APPLIED STATISTICS
FOR ECONOMISTS

APPLIED STATISTICS FOR ECONOMISTS

By

P. H. KARMEL

Vice-Chancellor, The Flinders University of South Australia

and

M. POLASEK

Senior Lecturer in Economics, The Flinders University of South Australia

THIRD EDITION

PITMAN PUBLISHING

First published 1957
Reprinted 1959
Second edition 1963
Reprinted 1963, 1965, 1967
Third edition 1970

SIR ISAAC PITMAN AND SONS LTD.
Pitman House, Parker Street, Kingsway, London, W.C.2
P.O. Box 6038, Portal Street, Nairobi, Kenya

SIR ISAAC PITMAN (AUST.) PTY. LTD.
Pitman House, Bouverie Street, Carlton, Victoria 3053, Australia

PITMAN PUBLISHING COMPANY S.A. LTD.
P.O. Box 9898, Johannesburg, S. Africa

PITMAN PUBLISHING CORPORATION
6 East 43rd Street, New York, N.Y. 10017, U.S.A.

SIR ISAAC PITMAN (CANADA) LTD.
Pitman House, 381–383 Church Street, Toronto, 3, Canada

THE COPP CLARK PUBLISHING COMPANY
517 Wellington Street, Toronto, 2B, Canada

© P. H. KARMEL and M. POLASEK
1970

SBN: 273 40295 1

MADE IN GREAT BRITAIN AT THE PITMAN PRESS, BATH
G0—(B.837)

PREFACE
TO THE THIRD EDITION

THE major addition relates to the sections on probability and significance, where a new chapter on Probability Distributions has been introduced. A section on ogives has been added to Chapter V, and a more detailed treatment of the significance of regression coefficients has been included in Chapter X. Apart from these additions, a number of minor amendments have been made and all the examples in the book have been brought up to date.

The authors are indebted to the Commonwealth Statistician and to members of the staff of the Commonwealth Bureau of Census and Statistics for their help in revising Appendix B; and to Mrs. G. Dunstan for her assistance with Chapter XII. The authors' thanks are also due to Miss H. S. Matters who assisted with the computations and Mrs. C. D. Clark who helped prepare the manuscript.

ADELAIDE
P. H. K.
M. P.

PREFACE
TO THE SECOND EDITION

THE major revisions, which I have made, relate to Chapter XI on Social Accounts. The treatment of the government sector has been modified to accord more with international practice and I have included a much fuller discussion of the uses of social accounting procedures and the extension of such procedures into the fields of national budgeting, flow-of-funds accounting, and inter-industry analysis. I am grateful to my colleague Professor R. L. Mathews for his advice and assistance in this connexion. Apart from this, I have included a brief section on one-tailed tests of significance in Chapter VI, and I have removed detailed references to Australian statistical practice from the text to the Appendix on the Sources of Australian Statistics. This Appendix has been completely rewritten and brought up to date. I am again indebted to the Commonwealth Statistician and to members of the staff of the Commonwealth Bureau of Census and Statistics for their help. Finally, where necessary, charts, tables and examples have been amended to incorporate up-to-date data, and my thanks are due to Miss J. M. Higgins who assisted me greatly in this.

ADELAIDE
P. H. K.

PREFACE
TO THE FIRST EDITION

THIS book, which is based on a course of about fifty lectures which I have been giving annually in the University of Adelaide, is designed to cater for the needs of economics and commerce students. It contains a treatment both of basic statistical methods (Chapters I, II, III, IV, V, VI and IX) and of those aspects of statistics which are of special interest to economists (Chapters VII, VIII, X, XI, XII, XIII and XIV). The basic statistical methods are treated in quite general terms, but where possible their relevance to economic problems has been indicated. Methods which do not find much application in the field of economics have been omitted. The examples and illustrations are drawn from Australian data.

The text provides the framework for a full one-year course at second- or third-year undergraduate level and covers those topics which in my view ought to be included as minimum requirements for economics and commerce graduates. In this respect the book can be regarded as complete in itself. However, for students intending to undertake work in econometrics the text provides an appropriate preliminary course, in which the earlier chapters can be covered fairly rapidly. The general emphasis of the text is on the understanding of the reasoning lying behind the various techniques, rather than on the techniques themselves. The danger of using the techniques without such an understanding is stressed.

Although mathematical formulations are used freely, they are for the most part of an elementary character, and verbal explanations are given throughout. Mathematics has been used wherever I have considered it helpful for an understanding of the methods given, but I have not attempted to provide a rigorous mathematical presentation, which would indeed be beyond the capacity of the student for whom the text is intended. In the few places where the mathematics is difficult, it can be omitted by students without the necessary mathematical equipment; my personal experience is that the course set out in this book can be handled quite well by students with little mathematics. Statistical methods cannot be taught satisfactorily by presenting them as if they were not of a mathematical nature, although they can be taught without placing too much reliance on any but the most elementary algebraic processes. Some facility in working in terms of symbols and in handling formulae is quite essential, and the non-mathematical student is likely to make better progress once he has recognized this.

PREFACE

I am indebted to Professor Sir Ronald A. Fisher, Cambridge, to Dr. Frank Yates, Rothamsted, and to Messrs. Oliver and Boyd, Ltd., Edinburgh, for permission to reproduce material from their book *Statistical Tables for Biological, Agricultural, and Medical Research*; this appears in my text as Table 7.1 and Tables II and III in Appendix A. I am indebted also to the Imperial Chemical Industries, Ltd., and to Messrs. Oliver and Boyd, Ltd., for permission to base my Chapter VIII on Chapter X of their book *Statistical Methods in Research and Production*. My thanks are due also to the Houghton Mifflin Company of Boston, U.S.A., for permission to use in my Table I of Appendix A material from Rugg's *Statistical Methods Applied to Education*.

I wish to thank Dr. F. G. Jarrett of the University of Adelaide who has helped me considerably in the preparation of the text with comment and criticism, and Mr. H. P. Brown of the Australian National University and Mr. R. R. Hirst and Mr. R. L. Mathews of the University of Adelaide who have read and commented on certain sections of the text. The Acting Commonwealth Statistician and members of the staff of the Commonwealth Bureau of Census and Statistics have been most co-operative in providing me with information on the activities and publications of the Bureau. I am much in the debt of Associate-Professor Jean Polglaze, M.B.E., of the University of Melbourne, not only for her helpful criticisms of the text, but also for first introducing me to statistical methods in the field of economics during a course of lectures at the University of Melbourne which in many respects has been the foundation upon which I have built. Finally, my thanks are due to Miss L. M. Sutton who has assisted me greatly in the preparation of the manuscript and who drew the diagrams, and to Miss J. Hanson who typed the manuscript.

ADELAIDE   P. H. K.

CONTENTS

	PAGE
Preface to the Third Edition	v
Preface to the Second Edition.	v
Preface to the First Edition	vi

CHAPTER I
INTRODUCTION

1.1. The nature of statistics	1
1.2. Economics and statistics	3
1.3. Need for care in statistical work	4

CHAPTER II
TABULAR PRESENTATION

2.1. Drawing up a table	7
2.2. Classification	7
2.3. Types of tables	9
2.4. Comparisons in tables	10

CHAPTER III
COLLECTION OF DATA

3.1. Method of collection	14
3.2. Questionnaires	15
3.3. Tabulation	21
3.4. Sources of economic and social statistics	23

CHAPTER IV
GRAPHICAL PRESENTATION

4.1. Object of graphical presentation	25
4.2. Bar charts	25
4.3. Graphing time series	28
4.4. Semi-logarithmic (or ratio) charts	32
4.5. Lorenz curves	39

CHAPTER V
THE FREQUENCY DISTRIBUTION AND ITS DESCRIPTION

5.1. Forming a frequency distribution	42
5.2. Cumulative frequency distributions	48
5.3. Types of frequency distributions	50
5.4. Measures of central value	51
5.5. Measures of dispersion	67

CHAPTER V (contd.)

5.6. Measures of skewness 76
5.7. Description of frequency distributions 78

CHAPTER VI
PROBABILITY AND PROBABILITY DISTRIBUTIONS

6.1. Statistical methods and probability 80
6.2. Measurement of probability 80
6.3. Axioms of probability 84
6.4. Multiplication of probability and statistical independence 86
6.5. Expected value 89
6.6. Populations and samples 90
6.7. Forming a probability distribution 91
6.8. The binomial distribution 95
6.9. The binomial distribution and the normal curve . . 103
6.10. The normal distribution 104

CHAPTER VII
SAMPLING AND SIGNIFICANCE

7.1. The general problem of statistical inference . . . 115
7.2. Mean and variance of a linear combination . . . 116
7.3. Sampling distribution of the mean 119
7.4. Tests of significance 122
7.5. Significance of the difference between $\bar{X}$ and μ, σ known 124
7.6. Significance of the difference between two sample means, $\bar{X}_1$ and $\bar{X}_2$, σ known 127
7.7. Significance of the difference between $\bar{X}$ and μ, σ not known 128
7.8. Significance of the difference between two sample means, $\bar{X}_1$ and $\bar{X}_2$, σ not known 131
7.9. Application to proportions 133
7.10. Significance of the difference between p and π . . 135
7.11. Significance of the difference between two sample proportions, p_1 and p_2 136
7.12. Goodness of fit 138
7.13. Independence of classification 144
7.14. Two types of errors 146
7.15. One-tailed tests of significance on the mean . . 148
7.16. Estimation 151
7.17. Confidence limits for a mean 153
7.18. Confidence limits for a proportion 154

CHAPTER VIII
SAMPLE SURVEYS

		PAGE
8.1.	Advantages of sample surveys over full counts	156
8.2.	The adequacy of a sample	158
8.3.	Selecting a random sample	160
8.4.	Sampling from a finite population	164
8.5.	Simple random sampling	165
8.6.	Stratified random sampling	170
8.7.	Other designs for sample surveys	182
8.8.	Sample surveys in practice.	184

CHAPTER IX
QUALITY CONTROL

9.1.	Statistical control of quality	187
9.2.	Control charts	188
9.3.	Advantages of quality control	193

CHAPTER X
REGRESSION AND CORRELATION

10.1.	Descriptive measures of regression and correlation for ungrouped data	195
10.2.	Descriptive measures of regression and correlation for grouped data	212
10.3.	Normal probability distribution of two variables	215
10.4.	General discussion on regression and correlation	219
10.5.	Tests of significance in regression and correlation analysis	226
10.6.	Prediction from a regression equation	233
10.7.	Regression in empirical economic investigations	236
10.8.	Multiple regression	240
10.9.	Relation between simple and multiple regression	249

CHAPTER XI
TIME SERIES

11.1.	Objectives of the analysis of time series	251
11.2.	Characteristic behaviour of time series.	252
11.3.	Basic assumptions in the analysis of time series	253
11.4.	Measurement of trend	257
11.5.	Fitting a mathematical trend	259
11.6.	Interpretation of a mathematical trend	268
11.7.	Use of time series in correlation and regression	269
11.8.	Measurement of seasonal variation	271

CHAPTER XII
SOCIAL ACCOUNTS AND THE MEASUREMENT OF NATIONAL INCOME

		PAGE
12.1.	The nature of social accounts	279
12.2.	Classification of economic transactions	280
12.3.	Content of national income accounts	285
12.4.	Consolidation of sector accounts	298
12.5.	Gross national product and allied aggregates	302
12.6.	The problem of imputation	305
12.7.	Estimation of items in social accounts	308
12.8.	Uses of the social accounting framework	312
12.9.	Flow-of-funds accounting	318
12.10.	Inter-industry analysis	321

CHAPTER XIII
PRICE INDEX NUMBERS

13.1.	The concept of an index number	327
13.2.	Price index numbers	327
13.3.	Index number formulae—aggregative type	329
13.4.	Relation between Laspeyres's and Paasche's index numbers	333
13.5.	Tests of adequacy of index number formulae	336
13.6.	Index number formulae—average type	339
13.7.	Relation between aggregative and average types	342
13.8.	Comparisons between more than two points of time	347
13.9.	Choice of an index number—a cost of living index	351
13.10.	Constructing an index number	358
13.11.	Changes in the regimen and quality changes	360
13.12.	Index numbers in practice	364

CHAPTER XIV
REAL NATIONAL PRODUCT AND INDEXES OF PRODUCTION

14.1.	Money national product and real national product	367
14.2.	Measurement of real national product	370
14.3.	Measurement of movements in the quantum of industrial production	376
14.4.	Computation of an index of industrial production by the method of deflation	381
14.5.	Computation of an index of industrial production by the method of indicators	386
14.6.	Comparisons over short and long periods	391
14.7.	Measurement of labour productivity	392

CONTENTS

CHAPTER XV
DEMOGRAPHY
15.1. Demography and demographic data 395
15.2. Measurement of total population 397
15.3. Sex and age distribution 401
15.4. Birth and death rates and rates of increase . . . 413
15.5. Measurement of mortality 424
15.6. Applications of the life table 433
15.7. Measurement of fertility 445
15.8. Measurement of population replacement . . . 449

APPENDIX A
STATISTICAL TABLES
I. Areas under the normal probability curve . . . 466
II. Values of t 468
III. Values of χ^2 470

APPENDIX B
SOURCES OF AUSTRALIAN STATISTICS 472

APPENDIX C
A SHORT LIST OF REFERENCES 505

Index 509

CHAPTER I

INTRODUCTION

1.1. The Nature of Statistics

The subject "statistics" is concerned with the *collection, presentation, description* and *analysis* of data which are measurable in numerical terms. The word "statistics" is used sometimes to refer to the data themselves, as when we speak of "statistics of national income" or "vital statistics," but it is also used to refer to the whole field of study of which statistics in the narrower sense of statistical data are the subject-matter. The subject "statistics" has a very wide application; and the basic principles and practice are the same irrespective of the field of application. This text is concerned with the use of "statistics" in the field of economics.

When data are obtained as a result of an experiment, as occurs in the physical or biological sciences, the collection of the data is part and parcel of the experiment itself. The experiment must be designed to produce data from which meaningful results can be derived; and the main problems associated with the collection of data are the problems of the design of the experiment. However, in the social sciences, and particularly in economics, experimentation is seldom possible, and data must for the most part be collected by requiring people to fill in questionnaires. First, the design of questionnaires is itself a matter raising considerable statistical problems. Secondly, it is often impossible or impracticable to make a survey of all the data in which one is interested. There then arises the question of how a fraction of the field can be surveyed in a way which will provide meaningful results about the whole. These are the two major problems which occur in the collection of economic data. Thus, if we wish to collect data on peoples' spending habits, we have to decide not only the sort of information we want, including precise definitions of the facts we wish to record and hence the sort of questions we must ask people, but also how we shall select particular people to answer our questions.

Once collected, data must be assembled into a useful form. This process is the statistical presentation of data. Data are usually presented in tabular form and are frequently made more readily comprehensible by being presented pictorially in charts and graphs. But presentation involves much more than this. The raw data must be classified into forms appropriate to the purpose in hand. Classification cannot proceed in a vacuum and must depend upon theoretical categories which determine whether one or another classification is meaningful.

Thus, if one had a great list of all the economic transactions which took place in an economy over a year, one would have to know the purpose for which the classification was to be used before classifying them. If this purpose were an investigation into the levels of consumption, investment, savings, etc., in the economy, the classification would have to be designed to take into account the theoretical meaning of these concepts and their relationships.

The description of statistical data involves the computation of measures to summarize the data. There are a host of such measures of wide application, of which the commonest perhaps is the average. However, in certain fields highly-specialized and complicated measures are required, as, for example, when we are concerned with measuring the rate of mortality of a human population. Moreover, it is sometimes impossible to fabricate a measure which will measure precisely what we are after, as, for example, when we want to measure changes in the cost of living.

The methods by which statistical data are analysed are called *statistical methods*, although the term is sometimes used more loosely to cover the subject "statistics" as a whole. The mathematical theory which is the basis of these methods is called the *theory of statistics* or *mathematical statistics*. Statistical methods have a very wide application and have had perhaps their greatest development in the biological and agricultural sciences. They are specially appropriate for handling data which are subject to variations that cannot be fully controlled by experimental method and for which we can observe only a fraction of the totality of observations which may exist. Thus, if a particular phenomenon were quite uniform, a single observation on it could be used to discover its nature, but when the phenomenon exhibits variability which cannot be controlled experimentally, techniques must be devised by which we can make inferences about the nature of the totality and by which we can test hypotheses about it from the particular observations we have. One of the main concerns of statistical methods is with such techniques, so that, for example, by observing a fraction of all rents in a particular city we can make inferences about the level of rents or test hypotheses about the nature of rents in the city as a whole.

The situation becomes more complex when instead of being interested in the characteristics of one particular phenomenon, like rent, we are concerned with relationships between phenomena. Suppose we wished to ascertain the numerical effect of a percentage change in the price of tea on the quantity of tea demanded, other things being equal. If we could conduct a controlled experiment we should hold all relevant factors constant, vary the price of tea and observe the resulting changes in the quantity demanded. But this is impossible, and all we are likely to have is a series of varying tea prices observed over a period of time with the corresponding amounts of tea bought. Over this period many factors which affect the demand for tea will have varied, as, for example, the level of income or the price of coffee, and it will be necessary to isolate the effect of the changes in the price of tea from the effects of these other factors. This can be attempted only by means of statistical methods. Even so it will not be possible to take into account the whole multiplicity of factors which may have affected the price of tea.

1.2. Economics and Statistics

As far as the economist is concerned, he needs a knowledge of the relevant statistical material and of how to handle it. This requires an understanding of statistical methods as such, although some methods are more appropriate than others to the field in which he works. Broadly speaking, the first half of this book is concerned with methods which have very general application, although naturally the emphasis is on their application in the field of economics. The second half is concerned with techniques which are more specialized. For example, the problems which arise in measuring growth of population or movements in price levels call for the development of special sorts of methods.

There is little doubt that a facility with statistics and statistical methods is an essential qualification for the student of economics. This does not mean that he has to have any intimate acquaintance with the mathematical theory of statistics, however desirable that may be. But he ought to know how to handle numerical data within his field— how to collect it, present it, describe it and analyse it. Above all he should be able to discriminate between valid and invalid inferences from such data.

Statistical methods make possible the development of the empirical side of economics. Their use is necessary to give real content to theoretical formulations. Many concepts which are commonplace in economic theory present major problems when we come to measure them in numerical terms. One of the major tasks of economic statisticians is that of measurement. The difficulty in measuring movements in "real income" or in the "general level of prices" is well known, but even such a deceptively simple concept as "number unemployed" conceals intricate problems of statistical measurement.

Empirical investigations involving statistical data are needed to test the hypotheses of pure economic theory and the conclusions drawn from these hypotheses. Thus the proposition that consumers' expenditure depends on persons' disposable income can be tested in the sense of whether observed data are or are not consistent with it; and the nature of the dependence and whether any variables other than income influence consumers' expenditure can be examined. At the same time empirical investigations may suggest new hypotheses for theoretical speculation—for example, the apparent stability in labour's share of national income evident in some economies has led to a number of important contributions to economic theory.

1.3. NEED FOR CARE IN STATISTICAL WORK

Statistical data are a powerful aid in economic analysis, but they must be used with care. You cannot "prove" an hypothesis with statistics; you can show only that the hypothesis is not inconsistent with the known facts. Indeed it is nearly always possible by a careful selection of data to bolster up an argument, and this makes it essential that in considering any problem *all* the relevant material should be examined. Thus, in comparing the current level of production in Australia with the pre-war one, it would be possible to use some figures (say gold, lead, clay bricks) to indicate stagnation and to use others (say electricity, refrigerators, brown coal) to indicate rapid growth. But what we require, of course, is the picture as a whole. Furthermore, since much of the data used by economists is published material which they have not collected themselves, it is important to find out the exact nature of the data, and whether or not there have been any changes in definition. To interpret a particular body of data without some knowledge of the background to it is a dangerous procedure. Thus it would be foolish to use the published budget surpluses or deficits of a Government as an indication of the extent to which the Government was becoming indebted to the rest of the economy, without first inquiring whether there are any funds through which the Government operates which are not included in the published figures.

The points made in the preceding paragraph are obvious. What is not so obvious is the need for a careful examination of all statistical material to ensure that inappropriate measures are not used and invalid inferences are not drawn. An example of the careless use of statistics occurred in a press report which stated that "more working-class than middle-class homes have television sets in the United Kingdom," the inference being that the tendency to possess television sets was greater in working-class than in middle-class homes. The facts may be correct, but the inference is invalid. To test the inference we should compare the proportion of all working-class homes which possess television sets with the proportion of middle-class homes which have them.

A more complex example may be worth quoting. In South Australia there were 9·01 and 8·06 deaths per 1,000 of the population in 1954 and 1961 respectively. In Queensland the corresponding rates were

INTRODUCTION

7·63 and 8·42. This might lead one to conclude that whereas Queensland was the healthier State in 1954, the situation had been reversed by 1961. But the rate at which a group of persons die varies directly with their age (apart from the very young); and the age distribution of the two States differed (and in different ways) in the two years under consideration. South Australia contained relatively more old people in 1954, but, through a higher rate of population growth over the period 1954 to 1961, contained relatively more young people in 1961. In fact in both years the rate at which people died was lower at almost every age in South Australia. Had the two States possessed the same age distributions, the death rates in South Australia would have been 4 per cent and 6 per cent lower than in Queensland for 1954 or 1961 respectively—results quite different from those obtained by observing the crude death rates per 1,000 of the population.

In this case the measures used in making the comparison were inappropriate. More complicated measures are needed. One such measure is the average age to which people could expect to live if they were subject to the mortality conditions under consideration. If such a measure were calculated for the two States, South Australia would be shown to have a higher average expectation of life than Queensland.

The need to look beneath the surface appearance of statistics can be further illustrated. The 1961 Census in Australia, for example, revealed that the average issue of married women aged 50 years and under 55 was 2·11 in metropolitan areas, 2·65 in other urban areas, and 2·99 in rural areas. *Prima facie*, this suggests that where people live affects family size. But it would be dangerous to infer this without further examination. For example, it might be the case that family size and educational attainment are inversely related and that city dwellers on the average enjoy higher educational attainments than country dwellers. In this case an apparent relationship between place of residence and family size would occur, even though place of residence has no direct influence whatsoever on family size. To discover whether this was so, it would be necessary to compare family size of people of the same educational attainments living in urban and rural areas. Place of residence might indeed influence family size, but that cannot be concluded from the two figures given above without examining carefully the influence of factors such as educational attainment, income, occupation and perhaps many others.

Attention should also be drawn to the danger of inferring a relationship between two phenomena because they happen to vary together. Thus over the past few decades the number of radios in use and the number of criminal convictions have both increased in Australia. On the basis of this covariation it might be suggested that the spread of radio has encouraged crime, i.e. that there is a causal relationship between radio and crime. However, the increase in the number of radios could be accounted for by the increase in the population and by the spread of the radio habit, and the increase in the number of criminal convictions by the increase in the population also and perhaps by the increased efficiency of the police force. Consequently radios and criminal convictions would exhibit a high degree of covariation without there being any direct relationship between them. Thus, although such a relationship as the one postulated above might obtain, the existence of covariation itself could not be taken as evidence of it. Further analysis would be required.

CHAPTER II

TABULAR PRESENTATION

2.1. Drawing up a Table

Although it would be logical to discuss the collection of data first, it is convenient to start with a brief discussion of statistical tables. Data when collected are summarized and set out in tabular form. The problems of collection are more readily appreciated when one has some idea of how the data have to be organized in tables.

A table summarizes the data by using columns and rows and entering figures in the body of the table. An example is given on page 8.

Drawing up a good table is an art which requires practical experience. The following points should be noted:

1. The table as a whole includes its title and all explanatory notes. As such it should be *self-explanatory*.
2. The title and the column and row headings should be clear, concise, unambiguous and as brief as possible.
3. The title should state what is being classified, the nature of the classification and the place and time concerned. See, for example, the title of Table 2.1.
4. Units of measurement should always be given.
5. Terms should be defined, in footnotes if necessary.
6. If the table is quoted or derived from other tables, the source must be given at the foot of the table.
7. Figures may be rounded to avoid unnecessary detail in the table. But, if necessary, a footnote should be added to the effect that figures do not necessarily add to totals because of rounding.
8. Tables should be adjusted to the space available, but should not be too narrow or too wide. Columns should be carefully planned and unnecessary variations in width avoided. Extensive ruling is frequently unnecessary.

2.2. Classification

All tables involve the classification of the subject-matter in some way or another. The classification may be through a characteristic of the subject-matter, say houses, which varies *qualitatively* or *quantitatively*, e.g. materials of outer walls (qualitative), number of rooms (quantitative). If the variation is qualitative we speak of the characteristic as an *attribute*, if quantitative as a *variable*. A variable may be *discrete* or *continuous*. A discrete variable is one which can vary only by finite "jumps," e.g. number of rooms. A continuous variable can vary

Table 2.1

OCCUPIED AND UNOCCUPIED DWELLINGS ACCORDING TO GEOGRAPHICAL AREA BY CLASS OF DWELLING
AUSTRALIA, CENSUS 30TH JUNE, 1966

Class of Dwelling	Number of Dwellings			
	Urban		Rural	Total
	Metropolitan	Other		
OCCUPIED DWELLINGS				
Private—				
Private house	1,529,037	692,504	460,146	2,681,687
Share of private house	20,940 (a)	3,678 (a)	1,296 (a)	25,914
Self-contained flat	274,328	63,325	7,893	345,546
Share of self-contained flat	956 (b)	193 (b)	19 (b)	1,168
Shed, hut, tent, etc.	5,085	9,526	16,445	31,056
Other	55,650	9,184	1,644	66,478
Total private dwellings	1,885,996	778,410	487,443	3,151,849
Non-private—				
Licensed hotel	1,760	2,390	1,788	5,938
Motel	188	717	219	1,124
Boarding house	9,070	3,347	696	13,113
Educational institution	224	221	80	525
Religious institution	620	450	204	1,274
Charitable institution	350	120	92	562
Hospital (c)	687	526	287	1,500
Staff barracks	463	1,290	6,321	8,074
Other	923	431	264	1,618
Total non-private dwellings	14,285	9,492	9,951	33,728
Total occupied dwellings	1,900,281	787,902	497,394	3,185,577
UNOCCUPIED DWELLINGS (d)	86,828	81,659	95,382	263,869

(a) Represents 10,077 private houses in Metropolitan areas, 1,803 in other urban areas and 638 in rural areas.
(b) Represents 464 self-contained flats in Metropolitan areas, 95 in other urban areas and 9 in rural areas.
(c) Includes mental hospitals.
(d) Includes vacant dwellings for sale or renting, holiday homes, seasonal workers' quarters unoccupied on Census night, unoccupied newly completed buildings, dwellings to be demolished, etc.

Source: Commonwealth Bureau of Census and Statistics, Australia: *Census Bulletin No.* 9.2: *Summary of Dwellings,* 1966, p. 5.

continuously in the sense that, if a and b are two values of the variable, a third value c can be located between a and b, no matter how close a and b are, e.g. age.

TABULAR PRESENTATION 9

When we make a classification, we break up the subject-matter into a number of classes. It is important that the classification should be *exhaustive* and *mutually exclusive*—

1. *Exhaustive*

There must be no items which cannot find a class. For example, a classification of persons by conjugal condition is not exhaustive if only the two classes "married" and "single" are included, since there are no classes in which widowed, divorced or separated persons can be placed. An exhaustive classification would require "never married," "widowed," "divorced," "separated" and "married" categories.

2. *Mutually Exclusive*

There must be no item which can find its way into more than one class. For example, a classification of houses by rent is not mutually exclusive if it contains the classes "$10 to $15," "$15 to $20," etc., since a rent of $15 can be placed in two classes. A mutually exclusive classification would require classes "$10 and under $15," "$15 and under $20," etc.

2.3. TYPES OF TABLES

Tables are broadly of two types—

1. *Frequency Type*

The subject-matter is classified according to some characteristic, and the frequencies of occurrence of the subject-matter in the various classes are recorded. The classification may be single or multiple. Thus "houses according to geographical area" or "houses according to class of dwelling" are single classifications, whereas "houses according to geographical area by class of dwelling" is a double or two-way classification. In Table 2.1 above the two sets of marginal totals each give a single classification, and the body of the table gives a double classification.

When the characteristic concerned is a variable we have what we call a *frequency distribution* (see Table 2.2).

2. *Aggregative Type*

The subject-matter is some aggregate or average, e.g. national income, value of exports, production of steel, average price of wheat. The aggregate may itself be broken up into a number of classes, e.g. exports according to country of destination, or may be shown to vary according to different situations, e.g. exports of various countries,

Table 2.2
Occupied Private Dwellings According to Number of Motor Vehicles
Australia, Census 30th June, 1966

Number of Vehicles per Occupied Private Dwelling (a)	Number of Occupied Private Dwellings
No vehicles	742,567
1 vehicle	1,608,321
2 vehicles	552,930
3 vehicles	120,500
4 or more vehicles	40,658
Not stated	86,873
Total Occupied Private Dwellings	3,151,849 (b)

(a) Motor Vehicles (excluding Motor Cycles and Scooters) used by members of the household and garaged or parked at or near the dwelling on Census night.
(b) Total number of vehicles was 3,257,180.

Source: Commonwealth Bureau of Census and Statistics, Australia: *Census Bulletin No. 9.2: Summary of Dwellings*, 1966, p. 9.

exports in various years. When the aggregate is shown to vary over *time* we have what is called a *time series*. For example—

Table 2.3
Exports and Imports of Merchandise
Australia, 1960–61 to 1966–67
(f.o.b. Port of Shipment)

Monthly Average	Value of Exports $A.'000	Value of Imports $A.'000
1960–61 (a)	161,474	181,263
1961–62	179,547	147,458
1962–63	179,318	180,223
1963–64	231,872	197,722
1964–65	220,954	242,059
1965–66	226,746	244,958
1966–67	251,993	253,777

(a) Australian financial years are 1st July to 30th June.
Source: Commonwealth Bureau of Census and Statistics, Australia: *Monthly Review of Business Statistics*, No. 362, 1968, pp. 24–25

2.4. Comparisons in Tables

Frequently tables are drawn up in order to make comparisons. In these cases it is often convenient to include figures *derived* from the basic statistics, such as *ratios* and *percentages*.

Table 2.4
Exports and Imports of Merchandise
Australia, 1960–61 to 1966–67
(Base: 1960–61 = 100)

	Value of Exports	Value of Imports
1960–61	100	100
1961–62	111	81
1962–63	111	99
1963–64	144	109
1964–65	137	134
1965–66	140	135
1966–67	156	140

1. *Relatives*

The movements in a time series are often made easier to comprehend by selecting one year as base and converting the other years to ratios to that base. In relative form, Table 2.3 appears as Table 2.4.

From this table we can see that in 1966–67 exports had increased, in value terms, by 56 per cent and imports by 40 per cent on their 1960–61 levels. The absolute values of the relatives depend on the year which has been selected as base. If 1966–67 were used as the base instead of 1960–61, Table 2.4 would appear as follows—

Table 2.5
Exports and Imports of Merchandise
Australia, 1960–61 to 1966–67
(Base: 1966–67 = 100)

	Value of Exports	Value of Imports
1960–61	64	71
1961–62	71	58
1962–63	71	71
1963–64	92	78
1964–65	88	95
1965–66	90	96
1966–67	100	100

Naturally the base is selected with a view to the purpose of the comparison.

Relatives show relative movements, not absolute levels. From Table 2.4 we see that the relative figures for 1965–66 are 140 and 135 for exports and imports respectively. This does not mean that in that year exports were higher than imports, but that compared with the base 1960–61 exports rose more than imports. In absolute terms exports

Table 2.6
EXPORTS AND IMPORTS OF MERCHANDISE
AUSTRALIA, 1960–61 TO 1966–67

	Ratio of Value of Exports to Value of Imports
1960–61	0·89
1961–62	1·22
1962–63	0·99
1963–64	1·17
1964–65	0·91
1965–66	0·93
1966–67	0·99

in 1965–66 were in fact lower than imports, having also been lower by a similar absolute amount in the base year 1960–61. The value of exports relative to imports in each year can be shown by taking the ratios of exports to imports for each year. This is done in Table 2.6.

2. *Percentage Distributions*

The percentage rather than the absolute distribution of a classification is often very useful. For example—

Table 2.7
SETTLER ARRIVALS ACCORDING TO SEX BY AGE, AUSTRALIA, 1967

Age Group (years)	Total Overseas Arrivals		Percentage Distribution	
	Males	Females	Males	Females
Under 15	21,634	20,093	29·62	32·42
15 and under 45	45,026	34,582	61·64	55·79
45 „ „ 60	4,541	4,365	6·22	7·04
60 and over	1,838	2,940	2·52	4·75
Total	73,039	61,980	100·00	100·00

Source: Department of Imigration: *Australian Immigration, Quarterly Statistical Summary*, Vol. 3, No. 6, 1967, pp. 14–15

3. *Ratios*

Frequently it is convenient to express a series of figures as ratios to another series, in particular as values per head of the population or of part of the population. For example—

Table 2.8

ESTIMATED QUANTITY OF ANNUAL CONSUMPTION OF BEER AUSTRALIA, 1936–37 TO 1965–66

	Annual Consumption (millions of gallons)	Annual Consumption per head of population (gallons)
1936–37 to 1938–39	80·1	11·7
1946–47 to 1948–49	129·5	16·9
1956–57 to 1958–59	221·0	22·7
1965–66	278·5	24·3

Source: Commonwealth Bureau of Census and Statistics, Australia: *Report on Food Production and the Apparent Consumption of Foodstuffs and Nutrients in Australia*, No. 15, 1959–60, pp. 45–46 and No. 21, 1965–66, p. 37

CHAPTER III

COLLECTION OF DATA

3.1. METHOD OF COLLECTION

ECONOMIC and social statistics are collected *incidentally* or *intentionally*. A considerable body of data is available as a result of administrative acts and is collected only incidentally to them, e.g. statistics of crime, car accidents, numbers of wage- and salary-earners incidental to payroll tax collections, details of imports incidental to collection of customs duties. These statistics are not collected primarily for research purposes, although they may be very useful in the field of research. Since they are not collected primarily for research, they may not be in a form completely appropriate to research. On the other hand, some statistics are collected *for their own sake*, i.e. to give general information and/or for use by research workers. It is these intentionally collected statistics with which we shall be mainly concerned, e.g. the census, statistics of prices, production, etc. Frequently statistics are collected incidentally to some administrative act but are also used for general information purposes. But whatever happens, it should be clear that when there is some intention of using the data collected for some purpose or other, then much more attention can and will be paid to the method of collection and to the precise nature of the statistics collected.

The collection of economic and social statistics generally requires the *filling in of forms*. These forms are called *questionnaires*. The questionnaire asks for certain information and has spaces on it for the answers. The answers may be filled in either by the *respondent*, i.e. the person or institution to whom the questionnaire is addressed, or by an *enumerator*, i.e. by a person, employed by the organization running the inquiry, who interviews the respondent and notes his answers to the questions. The enumerator must make personal contact with the respondent and, since his work carries him out into the field of inquiry itself, he is often called a *field-worker*.

A questionnaire is the operational instrument of any inquiry and inquiries can be of two broad types. They may be *general purpose* or *special purpose*. A general purpose inquiry attempts to obtain data which may be useful for many purposes, and it does not try necessarily to obtain data to answer specific problems. The best example of a general purpose inquiry is a population census. A special purpose inquiry tries to obtain information in a form suitable for analysing a specific problem or problems, e.g. an inquiry into the present level of rents in Adelaide, an inquiry to discover whether there is any difference between the

COLLECTION OF DATA 15

intelligence of children from large and small families. An example of a questionnaire is given on the next page.

3.2. QUESTIONNAIRES

We now consider the drawing up of questionnaires. Several factors require careful consideration.

1. *The Object of the Inquiry*

Specification of the general object of an inquiry is not enough in framing the questions which are to enter into the questionnaire. We must know the uses to which the answers are subsequently going to be put, in order to get the questions (and hence the answers) into the right form. It is for this very reason that administrative statistics are often not as useful for research workers as they would be had they been collected intentionally. It is also for this reason that a general purpose survey is not likely to be as useful for research into specific problems as a survey undertaken for a specific purpose, since the object of a general purpose survey is to provide information *in general*.

Let us suppose that we are concerned in drawing up a questionnaire for a specific purpose inquiry into "rents in Adelaide." That is the subject-matter of the inquiry, but we cannot attempt to draw up the questionnaire unless we know the questions which the results of our inquiry are supposed to answer. Suppose one of these questions is: "What are the factors governing the rent of a particular house?" We must now ask: "What are the ideal data necessary for answering that question?" These data might include the locality of the house, its construction, its age, the size of the block, etc. Given the ideal data, we can then set about framing the questions to obtain them. Alternatively, if another question is: "What are the factors governing the rent which a particular breadwinner pays for his house?" our ideal data might include his income, age, conjugal condition, number of dependants, occupation, etc.

We may generalize from the above. Before attempting to draw up the questionnaire, *we should set out in detail the ideal data* which we desire from the answers to the questionnaire. It might be wise to go even a stage further and actually construct the sorts of tables which we should like to emerge from our inquiry. Of course, it is not always possible to set out all the ideal data we should like in advance, since we may learn things in the course of the inquiry itself and may find that what we believed to be ideal does not quite meet the bill. Furthermore, it may be impossible to obtain all the ideal data from the inquiry. But detailed consideration of the ideal data is essential before drawing up the questionnaire. For this reason those who will be concerned with analysing the results of the inquiry should be called in at the very

Form D21
Section 29(2)

SOUTH AUSTRALIA
BIRTHS, DEATHS AND MARRIAGES REGISTRATION ACT, 1966
Seventh Schedule
INFORMATION STATEMENT FOR DEATH REGISTRATION

PARTICULARS OF DECEASED

1. Name—(a) Surname ....
    (b) Christian names ....
2. Date of death ....
3. Place where death occurred ....
4. Date of burial or cremation ....
5. Place of burial or cremation ....
6. Sex ....
    Date of birth ....
    Age last birthday ....
7. Usual profession or occupation ....
8. Usual residence ....
9. Birthplace ....
10. Length of residence in Australia ....
11. To be completed where deceased is **under** 21 years at date of death—
    (a) Father of deceased—
        Name and surname ....
        Profession or occupation ....
    (b) Mother of deceased—
        Name and surname ....
        Maiden surname ....
12. State whether deceased was bachelor, spinster, married, widowed or divorced ....

PARTICULARS OF MARRIAGE AND ISSUE

(This part is to be completed if the deceased has been married at any time, whether or not the deceased was married at date of death.)

	First Marriage	Each subsequent marriage	
		Second	Third, etc.

13. Date of or age at first marriage ....
14. To whom married ....
    Note: In the case of a deceased male, please quote the former surname of his wife.
15. Issue living at death of deceased (insert names and date of birth)—
    (a) Issue of first marriage—

MALES		FEMALES	
Name	Date of Birth	Name	Date of Birth
....	....	....	....
....	....	....	....

(b) Issue of subsequent marriages. Specify second, third, etc.—

....	....	....	....

16. Issue not living (dates of birth are not necessary—insert names only)—
....

OTHER PARTICULARS

17. (a) Name and address of Medical Practitioner
    or
    (b) Name and address of Coroner ....
18. Name and address of Funeral Director ....

CERTIFICATION OF INFORMANT

I certify that I have read the foregoing particulars and that the information is, to the best of my knowledge and belief, correct for the purpose of being inserted in the register of deaths.
Signed by me this .... day of .... 19 ....
Signature of informant ....
Description (occupier of building where death occurred, relationship to deceased, etc.) ....
Address of informant ....

earliest stage. For example, it would be foolish for a Government Statistician to collect some data on, say, retail trade and then hand them over to an economist to analyse. The economist should have first been consulted on what data were desirable. This point cannot be overstressed. But it should be noted that with a general purpose inquiry like a population census, it is impossible to anticipate in detail all the problems to the solution of which the results of the census may be put.

2. *The Practicability of the Questions*

Having determined our ideal data, we have in effect determined our ideal questionnaire, although we have not considered the precise form of the questions. Before doing this, we must consider whether certain information can or cannot be obtained. It may not be practicable to ask questions to ascertain some of the ideal data required. It may well be that the answers to some questions are not likely to be accurate enough to make them worth asking. Questions may be inaccurately answered either because the answers are deliberately falsified by the respondents or because the respondents do not really know the answers accurately enough themselves. In the former category fall, for example, questions about income and wealth. There is always reluctance to reveal one's private financial situation. It is probably better not to ask a question about income at all than to ask one which you feel sure will be falsely answered. We can often infer the approximate level of a person's income from other sources, e.g. the rent he pays for his house or the car he drives, so that it may be better to ask such questions rather than a question on income directly.

Questions which require a recollection are frequently inaccurately answered, e.g. "How old were your grandparents when they died?"; "How much did you spend on holidays in the past three years?"; "How many pairs of socks do you buy in a year?"; and so on. This does not mean that such questions should never be asked, only that the resulting data be treated with great caution.

To some questions we will never get the right answer, not because people deliberately falsify the answers or their answers are inaccurate in the everyday sense, but because subconscious factors, of which the respondents are not aware or are only vaguely aware, enter into the matter. Questions concerned with people's *motives* or *opinions* are of this nature. If we ask women the main reason for the limitation of their families, the economic factor will inevitably be stressed. But we cannot be certain that this is the real factor; it is quite likely that other motives are the fundamental ones and that women rationalize these into the rather neutral economic motive, e.g. the main motive may be fear of pain or dislike of children. Or if we ask people their opinions, we can never be quite certain what we are getting. Many people feel that there are certain opinions which they "ought" to hold, and they may give these opinions rather than the ones they really hold, e.g. if we ask people how they will vote at an election, they may say "Party A" because they expect most other people to say "Party A" or because they expect most other people will in fact vote for Party A or because it is considered respectable to vote for Party A.

This latter discussion leads to a division of questions into two types: those the answers to which are *refutable*, and those the answers to which are *non-refutable*. With the non-refutable type of question we can never be certain that the answers are correct, because we can never prove that they are wrong. This type includes questions about motives and opinions. These questions may be well worth asking, but the answers do *not* tell us people's motives and opinions, only what they *say* are their motives and opinions. With the refutable type of question, the answers may be inaccurate, but we can theoretically check up on them, e.g. a person's age or income. Some questions may be likely to result in more inaccurate answers than others. But usually we have a fairly good idea of the accuracy of the answers *a priori*, and we should avoid asking any question the answers to which are likely to be too inaccurate.

3. *The Form of the Questions*

Having decided on what questions to include, one must now decide on the form which the questions are to take. This is by no means easy. To some extent the form of the questions depends on whether they are to be answered by the respondent or by an enumerator. If the questionnaire is to be filled in by an enumerator in a personal interview, the questions need not be quite as carefully worded, since the enumerator can be given instructions as to how they are to be answered. But in general the following considerations apply to all types of questionnaires:

(i) The questions should be *clear*, *unambiguous* and *precise*. They should be capable of being answered in only a limited number of ways. For example, "What is the size of your house?" This is an unsatisfactory question, since it may be answered in terms of the dimensions of the house or the number of rooms. "Do you like living in a brick house or would you prefer to live in a wooden one?" This may be answered ambiguously by stating *yes* or *no*.

(ii) Most questions ask the respondent to place himself in a category. When the categories are few in number, the best method is to have a square for each category and for the respondent to place a cross in the appropriate square. When this is done, it is essential that the listed categories should be *exhaustive* and *mutually exclusive*.

COLLECTION OF DATA

NOT   Are you married? ☐ ⎫
            never married? ☐ ⎬ not exhaustive
BUT   Are you married? ☐
            widowed? ☐
            divorced? ☐
            separated? ☐
            never married? ☐

NOT   Do you own your farm? ☐ ⎫
            rent your farm? ☐ ⎬ not mutually exclusive
BUT   Do you own your farm? ☐
            rent your farm? ☐
            part own and part rent your farm? ☐

When categories are large in number and cannot be listed, it should be made clear to the respondent what sort of answer is required. For example—

     State present age .... years .... months.

     Occupation ....
                  (e.g. carpenter, labourer, school-teacher, etc.)

The categories into which the answers to a question are to be placed should be carefully selected, bearing in mind the nature of the data and the purpose for which they will be used. The person framing the question needs a detailed knowledge of the field of inquiry. Thus the question on farm-tenure quoted above may be quite unsatisfactory if, unknown to the framer of the question, other forms of tenure than those provided for exist, e.g. share cropping.

(iii) The question should be framed to get the desired information. Suppose it is desired to obtain information about the age distribution of the respondents, if we ask "present age?" or "age last birthday?" we shall get the age distribution at the date of the inquiry. But if we ask "year of birth?" we shall not be able to get an age distribution at all, although for some purposes a distribution by year of birth may be more useful than one by age.

(iv) Avoid leading questions, i.e. questions which suggest the answers. For example, in a depression year do not ask "By how much has your income fallen?" but "Has your income fallen? $\frac{\text{YES}}{\text{NO}}$. If yes, by how much?"

(v) Questions should be capable of objective answers, i.e. avoid questions of opinion and keep to questions of fact. For example, instead of asking a worker "whether he is content with his present job," ask him "if he desires to change his job and, if so, to what sort of job would he like to shift?"

(vi) Questions should be arranged in some logical order. For example, in an inquiry into the size of families, do not ask women the size of their family before you ask them whether they are married or single. Start with the simplest questions first.

(vii) Give precise and definite instructions and (if necessary), examples of how to fill in the questionnaire. It is better to include the instructions in the body of the questionnaire than to make them footnotes or put them all together on the other side of the page. But if instructions are very lengthy, this may not be possible. It is always necessary to define carefully all terms. For example, in the question "Do you own your own house?" does "house ownership" include purchasing by instalments or owning a share of a house? Precise definition is essential. Instructions should be simple and not too legalistic.

(viii) Some care should be taken in the actual setting out of the questionnaire. It should be made to look as attractive as possible. Plenty of room should be given for answers.

(ix) Having drawn up the actual questions and arranged them on a form, it is wise to try out the questionnaire by filling it in oneself and getting acquaintances to fill it in or, better still, by using it in a pilot (trial) survey. This will reveal weaknesses in the questions and layout. In all this work no precepts can replace practical experience.

4. *The Field of the Inquiry*

Before the questionnaire itself has been finally drafted, it is necessary to decide the respondents to whom it is to be addressed, i.e. the field of the inquiry. For example, if the inquiry is on rents, will you send the questionnaires to house-owners or house-occupiers? If it is on juvenile employment, will you send it to households, to the juveniles themselves or to employers? This will largely depend on which form of contact you expect to lead to the best results. It will also be necessary to decide whether the questionnaire is to be addressed to all possible respondents or only to a sample of them. This matter is discussed in Chapter VIII below.

3.3. TABULATION

Once the questionnaires have been filled in and collected, the data on them must be tabulated, i.e. counted and abstracted into tables. This is usually done by mechanical means, unless the number of questionnaires collected is small.

In mechanical tabulation there are four distinct processes: *coding*, *punching*, *sorting* and *tabulating*. First, the answers in the original questionnaires are translated into a numerical code. This is quite straightforward in the case of answers which are themselves numbers, e.g. age, which can be given a two-digit code; thus the age of 9 years can be coded 09, and that of 32, 32, etc. Where answers are not themselves numerical, the various possible answers are each allotted a number, e.g. to code *conjugal condition* we can designate "never married" 0, "married" 1, "widowed" 2, etc. It frequently happens, however, that the answers to questions are neither in numerical form nor split up into simple categories. In these cases it is necessary to set up a classification into which the answers derived from the questionnaires can be fed. Such is the case, for example, with "occupation." A question on occupation may result in thousands of different answers, and a classification of occupations will have to be established. Such a task requires great skill, involving not only an understanding of the purposes to which the classification will be put, but also a detailed knowledge of the occupational structure of the economy under consideration. It is customary to establish orders and sub-orders of occupations with even finer gradings. These categories can then be allotted code numbers. If, as is likely, some hundreds of classes of occupations are distinguished, a three-digit code will be necessary. The coding of the individual answers to the question on occupation performs the classification. This coding may be quite difficult, and complicated instructions may have to be issued to the coders.

Once coded, the answers to each individual questionnaire are then punched on to a punch card. Punch cards vary somewhat, but the usual type has eighty columns, each containing positions for the digits 0 to 9, and two other positions above the 0. These extra positions enable combinations of two or more holes in the one column to represent letters of the alphabet or special characters. Sufficient columns are allotted to each question to accommodate the code of that question. Thus, if one question is on occupation and the answers have a three-digit code, three columns are allocated to occupation; and an answer coded 143 would be punched on to the punch card by punching holes over the digits, 1, 4, 3 respectively in the three columns allotted

PUNCH CARD USED IN THE TABULATION OF AUSTRALIAN DEATHS STATISTICS

COLLECTION OF DATA

to occupation. An example of a punch card used in the tabulation of Australian death statistics is shown on page 22.

After punching, there will be a card corresponding to each filled-in questionnaire. To ascertain the numbers of a particular answer falling into the various categories, the cards are first passed through a sorting machine, which sorts them (at a rate of up to 2,000 cards a minute) into the various codes. The sorted cards are then passed through a tabulating machine which counts them, adds items together if necessary and prints the results on sheets of paper. These tabulation sheets are, of course, in terms of the code numbers, and translation of the code enables ordinary statistical tables to be set up.

Naturally the answers to each question in the questionnaire are tabulated separately, but quite often complicated cross-tabulations are required, i.e. tabulations involving the answers to more than one question. For example, in a survey on family size directed to married females, there might be questions on age, duration of marriage, husband's occupation and family income, and a four-way classification might be required—married females according to age, by duration of marriage, by husband's occupation, by family income. Such complicated tabulations can be readily effected by mechanical tabulation. Electronic digital computers are now used to perform this kind of analysis. By writing an appropriate programme it is possible to instruct the machine to carry out elaborate cross-tabulations and to print out the complete results.

3.4. Sources of Economic and Social Statistics

Economic and social statistics are collected both by Government and private agencies. For the most part they are collected by the former, although banks, stock exchanges, trade associations, etc., do make some collections, and research organizations sometimes make special purpose collections.

Most statistical collections are continuous, i.e. they cover, say, all months or all years without gaps, e.g. vital statistics, employment statistics. Some, however, are made only every so often. Of these the most important is the *population census*. The population census aims at making a complete count of the whole population and ascertaining information about how the population is distributed according to sex, age, conjugal condition, family size, occupation, industry, employment status, racial origin, birthplace, housing accommodation, etc. Some of the information derived from the census can be obtained nowhere else, but some is also obtained through continuous collection and the census is used as a periodic check on it. In addition to the population census, periodic censuses of production, retail trade, etc., may also be undertaken.

In most countries there is a Government department charged with the collection and publication of statistics. In addition to the statistical publications of this department, other governmentally collected statistics are to be found in the reports of various Government departments, such as the Taxation Department, in the budget papers and public accounts, in the reports of nationally operated business undertakings, etc. There are also usually some private statistical publications sponsored by banks and other financial institutions. Useful international summaries of statistics, containing a great deal of comparative material, are issued by the various agencies of the *United Nations*.

CHAPTER IV

GRAPHICAL PRESENTATION

4.1. Object of Graphical Presentation

A GRAPH is a simple and effective way of illustrating and comprehending a table. It gives pictorial effect to what would otherwise be just a mass of figures. Drawing charts and graphs is an art which can be acquired only through practice, but there are a number of simple rules,

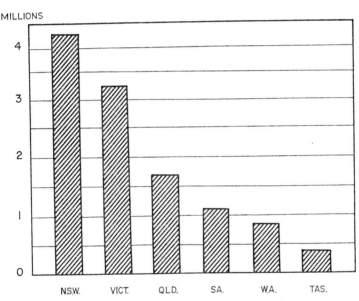

Fig. 4.1. Population of Australian States, 30th June, 1967
Source: Commonwealth Bureau of Census and Statistics, Australia:
Monthly Review of Business Statistics, No. 362, 1968, p. 9

the adoption of which lends to the effectiveness of graphs. It is important to avoid drawing misleading diagrams, and one must always be on one's guard against graphs which are drawn to bring an unwarranted conclusion out of the figures on which they are based.

4.2. Bar Charts

The most elementary type of graph is the bar chart. Bar charts are used for making comparisons and the diagram above illustrates a simple bar chart.

The rules for drawing graphs can be most easily set down in relation to simple bar charts. More complicated graphs merely require the observance of additional rules.

1. *Simple Bar Charts*

These illustrate movements or comparisons in one variable only. The following procedure should be employed in drawing them—

1. Examine the data carefully, and then select the scales. This must be done with reference to the data, the size of the graph paper and the general pictorial effect. The appearance of the graph depends entirely on the scale. The graph must not be allowed to distort the data. Dimensions in the ratio of about 7 to 10 are found in practice to be most satisfactory. In working out the scale decide how much of an inch is to go to one unit, *NOT* how many units are to go to one inch. There is no need to fill up the whole piece of graph paper—it can be cut to suit. Spaces between bars should be equal to about half a bar and similar spaces should be provided at both ends.

2. Having decided on the scale, draw in the frame and print in the scale. The vertical scale should start from zero, otherwise the graph can be very misleading. Thus suppose the values 200, 190, 220, 230 have to be compared. If the scale were started at 140, the graph would appear as follows—

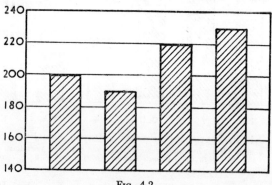

Fig. 4.2

This gives an impression of greater variability than is in fact the case. Using a zero base we should have the graph as shown in Fig. 4.3.

Sometimes, however, the range of the variable is so great as to make a full scale from a zero base impracticable. In that case the graph should be broken as shown in Fig. 4.4.

3. Draw in the bars.
4. Draw in horizontal reference lines. The grid on the graph

paper should be regarded as invisible. The most effective graphs are not drawn on graph paper.

5. Head the graph clearly and unambiguously, likewise the two scales and units used, observing the same rules as for tabular presentation (see p. 7). The source of the data should be printed at the foot of the graph. But avoid unnecessary verbiage. There is no need to head a diagram "*Bar Chart of. . .*" or to write in "*scale* £10 = $\frac{1}{10}$ *inch.*"

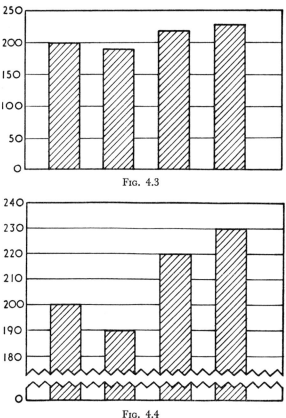

Fig. 4.3

Fig. 4.4

2. *Multiple Bar Charts*

These charts enable comparisons of more than one variable to be made at the same time. It is important not to overload the number of comparisons. The bars referring to each variable must be hatched differently. In doing this, do not use simply black and white, and avoid those diagonal hatchings that result in optical illusions. A key to indicate which bars are which must be included, but there is no need to

28   APPLIED STATISTICS FOR ECONOMISTS

label it "key" or "legend." An example of a multiple bar chart is given in Fig. 4.5.

3. *Composite Bar Charts*

These charts show the bars broken up into components. The segments of the bars must be hatched differently, with the darkest hatching at the bottom. Again one must be careful not to overload the diagram. An example is given in Fig. 4.6.

Sometimes it is helpful to show a percentage distribution by a composite bar chart. We then get what are called *100 per cent bars*.

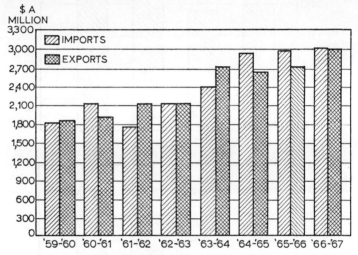

FIG. 4.5. VALUE OF EXPORTS AND IMPORTS—AUSTRALIA, 1959–60 TO 1966–67

Source: Commonwealth Bureau of Census and Statistics, Australia: *Oversea Trade*, No. 63, 1965–66, p. 749, and *Monthly Review of Business Statistics*, No. 362, 1968, pp. 24–25

This is illustrated in Fig. 4.7, where Fig. 4.6 is converted into a 100 per cent bar chart.

4. *Histograms*

The bar chart is also used to depict frequency distributions. In these cases no spaces are left between the bars.

A diagram such as Fig. 4.8 is called a *histogram*. An alternative method of presentation is the *frequency polygon*, shown in Fig. 4.9. Note that the points in the frequency polygon are joined by straight lines and not by curves.

4.3. GRAPHING TIME SERIES

Time series are graphed with time on the X-axis (horizontal) and the

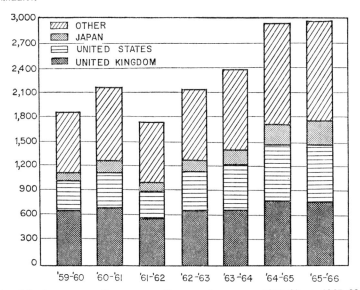

FIG. 4.6. IMPORTS BY COUNTRY OF ORIGIN—AUSTRALIA, 1959–60 TO 1965–66
Source: Commonwealth Bureau of Census and Statistics, Australia:
Year Book, Nos. 49–53, 1963–67, pp. 550, 528, 502, 408 and 392

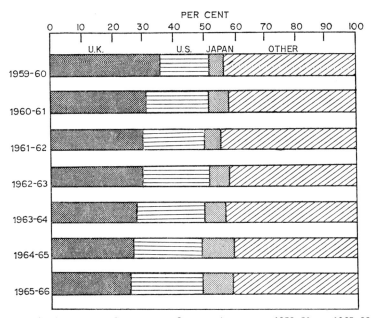

FIG. 4.7. IMPORTS BY COUNTRY OF ORIGIN—AUSTRALIA, 1959–60 TO 1965–66
(Percentage Distribution)
Source: Commonwealth Bureau of Census and Statistics, Australia:
Year Book, Nos. 49–53, 1963–67, pp. 551, 529, 503, 409 and 393

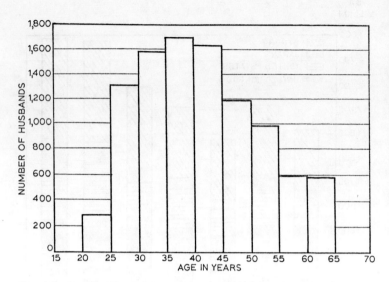

Fig. 4.8. Ages of Husbands at Time of Divorce—Australia, 1966
Source: Commonwealth Bureau of Census and Statistics, Australia:
Year Book, No. 53, 1967, p. 532

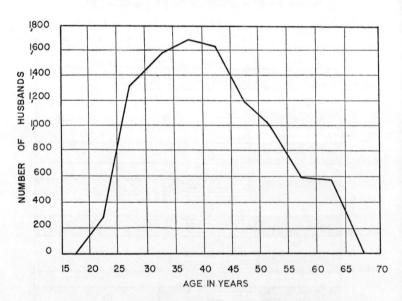

Fig. 4.9. Ages of Husbands at Time of Divorce—Australia, 1966
Source: Commonwealth Bureau of Census and Statistics, Australia:
Year Book, No. 53, 1967, p. 532

variable under consideration on the Y-axis (vertical). The same sort of rules as for graphing bar charts apply, but certain additional considerations should be noted—

1. Do not indicate plot points with circles or crosses. Use dots which disappear into the lines.

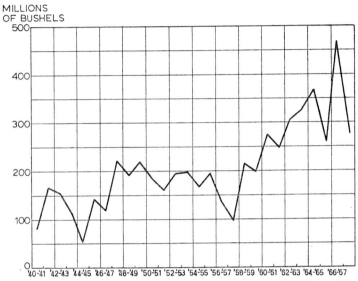

FIG. 4.10. PRODUCTION OF WHEAT FOR GRAIN—AUSTRALIA, 1940–41 TO 1967–68
Source: Commonwealth Bureau of Census and Statistics, Australia:
Rural Industries, No. 2, 1963–64, p. 33 and *Quarterly Summary of Australian Statistics*, No. 268, 1968, p. 23

2. Join points with straight lines, not curves.
3. The time scale should be fixed very carefully. Time is a continuous variable, so the following type of scale should be used as far as possible.

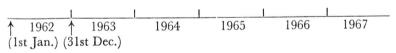

↑ 1962 ↑ 1963   1964   1965   1966   1967
(1st Jan.) (31st Dec.)

The variable under consideration can then be plotted at the appropriate places. For example, the value of exports in calendar years would be plotted at 30th June, the value of exports in financial years would be plotted at 31st December, the average price of wheat for quarters would be plotted at the middle of quarters, population as at 30th June would be plotted at 30th June. Sometimes it is convenient to draw the scale in terms of financial years as in Fig. 4.10 above.

4. The unit of time in which the variable under consideration is measured should be clearly stated in the title, e.g. an indication should be given as to whether the years are calendar or financial, or whether the variable is measured as at a date.

4.4. SEMI-LOGARITHMIC (OR RATIO) CHARTS

In graphing time series in the preceding section we graphed the data just as it stood. When the scale on the Y-axis is proportionate to the absolute values of the variable under consideration, it is said to be an *arithmetic* scale. If over three years a variable moved from, say, 100 to 110 and then from 110 to 120, these two movements would be shown on the graph as equal vertical distances and the slope of the lines joining the two pairs of points would be the same. Consequently, a graph drawn on an arithmetic scale, as well as showing us how the variable has changed, enables us to make comparisons of the *amounts* by which it has changed from time to time. Thus the two above movements each show an amount of increase of 10 and these two increases would occupy the same space on the graph. Furthermore, if a time series exhibited a constant amount of change per period, it would appear on the graph as a single straight line.

Frequently, however, the absolute amount of change in a variable is not of great interest to us, but rather the rate at which the variable is increasing or decreasing is of significance. For example, is an increment in the Australian population of 10,000 of as great significance when the population is 10 millions as when it was 100,000? Or is an increase in the population from 100,000 to 120,000 of less significance than one from 9 million to 10·8 million? Furthermore, if we wish to compare, for example, the Australian and the United States populations over time, the significant thing is the comparative rates at which the two populations have grown and not the fact that the United States population is eighteen times the Australian one. In an arithmetic scale, equal amounts of change are shown by equal vertical distances (or equal slopes). It would be very useful for some purposes if we could draw a graph in which equal vertical distances (or equal slopes) indicated equal *relative rates* of change, i.e. equal percentage changes.

Suppose that, instead of plotting the variable under consideration itself, we plot the logarithms of the variable. If three values of the variable are A, B and C, we shall be plotting $\log A$, $\log B$ and $\log C$ instead of A, B and C. The movement from A to B will now be shown on the graph as a movement from $\log A$ to $\log B$. The amount of this movement will equal $\log B - \log A$, i.e. $\log \frac{B}{A}$. Hence, if it happens that $\log B - \log A = \log C - \log B$, this will mean that $\log \frac{B}{A} = \log \frac{C}{B}$, i.e. $\dfrac{B}{A} = \dfrac{C}{B}$. In other words vertical distances now refer to relative rates of changes and not to amounts of change. Equal vertical distances (or equal slopes) mean equal relative rates of change (i.e. equal percentage changes). When we graph the logarithms of a variable rather than the original data themselves, we are plotting on a *logarithmic scale*. When we plot a time series in this way we have what is called a *semi-logarithmic* (or *ratio*) *chart*. It is called "semi-" because, although the variable under consideration is plotted logarithmically, time is still plotted arithmetically.

If a time series turns out to be a straight line on a semi-logarithmic scale, this means that the series exhibits a constant percentage rate of growth, e.g. 2 per cent per annum. Such a time series plotted arithmetically will result in a curve which is concave upwards, indicating an increasing *amount* of growth per period. Thus—

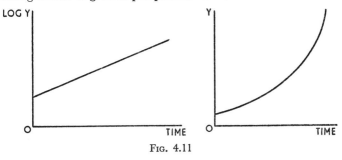

Fig. 4.11

On the other hand, a time series which is a straight line when plotted arithmetically, indicating a constant amount of growth per period, will yield a curve which is concave downwards when plotted logarithmically. Thus—

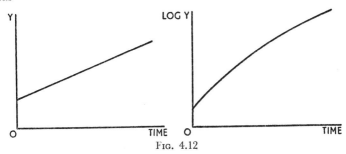

Fig. 4.12

As a numerical example, consider the series for five years: 100, 110, 120, 130, 140. The amount of growth is constant at 10 per annum.

Plotted arithmetically, this gives a straight line. The logarithms of these values are: 2·0000, 2·0414, 2·0792, 2·1139, 2·1461. When plotted, these will give a curve concave downwards. The percentage rate of growth is falling. As can be seen from the original figures, the rates are 10 per cent, 9·1 per cent, 8·3 per cent, 7·7 per cent respectively for the four spans. On the other hand consider the series: 100, 110, 121, 133·1, 146·41. The amount of growth per annum is here increasing, being 10, 11, 12·1 and 13·31 respectively, so that graphed arithmetically the curve will be concave upwards. But the logarithms of these values are: 2·0000, 2·0414, 2·0828, 2·1242, 2·1656. Plotted, these will be on a straight line, indicating a constant percentage rate of growth. As can be seen from the original figures, the rate is 10 per cent per annum.

To summarize: equal vertical distances or equal slopes mean equal amounts of change on an arithmetic chart, but mean equal relative rates of change on a semi-logarithmic chart.

The procedure for drawing a semi-logarithmic chart is as follows—
1. Convert the variable to logarithms.
2. Select the scales to be used. It should be noted that, since we are interested here in relative changes, there is no need to start the scale at zero. In any case there is no finite number corresponding to log 0. It is not possible to plot zero or negative numbers on a logarithmic scale. The logarithmic scale should start at the logarithm corresponding to a convenient natural number.
3. Complete the drawing of the graph.
4. Although the logarithms of the variable have been plotted, there is little value in attaching to the vertical scale these logarithmic values, since in themselves they do not convey very much. It would be preferable to include a scale of natural numbers so that the graph will portray absolute values as well as relative rates of growth. The logarithmic scale can readily be translated back into natural numbers. Consider the natural numbers 1 to 10. We have—

Natural Number	Logarithm
1	0·00
2	0·30
3	0·48
4	0·60
5	0·70
6	0·78
7	0·85
8	0·90
9	0·95
10	1·00

GRAPHICAL PRESENTATION

If we draw up a scale of the logarithms and write in the corresponding natural numbers, the result is as shown in Fig. 4.13. This gives us what is known as a complete *deck* or *cycle* in a logarithmic grid. We note that equally spaced natural numbers have diminishing spaces on the grid, and that the distance between, say, 1 and 2 is the same as that between 2 and 4 or 3 and 6 or 5 and 10, etc., because the *relative* relation between these pairs of numbers is the same. If we go on to draw up a scale of the numbers 10 to 100, it will simply duplicate

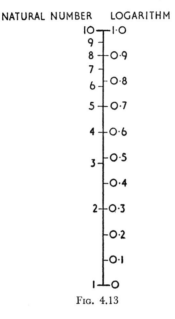

FIG. 4.13

the scale from 1 to 10 giving a second deck. Similarly the numbers 100 to 1,000 will give a third deck. The size of a deck is always a multiple of 10. We could have decks rising from 3 to 30, 30 to 300, etc.

Returning now to the marking-in of a vertical scale on our graph, we take convenient natural numbers, locate them on the scale and write them in. We then draw in a grid corresponding to these values. Suppose, for example, we have values of the variable under consideration ranging from 62,000 to 2,500,000. We could start our scale at the logarithm corresponding to 50,000. Reference lines could be drawn in at 50,000; 100,000; 150,000; 200,000; 250,000, etc. As the figures get higher rather less detail will be needed.

5. The graph is completed by writing in title, scales, source, etc. An indication should be given that the scale is semi-logarithmic.

The diagrams below show the number of broadcast listeners'

licences in force at 30th June in Australia for the years 1925 to 1967. Fig. 4.14 is arithmetic, Fig. 4.15 semi-logarithmic. It can be seen that whereas the arithmetic chart gives the impression of a steady upward

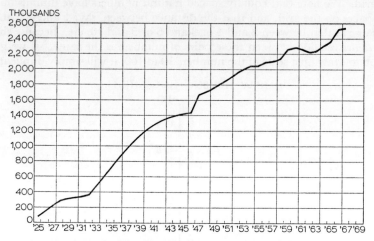

FIG. 4.14. BROADCAST LISTENERS' LICENCES IN FORCE AT
30TH JUNE—AUSTRALIA, 1925 TO 1967
(Arithmetic Scale)

Source: Commonwealth Bureau of Census and Statistics, Australia:
Transport and Communication Bulletin, No. 21, 1928–29, p. 58; No. 25, 1933–34, p. 58
No. 30, 1938–39, p. 58; No. 40, 1948–49, p. 83; No. 50, 1958–59, p. 71;
Monthly Review of Business Statistics No. 285, 1961, p. 47; and
Quarterly Summary of Australian Statistics, No. 267, 1968, p. 54

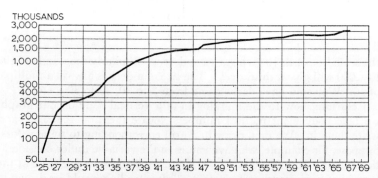

FIG. 4.15. BROADCAST LISTENERS' LICENCES IN FORCE AT
30TH JUNE—AUSTRALIA, 1925 TO 1967
(Semi-logarithmic Scale)

Source: Commonwealth Bureau of Census and Statistics, Australia:
Transport and Communication Bulletin, No. 21, 1928–29, p. 58; No. 25, 1933–34, p. 58
No. 30, 1938–39, p. 58; No. 40, 1948–49, p. 83; No. 50, 1958–59, p. 71;
Monthly Review of Business Statistics, No. 285, 1961, p. 47; and
Quarterly Summary of Australian Statistics, No. 267, 1968, p. 54

movement for most of the period under consideration, the semi-logarithmic chart shows that the relative rate of growth has flattened off and that it was greatest at the beginning of the period. In many respects the semi-logarithmic chart is the more interesting.

Semi-logarithmic charts are very useful when relative rates of growth are significant and when comparisons between the rates of growth of two or more series are required. The latter are rendered simple by the fact that if two series run parallel on a semi-logarithmic graph they must be changing at the same relative rates. Moreover, if we wish to plot data of very different magnitudes on the same chart, semi-logarithmic plotting is very convenient. Thus, consider a graph of the movement of the Australian population from 1788 to date. The variable ranges from about 700 to 12 million. If this were drawn arithmetically on a scale of 1 million to the inch, it would be almost impossible to graph the population at all in the earlier years. However, semi-logarithmically the same distance is given to the range 700 to 7,000 as to 700,000 to 7,000,000. Semi-logarithmically, there would be four and a bit decks, and the same relative detail could be included for earlier and later years.

By the same token, when two series, one of which is very large in magnitude compared with the other, have to be compared, a semi-logarithmic chart can be used to advantage. This is illustrated overleaf in the comparison of the growth of population in New South Wales and Western Australia. With an arithmetic scale it is only just possible to represent Western Australia on the same chart as New South Wales. However, the two States can be readily graphed on a semi-logarithmic chart by attaching different scales to the two sides of the grid and graphing New South Wales in relation to the left-hand scale and Western Australia to the right-hand scale. Once the left-hand scale has been determined, the right-hand one can be written in as any given multiple or fraction of it. Thus in Fig. 4.17 the Western Australia scale is one-fifth of the New South Wales scale. The former scale runs 10, 20, 40, 60, etc., and the latter 50, 100, 200, 300, etc. The scales are the same relatively and accordingly are represented by the same absolute distances on a semi-logarithmic grid. It will be noted that on the arithmetic chart (Fig. 4.16) Western Australia appears to have grown little and much less rapidly than New South Wales. This is of course true in absolute terms. However, by plotting the data on a semi-logarithmic chart (Fig. 4.17), it can be seen that in the latter half of the period Western Australia has had a relative rate of growth comparable with that of New South Wales and that for certain periods, including the past few years, it had a relative rate of growth which was in fact greater than that of New South Wales.

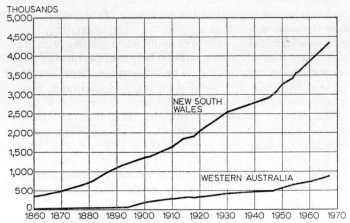

FIG. 4.16. POPULATION OF NEW SOUTH WALES AND WESTERN AUSTRALIA, AS AT 31ST DECEMBER, 1860 TO 1967
(Arithmetic Scale)

Source: Commonwealth Bureau of Census and Statistics, Australia:
Demography Bulletin, No. 67, 1949, pp. 154–55; No. 77, 1959, p. 174; No. 82, 1964, p. 119; and *Australian Demographic Review*, No. 248, 1968, p. 4

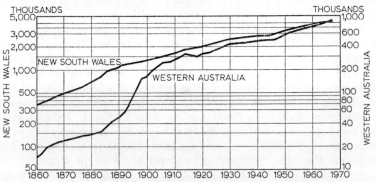

FIG. 4.17. POPULATION OF NEW SOUTH WALES AND WESTERN AUSTRALIA, AS AT 31ST DECEMBER, 1860 TO 1967
(Semi-logarithmic Scale)

Source: Commonwealth Bureau of Census and Statistics, Australia:
Demography Bulletin, No. 67, 1949, pp. 154–55; No. 77, 1959, p. 174; No. 82, 1964, p. 119; and *Australian Demographic Review*, No. 248, 1968, p. 4

4.5. Lorenz Curves

Economists are very often interested in the distribution of something by size, e.g. incomes, factories, etc. It is often important to know how far such a distribution departs from one of equality, or whether one distribution is more or less unequal than another. The table below gives details of the distribution of taxable income for Australia, 1964-65.

Table 4.1
Distribution of Taxable Income, Australia, 1964-65

Grade of Actual Income	Number of Taxpayers	Taxable and Contributable Income	Percentage in Income Grade		Accumulated Percentage	
			Number	Amount	Number	Amount
(1)	(2)	(3)	(4)	(5)	(6)	(7)
$		$'000	%	%	%	%
417-599	137,122	67,819	2·96	0·69	2·96	0·69
600-799	190,233	123,770	4·11	1·26	7·07	1·95
800-999	221,616	181,980	4·78	1·85	11·85	3·80
1,000-1,199	247,883	244,740	5·35	2·49	17·20	6·29
1,200-1,399	259,250	300,079	5·61	3·06	22·81	9·35
1,400-1,599	278,864	368,750	6·02	3·75	28·83	13·10
1,600-1,799	266,670	392,403	5·76	4·00	34·59	17·10
1,800-1,999	270,858	430,509	5·85	4·38	40·44	21·48
2,000-2,199	291,532	496,365	6·29	5·05	46·73	26·53
2,200-2,399	293,571	529,102	6·34	5·39	53·07	31·92
2,400-2,599	290,225	562,164	6·27	5·72	59·34	37·64
2,600-2,799	275,392	572,135	5·94	5·82	65·28	43·46
2,800-2,999	248,244	551,123	5·36	5·61	70·64	49·07
3,000-3,199	216,788	515,817	4·68	5·25	75·32	54·32
3,200-3,399	178,302	449,196	3·85	4·58	79·17	58·90
3,400-3,599	147,450	394,059	3·18	4·01	82·35	62·91
3,600-3,799	123,154	348,253	2·66	3·54	85·01	66·45
3,800-3,999	99,578	296,661	2·15	3·02	87·16	69·47
4,000-4,999	280,952	956,313	6·07	9·73	93·23	79·20
5,000-5,999	122,521	522,629	2·65	5·33	95·88	84·53
6,000-7,999	99,934	548,147	2·16	5·58	98·04	90·11
8,000-9,999	40,478	298,055	0·87	3·03	98·91	93·14
10,000-19,999	43,479	485,889	0·94	4·95	99·85	98·09
20,000-29,999	4,894	103,402	0·11	1·05	99·96	99·14
30,000 and over	2,021	84,526	0·04	0·86	100·00	100·00
Total	4,631,011	9,823,886	100·00	100·00		

Source: Commonwealth of Australia:
Taxation Statistics, 1965-66, pp. 26-27

Columns (1) and (2) taken together are a simple frequency distribution of income. In column (3) is given the amount of the taxable income earned by the individuals in the various grades of actual income, and columns (4) and (5) give the percentage distributions of columns (2) and (3). Columns (6) and (7) give columns (4) and (5) accumulated downwards. From these last two columns we can learn that the 2·96 per cent lowest income-earners earned between them 0·69 per cent of total taxable income, whereas the 95·88 per cent lowest income-earners earned 84·53 per cent of total taxable income. This latter implies, of course, that the 4·12 per cent highest income-earners earned 15·47 per cent of total taxable income. When columns (6) and (7) are graphed, one against the other, as in the diagram below, we have what is called a *Lorenz* curve. If income were exactly equally

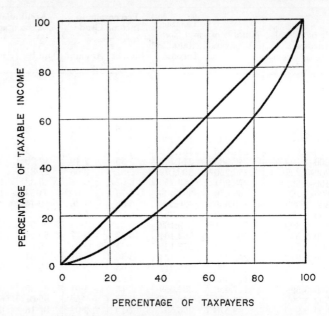

Fig. 4.18. Distribution of Taxable Income of Individual Taxpayers—Australia, 1964-65
(Lorenz Curve)

Source: Commonwealth of Australia: *Taxation Statistics*, 1965–66, pp. 26–27

distributed so that every individual had the same income, the first 10 per cent of income-earners would receive 10 per cent of total income, the first 20 per cent would receive 20 per cent, and so on. Strictly speaking, there would be only one grade of income and only two points on the diagram—0 per cent of income-earners would receive 0 per cent total income, and 100 per cent of income-earners would receive 100 per cent of total income. Consequently, in this situation the Lorenz curve would become a straight line drawn diagonally across the graph. The extent to which the actual curve diverges from the diagonal line is, therefore, illustrative of the degree of inequality of income. Naturally, Lorenz curves are most useful in making comparisons between different income distributions.

CHAPTER V

THE FREQUENCY DISTRIBUTION AND ITS DESCRIPTION

5.1. Forming a Frequency Distribution

HAVING considered the collection of data and their presentation in both tabular and graphical forms, we now proceed to the *description* of data. We shall consider in some detail the way in which we can describe statistical data when they consist of a large number of observations of a particular variable, e.g. rents of houses. The number of observations is designated by N, the variable observed (in this case, rent) by X.

Below are set out the rents of 200 houses of a particular type, e.g. furnished. The example is a hypothetical one.

Table 5.1
WEEKLY RENTS OF 200 TENANTED HOUSES
(Quoted to the last complete half-dollar)

30,	25,	25,	39,	11·5,	24,	34,	39,	16,	10,
25,	22·5,	8,	23,	36,	19·5,	26,	10,	35,	17·5,
29·5,	30,	25,	26,	27,	33,	19,	30,	21,	10,
25,	20,	9,	12,	25,	33,	14·5,	27·5,	12,	21,
37,	11,	12·5,	32,	15,	22·5,	34,	44,	27,	24,
40,	25,	33,	14,	25,	29·5,	12·5,	26,	24·5,	26,
16,	32,	15,	35,	29,	19,	25,	32,	40,	17·5,
17,	26·5,	20,	23·5,	25,	45,	29·5,	10,	15,	24·5,
22,	26,	28,	9·5,	28,	18,	31,	19,	24,	15,
20,	26,	15,	25,	14·5,	26,	30,	28,	19·5,	22,
31,	20,	35,	28,	24,	25,	19·5,	25,	30,	40,
29,	24,	30,	14,	19·5,	32,	24·5,	8,	22·5,	30,
26,	13,	30,	22,	15,	18,	20,	20·5,	24,	32,
23,	26,	24·5,	22,	29,	24,	14,	22,	24,	32,
25,	24,	25,	16,	18,	16,	23,	20,	27,	35,
34·5,	20,	19,	27·5,	31,	27,	24,	20,	24,	26,
29,	28,	25,	15,	28,	25,	21,	26,	19·5,	18,
19,	30,	19·5,	24,	20,	35,	17,	25,	25,	29·5,
20,	16,	20,	14,	21,	13,	20,	18,	23,	14,
22,	30,	22,	30,	20,	18·5,	25,	31,	19,	13.

From this unorganized data it is difficult to draw any conclusions at all about the distribution of rents in the given area. It would be somewhat more informative if we were to arrange the data into what is called an *array*, i.e. in their order of magnitude, but even so this would only indicate the range of the data and the way in which the individual items find their places within the range.

FREQUENCY DISTRIBUTION AND ITS DESCRIPTION 43

The most satisfactory procedure is to arrange the data into a *frequency distribution*, showing the frequency with which rents occur in certain specified intervals. The purpose of drawing up frequency distributions is to bring out the essential features in the data and to permit the use of analytical techniques in their description. One possible way of organizing the given raw data is shown in Table 5.2.

Table 5.2
WEEKLY RENT OF 200 TENANTED HOUSES

Weekly Rent (dollars) X	Number of Houses f
7·5 and under 12·5	12
12·5 ,, 17·5	26
17·5 ,, 22·5	45
22·5 ,, 27·5	60
27·5 ,, 32·5	37
32·5 ,, 37·5	13
37·5 ,, 42·5	5
42·5 ,, 47·5	2
All rents	200

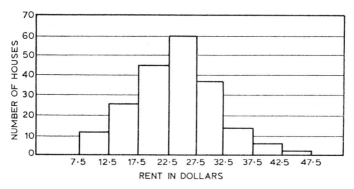

FIG. 5.1. WEEKLY RENT OF 200 TENANTED HOUSES

The left-hand column shows the *classes*, the right-hand column the *frequencies of occurrence*. Thus in the first class $7·5 is the *lower class limit*, $12·5 the *upper class limit* and the *class interval* is $5. The frequency distribution immediately gives us a picture of the form of the distribution when the results in Table 5.2 are plotted as a histogram or frequency polygon. (See Figs. 5.1 and 5.2.)

Several important considerations enter into the construction of frequency distributions. The first of these concerns the determination of an appropriate number of classes. Though formulae are available for this purpose, generally no hard and fast rules can be given and the ultimate decision will depend partly on personal judgment. The

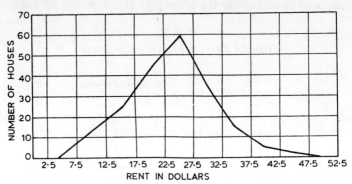

FIG. 5.2. WEEKLY RENT OF 200 TENANTED HOUSES

Table 5.3
WEEKLY RENT OF 200 TENANTED HOUSES

Weekly Rent (dollars) X	Number of Houses f	Weekly Rent (dollars) X	Number of Houses f
7·5 and under 8·5	2	27·5 and under 28·5	8
8·5 ,, 9·5	1	28·5 ,, 29·5	4
9·5 ,, 10·5	5	29·5 ,, 30·5	15
10·5 ,, 11·5	1	30·5 ,, 31·5	4
11·5 ,, 12·5	3	31·5 ,, 32·5	6
12·5 ,, 13·5	5	32·5 ,, 33·5	3
13·5 ,, 14·5	5	33·5 ,, 34·5	2
14·5 ,, 15·5	9	34·5 ,, 35·5	6
15·5 ,, 16·5	5	35·5 ,, 36·5	1
16·5 ,, 17·5	2	36·5 ,, 37·5	1
17·5 ,, 18·5	7	37·5 ,, 38·5	0
18·5 ,, 19·5	7	38·5 ,, 39·5	2
19·5 ,, 20·5	19	39·5 ,, 40·5	3
20·5 ,, 21·5	5	40·5 ,, 41·5	0
21·5 ,, 22·5	7	41·5 ,, 42·5	0
22·5 ,, 23·5	7	42·5 ,, 43·5	0
23·5 ,, 24·5	13	43·5 ,, 44·5	1
24·5 ,, 25·5	24	44·5 ,, 45·5	1
25·5 ,, 26·5	11	45·5 ,, 46·5	0
26·5 ,, 27·5	5	46·5 ,, 47·5	0
		All Rents	200

FREQUENCY DISTRIBUTION AND ITS DESCRIPTION 45

smaller the number of classes the more likely it is that important characteristics of the distribution will be concealed. On the other hand, too many classes will bring out more detail but summarize the distribution less well. Furthermore, the more classes the less the regularity of the frequency distribution. This can be easily seen by forming the given data into a frequency distribution with classes of, say, one dollar. The results are shown in Table 5.3 and Fig. 5.3.

Although the frequencies display a pattern similar to the results obtained with wider class intervals, they show great irregularity with a number of classes showing zero frequency. A compromise must be struck between too much detail and too little. In general, few distributions require more than sixteen to twenty or less than eight to ten

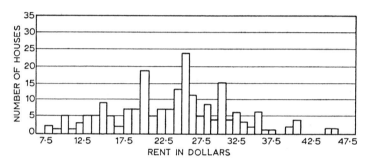

FIG. 5.3. WEEKLY RENT OF 200 TENANTED HOUSES

classes. Unless the frequency distribution is very highly skewed, all classes should have *equal intervals*.[1]

The second important consideration in drawing up frequency distributions lies in the choice of *class mid-points*. While the choice of class intervals is governed by convenience (e.g. in terms of multiples of 5 and 10), the classes must be correctly centred. The class mid-point is then regarded as representative of the cases occurring in the class, and should be half-way between the true upper and lower limits of each class. If there is a tendency in the original data for concentrations at particular values, the class limits should be so arranged that these values fall in the centre of the class or are evenly spaced throughout the class. This has in fact been the case in the present example. The data show strong concentrations at multiples of 5 dollars (i.e. at 10, 15, 20, etc., dollars), and if class intervals of 5 dollars are chosen, the classes should be $7·5 and under $12·5, $12·5 and under $17·5, etc., to ensure that each class is correctly centred.[2]

[1] For discussion of skewness in frequency distributions, see section 5.6 below.
[2] See p. 55 below.

For the purpose of selecting class limits, it is important to determine whether the variable under observation can be regarded as a *continuous* or *discrete* variable. The class limits can be stated in terms of either "*this* amount to *that* amount," or "*this* amount and *under that* amount." The latter form must be used in the case of continuous variables, where the upper class limit of one class and the lower limit of the next class should coincide in order to ensure an exhaustive classification.[1] Thus the variable age, being inherently a continuous variable, should be stated as, for example, "10 years and under 20," "20 years and under 30," etc. The other form should in general be used for discrete variables, although in the case of a variable expressed in a unit which moves by very small jumps, it may be convenient to use the "*this* and under *that*" form. In recording money amounts, discrete jumps of one cent may be regarded as sufficiently small to justify the latter form of stating class limits. Thus the rent of, say, $12·5 belongs to the class "$12·5 and under $17·5" the true limits of which are $12·5 and $17·49. When we use the discrete form of stating class limits, the upper class limit of one class and the lower limit of the next class must be appropriately differentiated in order to ensure mutually exclusive classification. Thus in classifying factories according to size, the limits should be stated as, for example, 100–199 employees, 200–299 employees, and not 100–200, 200–300, etc.

In selecting class limits, it is essential that the stated class limits clearly indicate the range of values falling within each class. When the variable is capable of being measured exactly, this requirement poses little difficulty. Thus, in classifying houses by number of rooms, we might have classes 1 to 2 rooms, 3 to 4 rooms, and these are clearly the true limits of the chosen classes; the mid-points are simply the averages of the true class limits, i.e. $1\frac{1}{2}$ rooms and $3\frac{1}{2}$ rooms. On the other hand, the true class limits of a money variable stated in the "*this* and under *that*" form have implicit class limits which differ by one cent between the adjacent classes, and strictly speaking the mid-points should be calculated by computing the averages of the true class limits and not of the stated limits. Thus, for instance, the mid-point of the class "$7·5 and under $12·5" should be $(7·5 + 12·49) \div 2$ and not $(7·5 + 12·50) \div 2$ which is the mean of the stated class limits. However, little accuracy will be lost by performing the latter calculation in computing class mid-points which then become $10, $15, $20, etc.

If a discrete variable is recorded only approximately, special care must be taken. Thus, for instance, if rent were quoted not exactly but to the *nearest half-dollar* (i.e. only in jumps of 50 cents), the true classes

---

[1] See p. 9 above.

FREQUENCY DISTRIBUTION AND ITS DESCRIPTION

would be "$7·25 to $12·24," "$12·25 to $17·24," etc., reflecting the general principle that the points lying to the left of the mid-point of the new unit of measurement (i.e. half-dollars) would be rounded downwards with the remaining points being rounded upwards. The mid-points for each class would be computed by averaging the true limits, i.e. $(7·25 + 12·24) \div 2 = 9$ dollars 74·5 cents, which would be rounded to $9·75. It should be noted that the class mid-point is not 10 dollars, i.e. that the effect of rounding each recorded amount to the nearest half-dollar, compared with exact measurement, is to lower all class mid-points by 25 cents.

On the other hand, rents may be quoted to the *last complete half-dollar*. If 5 dollar class intervals were required, the true class limits would be "$7·50 to $12·49," "$12·50 to $17·49," etc., with class mid-points $10, $15, etc. It was in fact on this principle that the frequency distribution in Table 5.2 was constructed.

In the case of a continuous variable (like age or weight) the observed data are always approximations and are usually quoted to the nearest ten or unit, or first or second decimal place, etc. Thus the variable age may be quoted to one decimal place, e.g., 51·3 years. This implies that the third digit is significant and we may assume that the correct value does not lie beyond the limits $51·3 \pm 0·05$. Accordingly, if we set up class limits, say 50 years and under 60, this really means 49·95 and under 59·95, and hence the true mid-point of that class is 54·95 and not 55. If age were quoted to the last full year, the class limits of 50 years and under 60 would be correct, and the mid-point of that class would lie at 55 years. The most important consideration in calculating class mid-points is that they lie halfway between the lowest true value and the highest true value which can occur in each class. This does not necessarily coincide with the halfway point between lower and upper class limits as stated. Thus, if factories are classified according to the number of their employees, a class "100 employees and under 200" has a mid-point of 149·5 and not 150.

The final point to remember is that the width of a class interval must be measured by the difference between the lower (or upper) class limits of two adjacent classes and not by the difference between the lower and upper class limit of a given class. The use of the latter measure may easily lead to error in the case of a discrete variable. Thus, if, in a distribution of houses by number of rooms, the classes are 1 to 2 rooms, 3 to 4 rooms, etc., the class interval is 2 rooms and not 1 room.

The actual mechanics of drawing up a frequency distribution are simple. The best plan is to select class intervals rather narrower than those which will be finally used and then go through the data allocating them as follows—

Weekly Rent (dollars)	Cases	Frequency
7·5 and under 8·5	8, 8	2
8·5 ,, 9·5	9	1
9·5 ,, 10·5	10, 10, 10, 9·5, 10	5
etc.	etc.	etc.

The narrow classes can then easily be rearranged and combined to form a frequency distribution which will combine sufficient smoothness with sufficient detail.

The frequency distribution can also be written down to show relative (percentage) frequencies rather than absolute frequencies. The relative frequencies column must add up to 100 per cent. This procedure is very useful for purposes of making comparisons. Table 5.4 expresses the frequency distribution of rents derived earlier on a relative basis.

Table 5.4

RELATIVE FREQUENCY DISTRIBUTION OF WEEKLY RENTS OF 200 TENANTED HOUSES

Weekly Rent (dollars)	Relative Frequency (per cent)
7·5 and under 12·5	6·0
12·5 ,, 17·5	13·0
17·5 ,, 22·5	22·5
22·5 ,, 27·5	30·0
27·5 ,, 32·5	18·5
32·5 ,, 37·5	6·5
37·5 ,, 42·5	2·5
42·5 ,, 47·5	1·0
All Rents	100·0

5.2. CUMULATIVE FREQUENCY DISTRIBUTIONS

For some purposes it is useful to show cumulative frequencies rather than frequencies of occurrence in any particular class. Thus, in our previous example we may wish to know what number (or proportion) of houses were rented at less than various amounts of rent. For quick reference we construct a cumulative frequency curve, or *ogive*.

The first step is to determine whether lower or upper class limits should serve as dividing points. In the case of a continuous variable, or a variable which can vary in small discrete jumps, this is a relatively simple matter. Thus the number of houses with rents smaller than, say,

FREQUENCY DISTRIBUTION AND ITS DESCRIPTION

22·5 will include all houses having rents ranging from $7·5 to $22·49. Similarly we can construct an ogive showing the number of houses with rents "equal to or more than" specified amounts. To distinguish between these two types of cumulative frequency curves, the first is often referred to as a "less than," and the second as an "or more" cumulative frequency curve.

In Table 5.5 the data on rents in Adelaide are shown on a cumulative basis. The table enables us to read off directly the aggregate frequencies of houses with rents lower than any particular value (middle column),

Table 5.5

CUMULATIVE FREQUENCY DISTRIBUTION OF
200 TENANTED HOUSES

Weekly Rent (dollars) X	Number of Houses with Rents *less than* X	Number of Houses with Rents *equal to or greater than* X
7·5	0	200
12·5	12	188
17·5	38	162
22·5	83	117
27·5	143	57
32·5	180	20
37·5	193	7
42·5	198	2
47·5	200	0

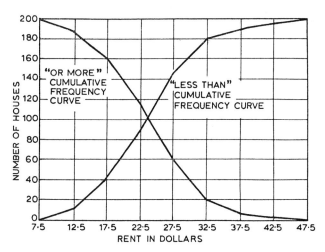

FIG. 5.4. WEEKLY RENT OF 200 TENANTED HOUSES

or the number of houses whose rents are equal to or exceed a given value (last column). In Fig. 5.4 this information is shown graphically.

If we translate frequency distributions into relative (percentage) frequency distributions, we may use the same principle to construct percentage ogives. These can then be used for comparing different sets of data. It may be noted here that Lorenz curves are an example of cumulative percentage frequency distributions constructed so as to bring out equality or inequality of distribution of two sets of data.[1]

5.3. Types of Frequency Distributions

Two tendencies are frequently observed in economic and business data, and indeed in most quantifiable data. One is a general tendency

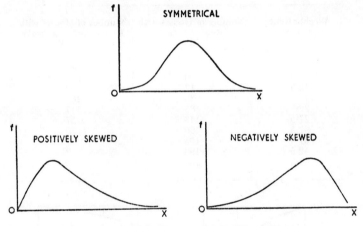

Fig. 5.5. Frequency Distributions

towards great diversity in observable phenomena. In our example, rents vary over a wide range depending on the age, location, size, etc., of the selected houses. Despite this diversity, however, most economic and business data have a tendency for clustering around particular values. Both tendencies are summarized by the shape of frequency distributions. As the variable under observation increases, frequency of occurrence increases to a maximum and then decreases. Graphed, the distributions most commonly found will be more or less bell-shaped, but they may be *symmetrical* or *skewed*. As seen in Fig. 5.5, the skewness may be *positive* (to the right) or *negative* (to the left).

Frequency distributions with only one peak are called *unimodal*.

[1] See section 4.5 above.

FREQUENCY DISTRIBUTION AND ITS DESCRIPTION

Two-peaked or *bimodal* distributions also occur, but their bimodality may often be caused by lack of homogeneity in the underlying data; when properly classified, such data often yield distributions with unimodal characteristics. Less commonly found distributions in the case of economic data are J- and U-shaped.

5.4. Measures of Central Value

Graphical representations of frequency distributions give us a good general picture of the "shape" of the original data. What we need now, however, are summary measures to describe frequency distributions. The characteristics we seek are those relating to—

(a) their central value,
(b) the degree of dispersion around the central value,
(c) their degree of skewness, if any.

The present section is concerned with measures of central value. Such measures may be looked upon as representative values about which a given distribution is centred. They will furthermore indicate the general level of magnitude of the distribution. Before developing such measures, we should first discuss the use of the summation sign in statistical work.

The Summation Sign

The symbol Σ (Greek capital sigma), which is sometimes called an *operator*, means the "sum of." Thus if X consists of three values, e.g. 4, 7, and 9, ΣX instructs us to sum all the values of X:

$$\Sigma X = 4 + 7 + 9 = 20$$

For clarity, we sometimes attach subscripts to the individual values of the variable X, i.e. $X_1 = 4$, $X_2 = 7$, $X_3 = 9$. Then we write

$$\sum_{i=1}^{3} X_i = X_1 + X_2 + X_3 = 20$$

This method of using Σ specifies the number of values of X which are being added. In general

$$\sum_{i=1}^{n} X_i = X_1 + X_2 \ldots X_n$$

But the choice of the first subscript is entirely arbitrary and we may equally well attach the subscript 0, or any other number, to the first value of X, e.g.

$$\sum_{i=0}^{n-1} X_i = X_0 + X_1 \ldots X_{n-1}$$

The following rules apply to operations with summation signs.

RULE 1: If X is a variable with n observations, and A a constant
$$\Sigma AX = A\Sigma X$$

RULE 2: If A is a constant number
$$\sum_{}^{n} A = nA$$

i.e. the constant A is added n times.

Combining these two rules, if
$$Z = A + BX,$$
$$\Sigma Z = nA + B\Sigma X$$

RULE 3: The sum of squares of a variable X is not equal to the square of its sum.

Suppose that there are n positive observations of X. Then
$$\Sigma X^2 = X_1^2 + X_2^2 \ldots + X_n^2$$

But
$$(\Sigma X)^2 = (X_1 + X_2 \ldots + X_n)(X_1 + X_2 \ldots + X_n)$$
$$= (X_1^2 + X_1 X_2 + X_1 X_3 \ldots + X_1 X_n)$$
$$+ (X_2 X_1 + X_2^2 + X_2 X_3 \ldots + X_2 X_n)$$
$$\cdot$$
$$\cdot$$
$$\cdot$$
$$+ (X_n X_1 + X_n X_2 + X_n X_3 \ldots + X_n^2)$$

Clearly
$$(\Sigma X)^2 > \Sigma X^2$$

RULE 4: The sum of products of n pairs of observations of variables X and Y is not equal to the product of their sums. For n such pairs of positive observations
$$\Sigma XY = X_1 Y_1 + X_2 Y_2 + \ldots X_n Y_n$$

But
$$\Sigma X \Sigma Y = (X_1 + X_2 + \ldots X_n)(Y_1 + Y_2 + \ldots Y_n)$$
$$= (X_1 Y_1 + X_1 Y_2 + \ldots X_1 Y_n)$$
$$+ (X_2 Y_1 + X_2 Y_2 + \ldots X_2 Y_n)$$
$$\cdot$$
$$\cdot$$
$$\cdot$$
$$+ (X_n Y_1 + X_n Y_2 + \ldots X_n Y_n)$$

Thus
$$\Sigma X \Sigma Y > \Sigma XY$$

RULE 5: $\Sigma(X + Y) = \Sigma X + \Sigma Y$. This rule merely states that the order of addition is immaterial, i.e. addition of numbers is commutative.

Arithmetic Mean

The simplest representative value of an array of numbers is the ordinary *average* or the *arithmetic mean*. For ungrouped data the mean is defined as

$$\bar{X} = \frac{X_1 + X_2 + \ldots X_N}{N} = \frac{\Sigma X}{N}$$

where N represents the number of observations of the variable X.

EXAMPLE 5.1

Given the following ten rents, calculate their mean.

X Rent (dollars)	
10·0	
10·5	
12·0	
14·5	
15·0	$\bar{X} = \frac{\Sigma X}{N} = \frac{181·5}{10} = 18·15$ dollars
18·0	
21·0	
24·5	
26·0	
30·0	

The arithmetic mean has two important properties. The first is that the sum of the deviations of values from their mean is zero.

Writing $x = X - \bar{X}$ as the deviation of a value from the mean in the above example, we have

X	x
10·0	−8·15
10·5	−7·65
12·0	−6·15
14·5	−3·65
15·0	−3·15
18·0	−0·15
21·0	2·85
24·5	6·35
26·0	7·85
30·0	11·85
$\Sigma X = 181·5$	$\Sigma x = 0$

Algebraically,
$$x = X - \bar{X},$$

$$\Sigma x = \Sigma X - N\bar{X}$$
$$= \Sigma X - N \cdot \frac{\Sigma X}{N}$$
$$= 0$$

This property of the mean can be used to facilitate computation. If we choose a value A such that $A \neq \bar{X}$, the sum of deviations will no longer be zero. Dividing this sum by N we obtain a correction factor by which A will need to be adjusted to yield the correct value of $\bar{X}$. Since A can be selected arbitrarily, it is known as *arbitrary origin*. Writing

$$x' = X - A$$
$$X = A + x'$$
$$\Sigma X = NA + \Sigma x'$$

and
$$\bar{X} = A + \frac{\Sigma x'}{N} \quad (\text{or } \bar{X} = A + \bar{x}')$$

where $\bar{x}'$ is the correction factor.

A second important property of the mean is that the sum of the squares of the deviations of values from any origin is a minimum when that origin is their mean. Using the above notation we shall require

$$\Sigma x'^2 = \Sigma(X - A)^2 \text{ to be a minimum.}$$

Since the values of X are given, the sum $\Sigma x'^2$ will depend only on the value of A as A is varied by small steps. This sum will be a minimum when its first derivative with respect to A is zero, i.e. when

$$-2\Sigma(X - A) = 0$$

i.e. when
$$-\Sigma X + NA = 0$$

i.e. when
$$A = \frac{\Sigma X}{N} = \bar{X}$$

Example 5.2

Calculate the mean rent in the above example, by using an arbitrary origin. Choosing a convenient A, e.g. $A = 20$, we have—

FREQUENCY DISTRIBUTION AND ITS DESCRIPTION 55

Rent (dollars) X	Arbitrary Deviations $A = 20$ x'
10·0	−10·0
10·5	−9·5
12·0	−8·0
14·5	−5·5
15·0	−5·0
18·0	−2·0
21·0	1·0
24·5	4·5
26·0	6·0
30·0	10·0
	$\Sigma x' = -18·5$

$$\bar{X} = A + \frac{\Sigma x'}{N} = 20 + \left(\frac{-18·5}{10}\right) = 18·15 \text{ dollars}$$

This method will prove very convenient in calculating the mean of a frequency distribution.

Computing the Mean from Grouped Data

The main point to remember is that we regard the class mid-point as representative of the observations in the class. For instance, if there are 12 rents in the class "$7·5 and under $12·5," we essentially treat these rents as though they were each 10 dollars. Evidently, the mean of a number of observations calculated from a frequency distribution of the observations will generally be only an approximation to the mean calculated from the original data. It is for this reason that it is of paramount importance to centre the classes at points which are typical of the whole classes. Consider our example of 200 rents. If instead of choosing intervals "$7·5 and under $12·5," etc., we were to define our class limits as "$5 and under $10," "$10 and under $20," etc., the mid-points would fall on $7·5, $12·5, etc. Since the rents tend to be concentrated on the "5's," the mean rent computed on this basis would be quite substantially out as compared with the mean computed directly from the original ungrouped data. So long then as the class mid-points are representative of the observations within classes, in the sense that they lie fairly close to the mean for each class, the arithmetic mean computed from grouped data will be a close approximation to the true mean, provided that the distribution is not too badly skewed. With a great mass of data, working with grouped observations is the only practicable way. In any case the nature of the data can best be appreciated when it is in the frequency distribution form.

Writing X for class mid-points and f for the frequencies in the classes, the mean for grouped data is defined as

$$\bar{X} = \frac{\Sigma fX}{\Sigma f} = \frac{\Sigma fX}{N}, \text{ where } N = \Sigma f$$

In calculating the mean for a frequency distribution, the work can be reduced by using an arbitrary origin located at the mid-point of a class. We shall have

$$\bar{X} = A + \frac{\Sigma fx'}{N}$$

where x' is the deviation of a class mid-point from an arbitrary origin. But the x''s in this formula are all multiples of the class interval. For instance, in the frequency distribution in Table 5.2, the class mid-points are 10, 15, 20, 25, 30, 35, 40, 45. If 20 is selected as arbitrary origin, the deviations of the mid-points will be −10, −5, 0, 5, 10, 15, 20, 25, respectively. These are all multiples of the class interval of 5 dollars and, in terms of the class interval, can be written −2, −1, 0, 1, 2, 3, 4, 5. When we express the arbitrary deviations in class interval units, the expression $\frac{\Sigma fx'}{N}$ will be in units of the width of the class interval, so that it must be multiplied by the width of the class interval to transform it into the original units. The formula then becomes

$$\bar{X} = A + \left(\frac{\Sigma fx'}{N} \times h\right)$$

where x' is in class interval units and h is the width of the class interval. Use of this formula simplifies the arithmetic considerably.

EXAMPLE 5.3

Calculate the mean rent of the frequency distribution set out in Table 5.2.

Rent (dollars)	Class Mid-point X	Frequency f	fX
7·5 and under 12·5	10	12	120
12·5 ,, 17·5	15	26	390
17·5 ,, 22·5	20	45	900
22·5 ,, 27·5	25	60	1,500
27·5 ,, 32·5	30	37	1,110
32·5 ,, 37·5	35	13	455
37·5 ,, 42·5	40	5	200
42·5 ,, 47·5	45	2	90
Total		200	4,765

$$\bar{X} = \frac{\Sigma fX}{N} = \frac{4,765}{200} = 23\cdot825, \text{ i.e. } 23\cdot 8 \text{ dollars (to one decimal place).}$$

FREQUENCY DISTRIBUTION AND ITS DESCRIPTION

We can reduce, however, the work in our calculations by using an **arbitrary origin**. Select as arbitrary origin the class mid-point, which from inspection appears to be near the mean, in this case $20.

Rent (dollars)	Class Mid-point X	Arbitrary Deviation $(A = 20)$ x'	Frequency f	fx'
7·5 and under 12·5	10	−10	12	−120
12·5 ,, 17·5	15	−5	26	−130
17·5 ,, 22·5	20	0	45	0
22·5 ,, 27·5	25	5	60	300
27·5 ,, 32·5	30	10	37	370
32·5 ,, 37·5	35	15	13	195
37·5 ,, 42·5	40	20	5	100
42·5 ,, 47·5	45	25	2	50
Total			200	765

$$\bar{X} = A + \frac{\Sigma fx'}{N} = 20 + \frac{765}{200} = 23 \cdot 8 \text{ dollars}$$

A further simplification can be made by working in **class interval units**.

Rent (dollars)	Class Mid-point X	Arbitrary Deviations in C.I. Units $(A = 20)$ x'	Frequency f	fx'
7·5 and under 12·5	10	−2	12	−24
12·5 ,, 17·5	15	−1	26	−26
17·5 ,, 22·5	20	0	45	0
22·5 ,, 27·5	25	1	60	60
27·5 ,, 32·5	30	2	37	74
32·5 ,, 37·5	35	3	13	39
37·5 ,, 42·5	40	4	5	20
42·5 ,, 47·5	45	5	2	10
Total			200	153

$$\bar{X} = A + \left(\frac{\Sigma fx'}{N} \times h\right) = 20 + \left(\frac{153}{200} \times 5\right) = 23 \cdot 8 \text{ dollars}$$

Weighted Arithmetic Mean

The mean for a frequency distribution, $\bar{X} = \dfrac{\Sigma fX}{\Sigma f}$, is in fact a mean of the X's (class mid-points), where each X is weighted by its importance. This is only a special case of the more general notion of a weighted mean $\bar{X} = \dfrac{\Sigma WX}{\Sigma W}$, where the W's are weights.

The concept of a weighted mean can be simplified by expressing the given weights W as relative weights $V = \dfrac{W}{\Sigma W}$ so that $\Sigma V = 1$. Then

$$\bar{X} = \Sigma VX$$

EXAMPLE 5.4

Suppose that bread is sold at three prices: 17 cents, 19 cents and 22 cents a loaf. The simple mean price is 19·3 cents, but a more useful measure is the mean which results from attaching to each price the quantity of bread sold at that price, i.e. a weighted mean. If the sales per week of the three types of bread are 50 thousand, 30 thousand and 20 thousand loaves respectively, we should have

$$\bar{X} = \frac{17 \times 50{,}000 + 19 \times 30{,}000 + 22 \times 20{,}000}{100{,}000}$$

$$= 18 \cdot 6 \text{ cents}$$

Using relative weights this reduces to

$$\bar{X} = (17 \times 0 \cdot 5) + (19 \times 0 \cdot 3) + (22 \times 0 \cdot 2) = 18 \cdot 6 \text{ cents}$$

Median

The mean is not the only measure of central value. Indeed, a simpler one is the *median*. The *median* is that value which exceeds, in magnitude, half the values and is exceeded by half the values. In the case of ungrouped data all we have to do is to arrange the data in order of magnitude and locate the median by inspection. If we have N items, the median will have to be located so that it exceeds $\dfrac{N}{2}$ items and is exceeded by $\dfrac{N}{2}$ items. By convention, when N is odd the median is located as the middle item, and when N is even the median is located as half-way between the two middle items.

EXAMPLE 5.5

What is the median of the following rents?

(*a*) 12, 12·5, 14, 15, 16, 18, 19·5
$Md = 15$ dollars.

(*b*) 12, 12·5, 14, 15, 16, 18
$Md = 14 \cdot 5$ dollars.

FREQUENCY DISTRIBUTION AND ITS DESCRIPTION 59

In the case of grouped data we have the frequencies in the various classes and not the detailed observations. Consequently the median cannot be located as a particular observation or midway between two observations. What we do is to find the value which will divide the total number of frequencies into halves. This can best be understood in terms of the histogram corresponding to the frequency distribution. We fix the median at a point on the X-axis such that a vertical line drawn at that point will divide the area of the histogram exactly in half. The area to the left of the median will then represent the lower half of the frequencies and the area to the right the upper half.

If the total number of the frequencies is N, the median will have to be such that $\frac{N}{2}$ values lie below it and $\frac{N}{2}$ above it. First, locate the median class by inspection, i.e. the class in which the median will lie. Let f_b be the number of frequencies in the classes below the median class, f_m the number in that class, and f_a the number above that class, so that $f_b + f_m + f_a = N$. If we assume that the distribution of observations in the median class is uniform over the class interval, then, by travelling a fraction $\left(\frac{N}{2} - f_b\right) / f_m$ along the class interval from the lower class limit, we shall reach a point below which $f_b + \left(\frac{N}{2} - f_b\right)$ $= \frac{N}{2}$ observations lie. This point will be the median, and it is given by the formula

$$Md = l_m + \left(\frac{\frac{N}{2} - f_b}{f_m} \times h\right)$$

where l_m is the lower limit of the median class and h is the class interval.

The same result is reached by travelling in the opposite direction. For, travelling a fraction $\left(\frac{N}{2} - f_a\right) / f_m$ along the median class interval from the lower class limit of the class immediately above the median class, we shall reach a point above which $f_a + \left(\frac{N}{2} - f_a\right) = \frac{N}{2}$ observations lie, and which is the same point as that derived from the above formula. This method holds irrespective of whether N is even or odd, because under the assumption of a uniform distribution within the median class no point other than the one given by the above method will divide the frequencies into halves.

EXAMPLE 5.6

Calculate the median rent of the frequency distribution set out in Table 5.2.

Rent (dollars)			f
7·5	and under	12·5	12
12·5	,,	17·5	26
17·5	,,	22·5	45
22·5	,,	27·5	60
27·5	,,	32·5	37
32·5	,,	37·5	13
37·5	,,	42·5	5
42·5	,,	47·5	2
All Rents			200

The median lies in the class $22·5 and under $27·5, for this class must contain the value which divides the frequencies into two parts containing 100 each. Since there are 83 items below this class, we shall have to travel another 17 items along this class to locate the median. We have

$$Md = l_m + \left(\frac{\frac{N}{2} - f_b}{f_m} \times h\right)$$
$$= 22·5 + \left(\frac{100 - 83}{60} \times 5\right)$$
$$= 23·9 \text{ dollars}$$

Quartiles

A knowledge of the median helps to characterize a frequency distribution by virtue of its central position. The quartiles are analogous measures. The first quartile (Q_1) is the value which exceeds, in magnitude, one-quarter of the values and is exceeded by three-quarters of the values. The second quartile is the median. The third quartile (Q_3) is the value which exceeds, in magnitude, three-quarters of the values and is exceeded by one-quarter of the values. The quartiles are calculated in a similar fashion to the median.

EXAMPLE 5.7

Calculate the first and third quartiles of the frequency distribution set out in Table 5.2.

$$Q_1 = 17·5 + \left(\frac{50 - 38}{45} \times 5\right)$$
$$= 18·8 \text{ dollars}$$
$$Q_3 = 27·5 + \left(\frac{150 - 143}{37} \times 5\right)$$
$$= 28·4 \text{ dollars}$$

FREQUENCY DISTRIBUTION AND ITS DESCRIPTION 61

Quintiles, deciles and percentiles can be defined and calculated similarly.

Mode

If we were to increase our number of observations very greatly and at the same time narrow the class interval, the histogram of our frequency distribution would approach a smooth curve. Such a smooth curve is shown in the diagram below.

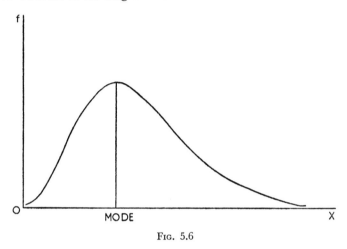

FIG. 5.6

The value of the variable at which the frequency curve reaches a maximum is called the *mode*. It is the value around which the items tend to be most heavily concentrated.

In a frequency distribution it is quite easy to locate the modal class, i.e. the class in which the mode lies. It is the most populous class and in Table 5.2 above is $22·5 and under $27·5. The location of an estimate of the mode itself is more difficult, for it is in a sense a hypothetical measure relating to the smooth curve underlying the observed frequency distribution. One procedure is to pass a smooth curve through the three central frequencies (a parabola) and to locate the mode in that fashion. This has the effect of locating the mode at a point away from the centre of the modal class towards the one of the two adjoining classes which has the greater frequency. Thus, if the class below the modal class contains more cases than the one above, the mode will be less than the mid-point of the modal class. This accords with common sense, as it is reasonable to suppose that, in these circumstances, the point of concentration lies to the left of the centre of the modal class. The method is illustrated in the diagram below.

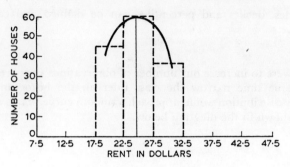

Fig. 5.7

This method yields the following formula—

$$Mo = l + \left(\frac{f_0 - f_{-1}}{(f_0 - f_{-1}) + (f_0 - f_1)} \times h \right)$$

where l is the lower limit of the modal class,

f_0 is the frequency of modal class,

f_{-1} is the frequency of next lower class,

f_1 is the frequency of next higher class.

EXAMPLE 5.8

Estimate the mode of the frequency distribution in Table 5.2.

$$\begin{aligned} Mo &= l + \left(\frac{(f_0 - f_{-1})}{(f_0 - f_{-1}) + (f_0 - f_1)} \times h \right) \\ &= 22 \cdot 5 + \left(\frac{60 - 45}{60 - 45 + 60 - 37} \times 5 \right) \\ &= 24 \cdot 4 \text{ dollars} \end{aligned}$$

Geometric Mean

The geometric mean is defined as

$$\text{G.M.} = \sqrt[N]{X_1 . X_2 \ldots X_N}$$

i.e. as the Nth root of the product of N observations of a variable X. Its meaning can perhaps be best understood with the aid of simple algebra.

It will be remembered that an arithmetic progression is a sequence of numbers each of which differs from the one which precedes it by a

FREQUENCY DISTRIBUTION AND ITS DESCRIPTION

constant amount called the common difference. A geometric progression is a sequence of numbers such that any term after the first is obtained from the preceding term by multiplying it by a fixed number called the common ratio. Consider two sequences 2, 4, 6 . . . and 2, 4, 8 . . . which from our definitions form an arithmetic progression and a geometric progression, respectively. We may now define the term of a progression, which lies between any two given terms, as the mean of those two terms. The middle term in the first sequence is equal to its arithmetic mean; the middle term of the second sequence, however, coincides with its geometric mean, i.e.

$$\text{G.M.} = \sqrt[3]{(2)(4)(8)} = 4$$

The arithmetic mean of any sequence of positive numbers will be greater than the geometric mean so long as the items averaged are not all of the same value.

In practice, the computation of the geometric mean is greatly facilitated by logarithms. Thus for ungrouped data

$$\log \text{G.M.} = \frac{1}{N} (\log X_1 + \log X_2 \ldots + \log X_N)$$

$$= \frac{\Sigma \log X}{N}$$

For grouped data

$$\log \text{G.M.} = \frac{1}{N} (f_1 \log X_1 + f_2 \log X_2 + \ldots f_n \log X_N)$$

$$= \frac{\Sigma f \log X}{N}$$

where f represents class frequencies and X the mid-point of each class. The logarithm of the geometric mean is the arithmetic mean (or, where appropriate, the weighted arithmetic mean) of the logarithms of the variable.

The most important property of the geometric mean is that the product of the ratio deviations of the observed values below and above the mean equals unity. This property makes the geometric mean a particularly suitable measure for averaging ratios and for computing average rates of change. In averaging ratios, the use of the arithmetic mean sometimes leads to inconsistent results, for instance in index number construction.[1] Such inconsistencies may be overcome by averaging relatives geometrically rather than arithmetically.

[1] See Ch. XIII, p. 340.

Harmonic Mean

The harmonic mean is not used a great deal, but it does occur in index-number formulae and should be mentioned here. It is defined as

$$\text{H.M.} = \frac{1}{\frac{1}{N}\sum \frac{1}{X}} = \frac{N}{\sum \frac{1}{X}}$$

i.e. the harmonic mean is the reciprocal of the arithmetic mean of the reciprocals of the variable. Thus, given the values 2, 4, 8 in the previous example

$$\text{H.M.} = \frac{3}{\frac{1}{2} + \frac{1}{4} + \frac{1}{8}} = 3\cdot 4$$

This is less than 4·7 and 4 which were the values of the arithmetic mean and the geometric mean, respectively. In general,

$$\text{H.M.} < \text{G.M.} < \bar{X}$$

unless the values being averaged are all of the same positive magnitude, in which case $\text{H.M.} = \text{G.M.} = \bar{X}$.

Quadratic Mean

Another useful measure in statistics is the quadratic mean. Although almost never used to average the values of actual observations, the quadratic mean is applied in the computation of measures of dispersion. It is defined as

$$\text{Q.M.} = \sqrt{\frac{X_1^2 + X_2^2 \ldots X_n^2}{N}} = \sqrt{\frac{\sum X^2}{N}}$$

For positive values of X, not all of equal value,

$$\text{H.M.} < \text{G.M.} < \bar{X} < \text{Q.M.}$$

Relations between the Mean, Median and Mode

The most commonly applied measures of central tendency in statistical data are the arithmetic mean, the median and the mode, and we now discuss further their characteristics and their respective advantages and disadvantages in description of statistical data.

If we interpret these three measures in terms of the smooth frequency curve which we should get if we had a very large number of observations, we can say that the mean is the value of the variable which is the point of balance or centre of gravity of the distribution, the median is the value which divides the distribution exactly in half, and the mode is the value at which the peak of the distribution occurs. In a symmetrical distribution all three measures coincide; but skewness pushes the measures apart. This can be seen by starting with a symmetrical distribution and then adding some high values at the upper end of the distribution. Positive skewness will result. The mode will remain unchanged, but the median will increase as the mode will now be exceeded by more than half the values. The mean will tend to increase even more. The reason for this can be appreciated when it is realized that the sum of the negative deviations, from the mean, of the values below the mean must equal in magnitude that of the positive deviations of the values above the mean, since the algebraic sum of the deviations

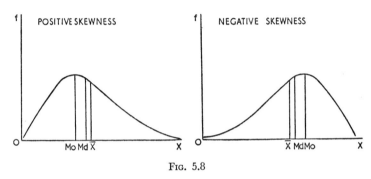

Fig. 5.8

from the mean must be zero. If the values are deviated from the new median, the sum will tend to be positive because that half of the values above the median will contain more extreme values than that half below the median. Hence the new mean will tend to be higher than the new median.

In practice, in moderately skewed distributions it is found that the median is usually one-third along the distance between the mean and the mode, the order being mode, median, mean in positively skewed distribution and mean, median, mode in negatively skewed distributions. This empirical relation can be expressed as

$$\bar{X} - Mo = 3(\bar{X} - Md)$$

This is illustrated in Fig. 5.8.

When a distribution is perfectly symmetrical, the three measures coincide and in cases of near symmetry they will be very close. However, in general, the mean possesses important advantages over the other two measures and is to be preferred in such cases as a measure of central value.

The first advantage is that the mean is defined algebraically, and is therefore capable of algebraic treatment. For example, if we have the means of several separate distributions of the same variable, we can combine them to obtain the mean of the combined distribution. This can be done by weighting the means of the separate distributions by their respective total frequencies, so that we have

$$\bar{X} = \frac{N_1\bar{X}_1 + N_2\bar{X}_2 + N_3\bar{X}_3 + \ldots}{N_1 + N_2 + N_3 \ldots}$$

where the subscripts refer to the separate distributions and $\bar{X}$ is the average of the means for the combined distribution. This cannot be done for the median or the mode. Similarly, if we have N values of a variable and we know the aggregate of the values (ΣX), we can calculate the mean even though we do not know the distribution in detail and hence cannot calculate the median or the mode. The mean is a key measure in statistical methods. Its two important properties, namely that the sum of the deviations of values from their mean is zero and that the sum of the squares of the deviations of values from an origin is a minimum when the origin is the mean, should be recalled at this point.

Secondly, when an attempt is made to estimate from sample data the central value of a totality of observations, the distribution of which is symmetrical, the mean will give a more reliable estimate in the sense that the means of repeated samples will show less dispersion than either the modes or medians. This is tantamount to saying that the sampling error of the mean is less than that of the median or mode. Sampling error is discussed in detail in Chapters VII and VIII.

On the other hand, the median and mode must be used when the mean cannot be satisfactorily calculated. This is the case when the distribution has open end classes, i.e. when the lower limit of the bottom class and the upper limit of the top class are not defined, and to a lesser extent when the class intervals are unequal. Evidently, one can still compute the mode, as well as the median, since the latter is not affected by the values of the variable but merely by their frequencies of occurrence.

When, however, the distribution is skewed, the mean, median and mode differ in their significance. They are no longer measures of *the* central value, and each has its use. The use of the median can be best illustrated by reference to distributions of income. These are usually positively skewed. This results in less than half of the income-earners receiving the mean income or more. Consequently the median income may, for some purposes, be regarded as a more representative figure, for half the income-earners must be receiving at least the median income. One can say that as many receive at least the median income as do not.

FREQUENCY DISTRIBUTION AND ITS DESCRIPTION 67

Whether the median or the mean should be used in analysis of income data often depends on the purpose of the exercise. Thus suppose that in a certain district there are five farmers with annual incomes of $3,000, $4,000, $6,000, $8,000 and $14,000. The mean is $7,000 and the median $6,000. If the last farmer's income were to increase substantially, e.g. to $29,000, the mean would increase to $10,000 whereas the median would be unaffected. The fact that the median is unaffected by extreme values, while the mean is, makes the mean a useful measure for some purposes, i.e. for measuring the overall level of income in the district, and the median for others, i.e. for inferring something about the division of incomes within the district.

The usefulness of the mode can be best illustrated by reference to certain types of distributions of discrete variables. Consider, for example, family size, which invariably is positively skewed. The mode will be the most common sized family and may for some purposes be regarded as more representative than the mean. It certainly is the "typical" family, in the everyday sense of the word. Mean family size may be, say, 2·8 children whereas the modal family is a two-child one. In the extreme case of reversed J-shaped distributions, the mean or median are in no sense "representative" values, whereas the mode may still be regarded as a representative or typical value, although it is clearly not a central value. Thus the distribution of number of cars owned by one household would be of this kind, with a modal value of one, and a mean higher than one.

Irrespective of whether or not the mean or the median can in certain cases be used as "representative" values, they always indicate important characteristics of the distribution. The mean gives a measure of the general level of magnitude of the variable under consideration in the sense that, if we multiply the mean by the total number of observations, we get their aggregate value (i.e. $N\bar{X} = \Sigma X$, by definition). Likewise the chances are even that any one observation in the distribution under consideration will be at least as large as the median.

5.5. MEASURES OF DISPERSION

The central value of a frequency distribution tells us something about the general level of magnitude of the distribution and, in the case of the usual bell-shaped distribution, indicates the point about which the observations tend to concentrate. This concentration can be more or less dispersed, however, as is illustrated in Fig. 5.9. Some measure of the dispersion about the central value is, therefore, required.

Range

One elementary measure is the *range*, i.e. the highest minus the lowest value. This is clearly unsatisfactory, since it depends on only the two extreme values and will, in general, get larger the larger the number of observations made.

Quartile Deviation

A more useful measure is the *quartile deviation*, defined as

$$Q.D. = \frac{Q_3 - Q_1}{2}$$

In a symmetrical distribution, $Md \pm Q.D.$ will contain 50 per cent of the cases, and in an asymmetrical distribution this will be approximately true. It clearly measures concentration of the observations around the median. The greater the quartile deviation the more the dispersion.

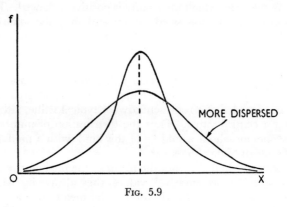

Fig. 5.9

EXAMPLE 5.9

Calculate the quartile deviation of the frequency distribution set out in Table 5.2 above.

$$Q.D. = \frac{Q_3 - Q_1}{2}$$

$$= \frac{28 \cdot 4 - 18 \cdot 8}{2}$$

$$= 4 \cdot 8 \text{ dollars}$$

In general the quartile deviation is not very satisfactory, because it is not based on all the values in the distribution and it is not capable of mathematical manipulation.

Mean Deviation

Ideally a measure of dispersion should take into account the deviations of all the observations from their central value. The obvious

FREQUENCY DISTRIBUTION AND ITS DESCRIPTION

suggestion is to calculate the mean of the deviations of all the items from their mean, i.e. if the deviation of an observation from the mean is written $x = X - \bar{X}$, calculate $\frac{\Sigma x}{N}$. But this always equals zero. If we ignore the minus signs in the deviations, we shall get a measure called the *mean deviation*—

$$\text{M.D.} = \frac{\Sigma |x|}{N}$$

where $|x|$ means the magnitude of x, irrespective of sign, i.e. the absolute value of a deviation. For grouped data, the equivalent formula would be

$$\text{M.D.} = \frac{\Sigma f |X - \bar{X}|}{N} = \frac{\Sigma f |x|}{N}$$

where X now represents class mid-points, and f the frequency weights.

In practice, however, the mean deviation of grouped data is awkward to calculate. It is furthermore incapable of mathematical manipulation. For these reasons, the mean deviation is hardly ever used in this form. Given that the data are not highly skewed, a variant of this method is based on the use of the median class mid-point, instead of the mean, as a measure of central value. This technique still gives us a measure of absolute dispersion in grouped data while greatly reducing computational labour, as illustrated by example 5.10.

EXAMPLE 5.10

Using the mid-point of the median class as measure of central value, calculate the mean deviation from the data set out in Table 5.2.

Weekly Rent	Class Mid-point X	Frequency f	Absolute Deviations in C.I. Units ($A = 25$) $|d|$	$f|d|$
7·5 and under 12·5	10	12	3	36
12·5 ,, 17·5	15	26	2	52
17·5 ,, 22·5	20	45	1	45
22·5 ,, 27·5	25	60	0	0
27·5 ,, 32·5	30	37	1	37
32·5 ,, 37·5	35	13	2	26
37·5 ,, 42·5	40	5	3	15
42·5 ,, 47·5	45	2	4	8
Total		200		219

$$\text{M.D.} = \frac{\Sigma f|d|}{N} \times h = \frac{219}{200} \times 5 = 5\cdot5 \text{ dollars}$$

Standard Deviation

If, instead of ignoring the signs of the x's, we square them, the result will be a series of positive squared deviations. The mean of these gives us what is called the *variance of X*—

$$\text{Var}(X) = \frac{\Sigma x^2}{\mathcal{N}}$$

However, when we calculate the variance for a group of observations, the group usually comes from a still larger group which is the totality of all possible observations of X. Generally we require the variance of our group as an estimate of the variance of the totality from which our observations have come. It can be shown[1] that when the variance of the larger group is defined as above, the variance of the smaller group should be defined as

$$\text{Var}(X) = \frac{\Sigma x^2}{\mathcal{N} - 1}$$

in order for it to be a good estimate of the variance of the larger group. Consequently we shall use $\mathcal{N} - 1$ rather than $\mathcal{N}$ as the divisor. The reason for doing this will be elaborated in a later chapter. Some texts on statistical methods use $\mathcal{N}$ as the divisor and make adjustments for the resulting bias at a later stage. So long as $\mathcal{N}$ is large, it is immaterial whether $\mathcal{N}$ or $\mathcal{N} - 1$ is used, but the matter is of some importance in small groups of observations of X.

The variance measures the variation or dispersion of the observations about their mean. It is a most important concept. The units in which the variance is expressed will be the original units squared and, to make the units of our measure of dispersion comparable with our original units, we take the square root. The result is akin to the concept of the quadratic mean mentioned earlier, and is known as the *standard deviation* or *root mean square deviation* of X—

$$s = \sqrt{\frac{\Sigma(X - \bar{X})^2}{\mathcal{N} - 1}} = \sqrt{\frac{\Sigma x^2}{\mathcal{N} - 1}}$$

Like the mean, the standard deviation is expressed in the same units as the original data. Both the standard deviation (s) and its square the variance ($Var(X)$ or s^2) are used as measures of dispersion. For some purposes, it is convenient to use the one, for other purposes the other.

EXAMPLE 5.11

Calculate the standard deviation of the following seven rents: $12, $12·5, $14, $15, $16, $18, $19·5.

The mean must first be calculated so that the deviation of the observations

---

[1] See pp. 121 and 122 below.

FREQUENCY DISTRIBUTION AND ITS DESCRIPTION

from their mean can be computed. If we require our standard deviation correct to, say, one decimal place, we shall have to work with three decimal places.

Rent (dollars) X	Deviation from Mean x	x^2
12·0	−3·286	10·798
12·5	−2·786	7·762
14·0	−1·286	1·654
15·0	−0·286	0·082
16·0	0·714	0·510
18·0	2·714	7·366
19·5	4·214	17·758
107·0		45·930

$$\bar{X} = \frac{\Sigma X}{N} = \frac{107}{7} = 15 \cdot 286, \text{ i.e. } 15 \cdot 3 \text{ dollars to one decimal place.}$$

$$s = \sqrt{\frac{\Sigma x^2}{N-1}} = \sqrt{\frac{45 \cdot 930}{6}} = 2 \cdot 8 \text{ dollars to one decimal place.}$$

It will be readily seen that the calculations in the above example are onerous and would be even more so if greater accuracy were desired. Fortunately a method is available which combines as high a degree of accuracy as one pleases with less arduous arithmetic. This "short method" makes use of an arbitrary origin. If we put

$$x' = X - A, \text{ where } A \text{ is an arbitrary origin,}$$

then $$\bar{X} = A + \bar{x}'$$

From this we have
$$X - \bar{X} = x' - \bar{x}'$$

i.e. $$x = x' - \bar{x}'$$

$$x^2 = x'^2 - 2x'\bar{x}' + \bar{x}'^2$$

and $$\Sigma x^2 = \Sigma x'^2 - 2\bar{x}'\Sigma x' + N\bar{x}'^2$$
$$= \Sigma x'^2 - 2N\bar{x}'^2 + N\bar{x}'^2$$
$$= \Sigma x'^2 - N\bar{x}'^2$$

Hence $$s = \sqrt{\frac{\Sigma x^2}{N-1}} = \sqrt{\frac{\Sigma x'^2 - N\bar{x}'^2}{N-1}}$$

For machine computation it is most convenient to work with the arbitrary origin of zero. Then

$$\Sigma x^2 = \Sigma X^2 - N\bar{X}^2$$

and

$$s = \sqrt{\frac{\Sigma X^2 - N\bar{X}^2}{N - 1}}$$

EXAMPLE 5.12

Calculate the standard deviation in the above example, using an arbitrary origin. Setting $A = 0$, we can work directly with the original observations.

Rent (dollars) X	X^2
12·0	144·00
12·5	156·25
14·0	196·00
15·0	225·00
16·0	256·00
18·0	324·00
19·5	380·25
	1,681·50

$$s = \sqrt{\frac{\Sigma X^2 - N\bar{X}^2}{N - 1}} = \sqrt{\frac{1{,}681 \cdot 5 - 7(15 \cdot 286)^2}{6}}$$

= 2·765, i.e. 2·8 dollars to one decimal place.

It should be noted that the same result would be achieved if we chose, say, $15 as our arbitrary origin:

Rent (dollars) X	Arbitrary Deviation ($A = 15$) x'	x'^2
12·0	−3·0	9·00
12·5	−2·5	6·25
14·0	−1·0	1·00
15·0	0	0
16·0	1·0	1·00
18·0	3·0	9·00
19·5	4·5	20·25
	2·0	46·50

$$\bar{x}' = \frac{\Sigma x'}{N} = \frac{2}{7} = 0 \cdot 286$$

$$s = \sqrt{\frac{\Sigma x'^2 - N\bar{x}'^2}{N - 1}} = \sqrt{\frac{46 \cdot 50 - 7(0 \cdot 286)^2}{6}}$$

= 2·8 dollars.

FREQUENCY DISTRIBUTION AND ITS DESCRIPTION 73

Note also that $\bar{X}$ can be readily calculated at the same time as s. Here $\bar{X} = A + \bar{x}' = 15 + 0.286 = 15.286$, i.e. 15·3 to one decimal place.

This method is simpler and more accurate than the original one, even though the formulae are somewhat more complicated. When we come to the case of grouped data, the deviations are deviations of class mid-points, and we have to weight these by their respective class frequencies. The formula becomes

$$s = \sqrt{\frac{\Sigma f(X - \bar{X})^2}{N-1}} = \sqrt{\frac{\Sigma f x^2}{N-1}} = \sqrt{\frac{\Sigma f x'^2 - N\bar{x}'^2}{N-1}}$$

where the X's refer to class mid-points and the f's to the corresponding frequencies. If we express the deviations in class interval units, we have

$$s = \sqrt{\frac{\Sigma f x'^2 - N\bar{x}'^2}{N-1}} \times h$$

EXAMPLE 5.13

Calculate the standard deviation of the frequency distribution set out in Table 5.2 above.

We first perform the calculation by using the original formula

$$s = \sqrt{\frac{\Sigma f x^2}{N-1}}$$

From example 5.3 above we know that the mean is 23·825 dollars. We then have

Rent (dollars)	Class Mid-point X	x	Frequency f	x^2	fx^2
7·5 and under 12·5	10	−13·825	12	191·131	2,293·572
12·5 ,, 17·5	15	−8·825	26	77·880	2,024·880
17·5 ,, 22·5	20	−3·825	45	14·631	658·395
22·5 ,, 27·5	25	1·175	60	1·380	82·800
27·5 ,, 32·5	30	6·175	37	38·131	1,410·847
32·5 ,, 37·5	35	11·175	13	124·880	1,623·440
37·5 ,, 42·5	40	16·175	5	261·631	1,308·155
42·5 ,, 47·5	45	21·175	2	448·380	896·760
Total			200		10,298·849

$$s = \sqrt{\frac{\Sigma f x^2}{N-1}} = \sqrt{\frac{10,298 \cdot 849}{199}} = 7 \cdot 194, \text{ i.e. } 7 \cdot 2 \text{ dollars to one decimal place.}$$

The computations are too burdensome, and once again by working from an arbitrary origin, the arithmetic can be greatly reduced. Thus we have

Rent (dollars)	Class Mid-point X	Arbitrary Deviations in C.I. Units ($A = 25$) x'	Frequency f	fx'	fx'^2
7·5 and under 12·5	10	−3	12	−36	108
12·5 ,, 17·5	15	−2	26	−52	104
17·5 ,, 22·5	20	−1	45	−45	45
22·5 ,, 27·5	25	0	60	0	0
27·5 ,, 32·5	30	1	37	37	37
32·5 ,, 37·5	35	2	13	26	52
37·5 ,, 42·5	40	3	5	15	45
42·5 ,, 47·5	45	4	2	8	32
Total			200	−47	423

$$\bar{x}' = \frac{\Sigma fx'}{N} = -\frac{47}{200} = -0.235$$

$$s = \sqrt{\frac{\Sigma fx'^2 - N\bar{x}'^2}{N-1}} \times h = \sqrt{\frac{423 - 200(-0.235)^2}{199}} \times 5$$

= 7·2 dollars to one decimal place.

Note that $\bar{X}$ can be readily calculated at the same time as s. Here

$$\bar{X} = A + (\bar{x}' \times h) = 25 + (-0.235 \times 5) = 23.8$$

Characteristics of the Standard Deviation

The mean and the standard deviation are fundamental measures in statistical theory. They play a key part in a special distribution of great importance known as the *normal distribution*. The normal distribution is a symmetrical bell-shaped distribution, and is completely determined by its mean and standard deviation. It will be discussed in detail in the following chapter.

Many actual distributions closely approximate the normal one. If we have a frequency distribution which is normally distributed and we lay off a distance equal to one standard deviation on both sides of the mean, the resulting range will contain 68·27 per cent of the items of the distribution, i.e. 68·27 per cent of the items will have values lying between the mean ± one standard deviation. Similarly, the range given by two standard deviations on both sides of the mean will contain 95·45 per cent of the items, and one given by three standard

FREQUENCY DISTRIBUTION AND ITS DESCRIPTION 75

deviations will contain 99·73 per cent of the items. Although the above holds exactly only for normal distributions, it holds approximately for distributions which are moderately skewed away from the normal. Incidentally, in the normal distribution the quartile deviation is always 0·6745 of the standard deviation, and the mean deviation is 0·7979 of the standard deviation.

The standard deviation measures the absolute dispersion or variability of a distribution in terms of the original units, i.e. in the case of rents, in dollars. Thus in example 5.13 the absolute dispersion is 7·2 dollars, and approximately two-thirds of the rents will be included within the range $23·8 \pm $7·2. The greater the amount of dispersion or variability in a frequency distribution, the greater the standard deviation, and the greater the absolute magnitude of the deviations of the values from their mean. This can be clearly seen from the properties of the standard deviation referred to in the preceding paragraph. The range given by one standard deviation marked off on both sides of the mean contains about two-thirds of the items. Hence the greater the standard deviation the greater will be the range required to include the middle two-thirds of the items.

Relative Dispersion

The standard deviation, being a measure of absolute dispersion, is expressed in the units in which the observations concerned have been made. This limits its use for comparative purposes in cases where we wish to compare the dispersion of two series of values, whose means are very different or which are expressed in different units. Suppose we are told that the weekly wages of males and females have the following properties—

$$\text{Male:} \quad \bar{X} = \$50 \quad s = \$5$$
$$\text{Female:} \quad \bar{X} = \$40 \quad s = \$4 \cdot 8$$

Male wages show greater dispersion or variability than female, but the general level of male wages is higher. In this case we may measure relative dispersion, by what is known as the *coefficient of variation*. It is defined by

$$V = \frac{s}{\bar{X}}$$

This V should not be confused with $\text{Var}(X)$, the variance of X. In our example,

$$\text{Male:} \quad V = \frac{5 \cdot 0}{50} = 0 \cdot 10, \text{ or } 10 \text{ per cent}$$

and

$$\text{Female:} \quad V = \frac{4 \cdot 8}{40} = 0 \cdot 12, \text{ or } 12 \text{ per cent}.$$

Accordingly, female wages show greater relative dispersion or variability than male. Similarly we may have two series expressed in different units. For example, the weights and heights of a group of persons:

Weight: $\bar{X} = 150$ lb $s = 10$ lb
Height: $\bar{X} = 69$ in. $s = 2$ in.

Is the variability in weight or height greater? This could only be answered by calculating V, which here is 6·7 per cent for weight and 2·9 per cent for height.

Standard Deviation Units

Sometimes we are interested in measuring how much above or below average a particular observation is. Suppose that a man's weight is 160 lb and his height is 72 in. Is he relatively more above average in weight or height? If the data in the previous example apply, we can calculate for weight,

$$\frac{X - \bar{X}}{s} = \frac{160 - 150}{10} = 1 \text{ standard deviation above average}$$

for height,

$$\frac{X - \bar{X}}{s} = \frac{72 - 69}{2} = 1\cdot5 \text{ standard deviations above average.}$$

Clearly he is relatively taller. The formula $\dfrac{X - \bar{X}}{s}$, known as *standard measure* or *standardized value*, expresses the deviation of an observation from the mean in terms of standard deviation units, i.e. the extent to which X is above (or below) average in terms of the mean variability of the data. Standardized values play an important role in statistical analysis and will be again referred to later.

5.6. Measures of Skewness

We have already discussed the nature of skewness, and it was shown that positive skewness results in the mean being greater than the mode and negative skewness in the mean being less than the mode. Pearson has suggested a measure of skewness, known as the *Pearsonian measure of skewness*—

$$Sk_P = \frac{\bar{X} - Mo}{s}$$

or

$$Sk_P = \frac{3(\bar{X} - Med)}{s}$$

The standard deviation is included as the denominator, so that the degree of skewness is measured relative to the dispersion of the distribution. This measure, being a ratio, is a pure number and it varies

FREQUENCY DISTRIBUTION AND ITS DESCRIPTION 77

between the limits ± 3. Values as large as ± 1 are quite unusual. It is positive for positive skewness, negative for negative skewness, and zero when the distribution is symmetrical.

Another measure of skewness is based on the quartiles. In a symmetrical distribution the third quartile is the same distance above the median as the first quartile is below it, i.e.

$$Q_3 - Md = Md - Q_1$$

If the distribution is positively skewed, the top 25 per cent of the values will tend to be further from the median than the bottom 25 per cent, i.e. Q_3 will be further from Md than Q_1 is from Md, and the reverse for negative skewness. Hence a possible measure is

$$Sk_Q = \frac{(Q_3 - Md) - (Md - Q_1)}{Q_3 - Q_1}$$

The denominator is in fact twice the quartile deviation, so that the degree of skewness is again measured relative to the dispersion of the distribution. This measure is called the *quartile measure of skewness*, and it varies between the limits of ± 1.

There is no relation between the magnitudes of Sk_P and Sk_Q. Sk_Q suffers from the defect that only the middle 50 per cent of the distribution is taken into account. It is possible to have Sk_P positive and Sk_Q negative. Both Sk_P and Sk_Q are empirical measures and are useful only for comparative purposes. More complicated analytical measures are available, but these will not be discussed here.

EXAMPLE 5.14

Calculate a measure of skewness for the frequency distribution in Table 5.2 above.

$$Sk_P = \frac{\bar{X} - Mo}{s}$$

$$= \frac{23 \cdot 8 - 24 \cdot 4}{7 \cdot 2} = -\frac{0 \cdot 6}{7 \cdot 2}$$

$$= -0 \cdot 08$$

and

$$Sk_Q = \frac{(Q_3 - Md) - (Md - Q_1)}{Q_3 - Q_1}$$

$$= \frac{(28 \cdot 4 - 23 \cdot 9) - (23 \cdot 9 - 18 \cdot 8)}{28 \cdot 4 - 18 \cdot 8}$$

$$= -0 \cdot 06$$

Both measures indicate a very slight degree of negative skewness. Our distribution of 200 rents is very nearly symmetrical, as already indicated by the mean, median and mode which are $23·8, $23·9 and $24·4 (to one decimal point), respectively.

5.7. Description of Frequency Distributions

We have now discussed the methods by which we can describe an unorganized mass of data. First, we form it into a frequency distribution. We can then draw a histogram which will give us a clear picture of the general shape of the distribution. Secondly, we calculate measures of central value, dispersion and skewness, which will characterize the distribution.

If we have two distributions to compare, it is useful to draw frequency polygons of the relative frequencies on the same scale and superimpose them. This immediately enables us to pick out the salient similarities and differences between the two distributions. We can then compute the various characteristic measures of the two distributions. Given the mean, the standard deviation and a measure of skewness, we shall be able to describe the main features of the distributions. A more detailed description can be made with the aid of the mode, median and quartiles.

Example 5.15

We are given a frequency distribution of metropolitan rents as set out in Table 5.2 above, and another distribution of 100 tenanted houses in a rural urban area, as set out in the third column of the table below. Compare and discuss the two distributions.

Rent (dollars)	Number of Houses		Percentage Distribution	
	Metropolitan	Rural Urban	Metropolitan	Rural Urban
2·5 and under 7·5	0	11	0	11·0
7·5 ,, 12·5	12	30	6·0	30·0
12·5 ,, 17·5	26	32	13·0	32·0
17·5 ,, 22·5	45	19	22·5	19·0
22·5 ,, 27·5	60	5	30·0	5·0
27·5 ,, 32·5	37	2	18·5	2·0
32·5 ,, 37·5	13	1	6·5	1·0
37·5 ,, 42·5	5	0	2·5	0
42·5 ,, 47·5	2	0	1·0	0
Total	200	100	100·0	100·0

The relative distributions of the rents are shown in the right-hand side of the above table and the corresponding frequency polygons are shown in Fig. 5.10.

FREQUENCY DISTRIBUTION AND ITS DESCRIPTION 79

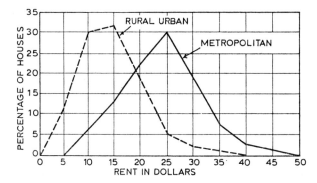

FIG. 5.10. WEEKLY RENT OF TENANTED HOUSES IN A METROPOLITAN
AREA AND A RURAL URBAN AREA
(Percentage Distribution)

Below are set out various characteristic measures of the two distributions—

Measure	Metropolitan	Rural Urban
$\bar{X}$	$23·8	$14·4
Md	$23·9	$13·9
Mo	$24·4	$13·2
Q_1	$18·8	$ 9·8
Q_3	$28·4	$18·0
s	$ 7·2	$ 6·1
V	30·2%	42·4%
Sk_p	−0·08	0·20
Sk_Q	−0·06	0

In comparison with metropolitan rents the distribution of which is very nearly symmetrical the distribution of rural urban rents appears slightly skewed to the right. The Pearsonian measure of skewness indicates a small degree of positive skewness, whereas the quartile measure suggests a symmetrical distribution. This apparent contradiction is due to the fact that the former measure takes the whole distribution into account; the latter concentrates on the middle of the distribution and ignores the tails, the influence of the positive tail with a small number of high rents thus being left out. The general level of rents in the rural urban centre is markedly lower than in the metropolitan area, as indicated clearly by all three measures of central tendency. Finally, although the metropolitan rents have a greater absolute dispersion, the variability relative to the general level of rents is greater in the rural urban district.

CHAPTER VI

PROBABILITY AND PROBABILITY DISTRIBUTIONS

6.1. Statistical Methods and Probability

So far we have dealt with the various techniques and measures by which a collection of data can be described and summarized. The next step is to introduce some elementary techniques of statistical analysis which will enable us to make valid deductions from the particular to the general. Thus, often we have a group of observations of a particular variable which covers only a fraction of all possible cases, and on the basis of this information we wish to infer something about the characteristics of the larger group from which the sample data had been drawn. The theory of statistics provides a scientific basis for making such judgments, and enables us to test hypotheses about the nature of the larger groups from which a particular set of data had presumably been obtained. The techniques employed in such analysis will be discussed in subsequent chapters.

Before embarking on this task, however, we need to introduce some basic ideas relating to probability and its measurement. Such ideas derive from a special branch of statistical methods known as the *theory of probability*. In general, the theory of probability provides a basis for the scientific analysis of uncertainty and its applications enable us to obtain a degree of predictability from uncertain states of nature. It has many specialized applications in economics and business, but these will not be considered here.

Although the term "probability" has a broad meaning with which we are all familiar, its definition and interpretation become the source of major difficulties when the term is to be defined strictly. There is in fact no single definition of probability. Probability statements can be derived both on an objective and subjective basis, experimentally as well as by *a priori* reasoning. Here we shall not be concerned with the nature of probability; and we shall limit our discussion to the case where probabilities are interpreted objectively. For the present purpose it will suffice to introduce concepts relating to the measurement of probabilities, and to derive a few simple rules for the algebraic manipulation of probability measures.

6.2. Measurement of Probability

Statistical Events

Uncertainty arises when any given situation has more than one possible outcome. Thus it may, or may not, rain later today, and we shall never be completely certain whether to take an umbrella or not. Nevertheless, we may wish to study uncertain events with a view to predicting which of the outcomes appears to be the most likely. The first step is to ascertain all the logical possibilities which may occur in any given situation. In statistical terminology, these possibilities form a collection or set of all *possible outcomes*.

In dealing with uncertain situations, it is often convenient to perform experiments, real or conceptual. Thus the toss of a coin, whether actually performed or not, is a statistical experiment with two possible outcomes, if we regard the possibility of the coin landing on its edge either as impossible or extremely unlikely. The drawing of a lottery ticket, the selection of ten persons with a view to ascertaining their political opinion, the calculation of means from randomly drawn samples, are all statistical experiments. In some cases the enumeration or listing of all possible outcomes of an experiment is quite a simple matter, as for instance in the experiment of tossing a six-sided die. In others, such as enumeration of various possible poker hands, it is a great deal more complex.

Suppose that a six-sided die is tossed. If we are interested in the face value of a particular toss, then there are six possible outcomes. If, on the other hand, we wish to know whether the number obtained is even or odd, then only two possibilities are relevant; these are made up of the face values 1, 3, 5 and 2, 4, 6, respectively.

We are able to distinguish between the *elementary outcomes* of the experiment, i.e. outcomes which cannot be decomposed any further, and *compound outcomes* or *events*, which are made up of several elementary outcomes, e.g. even number, or a number greater than 4.

In general, by the word event we mean a collection or set of elementary outcomes each of which satisfies the conditions for inclusion in the set. For instance, the event of obtaining a number greater than four in the die experiment consists of obtaining either 5 or 6, both of which are elementary outcomes in the experiment.

Often we are interested in events resulting from two or more experiments which are to be considered jointly. Thus we may toss a coin twice, or roll two dice simultaneously. In the first case, the possible outcomes are HH, HT, TH and TT, and these clearly cannot be decomposed any further if the experiment is to be a joint one. Consider, however, the event of obtaining "exactly one head." This is a compound event consisting of two elementary joint outcomes—the first toss resulting in heads and the second toss resulting in tails or *vice versa*, i.e. HT or TH. By similar reasoning, the experiment of rolling two dice would generate 6×6, i.e. 36 outcomes representing all the possible ways in which two dice may fall together. A general counting rule known as the *multiplication rule* may now be stated. If an experiment can have m distinct outcomes and another experiment can yield n distinct outcomes, the joint or combined experiment can result in $m \cdot n$ distinct outcomes.

EXAMPLE 6.1

Consider an experiment in which a coin and a six-sided die are tossed simultaneously. How many possible elementary outcomes are generated?

Using the multiplication rule, there are six distinct faces of the die and for each of these the coin can fall in two different ways. There are then 6×2 possible outcomes for the joint experiment. Enumerating these, we obtain

$$H1, H2, H3, H4, H5, H6$$
$$T1, T2, T3, T4, T5, T6.$$

If E now denotes the compound event of obtaining "heads and an even number," we may write
$$E = (H2, H4, H6)$$
which is a set of those elementary outcomes which satisfy the definition of E.

Probability Measures

We may introduce the important ideas which are to be discussed in this section by the general statement that probability measures are real numbers assigned to the occurrences of various possible outcomes in an experiment. This process can be mathematically complex, but here we shall be concerned only with relatively simple cases leading to the formulation of some elementary rules of probability measurement.

The simplest case arises when probability measures are to be assigned to *equiprobable events*. If an experiment consists of n possible elementary outcomes each of which is equally likely, it is natural to assign the ratio $1/n$ as a measure of probability associated with each outcome. In a compound event, say E, consisting of r such elementary outcomes, the probability measure associated with the event E will be the sum of the probability measures assigned to each element in E. This sum will be r/n since there are r such elements in E each having equal weight $1/n$. The ratio

$$\frac{\text{Number of ways in which event } E \text{ can occur}}{\text{Number of all possible outcomes in the experiment}}$$

when each outcome is equally likely, is defined as the *probability* of E occurring; this may be written as $P(E)$. In an experiment of tossing a die, there are six equally likely outcomes, and we assign the probability of $1/6$ to each of these outcomes. The probability of obtaining an even number on a single toss is $3/6$ since there are three possible ways of obtaining an even number, and each of these can be assigned an equal measure of probability.

If the statement defining an event is logically false (e.g. obtaining the number 7 from the toss of a six-sided die), the probability of that event occurring is zero. Conversely, if an event defined by a statement is true in every conceivable case (e.g. obtaining a number between 1 and 6 inclusive), the probability of its occurrence is unity; we say that the event is certain to occur. Thus probability measures cannot exceed the limits of 0 and 1.

EXAMPLE 6.2

In example 6.1, what probability should be assigned to the event "heads and a number greater than four"?

There are twelve possible elementary outcomes in the experiment, each equally likely to occur with unbiased die and coin. The event can occur in two possible ways. Hence the required probability is 2/12.

The statement, however, that two equiprobable events (i.e. two events which are agreed to be equally likely) have equal probability of occurrence is nothing more than a tautological definition of probability. This difficulty may be overcome by deriving probability measures empirically from actual experiments. Such experiments consist of a large number of repeated trials, e.g. tosses of a coin. Let there be n such trials, and let r indicate the number of successes, e.g. heads, in the n trials. We then observe the behaviour of the ratio r/n as n is increased indefinitely. If it is observed that with an increasing number of trials the ratio of relative frequency of success, r/n, approaches some number, say p, we accept this number p as the probability of an event (success) occurring. So long as the limiting ratio p exists, probability measures may be assigned to events on an experimental basis even though the likelihood of their occurrence cannot be ascertained *a priori*. Thus we may have a twisted coin with an unknown bias. If it is observed that the proportion of heads obtained in 100 tosses is 0·62, in 500 tosses 0·61, and in 1,000 tosses 0·604, we might accept 0·6 as the probability of heads occurring.

The experimental or frequency approach to probability rests on an important theorem of probability theory known as the *law of large numbers*. The theorem states that the probability of obtaining an absolute difference between the relative frequency of occurrence, r/n, and its limiting ratio p, such that this difference is less than some arbitrarily small value, say d, will approach unity as the number of trials n is increased indefinitely. In other words, a difference between r/n and p smaller than d is virtually certain to occur with a sufficiently large number of trials. Thus, in our example we can never be absolutely sure that the true value of p which we accept as the probability of heads occurring will be 0·6 if the ratio r/n appears to converge to 0·6. All we can say is that our estimate is not likely to be far out if the number of repeated trials is large enough. Conversely, if in an experiment consisting of 500 tosses of an unbiased coin we obtained, say, 300 heads, we would not conclude that the law of large numbers did not hold but that something rather unlikely had occurred.

The experimental approach to probability is not free of difficulties, but it enables us to obtain estimates of probabilities by the method of approximation. This approximation will be the more accurate the greater the number of trials. We have seen that by virtue of the definition of probability, probability measures cannot exceed the logical limits of zero and unity. However, some care is needed how these limits are interpreted in the experimental approach. If we assign 0 to the probability of an event E occurring, either E is impossible or it occurs so very rarely in a long stream of repeated trials that for practical reasons it may be regarded as impossible. Similarly, the upper limit of 1 is assigned to an event which is certain to occur or virtually certain to occur with a large number of trials.

6.3. Axioms of Probability

The discussion of probability measures in the preceding section may now be summarized by reference to three basic rules, or axioms, of probability.

1. If A_i is a set of all possible events, $P(A_i)$, i.e. the probability of occurrence of any one event in the set, is defined as a real number satisfying the requirement $(0 \leqslant P(A_i) \leqslant 1)$.

2. The sum of probabilities of all mutually exclusive events $A_1, A_2 \ldots A_n$ which may occur in a statistical experiment equals unity, i.e.

$$P(A_1) + P(A_2) + P(A_3) \ldots + P(A_n) = 1$$

3. For two mutually exclusive events A_1 and A_2 the probability of either A_1 occurring or A_2 occurring is equal to the sum of their respective probabilities, i.e.

$$P(A_1 \text{ or } A_2) = P(A_1) + P(A_2)$$

If an experiment consists of just two mutually exclusive possible events, A or B, the corollary holds

$$P(B) = 1 - P(A)$$

In the general case of m mutually exclusive events, the rule of probability addition becomes

$$P(A_1 \text{ or } A_2 \text{ or } A_3 \ldots \text{ or } A_m) = P(A_1) + P(A_2) + P(A_3) \\ \ldots + P(A_m)$$

Example 6.3

If two unbiased coins are tossed, what is the probability of obtaining exactly one head?

There are four equiprobable events in the experiment, i.e. HH, HT, TH and TT. The required event will occur if either HT or TH is obtained on a single toss. Thus

$$P(\text{HT or TH}) = P(\text{HT}) + P(\text{TH}) = 1/4 + 1/4 = 2/4$$

The addition rule given so far requires that the events considered are mutually exclusive events. The condition of mutual exclusiveness will be met if the occurrence of any particular event precludes the occurrence of all of the remaining possible events. Thus, suppose that we have a group of observations consisting of six rents: $7·50, $8·00, $10·00, $12·50, $13·00, $15·00. These rents are to be divided into two classes: "$7·50 to $12·50" and "$12·50 to $17·50." Let E be the event "the chosen rent belongs to the lower class," and F the event "the chosen rent belongs to the upper class." Is it possible to find an event, which we indicate by EF, which satisfies the condition that "the chosen rent belongs both to the lower class and the upper class"? Clearly, such event is the occurrence of the rent $12·50. The existence of an element which is common to both the event E and F implies that the classification is not mutually exclusive. If, on the other hand, we were to define the two classes as "$7·50 and less than $12·50," and "$12·50 and less than $17·50," there is no element (rent) belonging to both E and F, and the two categories are mutually exclusive.

What modification of the addition rule is necessary when the events to be considered are not mutually exclusive? Here we mention only the special case of two events, but a general method exists which extends the same principle to the general case of n events. Using the previous example of non-mutually exclusive classes E and F, we find $P(E) = 4/6$, and $P(F) = 3/6$. Addition of these probabilities violates the second axiom of probability. The reason is that one element, i.e. the rent of $12·50 belongs to both E and F and is thus counted twice. To eliminate double counting we use the general addition formula

$$P(E \text{ or } F) = P(E) + P(F) - P(EF)$$

The word "or" in this case signifies that it is possible for event E or F or both to occur. When, however, the two events to be considered are mutually exclusive, the "or" can only mean that either of the two events can occur but not both.

EXAMPLE 6.4

Of all the students at a university, one-half of students have access to a car to attend lectures, and one-third have access to a scooter. One-fourth have access to both. What is the probability of finding a student who has his own means of transport?

If A and B are the events relating to the first statement, and AB the event relating to the second statement, we have

$$P(A \text{ or } B) = P(A) + P(B) - P(AB)$$
$$= \tfrac{1}{2} + \tfrac{1}{3} - \tfrac{1}{4} = \tfrac{7}{12}$$

The events A and B are not mutually exclusive, and $P(A \text{ or } B)$ gives the probability of A or B or both occurring.

6.4. Multiplication of Probability and Statistical Independence

Having discussed mutual exclusiveness of statistical events, we now consider the conditions for their statistical *independence* or *non-independence*. Two events will be statistically independent if the outcome of one has no effect on the outcome of the other. Thus in tossing a coin twice, the probability of obtaining heads or tails on the second toss is completely unaffected by the outcome of the first toss.

Earlier we saw that experiments of this kind give rise to joint events which are the elementary outcomes in the experiment. In the present case of two coins, there are four such joint events, i.e. HH, HT, TH and TT, and these constitute all possible outcomes. Each of these is equally likely (if the coins are balanced) and is assigned equal probability of $\tfrac{1}{4}$. Thus it can be seen that the probability of obtaining a particular sequence of outcomes, for instance two heads, can be computed as the product of the probabilities associated with the outcomes of each separate toss, i.e. $P(\text{HH}) = P(\text{H})P(\text{H}) = \tfrac{1}{2} \cdot \tfrac{1}{2} = \tfrac{1}{4}$ which is the same as the probability measure assigned to each of the equiprobable events above.

Given two independent events A_1 and A_2, the probability of their joint occurrence is then defined by the *multiplication rule*

$$P(A_1 \cdot A_2) = P(A_1)P(A_2)$$

This rule holds for n independent events $A_1, A_2 \ldots A_n$ provided that the conditions for the multiplication of probabilities are satisfied for all combinations of two or more events, i.e. provided that for *all* such combinations the constituent events are independent. Then

$$P(A_1.A_2 \ldots A_n) = P(A_1)P(A_2) \ldots P(A_n)$$

Example 6.5

Given an experiment of tossing two unbiased coins and an unbiased die simultaneously, what is the probability of obtaining two heads and a six on a single toss?

The events are evidently independent, and the required probability is

$$P(\text{HH6}) = P(\text{H})P(\text{H})P(6) = \tfrac{1}{24}$$

If the condition of statistical independence is not satisfied, a more general formula involving *conditional probabilities* must be used. If we have two events E and F, the probability that event F will occur given that E has occurred is the conditional probability of F which can be written as $P(F/E)$. Similarly, we write the probability of occurrence of E which is conditional upon the occurrence of F as $P(E/F)$.

Suppose that there are ten paper tags which can be distinguished by number and colour, e.g. tags numbered 1, 2, 3 and 4 are red and the remainder are white. If these are placed in a hat and drawn at random, the probability of drawing a particular tag is 1/10. But if after drawing a tag at random we were told that it was red, how would we revise the probability that a particular tag, e.g. the tag which is numbered 1, had been drawn? Evidently, the number of possible outcomes is now reduced from ten to four; furthermore, it is given that tag number 1 is both the required number and red. We compute the required conditional probability as the ratio

$$P(\text{Tag No. 1/red}) = \frac{\text{Number of outcomes: Tag No. 1 and red}}{\text{Number of outcomes: red}} = \frac{1}{4}$$

For some purposes it is more convenient to express conditional probabilities as a ratio of probabilities. The latter are obtained by dividing the numerator and denominator in the above formula by the number of outcomes possible in the experiment. In the present case, there are ten tags altogether, and thus ten possible outcomes in the experiment. Then

$$P(\text{Tag No. 1/red}) = \frac{P(\text{Tag No. 1 and red})}{P(\text{red})} = \frac{1/10}{4/10} = \frac{1}{4}$$

In general, given two events which are not independent, the conditional probability of E given F is defined as

$$P(E/F) = \frac{P(EF)}{P(F)}$$

i.e. as a ratio of the probability of the joint event E and F occurring and of the probability of F occurring. The important point to notice is that since E is included in F and F itself includes only a fraction of all the outcomes possible in the experiment, the conditional probability E given F will be larger than the absolute probability associated with the occurrence of the event E. Thus, in our present example, the conditional probability of $\frac{1}{4}$ exceeds the previously computed probability of 1/10.

EXAMPLE 6.6

A card is drawn from a well-shuffled deck of cards. What is the probability of drawing a "black" king (spade or club) given that the card drawn was a "face" card (jack, queen or king)?

Let K be the event that the card drawn is a black king, and F that it is a face card. In the pack there are two black kings which are also face cards; hence $P(KF) = \frac{2}{52}$. Twelve of the cards are face cards; hence $P(F) = \frac{12}{52}$. We have

$$P(K/F) = \frac{P(KF)}{P(F)} = \frac{\frac{2}{52}}{\frac{12}{52}} = \frac{2}{12}$$

We may now similarly define the conditional probability of F given E as

$$P(F/E) = \frac{P(EF)}{P(E)}$$

Combining these results, we derive the general rule for probability multiplication—

$$P(EF) = P(F)P(E/F) = P(E)P(F/E).$$

This formula is applicable in computing the probability of two dependent events E and F occurring jointly when the conditional probabilities relating to E and F are known or can be ascertained.

EXAMPLE 6.7

An urn contains three white marbles and nine black. A marble is drawn at random without being replaced, then another marble is drawn. What is the probability that both are black? The first marble's being black evidently influences the probability of obtaining a second black marble: the two events are not statistically independent. We require

$P(\text{both black}) = P(\text{first black})P(\text{second black/first black})$

$$= \frac{9}{12} \cdot \frac{8}{11} = \frac{6}{11}$$

If the occurrence of event F has no influence on the event E, the two events are by definition independent and $P(E/F) = P(E)$; conversely $P(F/E) = P(F)$. The general multiplication formula then reduces to

$$P(EF) = P(E)P(F)$$

which is the result derived earlier.

EXAMPLE 6.8

In example 6.1 a coin and a die were tossed simultaneously. If E is the event "heads occur" and F is the event "3 occurs," show that the events E and F are independent.

$$P(F/E) = \frac{P(EF)}{P(E)} = \frac{1/12}{\frac{1}{2}} = \frac{1}{6}$$

But $P(F) = P(F/E) = \frac{1}{6}$, and the two events are independent.

6.5. Expected Value

An important concept in probability theory to which we shall refer frequently in subsequent sections is the *expected value of a variable* or *mathematical expectation*. If in an experiment, or perhaps a game of chance, it is possible to assign numerical values, X_i, to the various possible outcomes whose probabilities $P(X_i)$ can be ascertained, expected value is defined as

$$E(X) = X_1 P(X_1) + X_2 P(X_2) \ldots + X_n P(X_n)$$
$$= \Sigma X_i P(X_i)$$

Expected value is the weighted sum of the numerical values which can occur in an experiment and their respective probabilities.

Suppose we play a simple game of tossing one coin in which we receive $3 when heads turn up and lose $2 when tails turn up. If the coin is an unbiased coin

$$E(X) = (3)(\tfrac{1}{2}) + (-2)(\tfrac{1}{2}) = 0.5 \text{ dollar}$$

How should this result be interpreted? Evidently, it is not possible to win that amount in any single game since in order to participate we must contract to pay $2 if tails occur and receive $3 if heads occur, and this exhausts all possibilities. But we observe that the weights in the formula are the probabilities of heads and tails occurring, and by the law of large numbers the proportion of heads in a large number of tosses will tend to $\tfrac{1}{2}$, and the same will be true for the proportion of tails. Thus in a large number of games we can expect to win in 50 per cent of the cases, and lose in 50 per cent of the cases, and the net winnings per game can be expected to be 0.5 dollar. In general, mathematical expectation can be interpreted as the average value obtained in a long sequence of repeated experiments.

In our example, the game is evidently favourable to us since the expected value is positive. If the expected value were zero, the game would be a "fair game" enabling the promoter to break even in the long run. In life, we very frequently play games unfavourable to us, for instance, in taking out various insurance policies protecting us against contingencies such as fire, accident, etc. The word "unfavourable" should, however, not be interpreted that such games are in some ethical sense "unfair," but merely that they result in negative expected values; in any event, most people consider the unfavourable expectations to be outweighed by other advantages.

Example 6.9

A man wishes to insure his house against fire. The value of the house is $12,000. The annual premium which he must pay to insure this house is $4.

If the probability that a fire will destroy the house is 1/10,000, is his insurance contract a "fair game"?

$$E(X) = (12{,}000)(1/10{,}000) + (-4)(9{,}999/10{,}000)$$
$$\simeq -\$2 \cdot 80$$

The game must be considered "unfavourable."

The notion of "expected value" plays an important role in statistical theory. In practical applications, mathematical expectation serves as a basis for statistical decision theory. The concept is also widely used in describing the characteristics of various theoretical distributions some of which will be introduced in subsequent sections.

6.6. Populations and Samples

Before introducing the concept of probability distribution, it is necessary to draw a distinction between *populations* and *samples*. This distinction is of fundamental importance for all our future work.

If we have a group of observations of a variable which has been drawn from a still larger group, we refer to the former as a *sample* and to the latter as the *population* of that variable. The word "population" is used here in a technical sense and refers to the universe or totality of observations from which the sample has been drawn. Some populations are *finite*, e.g. a population of men's heights, a population of the ages of bridegrooms, or a population of rents. Others, sometimes known as theoretical or statistical populations, are *infinite*, e.g. a population of tosses of three coins, or a population of rolls of a six-sided die.

It is important to realize that even though a population, e.g. the population of rents in Adelaide at this moment, may be finite, the populations of samples of a certain size drawn from such a population may be infinite, because an infinite number of such samples can be drawn. Some finite populations are so large that they can be treated as infinite ones.

The distinction between populations and samples is a familiar common-sense distinction. We are all accustomed to the idea of picking a sample as an indication of the larger totality from which the sample has come, e.g. a sample of suiting, a sample of wheat, a sample of political opinion. In some situations when analysing the characteristics of a variable, a sample of observations must be used because the population is infinite or so very large as to render an examination of the whole population impossible. This would be the case in the analysis of a particular variety of wheat. In other situations, although an examination of the whole population would be possible, the magnitude of the task may be so great as to render a sample the only practicable solution. This question is discussed in detail in the following chapter.

If a sample is selected in such a way that each element in the population has an equal chance of selection, the sample is said to be a *random sample*. Since there are various kinds of random samples, it is more accurate to refer to such a sample as a *simple random sample*. The essential principle of random sampling is that it is based on chance selection, such as occurs when lottery marbles are taken from a properly mixed barrel. Random selection is a basic requirement of all sampling procedures; these will be discussed in more detail in the following chapter.

For the present, we shall consider populations of one variable only. Thus our concern will be, for instance, with the rents of houses in Adelaide, or the prices of potatoes and not, for example, with the way in which number of rooms and rents of houses vary conjointly.

6.7. Forming a Probability Distribution

We have already dealt with the question of how statistical data can be summarized and described. Such data are usually sample data in the sense that they represent only a portion of all the possible values of the variable. Certain measures such as the arithmetic mean and the standard deviation can then be computed. These measures are called *statistics*.

The description of the population can be performed by analogy with that of the sample. In the present section, we shall confine our attention to some simple cases in which we can rely on the rules and axioms of probability developed so far. Later we shall consider two theoretical distributions of great importance in applied statistics, the binomial and the normal distribution.

For the purpose of illustration, let us perform a statistical experiment in which tags are drawn at random from an urn containing four tags numbered 1, 2, 3 and 4; after each draw the tag is replaced. There are four possible outcomes in the experiment, each having the given numerical value. This set of possible values becomes the variable, X, of the experiment.

Since under our present assumptions each of the four outcomes is equally likely, we can assign an equal measure of probability to each of the possible values which X can assume. A variable to which probabilities can be assigned in accordance with the accepted axioms of probability is called a *random variable*. If we now plot the various probabilities against the appropriate values of the random variable X, we obtain a *probability distribution* of X which can be represented as in Table 6.1 and Fig. 6.1 below.

Table 6.1
PROBABILITY DISTRIBUTION OF RESULTS OF DRAWING ONE
TAG AT RANDOM WITH REPLACEMENT

Value (X)	$P(X)$
1	$\frac{1}{4}$
2	$\frac{1}{4}$
3	$\frac{1}{4}$
4	$\frac{1}{4}$
Total	1

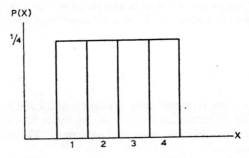

FIG. 6.1. PROBABILITY DISTRIBUTION OF RESULTS OF DRAWING ONE TAG AT RANDOM WITH REPLACEMENT

Some care is needed in interpreting these results. If we repeated the experiment of drawing a tag at random a large number of times we would *expect* to obtain each of the possible four numbers in approximately 25 per cent of all trials. By the law of large numbers this approximation would be the closer the greater the number of trials. Thus in one hundred such trials we would expect each tag to be drawn 25 times but in an actual experiment number 1 may occur only, say, 23 times, number 2 perhaps 28 times, etc. The first set of values are *theoretical* or *expected* frequencies of occurrence, obtained by multiplying the respective probabilities by the total number of trials, while the latter represent *actual* frequencies derived from experience. The actual frequencies are thus derived experimentally and can be summarized by frequency distributions of the kind that we studied earlier. The expected frequencies are computed from probability distributions based on theoretical considerations introduced earlier in this chapter.

In order to illustrate these important concepts, it may be useful to consider a more complicated statistical experiment. We now consider

PROBABILITY AND PROBABILITY DISTRIBUTIONS 93

two tags selected jointly by drawing one tag at random, recording its number, replacing it in the urn, and then drawing a second tag at random. All the possible combinations of numbers obtained in this manner can be tabulated as follows—

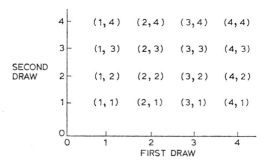

There are sixteen possible outcomes in drawing two tags together which under our assumptions are equally likely. We may then associate with each of these points equal probabilities of 1/16. This is verified by the multiplication rule which assigns to each point the probability of 1/16, since the results of the first draw and second draw are independent.

By using a simple rule we may now reduce the two values representing each point in the above diagram to a single value. One such rule may be "sum the results of the first and second draw"; another "find the mean of the results of the first and second draw." Let us now define Y as a random variable whose values are the possible sums obtained by drawing two tags at random. Using the second and third axioms of probability, we derive the following table—

Table 6.2
PROBABILITY DISTRIBUTION OF RESULTS OF DRAWING TWO TAGS AT RANDOM WITH REPLACEMENT

Sum of Two Tags (Y)	Events	$P(Y)$
2	(1, 1)	$\frac{1}{16}$
3	(2, 1), (1, 2)	$\frac{2}{16}$
4	(3, 1), (2, 2), (1, 3)	$\frac{3}{16}$
5	(4, 1), (3, 2), (2, 3), (1, 4)	$\frac{4}{16}$
6	(4, 2), (3, 3), (2, 4)	$\frac{3}{16}$
7	(4, 3), (3, 4)	$\frac{2}{16}$
8	(4, 4)	$\frac{1}{16}$
Total		1

The resulting probability distribution may be represented graphically—

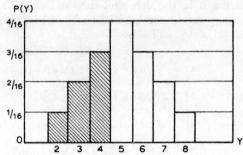

Fig. 6.2. Probability Distribution of Results of Drawing Two Tags at Random with Replacement

The result gives a functional representation of probabilities since for each value of the random variable Y there corresponds a point which is the probability of that value of Y occurring. But since each bar in the histogram represents the relative frequency with which a given value of Y occurs in a large number of trials, probabilities can be expressed as proportionate areas falling between the given values of Y; thus $P(2 \leqslant Y \leqslant 4)$ is given by the hatched area in Fig. 6.2. Furthermore, since the values of Y on the horizontal axis represent all the possible outcomes of the experiment, the area of the histogram will correspond to unity. In the present case, the variable under

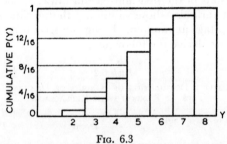

Fig. 6.3

consideration is discrete and its probability distribution is represented graphically by a histogram. Continuous variables give rise to smooth *probability curves*; an important example of a continuous probability distribution is the normal curve which will be introduced in section 6.9.

By analogy with ogives constructed from sample data, we may cumulate the probabilities associated with Y as Y varies over the given range; this is shown in Fig. 6.3. Such a cumulative distribution is

PROBABILITY AND PROBABILITY DISTRIBUTIONS   95

useful when it is of interest to know the probability that the value of the random variable is equal to or less than any particular value. When all the possible values of the variable are included in that range, the cumulative probability reaches unity, as can be seen from Fig. 6.3; thus a value of Y equal to or less than 8 is certain to occur.

In this section we have derived simple probability distributions from first principles by applying the axioms and rules of probability. As the next step, we shall introduce a special distribution, the binomial distribution, which finds many applications in statistical work. Other discrete probability distributions exist and have been extensively investigated, but these will not be discussed here.

6.8. The Binomial Distribution

The binomial distribution is a theoretical probability distribution applicable to situations with the following characteristics—

1. An experiment consists of a finite number of repeated trials each of which has only two possible outcomes which we denote as "success" and "failure."

2. The repeated trials are independent in the sense that the outcome of any one trial has no effect on the outcomes of the successive trials.

3. The probabilities associated with "success" and "failure" are known and remain unchanged throughout the experiment.

In experiments such as tossing an unbiased coin three times in succession, these conditions are strictly satisfied. The three repeated trials are independent of one another, each trial has two possible outcomes whose probabilities remain the same. Similarly in tossing a six-sided die if we are only interested in two possible outcomes such as "obtaining number six," and "not obtaining number six." Irrespective of the number of trials, i.e. tosses, the respective probabilities remain 1/6 and 5/6, and all the trials are independent. The second condition would however not be satisfied if, in drawing, say, two marbles from an urn containing three white marbles and nine black, the first marble drawn was not replaced. But the extent to which the condition of independence is violated by non-replacement depends also on the number of drawings in relation to the total number of marbles in the urn. If there were twelve hundred such marbles instead of twelve, it would evidently make little difference if the first marble drawn was replaced or not, and the condition of independence would be very nearly satisfied. This will be true when small samples are drawn at random from large groups of observations. In such cases the probability of selecting successive elements from the population may be assumed to remain unchanged.

The question which we now wish to answer may be stated as follows: if there are n independent trials, each having the same probability of success p and probability of failure $1 - p$, can we determine the probability with which exactly x successes will occur in the n trials? It may be agreed that for the purpose of the experiment the occurrence of heads constitutes "success"; then the occurrence of tails must be considered as "failure." Frequently the letter q is used to indicate the probability of failure instead of $1 - p$; this convention will be used for convenience here.

Let us now consider the experiments of tossing first one coin, then two coins and finally three coins each having the probability of heads (H) occurring equal to p and the probability of tails occurring equal to q. The possible results in these three experiments can be conveniently represented with the aid of a device known as a *tree diagram* as in Fig. 6.4. If only one coin is tossed, the leftmost region of Fig. 6.4

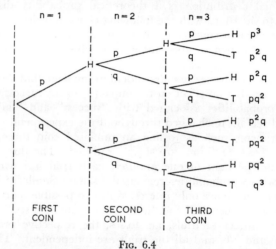

Fig. 6.4

is relevant. In this case X, i.e. the number of heads turning up, can assume only two values

$$X = \begin{cases} x_0 = 0 \\ x_1 = 1 \end{cases}$$

i.e. $x_0 = 0$ when tails occur, and $x_1 = 1$ when heads occur. Thus we have

No. of Heads X	$P(X)$
0	q
1	p

PROBABILITY AND PROBABILITY DISTRIBUTIONS 97

If $n = 2$, i.e. two coins are tossed simultaneously, we must consider the results obtained from the first coin and from the second coin jointly, and this may be represented by the four possible paths in the middle region of the tree diagram. For each of the two possible outcomes of the first toss, the second coin can fall in two distinct ways. But the two trials are independent, and the probabilities associated with the four possible outcomes may be obtained by multiplying the probabilities along the paths leading to these four outcomes—

$$pp$$
$$pq$$
$$qp$$
$$qq.$$

Denoting X as the number of heads obtained in the experiment we have

$$X = \begin{cases} x_0 = 0, \text{ when TT, i.e. two tails occur,} \\ x_1 = 1, \text{ when HT or TH, i.e. exactly one head occurs,} \\ x_2 = 2, \text{ when HH, i.e. two heads, occur.} \end{cases}$$

It should be noted that when $x_1 = 1$, we are only concerned with the event "exactly one head occurs" and not with the *order* in which heads can occur in the experiment. The required event is thus a compound event, and since the constituent events, HT and TH, are mutually exclusive, their probabilities can be added. The random variable "number of heads" is thus distributed with the following probabilities—

No. of Heads X	$P(X)$
0	q^2
1	$2pq$
2	p^2

By similar reasoning we find the probabilities associated with the eight possible outcomes of tossing three coins shown in the right-hand margin of Fig. 6.4. But there are three distinct ways in which the event "one head and two tails" and the event "two heads and one tail" can occur. Applying the addition rule we obtain—

No. of Heads X	$P(X)$
0	q^3
1	$3pq^2$
2	$3p^2q$
3	p^3

By mathematical induction, it is seen that in the general case of n independent trials the probabilities associated with the various possible values of X are the terms of the binomial expansion

$$(p+q)^n = \binom{n}{0}q^n + \binom{n}{1}pq^{n-1} + \binom{n}{2}p^2q^{n-2} \ldots + \binom{n}{n}p^n$$

where

$$\binom{n}{x} = \frac{n \cdot (n-1) \cdot (n-2) \ldots (n-x+1)}{x \cdot (x-1)(x-2) \ldots 2 \cdot 1}$$

for any given value of x and n.

Thus for any x within the range of X, the formula

$$P(x, n) = \binom{n}{x}p^x q^{n-x}, \qquad x = 0, 1, 2 \ldots n$$

gives the probability of x successes in n trials. The probability distribution which results from the application of this formula is known as the *binomial distribution*.

To illustrate the use of the binomial formula, we apply it to an experiment of tossing a coin three times in succession. Here $n = 3$ and $x = 0, 1, 2, 3$, and we have

$$\binom{3}{0} = 1, \binom{3}{1} = 3, \binom{3}{2} = 3 \quad \text{and} \quad \binom{3}{3} = 1$$

These values, known as *binomial coefficients*, measure the number of distinct ways in which x successes can be distributed among the n trials. In computing $P(x, n)$ we therefore calculate the probability of obtaining x successes in n trials regardless of the order in which the successes may occur. Suppose that the coin is not balanced, and $p = 0.4$ and $q = 0.6$. Substituting into the binomial formula $P(x, n)$, we obtain the following probability table—

Table 6.3

PROBABILITY DISTRIBUTION OF NUMBER OF HEADS IN THREE SUCCESSIVE TOSSES OF UNBALANCED COIN

($p = 0.4$; $q = 0.6$)

No. of Heads (X)	Binomial Coefficient $\binom{n}{x}$	$p^x q^{n-x}$	$P(X)$
0	1	0·216	0·216
1	3	0·144	0·432
2	3	0·096	0·288
3	1	0·064	0·064
Total	8	...	1·000

PROBABILITY AND PROBABILITY DISTRIBUTIONS 99

When the probabilities of "success" and "failure" are equal, as in the case of tossing unbiased coins, the binomial distribution formula reduces to

$$P(x, n) = \binom{n}{x} p^n$$

Applying this result to the previous example, we compute the required probabilities in Table 6.4.

Table 6.4

PROBABILITY DISTRIBUTION OF NUMBER OF HEADS IN THREE SUCCESSIVE TOSSES OF BALANCED COIN

No. of Heads (X)	Binomial Coefficient $\binom{n}{x}$	p^n	$P(X)$
0	1	0·125	0·125
1	3	0·125	0·375
2	3	0·125	0·375
3	1	0·125	0·125
Total	8	...	1·000

When graphed, these results evidently yield a symmetrical distribution. When $p < q$, the distribution is positively skewed, as seen from Table 6.3, and when $p > q$, it is negatively skewed.

The Mean of the Binomial Distribution

We have already explained that the summary measures used to describe sample frequency distributions are called *statistics*. Similar measures may also be derived for theoretical probability distributions. However, in order to specify a particular probability distribution we must know the equation by which such a distribution is defined. This requirement can be split up into two parts: (i) the general form of the equation, (ii) the particular member of the family of the general form. Thus $y = a + bx + cx^2$ is the general form known as the parabola, but a particular parabola depends on the actual values of the coefficients a, b and c. These coefficients are known as *parameters*. Similarly for probability distributions, given the general form of the distribution, a particular distribution depends on the value of the parameters. In statistical work, parameters are usually expressed in terms of the arithmetic mean, standard deviation, etc. Thus we have *parameters* in populations and *statistics* in samples. The statistics may often be used as estimates of parameters. The parameters are often indicated by Greek and the statistics by Roman letters. Thus, whereas $\bar{X}$ and s indicate the mean and standard deviation of a sample, we use μ and σ for the mean and standard deviation of the population.

Let us consider again the experiment of tossing three unbiased coins. Evidently, the experiment can be repeated a very large number of times. As the number of tosses increases, we can expect the observed relative frequencies of occurrence to approximate more and more closely a theoretical binomial distribution. In the present example, this distribution will be completely determined by the two parameters $n = 3$ and $p = \frac{1}{2}$. We now consider the mean and variance of binomial distributions with given parameters.

In computing the mean of a frequency distribution, the values of the variable are weighted by their frequencies of occurrence. Analogously, the expected value of a random variable was defined in section 6.5 as a sum of the values which can occur in the experiment weighted by their respective probabilities. Since the probabilities were defined earlier as limits of relative frequencies of occurrence in a large number of trials, it is seen that the mean value of a binomially distributed variable will be its expected value. Using the results of Table 6.4 and writing μ as the mean number of heads which can be expected in a long sequence of trials, we have

$$\mu = E(X) = \Sigma X \cdot P(X)$$
$$= (0)(0 \cdot 125) + (1)(0 \cdot 375) + (2)(0 \cdot 375) + (3)(0 \cdot 125)$$
$$= 1 \cdot 5 \text{ heads}$$

However, from previous results the required binomial probabilities are given by $P(x, n)$ for any specified value of n and $x = 0, 1, 2, 3, \ldots n$. Substituting these probabilities into the equation for $E(X)$[1], the expected value of X reduces to

$$\mu = n \cdot p$$

---

[1] This may be seen as follows—

Using the formula on p. 98, we have

$$\mu = E(X) = \sum_{x=0}^{n} x \cdot \binom{n}{x} p^x q^{n-x}, \quad \text{where} \quad \binom{n}{x} = \frac{n!}{x!(n-x)!}$$

In this summation, the first term vanishes since $x_0 = 0$, and we may write

$$\mu = \sum_{x=1}^{n} x \cdot \binom{n}{x} p^x q^{n-x}$$

Simplifying

$$x \cdot \binom{n}{x} = \frac{n!}{(x-1)!(n-x)!}$$

In the present example $n = 3$ and $p = \frac{1}{2}$; thus

$$\mu = (3)(\tfrac{1}{2}) = 1{\cdot}5 \text{ heads}$$

The common-sense interpretation of the mean of a binomial distribution is that if we, for instance, tossed three balanced coins 200 times we should expect to obtain 300 heads, i.e. 1·5 heads per experiment.

The Standard Deviation of the Binomial Distribution

In general, the variance of any population is derived analogously with the variance computed from a sample frequency distribution by weighting the mean square deviations of the variable by the relative frequencies of occurrence if the population is finite, or by the respective probabilities if we are dealing with theoretical probability distributions. The variance so derived, however, differs from the sample variance in that it is a measure of dispersion which we can expect *in a long sequence of experiments*, e.g. tosses of three coins. Accordingly, the population variance is defined as the expectation

$$\sigma^2 = E[X - \mu]^2 = \Sigma[(X - \mu)^2 \cdot P(X)]$$

In the case of the binomial distribution, the probability weights are known and given by the binomial formula $P(x, n)$.

The variance of a specific binomial distribution may be calculated directly from the above formula as in the table below.

Table 6.5
COMPUTATION OF VARIANCE OF NUMBER OF HEADS IN THREE SUCCESSIVE TOSSES OF BALANCED COIN

No. of Heads (X)	$x - \mu$	$(X - \mu)^2$	$P(X)$	$(X - \mu)^2 \cdot P(X)$
0	−1·5	2·25	0·125	0·28125
1	−0·5	0·25	0·375	0·09375
2	0·5	0·25	0·375	0·09375
3	1·5	2·25	0·125	0·28125
Total				0·75000

and factoring out n and one of the x factors of p, we have

$$\mu = np \sum_{x=1}^{n} \frac{(n-1)!}{(x-1)!(n-x)!} p^{x-1} q^{n-x}$$

Letting $m = n - 1$ and $y = x - 1$

$$\mu = np \sum_{y=0}^{m} \binom{m}{y} p^y q^{m-y}$$

where the summation term relates to a binomial distribution $P(y, m)$ for $y = 0, 1 \ldots m$, and is thus equal to unity. Hence

$$\mu = n \cdot p$$

The standard deviation of the number of heads in the experiment is thus 0·866, and this measures the amount of variation in the number of heads which we would expect in tossing three balanced coins. In this respect the measure is quite analogous to the sample standard deviation discussed earlier.

The general formula for the standard deviation of the binomial distribution may be derived in a similar manner to the formula for the mean in the previous section. This formula is

$$\sigma = \sqrt{np(1-p)}$$

For $n = 3$, and $p = \frac{1}{2}$, we obtain

$$\sigma = \sqrt{(3)(\tfrac{1}{2})(\tfrac{1}{2})} = \sqrt{0.75} = 0.866 \text{ heads}$$

It will be noted that the mean and standard deviation of the binomial distribution are determined by the parameters n and p, which specify a particular binomial distribution. Since n and p can be expressed in terms of μ and σ, the latter may, alternatively, be regarded as the parameters of the binomial distribution.

EXAMPLE 6.10

A company selling electronic equipment finds that of all the machines it installs 40 per cent require further adjustments after installation. If four machines were to be selected at random, what is the probability that at least two will require further work after installation?

We assume a large population of machines. We set $n = 4$, $p = 0.4$. Then

$$P(2 \leqslant x \leqslant 4) = P(2, 4) + P(3, 4) + P(4, 4)$$

where $\quad P(2, 4) = \binom{4}{2}(0.4)^2(0.6)^2 = \dfrac{4 \cdot 3}{2 \cdot 1}(0.4)^2(0.6)^2 = 0.3456,$

$$P(3, 4) = 0.1536, \quad \text{and} \quad P(4, 4) = 0.0256.$$

Thus the required probability is 0·5248.

The company now estimates that the cost of adjustment after installation is $100 per machine, and this cost must be borne by the company. What is the *expected* cost to the firm associated with any four independently installed machines?

Let us now make use of the theorem (see also section 7.2 below)

$$E(kX) = kE(X)$$

where k is a constant number. In the present case

$$E(X) = n \cdot p = 1.6$$

i.e. *on the average* 1·6 in every four independently installed machines will require adjustments after installation. But the cost of adjustment per machine is assumed to be constant, i.e. $k = \$100$. Thus

$$E(kX) = kE(X) = k \cdot n \cdot p = \$160$$

6.9. The Binomial Distribution and the Normal Curve

The binomial distribution is a theoretical discrete probability distribution which is most readily applicable when the number of independent trials is rather small. Even with moderately small n, however, the probability calculations can be quite cumbersome, and special binomial probability tables have been constructed to facilitate computation; we shall not be concerned with the use of such tables here.

Instead, let us consider for a moment the nature of the process involving n independent trials with a specified probability of success. Each of the trials represents an independent source of variation in the experiment, e.g. each of the coins in our three coin experiment contributes to the outcome of the experiment as a whole. The question then arises, what probability distribution can we expect when n, i.e. the number of independent sources of variation, is increased indefinitely. Intuitively, we can see that the discrete binomial distribution becomes more spread out as n increases, and also smoother. It can be shown with the use of more advanced techniques that as n becomes very large, the binomial distribution will approximate a curve whose general shape is shown in Fig. 6.5.

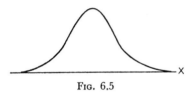

Fig. 6.5

We have already seen that when $p = \frac{1}{2}$ the binomial distribution will be symmetrical. The interesting fact is that even if p departs from $\frac{1}{2}$, the binomial distribution will closely approximate a symmetrical curve, provided that n is sufficiently large. This fact will be utilized in section 7.9 when we discuss sampling from dichotomous populations, i.e. populations classified according to only two attributes. The above curve, as illustrated in Fig. 6.5, is known as the *normal curve*. The properties of the normal curve and its use will be considered in detail in the next section; here we comment on its general characteristics.

The normal curve is often referred to as the "normal curve of errors." This stems from the notion that observations involve errors which are caused by a multitude of small and independent chance factors. The measurements of various natural phenomena may be affected in varying degrees by the objective conditions under which measurements are taken as well as by the skill of the observer. Whenever X is the sum of a true value plus a very large number of small independent errors each equally likely to be plus or minus, X will be distributed in the normal distribution. Quite apart from the distribution of errors of measurement, the distributions of many variables are normal or approximately normal. And more important, the distributions of many statistics, like the mean, are normal or nearly normal for large samples, even if the populations from which they have been drawn are not normal. In practice one might seldom find an exactly normal distribution; thus economic variables are frequently positively skewed, but this does not necessarily detract from the importance of the normal distribution which we must now consider in detail.

6.10. THE NORMAL DISTRIBUTION

If a variable X, for instance heights of Australian men, is normally distributed, the general shape of its distribution will be as in the diagram below.

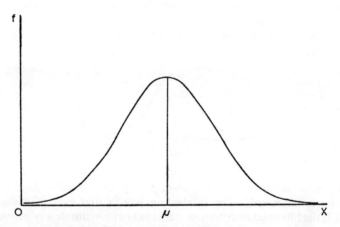

FIG. 6.6. THE NORMAL DISTRIBUTION

This distribution is defined by the equation

$$f(X) = k \cdot e^{-\frac{1}{2}\left(\frac{X-\mu}{\sigma}\right)^2}$$

where μ is the mean value of X in the population, σ is the standard deviation of X in the population, and k is a constant. It should be noted that the equation $f(X)$ above represents a probability distribution which is a *continuous* distribution. This distribution must also satisfy all the axioms of probability discussed in the preceding sections. In particular, the probabilities associated with all the possible observations of X in the population must sum to unity. By analogy with discrete probability distributions discussed earlier, this simply means that the area under the curve depicted in Fig. 6.6 must be unity. It can be shown mathematically that this condition will be satisfied when

$$k = \frac{1}{\sqrt{2\pi}\sigma}$$

where $\pi = 3\cdot14159$ and $e = 2\cdot71828$.

The normal distribution

$$f(X) = \frac{1}{\sqrt{2\pi}\sigma} e^{-\frac{1}{2}\left(\frac{X-\mu}{\sigma}\right)^2}$$

is bell-shaped, symmetrical and asymptotic in both directions to the X axis. It encloses unit area and depends on the two parameters μ and σ only, which are the mean and standard deviation of the distribution respectively. The fact of normality dictates the general shape

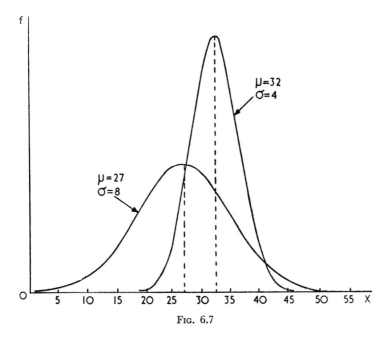

Fig. 6.7

of the distribution, but there is a whole family of normal distributions, depending on the particular values which the parameters take. This is illustrated in Fig. 6.7.

The parameter μ indicates the central value of the distribution and σ the dispersion about it, so that μ fixes the general level of the distribution and σ the extent to which it is spread out. Diagrammatically, a change in μ shifts the curve along without changing its shape, and a change in σ changes its spread. Since the distribution is symmetrical, the mean μ is unambiguously the central value and coincides with the mode and median.

If a variable X is normally distributed, we say that it is "normally distributed about a mean μ, with a standard deviation σ." Suppose we have a particular normal distribution with given μ and σ and we wish to find out the probabilities of X falling within certain *values* of the variable X which we shall indicate by subscripts, e.g. X_1 and X_2. How shall we do this?

From our discussion of probability and the properties of the normal curve, the probability of X falling within the required range, $P(X_1 < X < X_2)$, is represented by the hatched area as in Fig. 6.8.

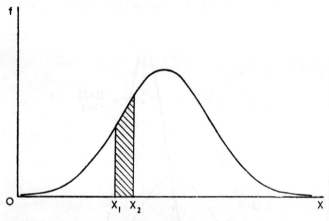

Fig. 6.8

The numerical value of this area is the integral of $f(X)$ between X_1 and X_2, i.e.

$$P(X_1 < X < X_2) = \int_{X_1}^{X_2} \frac{1}{\sqrt{2\pi}\,\sigma} e^{-\frac{1}{2}\left(\frac{X-\mu}{\sigma}\right)^2} dX$$

This integration cannot be performed analytically, and the integral must be computed by numerical methods. In practice it would be very convenient if we could look up the numerical values of areas like the hatched one in tables, but it looks at first sight as if we should want a different table for every particular combination of μ and σ. Fortunately this is not the case, for the normal distribution can be transformed into a standard form.

To do this, express the variable X as a deviation from its mean in terms of its standard deviation, thus—

$$\frac{X - \mu}{\sigma}$$

For example, if $\mu = 70$ and $\sigma = 5$, a particular value of X, say 80, is expressed as a deviation of two standard deviation units above the mean. We write

$$T = \frac{X - \mu}{\sigma}, \; T_1 = \frac{X_1 - \mu}{\sigma} \text{ and } T_2 = \frac{X_2 - \mu}{\sigma}$$

When $X_1 < X < X_2$, we must have

$$T_1 < T < T_2$$

so that

$$P(X_1 < X < X_2) = P(T_1 < T < T_2)$$

We now derive an analytical expression for $P(T_1 < T < T_2)$. We have

$$P(X_1 < X < X_2) = \int_{X_1}^{X_2} \frac{1}{\sqrt{2\pi}\,\sigma} e^{-\frac{1}{2}\left(\frac{X-\mu}{\sigma}\right)^2} dX$$

Changing the variable of integration from X to T and noting that since $X = \mu + \sigma T$, $dX = \sigma dT$, we shall obtain

$$P(T_1 < T < T_2) = \int_{T_1}^{T_2} \frac{1}{\sqrt{2\pi}} e^{-\frac{1}{2}T^2} dT$$

But

$$T = \frac{X - \mu}{\sigma}$$

and consequently has a mean of 0 and a standard deviation of 1. Thus the integrand in T is a normal distribution with mean of 0 and standard deviation of 1. Consequently we have shown that, when X is normally distributed about a mean μ with a standard deviation σ, the variable $T = \dfrac{X - \mu}{\sigma}$ is normally distributed about a mean of 0 with a standard deviation of 1. The distribution of T is unique in the sense that no parameters which can vary are contained in the integrand. Consequently it can be uniquely tabulated. Thus, in order to ascertain $P(X_1 < X < X_2)$ for given μ and σ, it is only necessary to transform the X's into T's and to evaluate $P(T_1 < T < T_2)$.

The distribution of T is tabulated in Table I of Appendix A. This table is known as a table of "areas under the normal curve." It gives, in the body of the table, the proportion of area under the normal curve lying between the central ordinate and any value of the variable $T = \frac{X - \mu}{\sigma}$ to the right of the central ordinate, such values being tabulated in steps of 0·01 in the margin of the table. Diagrammatically the table gives proportions of area like the hatched segment in Fig. 6.9.

The hatched area corresponds to the probability of T lying between 0 and T_1, i.e. $P(0 < T < T_1)$. But $P(0 < T < T_1)$ is the same thing as $P\left(0 < \frac{X - \mu}{\sigma} < \frac{X_1 - \mu}{\sigma}\right)$, so that the hatched area gives the probability that X will lie within T_1 standard deviations of its mean, or, given its standard deviation, that X will lie between X_1 and its mean. Given the table of areas, it is a perfectly straightforward matter to obtain the area between any two values of T, and hence probabilities of the form $P(T_1 < T < T_2)$, by adding or subtracting the areas given in the table. In doing this, it is helpful to sketch a small diagram, as in example 6.11

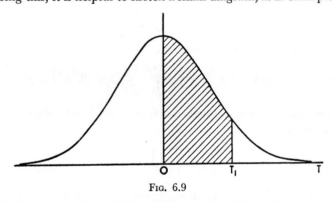

Fig. 6.9

on p. 109. It should be remembered that since the distribution is symmetrical, each half of the curve covers 50 per cent of the total area.

From the table we can always find out the probability that a variable will lie within a certain distance of its mean in terms of its standard deviation, for the area between the central ordinate and any given deviation of X from μ in terms of σ is always the same. Thus, if a variable has a mean of 100 and a standard deviation of 7, the probability that one observation of the variable will lie between 100 and 110 is the probability that an observation of a variable will lie within $\frac{110-100}{7} = 1\cdot43$ standard deviations from its mean. From the table this probability is found to be 0·4236. This probability means that if we were to draw at random, say, 1,000 observations of the variable under consideration from the population of the variable, we should expect on

PROBABILITY AND PROBABILITY DISTRIBUTIONS 109

the average approximately 424 of the observations to have values between 100 and 110. The qualification "on the average" implies that, whilst we do not necessarily expect to get exactly 424 observations in the specific range out of every 1,000, we should, with repeated sampling, expect the average number to be 424.

EXAMPLE 6.11

From the table of areas under the normal curve we have

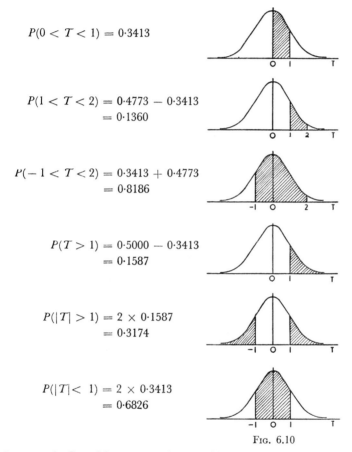

$P(0 < T < 1) = 0.3413$

$P(1 < T < 2) = 0.4773 - 0.3413$
$= 0.1360$

$P(-1 < T < 2) = 0.3413 + 0.4773$
$= 0.8186$

$P(T > 1) = 0.5000 - 0.3413$
$= 0.1587$

$P(|T| > 1) = 2 \times 0.1587$
$= 0.3174$

$P(|T| < 1) = 2 \times 0.3413$
$= 0.6826$

FIG. 6.10

To interpret the first of these we say the probability that a normally distributed variable will lie within one standard deviation above its mean is 34·13 per cent.

From the table of areas we see that the area between the central ordinate and one standard deviation to its right is 0·3413. Consequently, the range $\mu \pm \sigma$ will, in any normal distribution, contain

68·27 per cent of the population. Similarly, the range $\mu \pm 2\sigma$ will contain 95·45 per cent, and the range $\mu \pm 3\sigma$, 99·73 per cent. It is also useful to note at this stage that 95 per cent of the population will be contained in the range $\mu \pm 1\cdot96\sigma$ and 99 per cent in the range $\mu \pm 2\cdot58\sigma$. These facts about the normal distribution can be expressed in the following form and are illustrated in Fig. 6.11.

$$P(|T| < 1) = P(\mu - \sigma < X < \mu + \sigma)$$
$$= 0\cdot6827$$

$$P(|T| < 2) = P(\mu - 2\sigma < X < \mu + 2\sigma)$$
$$= 0\cdot9545$$

$$P(|T| < 3) = P(\mu - 3\sigma < X < \mu + 3\sigma)$$
$$= 0\cdot9973$$

$$P(|T| < 1\cdot96) = P(\mu - 1\cdot96\sigma < X < \mu + 1\cdot96\sigma)$$
$$= 0\cdot95$$

$$P(|T| < 2\cdot58) = P(\mu - 2\cdot58\sigma < X < \mu + 2\cdot58\sigma)$$
$$= 0\cdot99$$

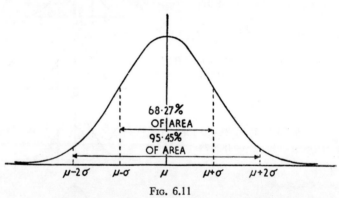

Fig. 6.11

By this stage it should be clear that given the mean and standard deviation of a particular normal distribution, the table of areas tells us all there is to know about the distribution.

The use of the table of areas of the normal curve is further illustrated in the examples below.

EXAMPLE 6.12

The weekly wages of tradesmen are normally distributed about a mean o $50 with a standard deviation of $4.

(a) Find the probability of a tradesman having a weekly wage lying
   (i) between $50 and $52;
   (ii) between $48 and $50;
   (iii) between $49 and $52;
   (iv) over $55;
   (v) under $44;
   (vi) more than $8 from the mean.

(b) What is the wage which 10 per cent of the tradesmen will earn more than?

(c) Within what deviation on both sides of the mean will 95 per cent of the tradesmen's wages lie?

(a) (i) $P(50 < X < 52) = P(0 < T < 0.5)$
$$= 0.1915$$

(ii) $P(48 < X < 50) = P(-0.5 < T < 0)$
$$= 0.1915$$

(iii) $P(49 < X < 52) = P(-0.25 < T < 0.5)$
$$= P(-0.25 < T < 0) + P(0 < T < 0.5)$$
$$= 0.0987 + 0.1915$$
$$= 0.2902$$

(iv) $P(X > 55) = P(T > 1.25)$
$$= 0.5000 - 0.3944$$
$$= 0.1056$$

(v) $P(X < 44) = P(T < -1.5)$
$$= 0.5000 - 0.4332$$
$$= 0.0668$$

(vi) $P(|X - \mu| > 8) = P(|T| > 2)$
$$= 2[0.5000 - P(0 < T < 2)]$$
$$= 2(0.5000 - 0.4773)$$
$$= 2 \times 0.0227$$
$$= 0.0454$$

These probabilities should be interpreted as follows—

Taking (i) as an example. If weekly earnings are normally distributed about a mean of $50 with a standard deviation of $4 and if a sample of tradesmen is drawn at random, we should expect on the average approximately 19.15 per cent of the sample to have earnings lying between $50 and $52.

In ascertaining the above probabilities it is often helpful to use a diagram. Taking (iii) as an example, we should have

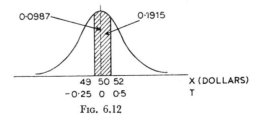

Fig. 6.12

(b) We require X_1 such that
$$P(X > X_1) = 0.1$$
i.e. $P(T > T_1) = 0.1$, where $T_1 = \dfrac{X_1 - \mu}{\sigma}$

i.e. $P(0 < T < T_1) = 0.4$

Referring to the table of areas we find that an area of 0·4 is included between the central ordinate and $T = 1.28$ approximately. Hence $T_1 = 1.28$ and $X_1 = \$55 \cdot 1$.

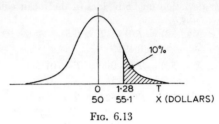

Fig. 6.13

(c) We require D such that
$$P(\mu - D < X < \mu + D) = 0.95$$
i.e. $P\left(-\dfrac{D}{\sigma} < \dfrac{X - \mu}{\sigma} < \dfrac{D}{\sigma}\right) = 0.95$

i.e. $P\left(-\dfrac{D}{\sigma} < T < \dfrac{D}{\sigma}\right) = 0.95$

i.e. $2 \times P\left(0 < T < \dfrac{D}{\sigma}\right) = 0.95$

i.e. $P\left(0 < T < \dfrac{D}{\sigma}\right) = 0.475$

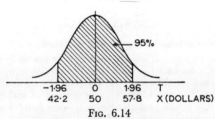

Fig. 6.14

Referring to the table of areas we find that an area of 0·475 is included between the central ordinate and $T = 1.96$. Hence
$$\dfrac{D}{\sigma} = 1.96$$
and $D = \$7 \cdot 8$

i.e. we can expect 95 per cent of tradesmen to have wages within the range $\$50 \pm \$7 \cdot 8$.

PROBABILITY AND PROBABILITY DISTRIBUTIONS 113

EXAMPLE 6.13

A random sample of 200 rents yields a mean of $23·82 and a standard deviation of $7·19. The rents of these houses are arranged into a frequency distribution with classes $2·5 and under $7·5, $7·5 and under $12·5, $12·5 and under $17·5, etc., as in Table 5.2 above. What frequency would we expect in each class if rents were distributed normally with the same mean and standard deviation as our data? This problem can be dealt with most simply in tabular form.

Rent (dollars) (1)	Lower Class Limit X_1 (2)	$\dfrac{X_1 - \mu}{\sigma}$ (3)	Area between Lower Limit and Central Ordinate (4)	Area within Classes (5)	Expected Frequency (6)
	0	−3·31	0·5000	—	—
2·5 and under 7·5	2·5	−2·97	0·4985	0·0116	2·3
7·5 ,, 12·5	7·5	−2·27	0·4884	0·0466	9·3
12·5 ,, 17·5	12·5	−1·57	0·4418	0·1312	26·3
17·5 ,, 22·5	17·5	−0·88	0·3106	0·2392	47·8
22·5 ,, 27·5	22·5	−0·18	0·0714	0·2664	53·3
27·5 ,, 32·5	27·5	0·51	0·1950	0·1919	38·4
32·5 ,, 37·5	32·5	1·21	0·3869	0·0844	16·9
37·5 ,, 42·5	37·5	1·90	0·4713	0·0240	4·8
42·5 ,, 47·5	42·5	2·60	0·4953	0·0047	0·9
Total				1·0000	200·0

Column (3) expresses the lower class limits given in column (2) as deviations from the mean in standard deviation units. Column (4) is obtained directly from the table of areas. Column (5) is obtained from column (4) by successive subtraction of the areas except for that class in which the mean lies for which the two areas are added. Column (5) must add to 1, since the probability of a

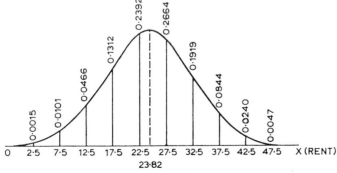

FIG. 6.15

variable falling anywhere must be unity. Any small fraction of area left over at either end of the scale should be included in the end classes. Column (6) is column (5) multiplied by 200. Since it contains the theoretically expected and not actually observed frequencies, the frequencies can be expressed with decimal places. Column (6) tells us how we can expect a sample of 200 houses drawn from a normally distributed population with a mean of $23·82 and a standard deviation of $7·19 to be distributed among the stated classes, i.e. we should expect repeated sampling of 200 to have these frequencies on the average.

The process outlined above is illustrated diagrammatically in Fig. 6.15.

CHAPTER VII

SAMPLING AND SIGNIFICANCE

7.1. The General Problem of Statistical Inference

If we have a sample number of observations of a particular variable, we can describe the sample quite fully by the techniques which have been set out in Chapter V. In general, however, we are not interested in the sample information for its own sake, but rather do we want it either—

(i) to use it to test hypotheses about the parent population from which it was drawn, or

(ii) to use it to make inferences about the nature of that parent population.

These two uses, both part of the general problem of statistical inference, can be referred to as (i) testing hypotheses, and (ii) estimation. However, it is only possible to use samples for these purposes if the samples are random samples. It will be recalled from section 6.6 that the basic sampling design is the simple random sample which allows each element in the population an equal chance of selection. In this chapter we discuss sampling procedures based on simple random sampling. Furthermore, we restrict our attention to sampling problems involving only one variable.

Our general problem is, given the sample, to decide whether it could have come from a population of a certain kind with certain parameters, i.e. to test an hypothesis about the population from the sample information. Alternatively, given the sample, what are the values of the population parameters likely to be? Since a sample covers only a fraction of the population, it cannot be expected to be an exact replica of the population. Consequently, it will be a pure fluke if a sample mean exactly equals the corresponding population mean. The discrepancies between sample values (statistics) from random samples and population values (parameters) are known as *sampling errors*, and these arise due to chance. Since all our sample statistics will be subject to sampling errors we shall never be able to be quite certain about our results, for just due to chance alone a population with, say, a certain μ may throw up a sample with a very different $\bar{X}$. Thus we want to know the way in which $\bar{X}$ would be distributed if we made repeated samplings from a given population. This probability distribution is known as the *sampling distribution* of the mean. Knowing the sampling distribution, we could compute the probability of $\bar{X}$ being different from μ by any given amount—or, in other words, we could decide whether a particular

$\bar{X}$ could reasonably have come from a population with a particular μ. The essential feature of such inferential reasoning is that we attempt to make probability judgments about the parent population on the basis of sample data.

As the first step, we must derive the sampling distribution of statistics from given populations. Thus, suppose a sample of 200 rents drawn at random from Adelaide has a mean rent of $13. It may be suggested that the mean rent of all rents in Adelaide is, say, $18. We may then ask, could this sample have reasonably come from a population in which the mean rent is $18? To be able to answer this we should want to know the probability of drawing a sample of 200 rents with a mean as different from the population mean as $13 is from $18. To know this we would want to know the way in which the means of samples of 200 drawn from the given population would be distributed, i.e. the sampling distribution of the mean. If we can show that the sample with the mean rent of $13 could not have reasonably come from a population with a mean rent of $18 because the probability of such a large discrepancy occurring is very small, then doubt is cast on the validity of the hypothesis about the nature of the parent population.

7.2. Mean and Variance of a Linear Combination

Before proceeding with sampling distributions, we must draw attention to some important theorems concerning the means and variances of variables which are expressed as linear functions of other variables. Since in economic applications the populations with which we deal are generally finite, we derive the means and variances of linear combinations with reference to finite populations. For such populations, the mean and variance are defined as

$$\mu = \frac{\Sigma X}{P}$$

and

$$\sigma^2 = \frac{\Sigma(X - \mu)^2}{P}$$

where the summation extends over the P elements in the population.

1. Suppose we have a variable X distributed about a mean μ_X with a variance σ_X^2 and we multiply every possible observation of X by a constant A, to form a new variable

$$W = AX$$

then

$$\Sigma W = A \Sigma X$$

Dividing by P, where P is the number in the population, we shall have

$$\mu_W = A\mu_X$$

where μ_W is the mean of W, i.e. the mean of a constant times a variable is the constant times the mean of the variable.

2. It follows from this that
$$W - \mu_W = A(X - \mu_X)$$
Hence
$$\Sigma(W - \mu_W)^2 = \Sigma[A(X - \mu_X)]^2$$
$$= A^2\Sigma(X - \mu_X)^2$$
Again dividing both sides by P, we shall have
$$\sigma_W^2 = A^2\sigma_X^2$$
where σ_W^2 is the variance of W, i.e. the variance of a constant times a variable is the constant squared times the variance of the variable.

3. Now, suppose we also have a variable Y distributed about a mean μ_Y with a variance σ_Y^2. We select an X at random and at the same time a Y at random and then combine the X and Y values to make a new value Z, such that $Z = AX + BY$, where A and B are constants. Then Z is said to be a linear combination of X and Y. We repeat this process a large number of times and generate a population of Z's. We now show how to express the mean and variance of Z in terms of the means and variances of X and Y.

We have
$$Z = AX + BY$$
$$\therefore \quad \Sigma Z = A\Sigma X + B\Sigma Y$$
and, dividing by P, we get
$$\mu_Z = A\mu_X + B\mu_Y$$
where μ_Z is the mean of Z, i.e. the mean of a linear combination is the linear combination of the means.

4. It follows from the above that
$$Z - \mu_Z = A(X - \mu_X) + B(Y - \mu_Y)$$
Writing small letters for deviations from means, this becomes
$$z = Ax + By$$
$$\therefore \quad z^2 = A^2x^2 + B^2y^2 + 2ABxy$$
$$\therefore \quad \Sigma z^2 = A^2\Sigma x^2 + B^2\Sigma y^2 + 2AB\Sigma xy$$

The sums of squares Σz^2, Σx^2 and Σy^2 are often referred to as the *variation* of Z, X and Y respectively; the term Σxy is then called the *co-variation* of X and Y, and will occur again when later we discuss regression and correlation analysis. Provided that X and Y are completely independent, so that the getting of a particular value of X has no influence on the value of Y with which it is paired (i.e. provided that

118   APPLIED STATISTICS FOR ECONOMISTS

higher than average X's do not tend to be paired mainly with higher or mainly with lower than average Y's), then in a large population of Z's the co-variation term Σxy will tend to zero; for the paired x's and y's will have the same signs as often as they have different signs and the plus xy's will cancel out the minus xy's.

$$\therefore \quad \Sigma z^2 = A^2 \Sigma x^2 + B^2 \Sigma y^2$$

and, dividing by P we get

$$\sigma_Z^2 = A^2 \sigma_X^2 + B^2 \sigma_Y^2$$

where σ_Z^2 is the variance of Z, i.e. the variance of a linear combination of two independent variables is a linear combination of the variances of the variables having as their coefficients the squares of the coefficients of the original linear combination. This result holds generally. Thus if $X_1, X_2 \ldots X_n$ are independent, the variance of their sum (Z) will be equal to the sum of their variances, i.e.

$$\mathrm{Var}\,(Z) = \mathrm{Var}\,(X_1) + \mathrm{Var}\,(X_2) + \ldots \mathrm{Var}\,(X_n)$$

Furthermore, this result holds for any linear combination, i.e. irrespective of the signs which bind the variables together. Thus if, for instance,

$$Z = X - Y$$

then

$$\Sigma z^2 = \Sigma x^2 + \Sigma y^2 - 2\Sigma xy$$

and the co-variation term vanishes if X and Y are independent. Thus we have the simple but important case that if $Z = X \pm Y$, and X and Y are independent, $\sigma_Z^2 = \sigma_X^2 + \sigma_Y^2$.

Another important result, which will not be proved here, is that if the independent variables are themselves normally distributed, a linear combination of them will be normally distributed, with mean and variance as above. In the remainder of this chapter, frequent use will be made of these results.

Example 7.1

The weights of packages (X) handled by a delivery service are found to be normally distributed with a mean weight of 300 pounds and a standard deviation of 50 pounds. If the packages are loaded onto trucks in lots of 25 packages chosen at random, what is the mean and standard deviation of the truck-load weights?

Let T be the new variable representing weights of 25 packages. With $N = 25$, and $\mu_X = 300$

$$\mu_T = N\mu_X = (25)(300) = 7{,}500 \text{ lb.}$$

and

$$\sigma_T^2 = N^2 \sigma_X^2 = (25)^2 (50)^2$$

$$\therefore \quad \sigma_T = 1{,}250 \text{ lb}$$

SAMPLING AND SIGNIFICANCE 119

The weights of 25 packages will also be distributed normally with a mean of 7,500 pounds and a standard deviation of 1,250 pounds.

7.3. Sampling Distribution of the Mean

Suppose we have a variable X normally distributed about a mean μ with a standard deviation σ. We draw a random sample of N observations from this population and calculate $\bar{X}$ for the sample. If we repeat this experiment, a whole series of $\bar{X}$'s will be generated. These will evidently vary between themselves and from μ, just due to chance. In other words, sampling errors will occur. However, we should expect in the long run that the mean of the $\bar{X}$'s would itself tend towards μ and that the variability of the $\bar{X}$'s would be less than the variability of the X's and would be smaller the greater N. It can be proved that if X is normally distributed about a mean μ with a standard deviation σ then $\bar{X}$, the mean of a sample of size N, will, with repeated sampling, be normally distributed about a mean μ with a standard deviation of $\dfrac{\sigma}{\sqrt{N}}$.

It is not difficult to prove that with repeated sampling the mean of the $\bar{X}$'s will tend to μ and the standard deviation to $\dfrac{\sigma}{\sqrt{N}}$. We write $\mu_{\bar{X}}$ and $\sigma_{\bar{X}}$ for the mean and standard deviation of the infinite population of $\bar{X}$'s which can be generated by repeated sampling, and μ and σ for the mean and standard deviation of the parent population of X's.

The mean of a sample of size N is defined by

$$\bar{X} = \frac{\Sigma X}{N}$$
$$= \frac{X_1 + X_2 + \ldots + X_N}{N}$$
$$= \frac{X_1}{N} + \frac{X_2}{N} + \ldots \frac{X_N}{N}$$

If we have a number of random samples of size N, we shall have

$$\bar{X}^{(1)} = \frac{X_1^{(1)}}{N} + \frac{X_2^{(1)}}{N} + \ldots \frac{X_N^{(1)}}{N}$$
$$\bar{X}^{(2)} = \frac{X_1^{(2)}}{N} + \frac{X_2^{(2)}}{N} + \ldots \frac{X_N^{(2)}}{N}$$
$$\bar{X}^{(3)} = \frac{X_1^{(3)}}{N} + \frac{X_2^{(3)}}{N} + \ldots \frac{X_N^{(3)}}{N}$$

and so on,
where the superscripts refer to the first, second, third sample, etc. Since the X's are all drawn from the same population, the mean of the

population of X_1's (i.e. of the first items in each sample) will be μ and the standard deviation σ (i.e. variance σ^2); likewise for the X_2's, X_3's, etc. Now it can be seen from the above that $\bar{X}$ is a linear combination of N variables. Hence, by virtue of section 7.2 above we shall have

$$\mu_{\bar{X}} = \frac{\mu}{N} + \frac{\mu}{N} + \ldots + \frac{\mu}{N}$$
$$= \mu$$

and
$$\sigma^2_{\bar{X}} = \frac{\sigma^2}{N^2} + \frac{\sigma^2}{N^2} + \ldots + \frac{\sigma^2}{N^2}$$
$$= \frac{\sigma^2}{N}$$

i.e.
$$\sigma_{\bar{X}} = \frac{\sigma}{\sqrt{N}}$$

The standard deviation of a statistic is called the *standard error* of that statistic, e.g. $\sigma_{\bar{X}}$ is called the standard error of the mean. These two important results are quite general irrespective of the nature of the

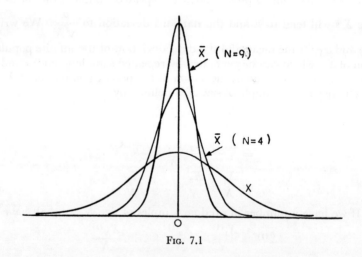

Fig. 7.1

parent population of the X's, and the first of them shows that a sample $\bar{X}$ is an unbiased estimate of the population μ, in the sense that with repeated sampling the mean of the $\bar{X}$'s tends to μ. If, however, X itself is normally distributed, $\bar{X}$ will also be normally distributed. But even if X is not normally distributed, the distribution of $\bar{X}$ will tend to normality as the size, N, of the samples chosen increases. Accordingly $\bar{X}$ can be treated as if it were normally distributed about μ with a standard error $\dfrac{\sigma}{\sqrt{N}}$, provided N is reasonably large, say, over 50. This gives us the sampling distribution of the mean, i.e. the way in which means of samples are distributed. From this distribution we can ascertain the probability that an $\bar{X}$ will lie within any particular range, e.g. the probability that a sample mean will lie within the range $\mu \pm 1 \cdot 96 \sigma_{\bar{x}}$, where $\sigma_{\bar{x}} = \dfrac{\sigma}{\sqrt{N}}$, is 0·95; in other words, in the long run 95 out of every 100 random samples of size N will have means within this range.

Since $\sigma_{\bar{x}} = \dfrac{\sigma}{\sqrt{N}}$, the standard error of the mean is smaller the larger the size of the sample, i.e. the variability of sample means is smaller the larger the size of the sample. Consequently, sampling errors will be smaller the larger is sample size. This is illustrated in Fig. 7.1, where the distributions of means of samples of different sizes are compared with the distribution of the parent population. It can be seen that for larger samples the distribution is tighter. A few moments' thought will show that the standard error of a sample mean must always be less than the standard deviation of the population, since, for each sample, the sample mean averages out the variability of the observations within the sample.

Before proceeding with the application of the results of this section to the testing of statistical hypotheses, we can make use of the result $\sigma_{\bar{x}} = \dfrac{\sigma}{\sqrt{N}}$ to show why we defined the standard deviation of a sample in section 5.5 above with $N - 1$ as the denominator rather than N. The variance of the population of the variable X is defined as

$$\sigma^2 = \frac{\Sigma(X - \mu)^2}{P}$$

where P is so large as to cover the whole population. The numerator of the variance of a sample of size N is given by $\Sigma(X - \bar{X})^2$. We have shown (page 71) that this can be written

$$\Sigma(X - \bar{X})^2 = \Sigma(X - A)^2 - N(\bar{X} - A)^2$$

where A is an arbitrary origin. We now put $A = \mu$, where μ is the mean of the population of X's, so that

$$\Sigma(X - \bar{X})^2 = \Sigma(X - \mu)^2 - N(\bar{X} - \mu)^2$$

If we take M such samples of size N, we shall have the above relation holding for each sample, and summing them we shall have

$$\sum_{}^{M}\sum_{}^{N}(X-\bar{X})^2 = \sum_{}^{M}\sum_{}^{N}(X-\mu)^2 - N\sum_{}^{M}(\bar{X}-\mu)^2$$

where $\sum^{N}$ refers to the summation within each sample and $\sum^{M}$ refers to the summation of samples. This can be rewritten

$$\frac{1}{M}\sum_{}^{M}\sum_{}^{N}(X-\bar{X})^2 = N\left[\frac{\sum^{M}\sum^{N}(X-\mu)^2}{MN} - \frac{\sum^{M}(\bar{X}-\mu)^2}{M}\right]$$

As M becomes larger the first term in the square brackets on the right-hand side tends towards σ^2 the variance of the population and the second towards $\sigma^2_{\bar{x}}$, the variance of the mean of samples of size N, so that the right-hand side tends towards

$$N(\sigma^2 - \sigma^2_{\bar{x}}) = N\left(\sigma^2 - \frac{\sigma^2}{N}\right)$$
$$= (N-1)\sigma^2$$

It follows that, as M becomes larger,

$$\frac{1}{M}\sum_{}^{M}\sum_{}^{N}\frac{(X-\bar{X})^2}{N-1} \text{ tends towards } \sigma^2.$$

But this expression is the average value of $\dfrac{\Sigma(X-\bar{X})^2}{N-1}$ for the M samples, so that $s^2 = \dfrac{\Sigma(X-\bar{X})^2}{N-1}$ will be an unbiased estimate of σ^2, in the sense that with repeated sampling the average of s^2 will tend towards σ^2. On the other hand, were s^2 defined as $\dfrac{\Sigma(X-\bar{X})^2}{N}$, the average of s^2 would in the long run tend to be lower than σ^2, so that it would be a downwardly biased estimate of σ^2. The reason for this is that the variance of the population is defined in terms of deviations from the population mean μ, whereas the deviations in the expression for s^2 must be from the sample mean $\bar{X}$. It has already been proved that the sum of the squares of deviations of values is least when the deviations are from the mean of the values, hence

$$\sum_{}^{N}(X-\bar{X})^2 < \sum_{}^{N}(X-\mu)^2$$

Consequently the numerator in s^2 will tend to be downwardly biased, and dividing by $N-1$ rather than N corrects this.

7.4. Tests of Significance

Tests of significance are tests of the significance of statistical hypotheses. For example, we have a certain sample taken from a population. Suppose the sample mean is $\bar{X}$ and we set up the hypothesis that it has come from a population with a mean of μ. This hypothesis implies that the discrepancy between $\bar{X}$ and μ is only due to chance, i.e. in the long run repeated sampling would produce data which would result in a mean discrepancy between $\bar{X}$ and μ of zero. We now ask what is the probability of our getting a discrepancy between $\bar{X}$ and μ as great as or greater than the actual one, *if the hypothesis were true*. We can answer this question from our knowledge of the sampling distribution of $\bar{X}$, for we can ascertain the probability of obtaining from a population, in which μ is the mean, an $\bar{X}$ further away from μ than the one under consideration. Suppose this probability turns out to be small ("small" to be defined later), we can then conclude:

(i) Hypothesis true and we have struck a fluke, or
(ii) Hypothesis true but sampling has not been random, or
(iii) Hypothesis false.

The choice lies between either (i) and (iii) if we can be certain that the sampling has been random, or between (i) and (ii) if we are certain that the hypothesis is true. In this latter case we are testing the randomness of the sampling. Let us assume, however, for the moment, that the sampling is random so that the choice is between (i) and (iii). By convention, we always conclude (iii), for a fluke is by its nature unlikely to occur. If the probability that a difference equal to or greater than the actual one between $\bar{X}$ and μ will occur just due to chance is small, we say that $\bar{X}$ is *significantly different* from μ, and we reject the hypothesis that $\bar{X}$ comes from a population with mean μ. Similarly, if the choice is between (i) and (ii), we conclude (ii).

The vital point, is, of course, just how small the probability referred to above has to be, before it can be called "small." The choice here is quite arbitrary. Usually we take, by convention, 5 per cent as the critical level. This is called the 5 *per cent level of significance*. Sometimes a 1 per cent level of significance is used.

The procedure is as follows—

1. *Set up an hypothesis* about the population from which the sample has been drawn. This is called the *null hypothesis*, because it asserts that there is no difference between the sample and the population in the particular matter under consideration. According to this hypothesis, the true population mean is some particular value, say, μ_o. But the null hypothesis can be either true or false, and if our test leads us to reject it, some other hypothesis, known as the alternative hypothesis, must be accepted. Thus we have

Null hypothesis: The population mean is μ_o

Alternative hypothesis: The population mean is not μ_o

For convenience and brevity, this may be written as

$$H_o: \mu = \mu_o$$
$$H_a: \mu \neq \mu_o$$

The alternative hypothesis is here stated in a general form which permits the testing of hypotheses about the true population mean lying in either direction of the assumed population mean μ_o. In some circumstances, the appropriate alternative hypothesis may be in the form $\mu > \mu_o$ or in the form $\mu < \mu_o$. The procedure to be followed in such cases will be considered in section 7.15 below.

2. *Test the null hypothesis* by determining what is the probability of getting a discrepancy as great as or greater than the actual discrepancy between the sample and population just due to chance if the null hypothesis is true.

3. *Draw a conclusion* about the reasonableness of the null hypothesis. If the probability under (2) is greater than 5 per cent, do not reject the null hypothesis; if it is 5 per cent or less, reject it. Actually if the probability is about 5 per cent, the wisest course may be to reserve judgment and draw another sample if possible. We can never prove or disprove an hypothesis. We can merely indicate that the sample is not inconsistent with it or, on the other hand, we can cast doubt on its reasonableness. For when we reject an hypothesis, there is always the possibility that the hypothesis is true and we have struck a fluke and, when we accept one, we cannot be certain that it holds, our data merely being not inconsistent with it.

7.5. Significance of the Difference Between $\bar{X}$ and μ, σ known

We now illustrate the above with a simple case. We have a random sample of N items with mean $\bar{X}$. Could this have come from a normal population with a mean μ and a standard deviation σ? We set up the null and the alternative hypothesis

$$H_o: \mu = \mu_o$$
$$H_a: \mu \neq \mu_o$$

We know that the means of such samples drawn from such a population are normally distributed about a mean μ with a standard error $\sigma_{\bar{X}} = \dfrac{\sigma}{\sqrt{N}}$. Consequently we can easily find the probability that $\bar{X}$ will differ from μ by an amount as great as or greater than any given amount. Thus we know that $\bar{X}$ will differ from μ by $1\cdot 96\ \sigma_{\bar{X}}$ or more with a probability of 5 per cent.

In the diagram, the two tails each cover $2\frac{1}{2}$ per cent of the area, the probability of $\bar{X}$ lying further from μ by more than $1\cdot 96$ $\sigma_{\bar{X}}$ in either direction being $0\cdot 05$, or 5 per cent.

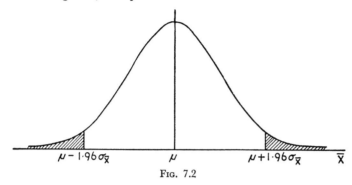

FIG. 7.2

Furthermore under the above conditions, $T = \dfrac{\bar{X} - \mu}{\sigma_{\bar{X}}}$ will be normally distributed in the standard fashion about 0 with unit standard deviation (see p. 107 above), and consequently T will exceed $1\cdot 96$ in magnitude with a probability of 5 per cent. Accordingly we now calculate

$$|T| = \frac{|\bar{X} - \mu|}{\sigma_{\bar{X}}}$$

where $|T|$ means the magnitude of T irrespective of sign and can be regarded as a measure of the discrepancy between the sample results and the null hypothesis.

The probability of $|T|$ being greater than $1\cdot 96$ is $0\cdot 05$, so that if $|T| > 1\cdot 96$ repeated sampling will yield a value of $|T|$ as great as or greater than the one obtained in 5 cases or less in 100 due to chance if the null hypothesis H_o is true. This means that, if H_o is true, repeated sampling will yield a discrepancy (as measured by $|T|$) between the sample results and H_o as great as or greater than the one actually obtained in 5 or less cases in 100 due to chance. It follows that either we have struck a rare case or H_o is not true. By its nature the former is unlikely, so we reject the null hypothesis H_o and accept the alternative hypothesis H_a. We say that $\bar{X}$ is significantly different from the hypothetical μ. Conversely, if $|T| < 1\cdot 96$, then, if H_o is true, repeated sampling will yield a discrepancy (as measured by $|T|$) between sample results and H_o as great as or greater than the one actually obtained in more than 5 cases in 100. Such being the case, the sample could reasonably have come from the hypothetical population, so we do not reject the null hypothesis. We say that $\bar{X}$ is not significantly different from the hypothetical μ.

Example 7.2

A random sample of twenty-five employees has a mean weekly wage of $52. Could this sample have been drawn from a population normally distributed about a mean of $48 with a standard deviation of $4?

Here we have

$$N = 25$$
$$\bar{X} = 52$$
$$\mu = 48$$
$$\sigma = 4$$
$$\sigma_{\bar{X}} = \frac{\sigma}{\sqrt{N}} = 0.8$$

Hypothesis. The sample is drawn from a normally distributed population with $\mu = 48$ and $\sigma = 4$, i.e.

$$H_o: \mu = 48$$
$$H_a: \mu \neq 48$$

Test—
$$|T| = \frac{|\bar{X} - \mu|}{\sigma_{\bar{X}}}$$
$$= \frac{|52 - 48|}{0.8}$$
$$= 5$$

From the normal table $P(|T| > 5) < 0.001$, which is less than 0.05.

Conclusion. If the null hypothesis H_o were true, repeated sampling would yield a value of $|T|$ as great as or greater than 5 in very many less than 1 case in 1,000 due to chance. This means that if H_o were true, repeated sampling would yield a discrepancy (as measured by $|T|$) between $\bar{X}$ and μ as great as or greater than the one actually obtained above in very many less than 1 case in 1,000 due to chance. It follows that either we have struck a rare case or H_o is not true. By its nature the former is unlikely. We thus regard the null hypothesis as being inconsistent with the sample result, and accept the alternative hypothesis H_a. The sample could not have reasonably come from a hypothetical population with $\mu = 48$. We say that $\bar{X}$ differs significantly from μ. In passing, we may note that, if we know that the sample has in fact been drawn from the specified population, the test in this instance would cast serious doubt on the randomness of the sample. For only very seldom would a random sample have a mean so different from the population mean. (See p. 143, below.)

Strictly speaking, this test and those which follow in sections 7.6 to 7.8 can be applied only if the population is assumed to be normal, but we know that $\bar{X}$ tends to a normal distribution if N is large, irrespective of the parent population. In practice, for $N > 50$ the test can always be applied.

7.6. Significance of the Difference between Two Sample Means $\bar{X}_1$ and $\bar{X}_2$, σ known

We may have two samples of size N_1 and N_2, giving means of $\bar{X}_1$ and $\bar{X}_2$. An obvious question is whether these two samples could have come from the same population, or, what is the same thing, whether they could have come from two assumed populations with equal means μ_1 and μ_2. In this case the hypotheses to be tested are

$$H_o: \mu_1 = \mu_2$$
and
$$H_a: \mu_1 \neq \mu_2$$

i.e. the null hypothesis that the true means of the two populations from which the samples have been drawn are equal against the alternative hypothesis that they are not.

First an important result must be stated. If X is normally distributed about a mean μ_X with a standard deviation σ_X, and Y is normally distributed about a mean μ_Y with a standard deviation σ_Y, and if X and Y are independent in the sense that particular values of X are not associated with particular values of Y, then $X \pm Y$ will be normally distributed about a mean of $\mu_X \pm \mu_Y$ with a standard deviation of $\sigma_{X \pm Y} = \sqrt{\sigma_X^2 + \sigma_Y^2}$. (See theorems 3 and 4, section 7.2 above.)

Suppose we have a normal distribution with mean μ and standard deviation σ. We draw one sample of N_1 items, with a mean $\bar{X}_1$ and a second of N_2 items, with a mean $\bar{X}_2$. This pair of samples will yield a difference of means of $\bar{X}_1 - \bar{X}_2$. If we repeated this experiment, we should find in view of the result stated above that $\bar{X}_1 - \bar{X}_2$ was normally distributed about a mean of $\mu_{\bar{X}_1 - \bar{X}_2} = 0$ with standard error

$$\sigma_{\bar{X}_1 - \bar{X}_2} = \sqrt{\sigma_{\bar{X}_1}^2 + \sigma_{\bar{X}_2}^2}$$
$$= \sqrt{\frac{\sigma^2}{N_1} + \frac{\sigma^2}{N_2}}$$
$$= \sigma \sqrt{\frac{1}{N_1} + \frac{1}{N_2}}$$

since each pair of samples is drawn from the same population. It follows that, if we have two sample means $\bar{X}_1$ and $\bar{X}_2$ and we set up the null hypothesis that they come from the same population with a standard deviation of σ, then we can test whether the difference between the means $(\bar{X}_1 - \bar{X}_2)$ differs significantly from zero, by writing

$$|T| = \frac{|(\bar{X}_1 - \bar{X}_2) - 0|}{\sigma_{\bar{X}_1 - \bar{X}_2}} = \frac{|\bar{X}_1 - \bar{X}_2|}{\sigma_{\bar{X}_1 - \bar{X}_2}}$$

where
$$\sigma_{\bar{X}_1 - \bar{X}_2} = \sigma \sqrt{\frac{1}{N_1} + \frac{1}{N_2}}$$

If $|T| > 1·96$, then it is unreasonable to suggest that the two samples come from the same population, because only rarely would two samples drawn from the same population have such different means. This test is precisely analogous to the preceding one.

EXAMPLE 7.3

A random sample of 40 male employees is taken at the end of a year and the mean number of hours of absenteeism for the year is found to be 63 hours. A similar sample of 50 female employees has a mean of 66 hours. Could these samples have been drawn from a population with the same mean and with $\sigma = 10$ hours?

Here we have
$$N_1 = 40 \quad N_2 = 50$$
$$\bar{X}_1 = 63 \quad \bar{X}_2 = 66$$
$$\sigma = 10$$

$$\sigma_{\bar{X}_1 - \bar{X}_2} = \sigma \sqrt{\frac{1}{N_1} + \frac{1}{N_2}}$$
$$= 10 \sqrt{\frac{1}{40} + \frac{1}{50}}$$
$$= 2·12$$

Hypothesis. The samples were both drawn from the same normal population with $\sigma = 10$, i.e.
$$H_o: \mu_1 = \mu_2$$
$$H_a: \mu_1 \neq \mu_2$$

Test—
$$|T| = \frac{|\bar{X}_1 - \bar{X}_2|}{\sigma_{\bar{X}_1 - \bar{X}_2}}$$
$$= \frac{|63 - 66|}{2·12}$$
$$= 1·42$$

From the normal table, $P(|T| > 1·42) = 0·16$, which is greater than 0·05.

Conclusion. If the null hypothesis H_o were true, repeated sampling would yield a value of $|T|$ as great as or greater than 1·42 in about 16 cases in 100 due to chance. This means that, if H_o were true, repeated sampling would yield a discrepancy (as measured by $|T|$) between $\bar{X}_1$ and $\bar{X}_2$ as great as or greater than the one actually obtained above in about 16 cases in 100 due to chance. Such being the case, the two samples could reasonably have come from the same hypothetical population, so we do not reject the null hypothesis. We say that $\bar{X}_1$ and $\bar{X}_2$ do not differ significantly. There appears to be no difference in the incidence of absenteeism between the two groups of employees.

7.7. SIGNIFICANCE OF THE DIFFERENCE BETWEEN $\bar{X}$ AND μ, σ NOT KNOWN

In the above two cases we have tested hypotheses about μ, where σ, the population standard deviation, is given. In practice we shall seldom know σ, for we shall have only sample results. If we wish to test whether $\bar{X}$ and a hypothetical μ are consistent or whether $\bar{X}_1$ and $\bar{X}_2$ could have come from the same population, we shall not in general know σ, and we shall have to make an estimate of it from the sample. We can, of course, do this from the formula

$$s = \sqrt{\frac{\Sigma(X-\bar{X})^2}{N-1}}$$

As has been shown above (section 7.3) the statistic, s, is an unbiased estimate of σ in the sense that with repeated samplings the mean of the s's will tend to σ.

Previously we have stated that $\bar{X}$ is normally distributed about μ with a standard error $\sigma_{\bar{X}} = \frac{\sigma}{\sqrt{N}}$, so that $T = \frac{\bar{X} - \mu}{\sigma_{\bar{X}}}$ is distributed in the standard fashion about 0 with unit standard deviation. This ceases to hold when we use a sample estimate of σ, instead of a known σ itself. If we write $t = \frac{\bar{X} - \mu}{s_{\bar{X}}}$ where $s_{\bar{X}} = \frac{s}{\sqrt{N}}$, ($s_{\bar{X}}$ is the sample estimate of $\sigma_{\bar{X}}$), then t is distributed not normally but in another distribution known as the *t-distribution*. Actually, for large N, the t-distribution tends to the normal distribution. We should expect t to be

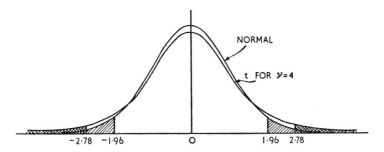

FIG. 7.3. THE NORMAL AND t-DISTRIBUTIONS

distributed differently from T, for in T, $\bar{X}$ is the only source of variation from sample to sample, whereas in t, $s_{\bar{X}}$ is an additional source. Hence the t-distribution will be more dispersed than the normal distribution and this dispersion will be the greater the smaller the size of the sample, because $s_{\bar{X}}$ will vary more from sample to sample the smaller the sample size. The t-distribution has a single parameter which is called the *number of degrees of freedom*. This number is related to the sample size and is indicated by ν. In the case of a single sample, $\nu = N - 1$.

The smaller ν, i.e. the smaller the sample size, the more dispersed is the *t*-distribution. Accordingly the smaller the size of the sample the greater is the value of *t* needed to cut off the two $2\frac{1}{2}$ per cent tails, i.e. the greater the value of *t* at the 5 per cent level of significance. As N gets larger, the distribution approaches the normal, and for $N > 30$ it is practically so. The distribution of *t* is tabulated in Table II of Appendix A. It gives in the body of the table, for particular numbers of degrees of freedom tabulated in the vertical margin, the value of *t* which cuts off tails covering certain proportions of area tabulated in the horizontal margin. Thus, for $\nu = 4$, $P(|t| > 2\cdot78) = 0\cdot05$, i.e. a distance equal to 2·78 from the centre of the distribution cuts off tails each equal to $2\frac{1}{2}$ per cent of the area, so that the probability of obtaining a value of *t* greater than 2·78 in magnitude is 5 per cent. This is shown in the diagram on p. 129. The figure of 2·78 is obtained from the table by reading across from $\nu = 4$ and down from $P = 0\cdot05$. This example can be interpreted as follows: If random samples of size 5 are taken repeatedly from a normal population with mean μ, and if $\bar{X}$, s and $t = \dfrac{\bar{X} - \mu}{s_{\bar{X}}}$ where $s_{\bar{X}} = \dfrac{s}{\sqrt{N}}$, are calculated from them, *t* will exceed 2·78 in magnitude in 5 per cent of the cases. If the distribution of *t* had been normal, the corresponding figure would have been 1·96. Again, from the *t*-table we can ascertain that, say with $\nu = 10$ and $t = 1\cdot2$, $P(|t| > 1\cdot2)$ lies between 0·3 and 0·2, so that with repeated sampling a value of *t* larger than 1·2 would occur in from 20 per cent to 30 per cent of the cases.

The *t*-distribution can be used to test an hypothesis about μ when σ is not known. If we have a random sample of N items with mean $\bar{X}$ and standard deviation s, we can test whether $\bar{X}$ differs significantly from an hypothetical μ by calculating

$$|t| = \frac{|\bar{X} - \mu|}{s_{\bar{X}}}, \quad \text{where } s_{\bar{X}} = \frac{s}{\sqrt{N}}$$

and referring it to the *t*-table with $\nu = N - 1$ degrees of freedom. We can thus ascertain the probability of getting a value of *t* as great as or greater than the actual value on the hypothesis that the sample was drawn from a population with the hypothetical μ.

EXAMPLE 7.4

A census of retail establishments in a particular year revealed that the mean annual turnover of suburban food stores was $12,500. A random sample of twenty-five such stores taken in the following year had a mean annual turnover of $13,135 and a standard deviation of turnover of $1,600. Could you conclude that the mean annual turnover had changed since the census?

Here

$N = 25$

$\bar{X} = 13,135$

$\mu = 12,500$

$s = 1,600$

$s_{\bar{X}} = \dfrac{1,600}{\sqrt{25}} = 320$

Hypothesis. The sample comes from a normal population with $\mu = 12,500$.

$H_o: \mu = 12,500$

$H_a: \mu \neq 12,500$

Test—

$$|t| = \dfrac{|\bar{X} - \mu|}{s_{\bar{X}}}$$

$$= \dfrac{635}{320}$$

$$= 1 \cdot 98$$

From the t-table with $\nu = 24$, $P(|t| > 1 \cdot 98) = 0 \cdot 06$, which is greater than $0 \cdot 05$.

Conclusion. If the null hypothesis H_o were true, repeated sampling with 24 degrees of freedom would yield a value of $|t|$ as great as or greater than $1 \cdot 98$ in about 6 cases in 100 due to chance. This means that, if H_o were true, repeated sampling would yield a discrepancy (as measured by $|t|$) between $\bar{X}$ and μ as great as or greater than the one actually obtained above in about 6 cases in 100 due to chance. Such being the case, we do not reject the hypothesis that $\mu = \$12,500$. We cannot conclude that there has been a statistically significant change in mean annual turnover of suburban food stores since the census. However, in this particular case, since we are close to the borderline of significance, it would be wise to withhold judgment until a further sample could be tested.

7.8. Significance of the Difference Between Two Sample Means $\bar{X}_1$ and $\bar{X}_2$, σ not known

In section 7.6 above we considered the case of two random samples of size N_1 and N_2, with means $\bar{X}_1$ and $\bar{X}_2$, and tested whether they could have come from the same normal population. We showed that, if they were from the same population, $\bar{X}_1 - \bar{X}_2$ would be normally distributed about a mean of 0 with a standard deviation

$$\sigma_{\bar{X}_1 - \bar{X}_2} = \sigma \sqrt{\dfrac{1}{N_1} + \dfrac{1}{N_2}}$$

so that

$$|T| = \dfrac{|\bar{X}_1 - \bar{X}_2|}{\sigma_{\bar{X}_1 - \bar{X}_2}}$$

could be used to test the hypothesis that the two samples come from the same population. But generally we shall not know σ. However, we shall know s_1 and s_2 from the two samples. Accordingly, we must use this information to make an estimate s of the population σ. The best estimate is given by

$$s = \sqrt{\frac{(N_1 - 1)s_1^2 + (N_2 - 1)s_2^2}{N_1 + N_2 - 2}}$$

$$= \sqrt{\frac{\Sigma(X_1 - \bar{X}_1)^2 + \Sigma(X_2 - \bar{X}_2)^2}{N_1 + N_2 - 2}}$$

This estimate is a pooled or weighted estimate.

Just as $t = \dfrac{\bar{X} - \mu}{s_{\bar{X}}}$ is distributed in the t-distribution with $N - 1$ degrees of freedom so $t = \dfrac{(\bar{X}_1 - \bar{X}_2) - 0}{s_{\bar{X}_1 - \bar{X}_2}}$ is distributed in the t-distribution with $N_1 + N_2 - 2$ degrees of freedom. Accordingly, by analogy with the preceding sections, to test the hypothesis that the two samples come from the same population, we compute

$$|t| = \frac{|\bar{X}_1 - \bar{X}_2|}{s_{\bar{X}_1 - \bar{X}_2}}$$

where $\quad s_{\bar{X}_1 - \bar{X}_2} = s\sqrt{\dfrac{1}{N_1} + \dfrac{1}{N_2}}$

and $\quad s = \sqrt{\dfrac{(N_1 - 1)s_1^2 + (N_2 - 1)s_2^2}{N_1 + N_2 - 2}}$

and refer it to the t-table with $N_1 + N_2 - 2$ degrees of freedom. If $N_1 + N_2 - 2 > 30$, the distribution can be taken as normal.

EXAMPLE 7.5

Ten plots of land are treated with fertilizer A and twelve with fertilizer B. The mean yield of the first plots is 6·00 bushels with a standard deviation of 0·03 bushels. The yields of the second plots have a mean of 5·95 bushels with a standard deviation of 0·04 bushels. Is there a difference in the effects of the fertilizers?

Here $\quad N_1 = 10 \quad \bar{X}_1 = 6{\cdot}00 \quad s_1 = 0{\cdot}03$

$\quad N_2 = 12 \quad \bar{X}_2 = 5{\cdot}95 \quad s_2 = 0{\cdot}04$

$$s = \sqrt{\frac{(N_1 - 1)s_1^2 + (N_2 - 1)s_2^2}{N_1 + N_2 - 2}}$$

$$= \sqrt{\frac{9 \times 0.0009 + 11 \times 0.0016}{20}}$$

$$= 0.036$$

$$s_{\bar{X}_1 - \bar{X}_2} = s\sqrt{\frac{1}{N_1} + \frac{1}{N_2}}$$

$$= 0.036\sqrt{\frac{1}{10} + \frac{1}{12}}$$

$$= 0.015$$

Hypothesis. The two samples come from the same normal population, i.e.

$$H_o: \mu_1 = \mu_2$$
$$H_a: \mu_1 \neq \mu_2$$

Test—

$$|t| = \frac{|\bar{X}_1 - \bar{X}_2|}{s_{\bar{X}_1 - \bar{X}_2}}$$

$$= \frac{0.05}{0.015}$$

$$= 3.3$$

From the *t*-table, with $\nu = 20$, $P(|t| > 3.3) < 0.01$, which is less than 0.05.

Conclusion. If the null hypothesis H_o were true, repeated sampling with 20 degrees of freedom would yield a value of $|t|$ as great as or greater than 3.3 in less than 1 case in 100 due to chance. This means that if H_o were true, repeated sampling would yield a discrepancy (as measured by $|t|$) between $\bar{X}_1$ and $\bar{X}_2$ as great as or greater than the one actually obtained above in less than 1 case in 100 due to chance. It follows that either we have struck a rare case or H_o is not true. By its nature the former is unlikely. We thus reject the null hypothesis H_o and accept the alternative hypothesis H_a. The two samples could not have reasonably come from the same hypothetical population. We say that $\bar{X}_1$ and $\bar{X}_2$ differ significantly. The samples are evidence of a difference in the effects of the two fertilizers.

7.9. Application to Proportions

Frequently we have a population which can be classified according to whether it does or does not possess a particular attribute, e.g. a population of persons according to whether male or female, an electorate of voters according to whether voters vote Labour or non-Labour. Populations which can be divided into two categories are called *dichotomous populations*, and they are completely specified by the proportion of the population falling into one of the categories. We shall call this proportion π, and the proportion falling into the other category $(1 - \pi)$. For convenience we can call the first category "successes" and the other one "failures."

A simple example of a dichotomous population is a barrel of black (successes) and white (failures) marbles. Now suppose that we draw a random sample of N marbles from this barrel. If the drawings are made with replacement or if the population is very large, each drawing may be regarded as an independent trial (see section 6.8, p. 95). Thus a sample of size N drawn from such a dichotomous population may be viewed as a sequence of independent trials with the probability of success and failure constant and equal to π and $(1-\pi)$ respectively. In section 6.8 we saw that the number of successes in N independent trials, X, was distributed in the *binomial distribution* about a mean of $N\pi$ and with a standard deviation $\sqrt{N\pi(1-\pi)}$, where π now represents the probability of success in a dichotomous population.

If we write $p = \dfrac{X}{N}$, we can convert the number of successes into the proportion of successes obtained in the experiment, i.e. in the sample.

If we repeat the sampling process, the statistic p will vary from sample to sample due to sampling errors. But it can be seen that the new variable $p = \dfrac{X}{N}$ is in fact a simple transformation of X, i.e. the number of successes in N independent trials. Hence by virtue of section 7.2[1]

$$E(p) = E\left(\frac{X}{N}\right) = \frac{1}{N}E(X)$$

But $E(X) = N\pi$, and hence

$$E(p) = \pi$$

and
$$\operatorname{Var}(p) = \frac{1}{N^2}\operatorname{Var}(X) = \frac{1}{N^2}[N\pi(1-\pi)]$$

$$= \frac{\pi(1-\pi)}{N}$$

We conclude that p, the sample proportion of successes, will in a large number of samples also be distributed in the binomial distribution with a mean π and a standard deviation $\sigma_p = \sqrt{\dfrac{\pi(1-\pi)}{N}}$. This gives the sampling distribution of a proportion in samples of size N drawn from a dichotomous population.

Since the sampling distribution of the p's follows the binomial distribution, it looks at first sight as if we should require special binomial probability tables to evaluate probabilities and to carry out significance tests. But it will be recalled from our discussion of the properties of the

---

[1] A change of notation in this section should be noted. While in section 6.8 the symbol p represented the probability of success, we now use p to indicate the proportion of successes found in samples of size N drawn from a dichotomous population; the proportion of successes in the population is indicated by π.

SAMPLING AND SIGNIFICANCE 135

binomial distribution in section 6.9 that if sample size, i.e. N, is sufficiently large, the binomial distribution can be very closely approximated by the normal curve. Accordingly, for large N, p will be distributed nearly normally about a mean π and with a standard deviation $\sigma_p = \sqrt{\dfrac{\pi(1-\pi)}{N}}$. The binomial distribution is near enough to normal for $N > 50$. Hence for large samples we can use the table of normal areas to calculate the probability that a sample of size N will have p lying within any specified range.

EXAMPLE 7.6

Under the mortality conditions of a certain country the proportion of births which survive to age 70 years is 0·6. What is the probability that out of 1,000 births taken at random, at least 630 will survive to age 70 years?

Here we have $\pi = 0\cdot6$. In samples of 1,000, p, the proportion of survivors, will be normally distributed about 0·6 with a standard deviation of

$$\sigma_p = \sqrt{\frac{\pi(1-\pi)}{N}} = \sqrt{\frac{0\cdot6 \times 0\cdot4}{1{,}000}} = 0\cdot0155$$

We require $P(p > 0\cdot63)$, i.e. $P\left(T > \dfrac{0\cdot63 - \pi}{\sigma_p}\right)$, i.e. $P(T > 1\cdot93)$. From the table of the normal distribution this probability is 0·0268. This is the required probability. It means that, if 1,000 such samples were selected, in only about 27 of them would more than 630 of the births survive to age 70 years.

7.10. SIGNIFICANCE OF THE DIFFERENCE BETWEEN p AND π

A random sample of N drawn from a dichotomous population has a proportion of successes of p. Could this sample have come from a population with a proportion of successes of π?

This problem is exactly analogous to the significance of the difference between $\bar{X}$ and μ, with σ known, described in section 7.5 above. The hypothesis is that the sample does come from such a population, i.e. we test the null hypothesis

$$H_0: \pi = \pi_0$$

against the alternative hypothesis that the true population proportion differs from the assumed value π_0

$$H_a: \pi \neq \pi_0$$

The test is

$$|T| = \frac{|p - \pi|}{\sigma_p}$$

where

$$\sigma_p = \sqrt{\frac{\pi(1-\pi)}{N}}$$

referred to the normal table.

Example 7.7

A sample of 400 electors selected at random gives a 51 per cent majority to the Labour Party. Could such a sample have been drawn from a population with a 50–50 division of political opinion?

Here $N = 400 \quad p = 0.51 \quad \pi = 0.50$

$$\sigma_p = \sqrt{\frac{\pi(1-\pi)}{N}}$$

$$= 0.025$$

Hypothesis. The sample has been drawn from a population in which $\pi = 0.50$, i.e.

$$H_o: \pi = 0.50$$
$$H_a: \pi \neq 0.50$$

Test—

$$|T| = \frac{|p - \pi|}{\sigma_p}$$

$$= \frac{0.01}{0.025}$$

$$= 0.4$$

From the normal table, $P(|T| > 0.4) = 0.69$, which is greater than 0.05.

Conclusion. If the null hypothesis H_o were true, repeated sampling would yield a value of $|T|$ as great as or greater than 0.4 in about 69 cases in 100 due to chance. This means that, if H_o were true, repeated sampling would yield a discrepancy (as measured by $|T|$) between p and π as great as or greater than the one actually obtained above in about 69 cases in 100 due to chance. Such being the case, the sample could reasonably have come from the hypothetical population, so we do not reject the null hypothesis. We say that p is not significantly different from π. The sample could have been drawn from a population with a 50–50 division of political opinion.

7.11. Significance of the Difference between Two Sample Proportions, p_1 and p_2

Two random samples of sizes N_1 and N_2 have proportions of successes p_1 and p_2. Could they have come from the same population? This problem is exactly analogous to the significance of the difference between $\bar{X}_1$ and $\bar{X}_2$ with σ known, dealt with in section 7.6 above. We set up the null hypothesis $H_o: \pi_1 = \pi_2$, and the alternative hypothesis $H_a: \pi_1 \neq \pi_2$, and test whether or not the former is consistent with our sample results.

If the two samples do come from the same population, then $p_1 - p_2$ will be normally distributed about a mean of 0 with a standard

SAMPLING AND SIGNIFICANCE

deviation $\sigma_{p_1-p_2}$. This standard deviation will be given by (see theorem 4, section 7.2 above)

$$\sigma_{p_1-p_2} = \sqrt{\sigma_{p_1}^2 + \sigma_{p_2}^2} = \sqrt{\frac{\pi(1-\pi)}{N_1} + \frac{\pi(1-\pi)}{N_2}}$$

$$= \sqrt{\pi(1-\pi)} \sqrt{\frac{1}{N_1} + \frac{1}{N_2}}$$

In practice we shall not know π. We must make an estimate of it, so that we can estimate $\sigma_{p_1-p_2}$. A good estimate of π is given by pooling the information from the two samples. We write

$$p = \frac{N_1 p_1 + N_2 p_2}{N_1 + N_2}$$

where p is an estimate of π. For large samples

$$s_{p_1-p_2} = \sqrt{p(1-p)} \sqrt{\frac{1}{N_1} + \frac{1}{N_2}}$$

may be used in conjunction with the normal distribution.

The test is $|T| = \dfrac{|p_1 - p_2|}{s_{p_1-p_2}}$ referred to the normal table.

EXAMPLE 7.8

A random sample of 100 business men in rural areas gives 54 per cent who expect business to improve next year, whereas one of 200 in city areas gives 48 per cent. Is this evidence of a difference in the expectations of business men in rural and city areas?

Here
$$N_1 = 100 \quad p_1 = 0.54$$
$$N_2 = 200 \quad p_2 = 0.48$$
$$p = \frac{54 + 96}{300} = 0.50$$
$$s_{p_1-p_2} = \sqrt{p(1-p)} \sqrt{\frac{1}{N_1} + \frac{1}{N_2}} = 0.061$$

Hypothesis. The two samples come from a population similar in respect of business expectations, i.e.
$$H_o: \pi_1 = \pi_2$$
$$H_a: \pi_1 \neq \pi_2$$

Test—
$$|T| = \frac{|p_1 - p_2|}{s_{p_1-p_2}}$$
$$= \frac{0.06}{0.061}$$
$$= 0.98$$

From the normal table, $P(|T| > 0.98) = 0.33$, which is greater than 0.05.

Conclusion. If the null hypothesis H_o were true, repeated sampling would yield a value of $|T|$ as great as or greater than 0·98 in about 33 cases out of 100 due to chance. This means that, if H_o were true, repeated sampling would yield a discrepancy (as measured by $|T|$) between p_1 and p_2 as great as or greater than the one actually obtained above in about 33 cases in 100 due to chance. Such being the case, the two samples could reasonably have come from the same hypothetical population, so we do not reject the null hypothesis. We say that p_1 and p_2 do not differ significantly. The two samples are not evidence of a difference in business expectations in rural and city areas.

7.12. Goodness of Fit

Suppose we have a sample of size N of a variable X classified into various classes, so that we have the frequencies of occurrence in each class, as, for example, we should have in a frequency distribution. We might then set up some hypothesis about the frequencies of occurrence in each class in the population and ask whether the sample data are consistent with such an hypothesis. To test this hypothesis we should have to compare the frequencies we observe in our sample, which we call the *observed frequencies*, with the frequencies which we should expect on the basis of the hypothesis, which we call the *expected frequencies*. This comparison involves what is known as a test of the *goodness of fit*, for we test how well the hypothesis fits the sample data. To take a trivial example: if we toss a die 120 times, we shall be able to record the number of ones, twos, threes, fours, fives and sixes we get. If the die is unbiased, we should expect that each face would be uppermost in one-sixth of the tosses, i.e. we should expect the frequencies of the ones, twos, etc., to be 20 each. By comparing the observed frequencies with these expected ones we can test the hypothesis that the die is unbiased. Naturally we should not expect our observed frequencies to equal exactly the expected ones even if the hypothesis were true, for just due to chance there will be discrepancies. The question is: Are the discrepancies too great to be attributed reasonably to chance? If so, doubt is cast on the hypothesis.

We write f_o for the observed frequency in any class and f_e for the corresponding expected frequency on the basis of some hypothesis. Summing the f's over the classes, we must have $\Sigma f_o = \Sigma f_e = N$, where N is the total number of observations. We first require some summary measure of the discrepancy between the f_o's and the f_e's. For each class $(f_o - f_e)$ measures the discrepancy. However, $\Sigma(f_o - f_e) = 0$, so that simple addition will not do. But by squaring each discrepancy we can get rid of the minus signs; accordingly, we work on $(f_o - f_e)^2$ as a measure of the discrepancy for each class. The importance of the magnitude of $(f_o - f_e)^2$ will depend on the size of the class frequency itself, so that we use $\dfrac{(f_o - f_e)^2}{f_e}$ as a measure of relative discrepancy.

SAMPLING AND SIGNIFICANCE 139

Summing these for all classes we get a measure of over-all discrepancy. This measure is called χ^2 (chi-square) and we have

$$\chi^2 = \sum \frac{(f_o - f_e)^2}{f_e}$$

Suppose we take repeated samples from a known population classified in a particular way. We shall have one set of f_e's derived from the

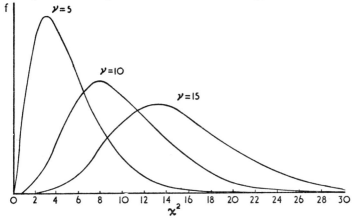

FIG. 7.4. THE χ^2 DISTRIBUTION

nature of the population itself and a large number of sets of f_o's, one from each sample. We can calculate χ^2 for each sample. χ^2 can then be formed into a frequency distribution. As the number of samples taken increases, this frequency distribution will tend towards a probability distribution, from which it will be possible to derive the probability of getting any particular value of χ^2 or larger just due to chance. This distribution is called the χ^2-*distribution*.[1]

The χ^2-distribution, like the t-distribution, depends on one parameter ν, the number of degrees of freedom. The number of degrees of freedom is the number of independent comparisons (i.e. independent comparisons between an f_o and the corresponding f_e) made in computing χ^2. The more comparisons the greater room for variability; and the higher and more dispersed should we expect χ^2 to be. The χ^2-distribution is shown in the diagram above. Since χ^2 is a sum of squares, it must always be positive, and the probability curve of χ^2 starts at the origin.

[1] More generally, the χ^2-distribution is generated by calculating

$$\chi^2 = \sum \frac{(f_o - f_e)^2}{f_e}$$

for repeated samples, even when the samples have been drawn from different populations, provided that the f_e's are the true population values in each case and that the number of independent comparisons made in each χ^2 is the same for all samples.

The distribution of χ^2 is tabulated in Table III of Appendix A. It gives in the body of the table, for particular numbers of degrees of freedom tabulated in the vertical margin, the value of χ^2 which cuts off a tail at the right-hand side of the distribution covering certain proportions of area tabulated in the horizontal margin. Such a tail is shown in Fig. 7.5. This tail measures $P(\chi^2 > \chi_1^2)$. Thus, for $\nu = 9$, $P(\chi^2 > 16\cdot92) = 0\cdot05$, i.e. in a case where $\nu = 9$, repeated sampling from known populations will throw up discrepancies between observed and expected frequencies which will result in $\chi^2 > 16\cdot92$ in 5 cases out of 100. Or if, with $\nu = 9$, on the basis of some hypothesis about the

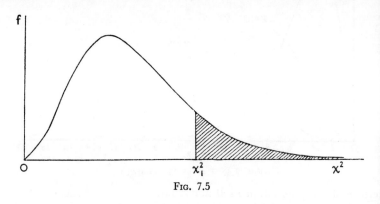

Fig. 7.5

population, we obtained for a particular sample a value of $\chi^2 = 12$, we could ascertain from the table that $P(\chi^2 > 12)$ lies between $0\cdot30$ and $0\cdot20$, so that we could say that a value of χ^2 as great as or greater than 12 would crop up just due to chance in from 20 to 30 cases out of 100.

We can test the goodness of fit of an hypothesis to sample data by computing the expected frequencies on the basis of the hypothesis, comparing them with the observed frequencies and calculating χ^2. By reference to the table of the χ^2 distribution we can ascertain the probability of obtaining just due to chance a value of χ^2 as great as or greater than the one calculated if the hypothesis were true. If this probability is small (i.e. less than 5 per cent), we say that the hypothesis is a bad fit and we reject it, for the discrepancy between the f_o's and the f_e's could not then be reasonably attributed to chance.

The number of degrees of freedom is given by the number of independent comparisons between the f_o's and the f_e's, i.e. the number of comparisons less the number of constraints. Broadly, a constraint, in this connexion, can be defined as a relation between the f_o's and the f_e's. Since $\Sigma f_o = \Sigma f_e$, there must always be one constraint but, as will be illustrated in the case below, there may be more. Another way of looking at the number of degrees of freedom is to imagine that the f_e's have been set down and to ask in how many of the classes can we freely set down f_o's. Clearly, if there are M classes and $\Sigma f_o = \Sigma f_e$, we can freely fill in only $M - 1$ classes, since the Mth class frequency will follow after the $M - 1$ classes have been set down; consequently there would be $M - 1$ degrees of freedom in this case. It should be noted that the χ^2-test should not be applied when any one f_e is very small, certainly not when an f_e is less than 5 and preferably not when one is less than 10. In such cases classes should be combined to produce higher f_e's. This is also illustrated below.

By way of illustration of the application of the χ^2-test we consider the case of a sample of size N of a variable X classified into a frequency distribution. One question which arises is: is the distribution of the values of X in the sample such that the sample could have come from a population which is normally distributed? This question can be asked in two ways—

(a) Could the sample have come from a normally distributed population with given mean μ and given standard deviation σ; or

(b) Could it have come from any normal population? In this case μ and σ are unspecified, and the best we can do is to estimate them from the sample $\bar{X}$ and s.

In this situation we have to compare the f_o's of our sample frequency distribution with the class frequencies which we should expect to obtain on the average if we repeatedly sampled the hypothetical normal distribution. These expected class frequencies can be obtained quite easily from the normal table (see example 6.13 in section 6.10 above). The number of degrees of freedom depends upon whether we are asking question (a) or (b) above. If the expected frequencies have been derived on the basis of a hypothetical μ and σ, there will be only one constraint, namely

$$\Sigma f_o = \Sigma f_e$$

so that if there are M classes there will be $M - 1$ degrees of freedom. On the other hand, if the expected frequencies have been derived on the basis of the sample $\bar{X}$ and s, there will be only $M - 3$ degrees of freedom, for there will be three constraints:

$\Sigma f_o = \Sigma f_e$ (observed and theoretical frequencies equal)

$\dfrac{\Sigma X f_o}{\Sigma f_o} = \dfrac{\Sigma X f_e}{\Sigma f_e}$ (observed and theoretical means equal)

$\sqrt{\dfrac{\Sigma x^2 f_o}{\Sigma f_o}} = \sqrt{\dfrac{\Sigma x^2 f_e}{\Sigma f_e}}$ (observed and theoretical standard deviations equal)

Having ascertained the f_e's, we set up the hypothesis that the sample comes from a normal population, calculate χ^2 and ascertain whether the value of χ^2 is such that it could or could not be reasonably attributed to chance. In the former case the hypothesis stands, in the latter case it is rejected. Two examples are given below.

Example 7.9

A sample of 200 rents is taken and their distribution is as below. Could this sample have come from a normal population?

Class (dollars)	Observed Frequencies f_o	Expected Frequencies f_e	$(f_o - f_e)^2$	$\dfrac{(f_o - f_e)^2}{f_e}$
2·5 and under 7·5	—⎱ 12	2·3⎱ 11·6	0·16	0·014
7·5 ,, 12·5	12⎰	9·3⎰		
12·5 ,, 17·5	26	26·3	0·09	0·003
17·5 ,, 22·5	45	47·8	7·84	0·164
22·5 ,, 27·5	60	53·3	44·89	0·842
27·5 ,, 32·5	37	38·4	1·96	0·051
32·5 ,, 37·5	13	16·9	15·21	0·900
37·5 ,, 42·5	5⎱ 7	4·8⎱ 5·7	1·69	0·296
42·5 ,, 47·5	2⎰	0·9⎰		
All Rents	200	200·0		2·270

Since the mean and standard deviation of the hypothetical normal population are not specified, the sample mean and standard deviation must be used as estimates of the population parameters. These statistics can be calculated from the f_o column and are $\bar{X} = \$23 \cdot 82$ and $\$7 \cdot 19$ (see examples 5.3 and 5.13 on pp. 56 and 73). We now set up the hypothesis that the sample has been drawn from a normally distributed population. On the basis of this hypothesis we calculate the class frequencies which we should expect in the long run if the hypothesis were true. These expected frequencies are given in the f_e column (see example 6.13 on p. 113 above). χ^2 can be calculated. Note that the end classes have been combined so that no class contains less than 5 expected frequencies. We have

$$\chi^2 = \sum \frac{(f_o - f_e)^2}{f_e} = 2 \cdot 270$$

There are seven classes yielding seven comparisons, but there are three constraints—the total number of observations in the observed and expected frequency distributions are the same, as are their means and standard deviations. Hence there are 4 degrees of freedom.

From the table of χ^2, with $\nu = 4$,

$P(\chi^2 > 2 \cdot 270)$ is about 0·7, which exceeds 0·05.

If the hypothesis were true, repeated sampling with 4 degrees of freedom would yield a value of χ^2 as great as or greater than 2·270 in approximately 70 cases in 100 due to chance. This means that, if the hypothesis were true, repeated sampling would yield a discrepancy (as measured by χ^2) between the f_e's and the f_o's as great as or greater than the one actually obtained in about 70 cases in 100 due to chance. This being the case, we do not reject the hypothesis. The sample could reasonably have come from a normal population. We say that the value of χ^2 obtained is not significant and that the f_e's are a good fit to the f_o's. Incidentally, we should note that the χ^2-test as used here is only one way of testing normality. Other tests are available and some of them are more powerful.

EXAMPLE 7.10

A sample of 500 factories was taken to investigate certain aspects of the working conditions obtaining in them. The method used was to select names at random from trade lists. The classification of the sample by size of factory in terms of number of employees is given in the first two columns of the table below. The classification of all factories by size is known, and the percentage distribution is shown in the third column. From this column the frequencies which we should expect to get in the long run with repeated random sampling of 500 factories from the population can be calculated, and these are shown in the fourth column. Can the contention that the sample is random be supported?

We set up the hypothesis that the sample is a *random* sample drawn from the given population. On the basis of this hypothesis, $\chi^2 = \sum \dfrac{(f_o - f_e)^2}{f_e}$ is found to be 14·89. There are seven comparisons, but the totals of expected and observed frequencies must agree; accordingly there are 6 degrees of freedom. From the table of χ^2 we find that with $v = 6$, $P(\chi^2 > 14 \cdot 89) < 0 \cdot 05$. If the

Size of Factory (persons employed)	Observed Sample Frequencies f_o	Distribution of All Factories (per cent)	Expected Sample Frequencies f_e
under 5	180	39·3	197
5–10	123	27·0	135
11–20	92	14·6	73
21–50	51	11·5	57
51–100	29	4·0	20
101–200	14	2·0	10
over 200	11	1·6	8
Total	500	100·0	500

hypothesis were true, repeated samples with 6 degrees of freedom would yield values of χ^2 as great as or greater than 14·89 in less than 5 cases in 100 due to chance. This means that, if the sampling were random, a discrepancy between observed and expected frequencies (as measured by χ^2) as great as or greater than the one actually obtained would occur due to chance in less than 5 cases in 100 with repeated samples. It follows that either we have struck a rare case

7.13. Independence of Classification

An important application of the χ^2 test occurs when we wish to test hypotheses about the relation between two attributes. Suppose we have N observations classified according to two criteria. We may ask whether the criteria are related or independent. This problem can best be expressed by following through an example.

Suppose we have a random sample of 400 dwellings, classified according to area and nature of occupancy as below. The question is, are the two classifications independent?

Table 7.1

Nature of Occupancy	Area			Total
	Metropolitan	Urban Provincial	Rural	
Owner-occupied	102	41	100	243
Tenanted	82	30	45	157
Total	184	71	145	400

This type of table is called a *contingency table*. The above one is a 3×2 contingency table, the horizontal classification containing three categories, the vertical one two categories, and the table six distinct cells.

If area and nature of occupancy were unrelated we should expect, on the basis of the marginal totals, $\frac{184}{400} \times \frac{243}{400}$ of the cases to fall in the cell "metropolitan and owner-occupied." This is so because the probability of a dwelling being metropolitan can be estimated from the sample as $\frac{184}{400}$ and the probability of its being owner-occupied as $\frac{243}{400}$, and the probability of two independent events occurring together is the product of the probabilities of their occurring separately (see section 6.4 above). It follows that on the basis of the hypothesis of independence we should expect that repeated samples with the same marginal totals as the above would, in the long run, have $\frac{184}{400} \times \frac{243}{400} \times 400 = 111 \cdot 8$ cases in the cell under consideration. The expected frequencies of the other cells can be computed in the same way and are given in brackets in the table below. The expected frequencies must of course add to the marginal totals.

Table 7.2

Nature of Occupancy	Area			Total
	Metro-politan	Urban Provincial	Rural	
Owner-occupied	102 (111·8)	41 (43·1)	100 (88·1)	243
Tenanted	82 (72·2)	30 (27·9)	45 (56·9)	157
Total	184	71	145	400

It is now a simple matter to compare observed with expected frequencies by means of the χ^2 test in the same way as in the preceding section. The simplest way of determining the number of degrees of freedom in this case is to ask how many cells can be freely filled in, given the marginal totals. There are six cells, but, if two are filled in, the other four follow. Accordingly there are 2 degrees of freedom. Alternatively, we may say that there are six comparisons and four constraints—the latter being the equality of total observed and expected frequencies and three marginal probabilities (the other two follow given those three)—and hence 2 degrees of freedom. In general, if we have an $A \times B$ contingency table, there will be $(A-1)(B-1)$ degrees of freedom.

Calculating χ^2 for the above we have

$$\chi^2 = \sum \frac{(f_o - f_e)^2}{f_e}$$
$$= \frac{(9\cdot8)^2}{111\cdot8} + \frac{(2\cdot1)^2}{43\cdot1} + \frac{(11\cdot9)^2}{88\cdot1} + \frac{(9\cdot8)^2}{72\cdot2} + \frac{(2\cdot1)^2}{27\cdot9} + \frac{(11\cdot9)^2}{56\cdot9}$$
$$= 6\cdot55$$

From the table of χ^2 with $\nu = 2$ we have $P(\chi^2 > 6\cdot55) = 0\cdot04$, which is less than 0·05. If the hypothesis that the two classifications are independent were true, repeated sampling with 2 degrees of freedom would yield a value of χ^2 as great as or greater than 6·55 in about 4 cases in 100 due to chance. This means that, if the hypothesis were true, repeated sampling would yield a discrepancy (as measured by χ^2) between the f_o's and the f_e's as great as or greater than the one actually obtained in about 4 cases in 100 due to chance. It follows that either we have struck a rare case or the hypothesis is not true. By its nature the former is unlikely, and we reject the hypothesis. The sample could not reasonably come from a population in which the two classifications are independent. We say that the value of χ^2 obtained is significant. The sample suggests that area and nature of occupancy of dwellings are related in some way.

7.14. Two Types of Errors

When we set up a statistical hypothesis and test it, there are four possible results:

1. The hypothesis is true but our test rejects it.
2. The hypothesis is false but our test accepts it.
3. The hypothesis is true and our test accepts it.
4. The hypothesis is false and our test rejects it.

The first two of these possibilities lead to errors being made. When a hypothesis is true but rejected, it is called a *type one* error. When it is false but accepted, it is called a *type two* error. The probability that a true hypothesis will be rejected is given by the level of significance employed in the test. At the 5 per cent level, we know that we shall be rejecting on the average 5 per cent of all true hypotheses, because even if the hypothesis were true, $|T|$, for example, would be greater than 1·96 just due to chance 5 times out of 100. This 5 per cent is called the *size* of the type I error. Thus, suppose we have a random sample of N items with a mean $\bar{X}$ and we are testing the hypothesis that the sample comes from a population with a mean μ_o and a standard deviation σ_o. Using a 5 per cent level of significance, this means that, if $\bar{X}$ lies within the range $\mu_o - 1·96 \frac{\sigma_o}{\sqrt{N}}$ to $\mu_o + 1·96 \frac{\sigma_o}{\sqrt{N}}$, we accept the hypothesis; whereas if it lies outside this range, we reject it. But if the hypothesis is true (i.e. if the population really has a mean μ_o), 5 per cent of random samples drawn from the population will have $\bar{X}$'s lying outside the range, so that with repeated sampling we shall reject 5 per cent of true hypotheses. This is illustrated in Fig. 7.6, where the probability distribution of the means of samples of size N drawn from a population with mean μ_o and standard deviation σ_o is given. The two points, b and a, represent the values $\mu_o \pm 1·96 \frac{\sigma_o}{\sqrt{N}}$ respectively, and the shaded area represents the size of the type I error, 5 per cent in this case.

Now suppose that the hypothesis that the population mean is μ_o is in fact wrong, and that the correct population mean is μ_1, the standard

SAMPLING AND SIGNIFICANCE 147

deviation being σ_0 as before. In Fig. 7.7 the probability distribution of the means of samples of size N drawn from a population with mean

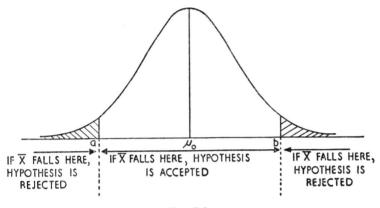

IF $\bar{X}$ FALLS HERE, HYPOTHESIS IS REJECTED | IF $\bar{X}$ FALLS HERE, HYPOTHESIS IS ACCEPTED | IF $\bar{X}$ FALLS HERE, HYPOTHESIS IS REJECTED

Fig. 7.6

μ_1 and standard deviation σ_0 is superimposed on the distribution given in Fig. 7.6 above.

If we are testing the hypothesis that the population mean is μ_0, but the correct population mean is μ_1, whenever an $\bar{X}$ falls to the left of b the hypothesis will be accepted, and a type II error will be made. If the population mean is μ_1, the probability of obtaining an $\bar{X}$ which is less than b will be given by the solidly-shaded area, so that the probability of accepting the hypothesis that the mean is μ_0 when this is incorrect will be given by that area. Accordingly that area measures the size of the type II error. Thus, if the hypothesis is that the population mean is μ_0, we shall wrongly reject a true hypothesis whenever the mean really is μ_0 but $\bar{X}$ lies outside the range a to b (i.e. make a type I error) and we shall wrongly accept a false hypothesis whenever the mean really is μ_1 but $\bar{X}$ lies inside the range a to b (i.e. make a type II error).

It should be plain from Fig. 7.6 above, that we can make the type I error as small as we please by lowering the level of significance, i.e. by increasing the range a to b in which we accept the hypothesis. If, for example, we work with a 1 per cent level of significance, the critical value of $|T|$ is raised from 1·96 to 2·58. By so doing we increase the risk of accepting a false hypothesis, i.e. of making a type II error. This can be seen from Fig. 7.7, for as we widen the range a to b, we reduce the lightly-shaded area, but increase the solidly-shaded area. It follows that it is impossible to minimize both errors simultaneously. What is usually done in practice is to fix the size of the type I error (say, at 5 per cent) and then select a test (if any) which will minimize the size of the type II error. This topic is mathematically complex and will not be pursued here.

It can be seen, however, that the appropriate balance to be given to the two types of error in designing tests of significance will depend upon

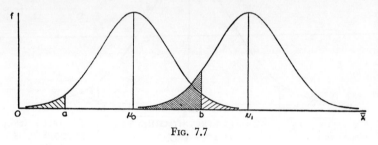

Fig. 7.7

the practical consequences which flow from making errors of both kinds. Thus a factory making wire ropes may be contemplating introducing a new process which is expected to increase their strength. A sample of ropes made under the new process is tested, the hypothesis being that the process results in no improvement. A type I error will occur if the process is really no improvement although the statistical test rejects the hypothesis, and a type II error will occur if the process is really an improvement although the statistical test accepts the hypothesis. If the cost of making the innovation is small, whereas the possible improvements in strength are considerable, clearly a type II error is more serious than a type I error, whereas, if the cost is substantial and the possible improvements slight, the reverse is true.

The case of testing Salk anti-polio vaccine for viability provides a more graphic example. Suppose a sample is drawn from a large batch and tested. The hypothesis is that the vaccine is not viable. If a type I error is made (vaccine not viable, but test rejects the batch) supplies of good vaccine will be discarded. People will go unvaccinated; and some may contract polio although had the vaccine been used they would have escaped it. If a type II error is made (vaccine viable, but test accepts the batch) vaccination may itself produce polio. The relative importance of the two types of error is in practice seldom as clear-cut as these examples might suggest, but at least they indicate the significance of the two types of error in a practical setting.

7.15. One-tailed Tests of Significance on the Mean

In the previous section it was pointed out that when we test a statistical hypothesis that a sample comes from a population with a given mean μ_o, we set up a range of acceptance and a range of rejection. If the sample mean $\bar{X}$ lies within the former range we accept the hypothesis; if it lies within the latter we reject it. This is illustrated in Fig. 7.6 above. In that figure the case is shown where the acceptance range is $\mu_o - 1\cdot 96 \frac{\sigma_o}{\sqrt{N}} < \bar{X} < \mu_o + 1\cdot 96 \frac{\sigma_o}{\sqrt{N}}$ and the rejection range $\bar{X} < \mu_o - 1\cdot 96 \frac{\sigma_o}{\sqrt{N}}$ or $\bar{X} > \mu_o + 1\cdot 96 \frac{\sigma_o}{\sqrt{N}}$. These ranges are those appropriate to a 5 per cent level of significance. In other words, the ranges are such that the size of the type I error (the probability of rejecting a true hypothesis) is 5 per cent.

Suppose that we are testing the hypothesis that the mean is μ_o, and that we know that if the mean is not μ_o it must be above μ_o. In this case it would be pointless to use the acceptance range as set out in Fig. 7.6. For if any $\bar{X}$ fell to the left of a, we should attribute this to chance and not to the true value of the mean being less than μ_o. In other words we should not wish to reject the hypothesis if $\bar{X} < a$. Thus, in order to keep the size of the type I error at 5 per cent, it would be necessary to move b to the left, until the shaded area under the right-hand tail became equal to 5 per cent, i.e. the acceptance range would be set at $\bar{X} < \mu_o + 1\cdot 645 \frac{\sigma_o}{\sqrt{N}}$ and the rejection range at $\bar{X} > \mu_o + 1\cdot 645 \frac{\sigma_o}{\sqrt{N}}$. Clearly, for testing the hypothesis that the mean is μ_o against alternative hypotheses that the mean is greater than μ_o, such an acceptance range results in a lower type II error than does the range shown in Fig. 7.7; for with b moved to the left the solidly-shaded area is diminished. A test of significance based on an acceptance range of the kind outlined in this paragraph is known as a *one*-*tailed* test. It contrasts with the *two*-*tailed* test illustrated in Fig. 7.6. Instead of testing hypotheses about the true population mean lying in either direction from μ_o, we now write

$$H_o: \mu = \mu_o$$
$$H_a: \mu > \mu_o$$

for tests of significance involving only the right-hand tail of the curve in Fig. 7.6, and

$$H_o: \mu = \mu_o$$
$$H_a: \mu < \mu_o$$

for tests involving the left-hand tail.

The tests set out in sections 7.5 to 7.11 are all two-tailed tests. Two-tailed tests are appropriate when there is no particular reason to believe that, if the hypothesis being tested is false, the true hypothesis involves a mean lying above rather than below the hypothetical mean (or vice versa); or, when the action flowing from the rejection of a hypothesis is the same whether the true hypothesis involves a mean lying above or below the hypothetical mean. One-tailed tests are appropriate in cases where these conditions do not hold. In general, two-tailed tests test whether, for example, $\bar{X}$ *differs* significantly from μ; whereas one-tailed tests test whether $\bar{X}$ is significantly *greater* than or *less* than μ. Two examples will illustrate the use of one-tailed tests.

Suppose that a census of city-dwellers reveals an average family size of 4·2 with a standard deviation of 0·5. A random sample of 100 country families is taken and this reveals a family size of 4·29. We wish to test whether family size in the country is the same as in the city. The null hypothesis is that the sample comes from a population in which the mean is 4·2. If we were to apply the usual two-tailed test we should calculate

$$|T| = \frac{|\bar{X} - \mu|}{\sigma_{\bar{X}}}$$
$$= \frac{|4·29 - 4·20|}{0·5/\sqrt{100}}$$
$$= 1·8$$

It follows that, if the hypothesis were true, repeated sampling would yield a discrepancy (as measured by $|T|$) between $\bar{X}$ and μ as great as or greater than the one obtained in rather more than 7 cases in 100 $[P(|T| > 1·8 = 0·0718]$. We should then say that the hypothesis should not be rejected at the 5 per cent level of significance, and conclude that $\bar{X}$ is not significantly different from μ. This would be perfectly all right if we had no reason to believe that, if the null hypothesis were wrong, the true hypothesis might as readily be that country families are smaller as that they are larger than city families. However, if we know, from other sources, that, whatever size country families are, they are certainly no smaller than city families we should apply a one-tailed test. Having calculated that $T = +1·8$, we should argue that, if the null hypothesis were true, repeated sampling would yield an excess of $\bar{X}$'s above μ as great as or greater than the one obtained in rather less than 4 cases in 100 $[P(T > 1·8) = 0·0359]$. We should then say that the null hypothesis H_o: $\mu = 4·2$ should be rejected at the 5 per cent level of significance and the alternative hypothesis H_a: $\mu > 4·2$ accepted. Thus our one-tailed test would lead us to conclude that $\bar{X}$ was significantly greater than the assumed μ.

Whether two-tailed or one-tailed tests are used the size of the type I error remains at 5 per cent; but for the former the critical value for

T is $T > 1\cdot 96$ or $< -1\cdot 96$, and for the latter it is $T > 1\cdot 645$. Where the t-test is used, the critical values are $t > t_{\cdot 05}$ or $< -t_{\cdot 05}$ and $t > t_{\cdot 10}$ respectively.

A second example illustrates another aspect of the one-tailed test. Suppose a large factory, producing a single product, has a normal output of 900 units per worker per day, with a standard deviation of 30 units. The management wishes to find out whether output will be affected by an alteration in the lay-out of the work benches. It modifies one small section of the factory and selects 25 workers at random to work under the new conditions. Let the mean output of these 25 workers be $\bar{X}$. There are now three possibilities: the new lay-out leaves output unchanged, it reduces output, it increases output. Only in the third case will the management wish to take action. Consequently the management is interested only in $\bar{X}$ being significantly greater than 900. If it were significantly smaller, no action would be taken, just as if there were no significant difference between $\bar{X}$ and 900. Hence, given that the management is prepared to take a 5 per cent risk of altering the lay-out when the new lay-out was ineffective, the range of rejection for the hypothesis that the sample comes from a population with a mean of 900 should be $\bar{X} > 900 + 1\cdot 645 \dfrac{30}{\sqrt{25}}$, i.e. $\bar{X} > 910$. With a two-tailed range of rejection, whereas the size of the type I error would, as in the case of the one-tailed range, be 5 per cent, the probability of the management altering the lay-out when the new lay-out had no effect would be only $2\frac{1}{2}$ per cent. If the management is prepared to tolerate 5 per cent for this probability, the use of the one-tailed range reduces the probability of the error of the management taking no action when the new lay-out has in fact improved production (type II error).

7.16. Estimation

So far we have used sample data to test hypotheses about the population. The second important object of taking samples is to be able to make inferences about the population parameters.

Suppose we have a random sample of N items with mean $\bar{X}$. What can we say about μ, the population mean? We should expect that $\bar{X}$ would be reasonably close to μ, but we should be very surprised if it exactly coincided. Actually, if we had to use a figure as an estimate of μ, we should use $\bar{X}$, and it would be a good estimate of μ in the sense that with repeated sampling the mean of the $\bar{X}$'s would tend to μ (i.e. $\bar{X}$ is an *unbiased* estimate of μ). However, ideally we should like to be able to ascertain a range within which we could be reasonably confident that μ would lie.

Consider the drawing of a random sample of N items from a normal population with mean μ and standard deviation σ. Let $\bar{X}$ be the sample mean. We know that $\bar{X}$ will be normally distributed about μ with a standard error of $\sigma_{\bar{X}} = \dfrac{\sigma}{\sqrt{N}}$. We know that with repeated sampling 95 per cent of sample means will lie within the range $\mu \pm 1{\cdot}96\sigma_{\bar{X}}$. Suppose now we do not know μ, but have calculated an $\bar{X}$, and further

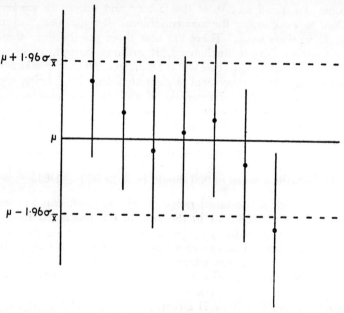

Fig. 7.8

suppose we know σ, what can we say? If $\bar{X}$ happens to be in the range $\mu \pm 1{\cdot}96\sigma_{\bar{X}}$, then μ must be contained in the range $\bar{X} \pm 1{\cdot}96\sigma_{\bar{X}}$. The sample mean $\bar{X}$ will be in the range $\mu \pm 1{\cdot}96\sigma_{\bar{X}}$ for 95 per cent of samples, so that μ will lie in the range $\bar{X} \pm 1{\cdot}96\sigma_{\bar{X}}$ for 95 per cent of samples. The sample mean $\bar{X}$ varies, of course, from sample to sample, so that the range $\bar{X} \pm 1{\cdot}96\sigma_{\bar{X}}$ will also vary. This is illustrated diagrammatically in Fig. 7.8. The population mean, μ, and the range $\mu \pm 1{\cdot}96\sigma_{\bar{X}}$ are scaled on the vertical axis. Each vertical line corresponds to a particular sample, is centred at $\bar{X}$ and ranges $\pm 1{\cdot}96\,\sigma_{\bar{X}}$.

It can be seen from the diagram that if $\bar{X}$ lies within $1{\cdot}96\,\sigma_{\bar{X}}$ of μ, then μ lies within $1{\cdot}96\,\sigma_{\bar{X}}$ of $\bar{X}$ also. This can easily be shown algebraically.

If $\mu - 1\cdot 96\, \sigma_{\bar{X}} < \bar{X} < \mu + 1\cdot 96\, \sigma_{\bar{X}}$

then adding $-\bar{X} - \mu$

we have $-\bar{X} - 1\cdot 96\, \sigma_{\bar{X}} < -\mu < -\bar{X} + 1\cdot 96\, \sigma_{\bar{X}}$

and multiplying by -1

$$\bar{X} - 1\cdot 96\, \sigma_{\bar{X}} < \mu < \bar{X} + 1\cdot 96\, \sigma_{\bar{X}}$$

It follows that if we make the statement "μ lies within $\bar{X} \pm 1\cdot 96\, \sigma_{\bar{X}}$" we shall be right in 95 per cent of such statements. This is evidently true for estimates of μ for samples drawn from the same population but, since there is a 95 per cent chance of being right in stating "μ lies within $\bar{X} \pm 1\cdot 96\, \sigma_{\bar{X}}$" for any one population, it must also be true for all such statements taken together. If we make such statements, we shall be right 95 per cent of the time irrespective of what the statements refer to. Accordingly, the probability is 95 per cent that we shall be correct in stating that "μ lies within $\bar{X} \pm 1\cdot 96\, \sigma_{\bar{X}}$." The limits $\bar{X} \pm 1\cdot 96\, \sigma_{\bar{X}}$ are known as the 95 per cent *confidence limits* for μ. In a particular case, where we have computed limits from the $\bar{X}$ of a given sample, we may say that we are 95 per cent confident that μ lies within the computed limits.

Similarly, by the same argument, $\bar{X} \pm 2\cdot 58\, \sigma_{\bar{X}}$ are the 99 per cent confidence limits. We can, of course, increase our confidence by widening the limits. But in practice the limits are not much use unless they are fairly narrow. Since $\sigma_{\bar{X}} = \dfrac{\sigma}{\sqrt{N}}$, we can narrow them for a given degree of confidence by increasing the sample size. Confidence limits give an indication of the sampling error involved in making inferences about a population from sample data. Naturally, the larger the sample the less the sampling error and the narrower the confidence limits.

7.17. Confidence Limits for a Mean

We have shown that the 95 per cent confidence limits for a mean are $\bar{X} \pm 1\cdot 96\, \sigma_{\bar{X}}$. The difficulty in practice is that we do not know $\sigma_{\bar{X}}$, and we have to estimate it from the sample. If $\bar{X}$ and s are the mean and standard deviation from a sample of size N, we know that $t = \dfrac{\bar{X} - \mu}{s_{\bar{X}}}$, where $s_{\bar{X}} = \dfrac{s}{\sqrt{N}}$, is distributed in the t-distribution with $N - 1$ degrees of freedom. Hence we know that in repeated sampling 95 per cent of the t's will be between $\pm t_{.05}$, where $t_{.05}$ is the value of t at the 5 per cent significance level (i.e. cutting off two 2½ per cent tails). Accordingly there is a 95 per cent probability that

$$-t_{.05} < t < t_{.05}$$

i.e.
$$-t_{.05} < \frac{\bar{X} - \mu}{s_{\bar{X}}} < t_{.05}$$

These inequalities can be rearranged to get

$$-t_{.05} s_{\bar{X}} < \bar{X} - \mu < t_{.05} s_{\bar{X}}$$

i.e. $\quad -\bar{X} - t_{.05} s_{\bar{X}} < -\mu < -\bar{X} + t_{.05} s_{\bar{X}}$

i.e. $\quad \bar{X} - t_{.05} s_{\bar{X}} < \mu < \bar{X} + t_{.05} s_{\bar{X}}$

By the same argument as in section 7.16 above, the last statement gives the 95 per cent confidence limits for μ, i.e. $\bar{X} \pm t_{.05} s_{\bar{X}}$. For large N this becomes $\bar{X} \pm 1 \cdot 96 s_{\bar{X}}$, since $t_{.05} = 1 \cdot 96$ for large N.

EXAMPLE 7.11

Find confidence limits for the mean height of Australians, given that the mean of a random sample of sixteen is 68 in. with a standard deviation of 2 in.

Here
$$s_{\bar{X}} = \frac{s}{\sqrt{N}} = \frac{2}{\sqrt{16}} = 0 \cdot 5$$

and $t_{.05}$ is obtained from the t-table with 15 degrees of freedom. It is the value of t such that $P(|t| > t_{.05}) = 0 \cdot 05$, and in this case is $2 \cdot 131$.

Confidence limits are

$$\bar{X} \pm t_{.05} s_{\bar{X}} = 68 \pm (2 \cdot 131 \times 0 \cdot 5)$$
$$= 68 \pm 1 \cdot 065$$

so that on the basis of our sample we can be 95 per cent confident that the mean height of Australians lies between about 67 and 69 inches.

7.18. CONFIDENCE LIMITS FOR A PROPORTION

If we have a sample of N from a dichotomous population with proportion of successes, p, within what range can we expect the proportion of successes in the population, π, to lie?

In large samples, these limits can be given by analogy with the mean, for p is then normally distributed about a mean π with standard deviation

$$\sigma_p = \sqrt{\frac{\pi(1-\pi)}{N}}$$

Hence confidence limits to π will be $p \pm 1.96\, \sigma_p$. However, σ_p itself is dependent on π, and π is unknown. For the calculation of σ_p we can use p as an estimate of π, and we get as an approximate expression for the confidence limits for π

$$p \pm 1.96 s_p$$

i.e.
$$p \pm 1.96 \sqrt{\frac{p(1-p)}{N}}$$

EXAMPLE 7.12

A random sample of 400 dwellings gives 30 per cent as being rented houses. Within what range can we be reasonably confident that the proportion for all dwellings lies?

Here $\qquad N = 400 \qquad p = 0.3$

and
$$s_p = \sqrt{\frac{p(1-p)}{N}}$$
$$= \sqrt{\frac{0.3 \times 0.7}{400}}$$
$$= 0.023$$

Confidence limits are $\qquad p \pm 1.96 s_p$

i.e. $\qquad 0.3 \pm 0.045$

so that on the basis of our sample we can be 95 per cent confident that the proportion of all dwellings which are rented houses lies between 25·5 and 34·5 per cent.

This chapter has covered the more important considerations in testing hypotheses and estimating population parameters. The tests discussed have been mainly concerned with means, but tests are available for other statistics, such as standard deviations and for handling much more complex situations. However, they involve the same principles as those outlined above.

CHAPTER VIII

SAMPLE SURVEYS

8.1. Advantages of Sample Surveys over Full Counts

In discussing the collection of data in Chapter III it was pointed out that in conducting a survey a decision would have to be made whether the survey should cover all units in the field of inquiry or whether only a sample of units should be examined.[1] In the present chapter this question is discussed. The use of sample surveys to elicit information about the populations from which the samples have been drawn is one of the most important applications of the theory of sampling and significance outlined in Chapter VII. The present chapter is intended to be an illustration of this application rather than a complete discussion of it, even at an elementary level. Indeed the subject of "sample surveys" has a substantial and growing literature of its own, and reference should be made to this literature, some of which is listed in Appendix C, 4. It is hardly necessary to add that considerable practical experience as well as extensive knowledge of the theoretical principles is essential for those who engage in this highly specialized work. But against the few who are responsible for conducting sample surveys there are the many who use their results, and these should have at least some broad understanding of the principles involved. Some of the more important aspects of these principles are discussed in the following pages.

An inquiry like a census is said to be a full count. Every unit in which we are interested is included in the survey. However, in order to find out something about a certain characteristic of a population it is not necessary to make a full count of all the units in the population, e.g. in order to determine the quality of the 1967-68 Australian wheat harvest it is not necessary to examine every grain of wheat, or in order to find out the average level of rents in Adelaide it is not necessary to find out the rent of every house. In practice, we can get along very well by taking a sample, e.g. a sample of wheat, a sample of houses, etc. By a sample we observe only a representative fraction of the whole field and from it calculate or infer something about the characteristics of the whole field. We have already distinguished between the notions of a *sample* and a *population* (see section 6.6 above). We repeat that the word "population" is here used in a technical sense to indicate the totality or universe of units from which the sample is drawn. If we take a

---

[1] See p. 20 above.

sample of rents in Adelaide, the population is all the rents in Adelaide; if we take a sample of South Australian wheat, the population is all South Australian wheat. When we take a *full count* we examine the whole population; when we take a sample we examine only a fraction of the population. The idea of taking a sample is a very simple everyday common-sense idea—there is nothing abstruse or particularly unusual about it—but it is fundamental to practical and theoretical statistics, and the task of selecting a reliable sample has many technical difficulties.

Why should we want to take a sample rather than a full count? There are several important reasons for doing so, and these are listed below.

1. *Practicability*

A full count may be impracticable because the job is simply too big, e.g. quality of wheat, timber resources, frequent surveys of retail establishments.

2. *Speed*

The data may be collected and summarized much more quickly with a sample than with a full count. This may be a vital consideration when the information is urgently needed or when it is to be used as a basis for government policy.

3. *Accuracy*

More accurate results can often be obtained when questionnaires are filled in by skilled enumerators instead of by the respondents themselves. Personal interviews may elicit more accurate information than sending out questionnaires by post and requesting respondents to fill them in themselves, although some people may be reticent when interviewed personally and give less accurate answers than they would in a private signed questionnaire. In cases where the number of enumerators required varies directly with the number of questionnaires to be filled in, the personal interview is much more of a practical proposition with a sample than with a full count. Even if it is possible to employ enumerators in a full count, for a sample a smaller staff will be necessary and it may be possible to have more efficient and better trained enumerators. It may also be possible to allow the enumerators to spend more time and take greater care over each individual questionnaire and to ask a greater number of questions. Thus, although a sample does not give the coverage which a full count does, the accuracy of that part of the total information which is collected may be greater in certain cases.

4. Cost

If data are secured from only a fraction of the population, expenditures will be much smaller than if a full count is taken. The costs of an inquiry can be broken down into

(a) overhead cost of staff organization,
(b) costs of collection of data,
(c) costs of processing and tabulating data,
(d) costs of publication.

Items (b) and (c) are in the nature of variable costs, (a) and (d) fixed costs. Items (b) and (c) will certainly be much smaller for a sample than a full count. On the other hand, the design of and the selection of the sample will involve expenditure. But in general a sample effects very considerable savings in cost.

Although the above considerations indicate that there are very real advantages in taking samples, this does not mean that full counts should never be taken. For some purposes it is not possible to select a satisfactory sample (e.g. to ascertain total population) and a full count must be taken. A sample if unreliable has no virtue simply because it is cheap. Apart from this, censuses are essential to provide certain information necessary for the selection of samples to give other information.

8.2. THE ADEQUACY OF A SAMPLE

A sample is a substitute for a full count of the population from which it is drawn. We require the information we derive from it not for its own sake, but to *infer* something about the population from which it has been drawn. For example, if we take a sample of houses in Adelaide and calculate the mean rent from it, we use that figure to make an inference about or as an estimate of the true mean rent of all houses in Adelaide. Naturally we should not expect the sample mean to equal exactly the population mean, and if we take repeated samples from the same population the sample means will all differ amongst themselves, even though the population mean is always in any particular instance the same. These discrepancies between sample statistics and population parameters are *sampling errors* (see section 7.1, p. 115). They must be carefully distinguished from the inaccuracies which arise in collecting information (see section 3.2, p. 17). These inaccuracies occur both in full counts and sample surveys, and they may, for reasons outlined above, be reduced in sample surveys; but full counts do not by their nature contain sampling errors.

Discrepancies between sample statistics and population parameters may arise for two reasons. First, the method of selection of the sample may lead to *bias* in the sample in the sense that with repeated sampling the mean of a sample statistic (mean, standard deviation, proportion, etc.) does not tend to the corresponding population parameter as the number of samples increases. Secondly, even where bias is absent, discrepancies will occur due to chance, since the sample will never exactly reproduce all the characteristics of the population. The first source of discrepancies can be eliminated by the use of proper sampling methods. Unless this is done, we can have no confidence in sample results as means of making inferences about the underlying population. With the elimination of bias the sampling errors which remain will be inevitable. They can be eliminated only by taking a full count, but they can of course be reduced by increasing the size of the sample.

It is not difficult to see how bias in the selection of a sample may arise. Thus, taking an extreme case, a sample of houses in Adelaide will be biased if the houses are all taken from inner suburbs, since houses in outer suburbs are not represented in the sample. The mean rent of the sample could in no way be taken as an estimate of the mean rent of all houses in Adelaide. It follows that bias can be avoided if each unit in the population has an equal chance of being selected for the sample. A sample drawn on this principle is a *simple random sample*. In a random sample the units for the sample are selected quite at random. If we have a barrel of marbles consisting of a very large number of white, black and red marbles and we desire to select a sample in order to infer the proportions of the different colours in the barrel, the best we can do is to mix thoroughly the marbles in the barrel and select a sample of them quite at random. Each marble then has the same chance as any other of being selected, and bias will be avoided. Random selection is selection by pure chance—lottery selection. Suppose, for example, we select 100 marbles from the barrel and find that there are 48 white, 23 black and 29 red marbles in our sample. This does not mean that these are the exact proportions in the barrel, but we should expect them to be close to them, and if we took a second sample we should not expect to get very different proportions. Moreover, here the sampling method (or the sample) is free from bias, for if we took repeated samples of the same type the mean of the sample results would in the long run get closer and closer to the population values. Thus, if 50 per cent of the marbles in the barrel are white, a series of random samples of 100 from the barrel might yield the following numbers of whites: 48, 51, 53, 49, 50, 46, 53, 51 . . . The mean of these figures will, in the long run, more and more closely approach 50 as the number of samples is increased. This illustrates the essence of random sampling. It is hardly necessary to point out that in practice the selection of a sample is a procedure much more complex than picking marbles out of a barrel.

The sampling error between a sample statistic and the corresponding population parameter for which it is to be used as an estimate cannot, of course, be specified for any one particular sample, for if it could there would be no point in having the sample estimate. But the magnitude of the sampling error can be measured in the sense of the dispersion which sample values would have about the population value with repeated sampling. Hence, when we speak of the sampling error of a statistic being smaller under one method of random sampling than another, we mean that with repeated sampling the one method would give sample values more tightly clustering around the population value than the other. Whenever a sample value is used as an estimate of a population value, it is of the first importance to attach to the sample value an indication of the magnitude of the sampling error which may be involved. This may be done by using the sample results to estimate confidence limits within which we may be reasonably certain that the population parameter under consideration will lie (see section 7.16 above). An estimate of a population parameter without confidence limits attached is of very doubtful utility, for then no indication is given of how reliable the estimate is likely to be, and consequently our confidence in the estimate must be uncertain. Thus, if a random sample of houses in Adelaide has a mean weekly rent of $13, we should not state that "our estimate of the mean rent in Adelaide is $13," but rather that "we can be 95 per cent confident that the mean rent in Adelaide lies in the range $13 $\pm$ $1," say. We are 95 per cent confident of this latter statement in the sense that, if we make a large number of such statements, we shall be right in about 95 per cent of them. In this connexion it must be emphasized that confidence limits cannot be ascertained unless the sample under consideration is a properly random one. Indeed, all the methods of statistical analysis set out in Chapter VII are valid only in reference to random samples.[1]

8.3. Selecting a Random Sample

In principle the method of selecting a simple random sample is quite easy. We take all the units in the population, mix them up thoroughly, and draw out the sample quite at random. In practice we are not usually dealing with barrels of marbles, and the selection of a random sample is not quite so easy. However, it is possible to simulate very closely the random process. We can make a list of all the units in the population, and number each unit. Then, if we transfer the numbers on to marbles, we can select numbers as in a lottery and, by reference back to our original list, identify our sample.

To aid in this process there are available tables of *random numbers*. These tables simply list the digits 0 to 9 at random. Quite complicated

---

[1] Strictly speaking, in order to be able to draw valid inferences about sampling errors, it is necessary for each unit in the population to have a known probability of selection other than zero, or for the probabilities of selection of each unit to be equal. This latter is the case of simple random sampling.

SAMPLE SURVEYS

methods have to be used to make sure the numbers are properly random. Such tables are to be found, among other places, in Fisher and Yates: *Statistical Tables for Biological, Agricultural and Medical Research* (Hafner Publishing Co., New York, 6th Edition, 1963) and a block of such numbers taken from page 134 of these tables is reproduced below by kind permission of the authors and publishers.

Table 7.1

RANDOM NUMBERS

03 47 43 73 86	36 96 47 36 61	46 98 63 71 62	33 26 16 80 45
97 74 24 67 62	42 81 14 57 20	42 53 32 37 32	27 07 36 07 51
16 76 62 27 66	56 50 26 71 07	32 90 79 78 53	13 55 38 58 59
12 56 85 99 26	96 96 68 27 31	05 03 72 93 15	57 12 10 14 21
55 59 56 35 64	38 54 82 46 22	31 62 43 09 90	06 18 44 32 53
16 22 77 94 39	49 54 43 54 82	17 37 93 23 78	87 35 20 96 43
84 42 17 53 31	57 24 55 06 88	77 04 74 47 67	21 76 33 50 25
63 01 63 78 59	16 95 55 67 19	98 10 50 71 75	12 86 73 58 07
33 21 12 34 29	78 64 56 07 82	52 42 07 44 38	15 51 00 13 42
57 60 86 32 44	09 47 27 96 54	49 17 46 09 62	90 52 84 77 27
18 18 07 92 46	44 17 16 58 09	79 83 86 19 62	06 76 50 03 10
26 62 38 97 75	84 16 07 44 99	83 11 46 32 24	20 14 85 88 45
23 42 40 64 74	82 97 77 77 81	07 45 32 14 08	32 98 94 07 72
52 36 28 19 95	50 92 26 11 97	00 56 76 31 38	80 22 02 53 53
37 85 94 35 12	83 39 50 08 30	42 34 07 96 88	54 42 06 87 98

These tables are used as follows. Suppose we have a population containing 700,000 items, and we wish to draw a sample of 1,000. We list the items in the population and number them 1 to 700,000. We then take any page of random numbers and, starting at any point, we write down the numbers in groups of six, e.g. 034743, 738636, 964736, 614698, 637162, 332616, etc. These will select the items in the population numbered 34,743; 614,698; 637,162; 332,616, etc. The second and third numbers do not correspond to any numbers on the list, and they are wasted. Methods are available to reduce this sort of wastage and hence to increase the speed of selecting the sample.

In order to draw a sample in this way, it is, of course, necessary to have a list of the units in the population to be sampled. Such a list is called the *frame*. Sometimes when the list is drawn up, the sample is drawn directly from the list, e.g. taking every 10th house on a list of houses to give a 1 in 10 sample. This is not strictly speaking random sampling. It is called *systematic sampling*, but, unless the list has periodic features (see below), it is usually quite satisfactory. It will be appreciated that in practice it may be very difficult to draw up a complete and accurate list of the population to be sampled. This applies particularly if the sampling is to be carried out periodically and the population is changing. For some purposes more or less satisfactory "ready-made" frames are available, e.g. street directories, electoral rolls, etc.

If a sample is not properly random, no estimate of the likely sampling errors involved can be made; and use of the sample may lead to biased results. There are various ways of selecting samples which do not involve random processes and hence which lead to invalid samples.

The extent to which a particular sample is properly random depends not only on the determination of the investigator to make it random but also on the accuracy of the frame used to select the sample and the extent to which all the units selected for the sample can actually be incorporated into the sample. Some of the more common sources of bias are set out below.[1]

1. *Deliberate Selection of a "Representative" Sample*

This is a form of *purposive selection*. The investigator selects the units for the sample so that they will be what *he regards* as representative of the population. For example, an investigator wishes to sample 100 rents in Adelaide, and he walks round the streets and selects houses which he regards as representative. In practice, this means either trying to select 100 average houses or else trying to select different types of houses according to the proportions he believes them to be in the total population of houses. Clearly, in the first case the sample will not be representative since it will contain a much lower degree of variability than obtains in the population. But in both cases the investigator is really begging the question, i.e. he has some preconceived ideas about rents, and he picks his sample to be representative of them. It is clear that such selection must be unreliable. Similarly, interviews with representative people are almost certain to reveal incorrectly true average public opinion. What is a representative person? When the personality of the selector enters into selection, the sample cannot be unbiased. The selection should be governed only by the impersonal laws of chance.

2. *Conscious or Unconscious Bias in the Selection of a "Random" Sample*

It may happen that, although the selection process is in the main random, there are points at which the personal judgment of the investigator is allowed to enter. For example, the investigator may reject certain units which have been sampled, on the grounds that they are extreme or atypical, or he may modify the frame before the sampling takes place.

[1] This list is based on Yates, F.: *Sampling Methods for Censuses and Surveys* (Hafner Publishing Co., New York, 3rd Edition, 1963), sections 2.2 and 4.22, by permission of author and publishers.

3. Selection Depending on Some Characteristic which Happens to be Related to the Properties of the Unit which are of Interest

For example, an investigator wishes to sample coal and he decides to take ten shovelfuls from the edges of coal stacks; this is impersonal, but it so happens that the largest lumps tend to fall to the edge. Alternatively, to sample opinion about liquor laws, it would be unwise to stand outside an hotel near closing time and interview every man who passes. Again, a selection of 1,000 names taken at random from the telephone book will not be a satisfactory method for obtaining information about people's spending habits, since people with telephones are probably of a different income and social status from those without.

4. Systematic Selection

Systematic selection occurs when a list of the population is compiled and every nth unit is selected. If there is any periodicity in the list, completely biased results may occur. For example, if a list of dwellings arranged in streets is sampled by taking every tenth dwelling, then if it happens that there are ten houses to a street block, the sample will contain either all corner houses or no corner houses. Most systematic selection, however, is equivalent to random selection.

5. Substitution

It sometimes happens that, given a properly selected sample, examination of all the units in the sample is not possible, and substitutions of other units are made. For example, in a budget inquiry certain households may have been selected. When the investigator calls, he may find no one at home in some houses, and he may substitute the house next door. This will necessarily lead to the over-representation in the sample of houses occupied all day, i.e. houses with families. To avoid this sort of bias, call-backs rather than substitutions are necessary.

6. Failure to Cover the Whole Sample Chosen

For example, investigators in a budget inquiry may not bother to visit the houses which are difficult to get to, and these houses may contain people who are significantly different from other people. Again, in a postal inquiry, only a proportion of the persons to whom the questionnaire has been posted generally replies. This may introduce bias, since the proportion which has answered may be of higher general intelligence than the average or may have an axe to grind. This is the problem of *non-response*, and, if possible, the non-respondents or a subsample of these should be interviewed personally.

7. *Quota Sampling*

This method is common in making surveys of public opinion. Interviewers are given definite quotas of persons in different social classes, different age groups, different regions, etc., and are then instructed to obtain the required number of interviews to fill each quota. The quotas ensure that the total sample includes approximately the right proportion of persons of the various categories which appear in the underlying population, but the actual persons sampled to fill each quota are not necessarily representative of the underlying population in that category. This is so because the quotas are not filled by a random selection, but by, for example, the first so many appropriate persons the interviewer strikes in a house-to-house survey or in a telephone survey. Quota sampling, however scientific it may be made to appear, is not equivalent to random sampling, unless the quotas are filled by proper random processes.

8.4. Sampling from a Finite Population

At this stage it is convenient to note that whereas the sampling theory developed in Chapter VII and referred to above was all based on the assumption that the populations from which samples are drawn are infinite, most of the populations with which we have to deal in economic and allied fields are in fact finite. When the population under consideration is infinite or when it is so large compared with samples drawn from it that it can be assumed so, the standard error of a sample mean is given by the well-known formula

$$\sigma_{\bar{x}} = \frac{\sigma}{\sqrt{N}}$$

Thus, given the population standard deviation σ, the sampling errors to which a sample mean is subject depend only on the absolute size of the sample and not on its size relative to the population. However, if the population is finite in size and the size of the sample is not very small compared with the size of the population, the relative coverage of the population by the sample must have some influence on the size of sampling errors. Thus a sample of 1,000 from a population of 5,000 must have greater reliability than the same-sized sample from a population of 5,000,000.

It can be shown that the standard error of a sample mean, when the sampling is from a finite population, is given by

$$\sigma_{\bar{x}} = \frac{\sigma}{\sqrt{N}} \sqrt{\frac{P - N}{P - 1}}$$

SAMPLE SURVEYS

where P is the size of the population. Except for very small populations this can be written

$$\sigma_{\bar{X}} = \frac{\sigma}{\sqrt{N}} \sqrt{1-F}$$

where F is the *sampling fraction*, i.e. the ratio of the sample size, N, to the population size, P. Hence the 95 per cent confidence limits for the population mean will, in this case, be given by

$$\bar{X} \pm 1\cdot 96 \frac{s}{\sqrt{N}} \sqrt{1-F}$$

The influence of the factor $\sqrt{1-F}$ in the above formula is to reduce the standard error of the mean and hence to narrow confidence limits based on this standard error. The higher the sampling fraction the lower the standard error, and, of course, when $F=1$, the standard error is zero, since such a sampling fraction implies a full count and not a sample at all. A sampling fraction of 1 in 20 will give a standard error of the mean equal to $\sqrt{1-\frac{1}{20}} = 97\cdot 5$ per cent of what it would have been if the sampling were from an infinite population; for a fraction of 1 in 10 the percentage is reduced to 94·9 per cent, for one of 1 in 5 to 89·4 per cent, and for one of 1 in 2 to 70·7 per cent. It is clear that so long as the sampling fraction is small, the refinement obtained by introducing the factor $\sqrt{1-F}$ does not amount to much, and in what follows it will be ignored. However, when the sampling is from small populations, it may be of considerable importance. Thus, for example, if two populations of a variable X, sized 20,000 and 2,000 respectively, have the same standard deviation, then it can easily be calculated that a sample of 690 from the smaller population will be as reliable as one of 1,000 from the larger population.

8.5. SIMPLE RANDOM SAMPLING

In designing a sample survey the two major considerations are the selection of the sample and the estimation of population values from the results of the sample. These are now discussed in relation to simple random sampling.

In simple random sampling, apart from the technicalities of the actual selection of the sample, the main matter to be determined is the size of the sample. Since the size of the sample directly affects sampling error, it can be determined by specifying the degree of sampling error that will be tolerated. The degree of sampling error involved in making an estimate of a population parameter from sample data can be indicated by ascertaining the confidence limits for the population parameter in question. Thus we know that, if we have a sample of size N of a variable X with sample mean $\bar{X}$ and sample standard deviation s, the 95 per cent confidence limits for the population mean, μ, are given by

$$\bar{X} \pm t_{.05} \frac{s}{\sqrt{N}}$$

When N is large, the appropriate value for $t_{.05}$ is 1·96, and the confidence limits are

$$\bar{X} \pm 1 \cdot 96 \frac{s}{\sqrt{N}}$$

In the remainder of this chapter we shall be concerned with large N, say $N > 30$. The confidence limits give us a range within which we can be reasonably certain the population mean lies. This range is based on the mean derived from the sample and takes into account the sampling error involved. It can, of course, be narrowed by reducing the desired level of confidence. For example, the 90 per cent confidence limits are given by

$$\bar{X} \pm 1 \cdot 64 \frac{s}{\sqrt{N}}$$

Such a reduction in range, however, is obtained at the expense of the confidence which we place in our results. If we adhere to the conventional 95 per cent level of confidence, the size of the range will evidently depend on the size of the sample and, as will be shown in the following section, on the design of the sample.

Consequently the essence of sampling design is first to determine the margin of sampling error which we shall tolerate and then to design a sample which will meet the requirements. In the case of sampling with the object of estimating the population mean, we might specify that the sample must be such that the population mean lies within a range, say, $\pm d$ of the sample mean with 95 per cent confidence. With repeated samples of size N drawn from a population with known σ, in 95 per cent of the cases μ will be correctly located in the range $\bar{X} \pm 1 \cdot 96 \frac{\sigma}{\sqrt{N}}$. Hence, given σ, we can ascertain the size of the sample which will make

$$1 \cdot 96 \frac{\sigma}{\sqrt{N}} = d$$

i.e.

$$N = \left(\frac{1 \cdot 96 \sigma}{d} \right)^2$$

SAMPLE SURVEYS

Clearly, for given σ, N must be greater the greater the degree of accuracy required (i.e. the smaller d), and for given d, N must be greater the greater the variability in the underlying population (i.e. the greater σ). In practice, we shall seldom know σ—the standard deviation of the population from which we are sampling—but we may have an estimate of it from a previous census or survey, or we may be able to get an estimate of it from a pilot survey. This will enable us to estimate approximately the required size of N.

Example 8.1

Given that the standard deviation of metropolitan rents derived from a pilot survey is $3·6, what sized sample should be taken to ascertain the mean level of metropolitan rents, so that we shall be 95 per cent confident that the population mean rent lies within 10 cents either way of the sample mean rent?

Here we require
$$1·96 \frac{\sigma}{\sqrt{N}} = 0·10, \quad \text{where} \quad \sigma = \$3·6$$

i.e.
$$\frac{1·96 \times 3·6}{0·10} = \sqrt{N}$$

i.e.
$$N = 4{,}979$$

Accordingly if we take a sample of about 5,000 rents, we shall be able to be 95 per cent confident that the population mean rent lies within 10 cents of the resulting mean rent, provided, of course, that the standard deviation is about $3·6. We shall be 95 per cent confident, in the sense that in the long run of every 100 similar attempts at estimation we shall be right 95 times in locating the population value within the appropriate confidence limits.

This example is based on the assumption that the population of rents is sufficiently large for it to be treated as infinite. If, however, the population were relatively small, a smaller sample would suffice. Thus, if we were told that the population consisted of 10,000 rents, we should require

$$1·96 \frac{\sigma}{\sqrt{N}} \sqrt{1 - \frac{N}{P}} = 0·10$$

i.e.
$$\frac{1·96 \times 3·6}{0·10} = \frac{\sqrt{N}}{\sqrt{1 - \frac{N}{10{,}000}}}$$

i.e.
$$N = 3{,}324$$

The above refers to sample size in relation to means, but a similar treatment can be used when other parameters are of interest. When more than one variable is to be investigated by the sample survey, the error requirements for one variable may conflict with those for others in the sense that the sample size to achieve the error requirement for one variable may differ from those necessary for another. In such cases either the sample size must be made large enough to meet all the requirements or some compromise in the various error requirements will be necessary.

As far as estimation from a simple random sample is concerned, the sample mean $\bar{X}$ provides an estimate of the population mean μ. Since with repeated random sampling the mean of $\bar{X}$'s will tend to the population mean μ, $\bar{X}$ is an unbiased estimate of μ.[1] However, no estimate is complete without attaching an indication of sampling errors. This is best done by determining confidence limits. In the case of a simple random sample the 95 per cent confidence limits are given by

$$\bar{X} \pm 1\cdot 96 \frac{s}{\sqrt{N}}$$

where s is the sample standard deviation. In this formula, s is used and not the estimate of σ which may have been used to determine N, for the latter would be only a rough estimate.

Frequently we are interested in totals rather than means, e.g. in total rent paid instead of mean rent paid. An estimate of the population total from a simple random sample is given by

$$P\bar{X}$$

where P is the number in the population. This can be written

$$\frac{P}{N}\Sigma X$$

where P/N is the reciprocal of the sampling fraction and ΣX the sample total. The ratio P/N is called the *raising factor*, and we sometimes speak of "blowing up" sample totals to population totals. Since the variance of $P\bar{X}$ is P^2 times the variance of $\bar{X}$ (see section 7.2 above) the standard error of an estimate of a population total is

$$P\frac{\sigma}{\sqrt{N}}$$

so that the 95 per cent confidence limits for a population total calculated from a simple random sample are

$$P\bar{X} \pm 1\cdot 96\, P \frac{s}{\sqrt{N}}$$

i.e.
$$\left(\bar{X} \pm 1\cdot 96 \frac{s}{\sqrt{N}}\right) P$$

[1] Not all sample statistics are unbiased estimates of the corresponding population parameters. Thus $\dfrac{\Sigma(X-\bar{X})^2}{N}$ is not an unbiased estimate of the population variance.

SAMPLE SURVEYS 169

For proportions and numbers possessing a certain attribute (e.g. proportion of dwellings which are made of brick, total number of dwellings which are made of brick) analogous formulae apply. Thus a population proportion can be estimated from a simple random sample as the sample proportion p, and the 95 per cent confidence limits are

$$p \pm 1.96 \sqrt{\frac{p(1-p)}{N}}$$

Similarly the number possessing a certain attribute in the population can be estimated from a simple random sample as Pp, with 95 per cent confidence limits

$$\left(p \pm 1.96 \sqrt{\frac{p(1-p)}{N}}\right) P$$

The preceding estimation procedures are quite straightforward. Rather more complicated ones can be devised which make use of supplementary information and which are more efficient in the sense that they give rise to smaller standard errors. Such a method which is commonly used is that of *ratio estimation*. Suppose we know that the variable X, of which we wish to estimate the population mean, is related to another variable Z, of which we know the population mean. We can estimate the population mean of X by multiplying the population mean of Z by the ratio of the means of the values of X and Z in the sample, i.e. we write

$$\text{estimate of } \mu_X = \mu_Z \frac{\bar{X}}{\bar{Z}}$$

For example, suppose we take a sample of wage-earner households to ascertain mean annual expenditure on food. We find the sample mean to be $920. We know from other sources that the annual mean earnings of all wage-earner households is in fact $2,210. If the mean earnings of the sample is $2,100, we can estimate the mean expenditure on food as $2,210 × $\frac{920}{2,100}$ = $968, instead of as $920, which is the direct sample result.

Provided the relation between X and Z is close, the standard error of an estimate performed in this way is lower than that of a direct estimate from the sample mean $\bar{X}$. This means that in these circumstances estimates from repeated samples will fluctuate less if they are made by the ratio method. Hence the range of confidence limits will be narrower. The formula for the standard error of ratio estimates is rather complicated and will not be given here. Ratio estimates are not, in general, unbiased estimates, although for large samples ($N > 30$) the bias is negligible.

Ratio estimation is particularly useful when samples of the same data are taken periodically. For example, we may have a census which gives us the mean rent of tenanted dwellings accurately for the census year. In subsequent years sample surveys are taken. We can estimate post-censal mean rent either by the means of the samples or by marking up the census figure by the movements in the mean rent of sampled houses from the census period onwards. In this case the formula

$$\mu_Z \frac{\bar{X}}{\bar{Z}}$$

is interpreted as follows—

μ_Z is the mean rent in the census,

$\bar{X}$ is the mean rent of a sample taken subsequently,

$\bar{Z}$ is the mean rent of the houses in that sample at the time of the census.

The possible advantage of ratio estimation can be seen in the extreme case where all rents, say, doubled. In this case the ratio method would give the exact population mean, even though $\bar{X}$ differed from it because of the chances of sampling; for since the same houses are involved in $\bar{X}$ and $\bar{Z}$, we shall have $\bar{X} = 2\bar{Z}$.

8.6. Stratified Random Sampling

The simple random sample is the simplest type of sampling design, but when supplementary information about the population to be sampled (i.e. information other than the mere list of units to be sampled) is available, other designs can be used. When alternative sample designs for a particular survey are available, we need criteria for discriminating between different designs which conform to the same error requirements. Since one of the main reasons for using samples rather than full counts is the saving in cost, the most workable criterion is the one of cost. The cost of making a survey will be roughly proportional to the size of the sample. Thus we can determine the margin of sampling error which we will tolerate and design the survey to minimize its cost. Alternatively, we can determine the amount we are prepared to spend on the survey, and design the sample so that the margin of sampling error is a minimum, i.e. so that the range of the confidence limits is as narrow as possible. Obviously these two criteria amount to the same thing. It is important to appreciate that there can be no "best" design in an absolute sense. There can only be a design which is "best" in relation to a certain criterion—for example, one which minimizes cost for a given error or minimizes error for a given cost. No meaning can be attached to the absolute minimization of sampling error for this can be achieved only by making the sample cover the whole population, i.e. by eliminating sampling altogether.

Of the more complicated designs the basic one is the *stratified random sample*. If the necessary information is available, the population can be divided into distinct *strata* and random samples taken from each stratum. The stratification is based upon some known characteristic of the population, e.g. rents may be stratified according to area, household budgets according to occupation of head of household, retail shops according to size of turnover, etc.

In certain circumstances a stratified random sample will give rise to smaller sampling errors than a simple random sample of the same size. This can best be seen by an example. Suppose our problem is to estimate the mean rent paid in South Australia. We divide the State into three areas—metropolitan, urban-provincial and rural. We may suppose that 50 per cent, 20 per cent and 30 per cent of houses are in these three strata respectively. We draw a simple random sample of 100 houses from the population. If rents vary within each stratum, but the distribution of rents is the same for the three strata, the way in which the three strata are represented in samples of 100 will not affect the pattern of variation of sample means resulting from repeated sampling. On the other hand, if rents are all the same within each stratum, but differ as between strata, the mean rent of a sample will depend entirely on the way in which the three strata are represented in the sample. Since with repeated random samples of 100 from the population the numbers of metropolitan, urban-provincial and rural houses in samples will fluctuate around 50, 20 and 30 respectively, the mean rent of samples will fluctuate although all rents within individual strata are the same. If, as will usually be the case in practice, rents vary within strata and the distributions of strata differ from each other, variations between mean rents of repeated simple random samples can be regarded as arising from two sources: first, from variations of rents within strata, and secondly, from variations in the proportions in which the strata are represented in different samples. By taking random samples from each stratum separately and combining the stratum sample means by the proportions in which the strata occur in the population, we can obtain a sample estimate for the population mean in which the second source of variation is controlled. This type of sample is a stratified random sample. With repeated sampling of this kind, the estimate of the population mean will vary only on account of the variation of rent within strata, and variations due to variations in the proportions in which the strata are represented in samples will be eliminated, i.e. the estimated mean rent from the sample will not differ from the population mean simply because the houses in the sample are distributed between strata differently from the way in which they are in the population. In the extreme case where the only variation in rents is between strata, all rents within a stratum being the same, all variation in the sample estimate of population mean rent will be eliminated by stratified sampling, and any one sample will correctly estimate the population mean. It follows that, with repeated sampling, the means of a stratified random sample will tend to vary less than those of a simple random sample of the same size (subject to certain restrictions about the allocation of the sample between the strata, to be discussed below) provided that stratum distributions differ. In these circumstances stratification reduces sampling error.

Suppose, we have a population of a variable X, which can be divided into various strata according to some characteristic, and we take a stratified random sample of size N. Let μ be the population mean, μ_s the mean of the sth stratum, P the number in the total population, P_s the number in the sth stratum, then

$$\mu = \frac{\Sigma P_s \mu_s}{P}$$
$$= \Sigma \alpha_s \mu_s$$

where $\alpha_s = P_s/P$ is the proportion of the population in the sth stratum, so that $\Sigma \alpha_s = 1$. Further, let N_s be the number in the sample drawn from the sth stratum, so that $\Sigma N_s = N$; and $\bar{X}_s$ be the mean of the sample drawn from the sth stratum. Then, by combining stratum sample means in the correct proportions, we shall have as an estimate of the population mean

$$\bar{X}^* = \alpha_1 \bar{X}_1 + \alpha_2 \bar{X}_2 + \alpha_3 \bar{X}_3 + \ldots$$
$$= \Sigma \alpha_s \bar{X}_s$$

where the star (*) indicates that $\bar{X}$ has been calculated from a stratified sample.

If we repeat the process of taking stratified random samples, then since $\bar{X}^*$ is a linear combination of the $\bar{X}_s$'s, by virtue of section 7.2 above, we shall have for the mean of $\bar{X}^*$

$$\Sigma \alpha_s \mu_s$$

i.e. the population mean, and for the standard error of $\bar{X}^*$

$$\sigma_{\bar{X}^*} = \sqrt{\Sigma \alpha_s^2 \sigma^2_{\bar{X}_s}}$$

where $$\sigma^2_{\bar{X}_s} = \frac{\sigma_s^2}{N_s}$$

$\sigma_{\bar{x}_s}$ being the standard error of the mean of the sth stratum and σ_s being the standard deviation of the sth stratum.

Hence
$$\sigma_{\bar{X}*} = \sqrt{\Sigma\left(\alpha_s^2 \frac{\sigma_s^2}{N_s}\right)}$$

The standard error of the mean computed from a simple random sample is

$$\sigma_{\bar{X}} = \frac{\sigma}{\sqrt{N}}$$

where σ is the standard deviation of the whole population, irrespective of stratification. The question arises: under what conditions will $\sigma_{\bar{X}*}$ be smaller than $\sigma_{\bar{X}}$? This is important, because it is only if $\sigma_{\bar{X}*}$ is appreciably smaller than $\sigma_{\bar{X}}$ that stratification is worth while.

Clearly the value of $\sigma_{\bar{X}*}$ will depend on how the sample size N is distributed between the different strata. One possibility immediately presents itself. The sample may be allocated as between strata in the same proportions as in the population. This is equivalent to having the same sampling fraction for each stratum and is called the method of a *constant sampling fraction*. We shall have

$$N_s = \alpha_s N$$

so that, in this case, $\bar{X}*$ can be calculated directly from the sample, without re-weighting, for

$$\bar{X}* = \Sigma \alpha_s \bar{X}_s$$
$$= \frac{\Sigma N_s \bar{X}_s}{N}$$
$$= \frac{\Sigma X}{N}$$

Here the sample is allocated to strata according to the importance of the strata in the population. In this case $\sigma_{\bar{X}*}$ reduces to

$$\sigma'_{\bar{X}*} = \sqrt{\frac{\Sigma \alpha_s \sigma_s^2}{N}}$$

where $\sigma'_{\bar{X}*}$ is the standard error of a sample mean calculated from a stratified random sample with a constant sampling fraction.

There are many ways in which the sample size might be allocated between the strata, of which the method of a constant sampling fraction is only one. Recalling that a criterion for sampling design is the minimization of error for a given cost, it will be appreciated that the allocation ought to be made with some such end in mind. Indeed, when a series of alternative designs presents itself, a major task of the statistician is to select the *optimum* one according to some criterion. This general principle can be applied here. If we assume that the size of the sample is given, then the problem is to select an optimum allocation between strata which minimizes sampling error. Accordingly we may fix the allocation of N between strata so as to minimize $\sigma_{\bar{x}*}$. This is called the method of *optimal allocation*.[1]

We require that $W = \Sigma\left(\alpha_s^2 \dfrac{\sigma_s^2}{N_s}\right)$ be a minimum, subject to $\Sigma N_s = N$. Bearing in mind that

$$N_1 = N - N_2 - N_3 \ldots,$$

we shall require $\dfrac{\partial W}{\partial N_s} = 0$ for all s, except $s = 1$,

i.e. we require $\dfrac{\alpha_1^2 \sigma_1^2}{(N - N_2 - N_3 - \ldots)^2} - \dfrac{\alpha_s^2 \sigma_s^2}{N_s^2} = 0$ for all s, except $s = 1$,

i.e. the expressions $\dfrac{\alpha_s^2 \sigma_s^2}{N_s^2}$ are the same for all s. It follows that N_s must be proportional to $\alpha_s \sigma_s$,

i.e. $$N_s = \lambda \alpha_s \sigma_s,$$

where λ is a constant.

Hence $$N = \lambda \Sigma \alpha_s \sigma_s$$

so that $$N_s = \dfrac{\alpha_s \sigma_s}{\Sigma \alpha_s \sigma_s} N.$$

Here the sample is allocated to strata according not only to the importance of the strata in the population but also to the variability within strata, so that the more variable strata are allocated a relatively larger part of the sample. The formula for optimal allocation reduces to $N_s = \alpha_s N$ when σ_s is constant for all strata. From the above, it can be seen that, for optimal allocation, $\sigma_{\bar{x}*}$ reduces to

$$\sigma_{\bar{x}*} = \dfrac{\Sigma \alpha_s \sigma_s}{\sqrt{N}}$$

[1] This procedure and its results are exactly analogous to the problem in economic theory of allocating a given money income between alternative uses. The equal marginal utility principle is here paralleled by equal reductions of error, i.e. the numbers are allocated such that the reduction in the variance from adding an additional unit to a stratum is the same for all strata.

SAMPLE SURVEYS

where $\sigma'_{\bar{x}*}$ is the standard error of a sample mean calculated from a stratified random sample with optimal allocation.

We now investigate the conditions under which $\sigma_{\bar{x}*}$ is smaller than $\sigma_{\bar{x}}$. We must first express the population variance σ^2 in terms of the stratum variances σ_s^2. It will be recalled (section 5.5 above) that

$$\Sigma(X-\bar{X})^2 = \Sigma(X-A)^2 - N(\bar{X}-A)^2$$

i.e.
$$\Sigma(X-A)^2 = \Sigma(X-\bar{X})^2 + N(\bar{X}-A)^2$$

where $\bar{X}$ is the mean of the X's, N is the number of X's and A is any arbitrary figure. Now consider the sth stratum. Writing X_s for the variable in this stratum, P_s for the number of the population in it, μ_s and σ_s for its population mean and standard deviation respectively, and μ for the mean of the population as a whole, we shall have

$$\Sigma(X_s - \mu)^2 = \Sigma(X_s - \mu_s)^2 + P_s(\mu_s - \mu)^2$$

i.e.
$$\Sigma(X_s - \mu)^2 = P_s[\sigma_s^2 + (\mu_s - \mu)^2]$$

This holds for each stratum. Adding all strata and dividing by P, the total population, we get

$$\frac{\Sigma\Sigma(X_s - \mu)^2}{P} = \Sigma\alpha_s\sigma_s^2 + \Sigma\alpha_s(\mu_s - \mu)^2$$

since $\alpha_s = P_s/P$,

i.e.
$$\sigma^2 = \Sigma\alpha_s\sigma_s^2 + \Sigma\alpha_s(\mu_s - \mu)^2$$

where σ^2 is the variance of the population as a whole. It can be seen that we have now split up the variability of X in the population into two components.[1] The first is a weighted average of stratum variances and measures variability within strata, and the second is a weighted variance of stratum means about the whole population mean and measures variability between strata. It follows from the above formula that we can write the standard error of the mean from a simple random sample of size N as

$$\sigma_{\bar{x}} = \frac{\sigma}{\sqrt{N}}$$

$$= \sqrt{\frac{\Sigma\alpha_s\sigma_s^2}{N} + \frac{\Sigma\alpha_s(\mu_s - \mu)^2}{N}}$$

[1] The breaking up of the variability of a variable, as measured by its variance, into components which can be attributed to different sources of variation is the central theme of a powerful tool of statistical analysis known as the *analysis of variance*. Although the notions of analysis of variance are of great generality, the application of its technique in the economic field is limited to more advanced problems than are covered by this text. The technique of analysis of variance is covered by most general texts on statistical methods (see Appendix C, 1).

Now if $\sigma'_{\bar{X}*}$ refers to the standard error of the mean from a stratified random sample of size N with a constant sampling fraction, we can write

$$\sigma_{\bar{X}} = \sqrt{\sigma'^2_{\bar{X}*} + \frac{\Sigma \alpha_s (\mu_s - \mu)^2}{N}}$$

It follows that

$$\sigma'_{\bar{X}*} < \sigma_{\bar{X}}$$

so long as all the μ_s's do not equal μ, i.e. so long as stratum means differ from each other.

Again, if $\sigma''_{\bar{X}*}$ refers to the standard error of the mean from a stratified random sample of size N with optimal allocation, we can write

$$\sigma_{\bar{X}} = \sqrt{\sigma''^2_{\bar{X}*} + \frac{1}{N}[\Sigma \alpha_s \sigma_s^2 - (\Sigma \alpha_s \sigma_s)^2] + \frac{\Sigma \alpha_s (\mu_s - \mu)^2}{N}}$$

$$= \sqrt{\sigma''^2_{\bar{X}*} + \frac{\Sigma \alpha_s (\sigma_s - \bar{\sigma}_s)^2}{N} + \frac{\Sigma \alpha_s (\mu_s - \mu)^2}{N}}$$

where $\bar{\sigma}_s = \Sigma \alpha_s \sigma_s$ is the mean of the σ_s's, weighted by the proportions α_s.[1] It follows that

$$\sigma''_{\bar{X}*} < \sigma_{\bar{X}}$$

so long as either stratum means differ from each other or stratum standard deviations differ from each other or both.

We can conclude from the above that, if we are taking a sample of size N to estimate the population mean, μ, of a variable X, the reliability of the estimate can be increased by taking a stratified random sample with optimal allocation if the population can be stratified into categories which themselves have different mean values of X, and/or which themselves have different standard deviations of X. If the strata all have the same standard deviation, optimal allocation in fact involves a constant sampling fraction. In any event optimal allocation gives a minimum sampling error for a stratified random sample of a given size.

The reasons for the gain in reliability obtained through stratification should be fairly obvious. By stratifying we do not allow variations in the proportions of the sample lying in any stratum to affect the estimate of the population mean. This eliminates one source of variability. Furthermore, with optimal allocation we take a relatively larger sample from those strata with relatively greater variability. This averages out the variability within the different strata.

[1] Note:
$$\Sigma \alpha_s (\sigma_s - \bar{\sigma}_s)^2 = \Sigma \alpha_s \sigma_s^2 - 2\bar{\sigma}_s \Sigma \alpha_s \sigma_s + \bar{\sigma}_s^2 \Sigma \alpha_s$$
but $\Sigma \alpha_s \sigma_s = \bar{\sigma}_s$ and $\Sigma \alpha_s = 1$

Hence $$\Sigma \alpha_s (\sigma_s - \bar{\sigma}_s)^2 = \Sigma \alpha_s \sigma_s^2 - (\Sigma \alpha_s \sigma_s)^2$$

SAMPLE SURVEYS 177

In order to determine whether or not a stratified sample is worth while in a particular case and, if so, what sized sample to take, and how to allocate it between strata, it is necessary to have estimates of the various σ's in the above formulae. These may be available from a prior investigation, or it may be possible to make rough estimates of them. Once the sample is taken, confidence limits to $\bar{X}^*$ should be based on the sample standard deviations, since the estimates of the σ's used in designing the sample will necessarily be fairly rough.

In general, a stratified random sample, appropriately allocated between strata, will enable a given margin of sampling error to be achieved with a smaller sample than a simple random sample. However, it does not follow that stratification is always worth while, for while stratification will reduce costs of collection and tabulation, the process of stratification may itself be costly, since the frame will have to be stratified and each stratum in the frame separately sampled. Furthermore stratification is impossible unless the proportions of the population falling into the different strata are known, i.e. unless we have supplementary information about the population. This can be seen by the fact that the α_s's occur in all the relevant formulae. Finally, the advantages of stratification are gained only when the sample is allocated between strata in one or another appropriate form. Any sort of allocation will not do. This becomes obvious when one considers what would happen if practically the whole sample were allocated to one stratum, the other strata being hardly sampled at all.

This section and the preceding one have dealt with the problem of sample design entirely in terms of means. Similar methods are available for handling proportions and totals. Strictly speaking, account should have been taken of the fact that a population which can be stratified is necessarily finite, although if the population is large relatively to the sample to be drawn from it, this does not matter a great deal. In any case the basic principles of stratification have been dealt with. The example below illustrates these principles in numerical terms.

EXAMPLE 8.2

A sample survey is to be undertaken to ascertain the mean annual income produced by farms in a certain area. The farms can be stratified according to their principal products. A census conducted several years earlier yielded the data set out in the table below.

Assuming that the proportions in the four strata have remained unchanged, and using the census standard deviations as approximations to the unknown current standard deviations, estimate:

(i) the margin of sampling error which can be expected to arise for a sample of 500 farms if it is (a) a simple random sample, (b) a stratified random sample with a constant sampling fraction, and (c) a stratified random sample with optimal allocation.

178 APPLIED STATISTICS FOR ECONOMISTS

(ii) the size of (*a*) a simple random sample, (*b*) a stratified random sample with a constant sampling fraction, and (*c*) a stratified random sample with optimal allocation, in order to be reasonably confident that the population mean will lie within $100 of the sample estimate.

Type of Farm	Proportion of Farms in Stratum	Mean Annual Income $	Standard Deviation $
Wool	0·16	10,946	2,236
Wheat	0·19	6,402	2,644
Dairying	0·27	2,228	606
Other	0·38	1,458	230
All Farms*	1·00	4,124	3,788

* This line gives values for the population as a whole. The mean of the population as a whole is related to the stratum means by the identity

$$\mu = \Sigma \alpha_s \mu_s$$

and the population standard deviation is related to the stratum standard deviations by the identity

$$\sigma^2 = \Sigma \alpha_s \sigma_s^2 + \Sigma \alpha_s (\mu_s - \mu)^2$$

These relations necessarily hold. (See pp. 172 and 175 above.)

The relation between the population variance and the stratum variances can, perhaps, be most readily understood by working in terms of an arbitrary origin of zero (see section 5.5, p. 72). Thus we have, for the population variance

$$\sigma^2 = \frac{\Sigma(X - \mu)^2}{P} = \frac{\Sigma X^2 - P\mu^2}{P}$$

and for the variance of the *s*th stratum

$$\sigma_s^2 = \frac{\Sigma(X_s - \mu_s)^2}{P_s} = \frac{\Sigma X_s^2 - P_s \mu_s^2}{P_s}$$

Since ΣX^2 in the formula for σ^2 is simply the sum of the squares of the original observations, it can be obtained by adding the ΣX_s^2's for the individual strata. Thus we can derive ΣX^2 from a knowledge of the σ_s^2's. Thus, from the formula for σ_s^2

$$\Sigma X_s^2 = P_s(\sigma_s^2 + \mu_s^2)$$

and hence

$$\sigma^2 = \frac{\Sigma P_s(\sigma_s^2 + \mu_s^2) - P\mu^2}{P}$$

$$= \Sigma \alpha_s (\sigma_s^2 + \mu_s^2) - \mu^2$$

where the summation is a summation of stratum results. By rearrangement this latter formula becomes

$$\sigma^2 = \Sigma \alpha_s \sigma_s^2 + \Sigma \alpha_s \mu_s^2 - \mu^2$$

$$= \Sigma \alpha_s \sigma_s^2 + \Sigma \alpha_s (\mu_s - \mu)^2$$

since $\Sigma \alpha_s \mu_s = \mu$ and $\Sigma \alpha_s = 1$.

SAMPLE SURVEYS

(i) (a) *Simple random sample*

Here $N = 500$, and we have $\sigma_{\bar{X}} = \dfrac{\sigma}{\sqrt{N}}$

$$= \$169\cdot 4$$

(b) *Stratified random sample with constant sampling fraction*

Here $N_s = \alpha_s N$

and with $N = 500$

we have $N_1 = 80, N_2 = 95, N_3 = 135, N_4 = 190$

The sampling error will be reduced by this form of stratification since the data indicate that stratum means are likely to differ. We have

$$\sigma'_{\bar{X}*} = \sqrt{\dfrac{\Sigma \alpha_s \sigma_s^2}{N}}$$

$$= \$67\cdot 0$$

(c) *Stratified random sample with optimal allocation*

Here $N_s = \dfrac{\alpha_s \sigma_s}{\Sigma \alpha_s \sigma_s} N$

and with $N = 500$

we have $N_1 = 161, N_2 = 226, N_3 = 74, N_4 = 39$

The sampling error will be further reduced by this form of stratification since the data indicate that stratum standard deviations differ. Note that the wool and wheat strata which have relatively high standard deviations are relatively more heavily sampled in this design than in design (b). We have

$$\sigma''_{\bar{X}*} = \dfrac{\Sigma \alpha_s \sigma_s}{\sqrt{N}}$$

$$= \$49\cdot 6$$

This example shows how stratification can reduce sampling error. With repeated samples of 500, sample means from stratified samples, calculated by the formula $\bar{X}* = \Sigma \alpha_s \bar{X}_s$, will show less variability than sample means from simple random samples, and design (c) will show less variability than design (b).

(ii) (a) *Simple random sample*

Using 95 per cent confidence limits we shall require

$$1\cdot 96 \dfrac{\sigma}{\sqrt{N}} = 100$$

i.e. $N = 5{,}512$

(b) *Stratified random sample with constant sampling fraction*

We shall require $1.96\sqrt{\dfrac{\Sigma\alpha_s\sigma_s^2}{N}} = 100$

i.e. $N = 863$, with $N_1 = 138$, $N_2 = 164$, $N_3 = 233$, $N_4 = 328$

(c) *Stratified random sample with optimal allocation*

We shall require $1.96\dfrac{\Sigma\alpha_s\sigma_s}{\sqrt{N}} = 100$

i.e. $N = 474$, with $N_1 = 153$, $N_2 = 214$, $N_3 = 70$, $N_4 = 37$

Provided the above σ_s's are close to the σ_s's currently obtaining in the population, we can be 95 per cent certain that the population mean lies within $100 of the sample mean for the three sample designs set out above. Clearly stratification here involves appreciable savings in sample size. Design (c) gives the same accuracy as design (a) with a sample less than one-tenth the size.

As far as estimation from a stratified random sample is concerned, the estimate of the population mean is given by

$$\bar{X}^* = \Sigma\alpha_s\bar{X}_s$$

and the 95 per cent confidence limits by

$$\bar{X}^* \pm 1.96\sqrt{\Sigma\left(\alpha_s^2\dfrac{s_s^2}{N_s}\right)}$$

Similarly the estimate of the population total is

$$P\bar{X}^*$$

and the 95 per cent confidence limits are

$$\left[\bar{X}^* \pm 1.96\sqrt{\Sigma\left(\alpha_s^2\dfrac{s_s^2}{N_s}\right)}\right]P$$

where s_s is the sample standard deviation of the sth stratum.

For proportions and numbers possessing a certain attribute analogous formulae apply. Thus we have as an estimate of the population proportion

$$p^* = \Sigma\alpha_s p_s$$

where p_s is the proportion in the sth stratum in the sample, and the 95 per cent confidence limits are

$$p^* \pm 1.96\sqrt{\Sigma\left(\alpha_s^2\dfrac{p_s(1-p_s)}{N_s}\right)}$$

SAMPLE SURVEYS 181

As an estimate of the number possessing a certain attribute we have

$$Pp*$$

with 95 per cent confidence limits

$$\left[p^* \pm 1{\cdot}96\sqrt{\Sigma\left(\alpha_s^2 \frac{p_s(1-p_s)}{N_s}\right)}\right]P$$

Sometimes it is not possible to select a stratified random sample because the frame cannot be readily stratified, although it may be possible to *stratify after selection* if the distribution of the population between the various strata is known. Thus, if dwellings are known to be divided between metropolitan, urban-provincial and rural regions in the ratios 50:20:30, but the full list of dwellings is not so classified, a simple random sample can be stratified after selection by dividing the sample into the three strata. The above formulae can then be applied, and provided that a sufficient number of dwellings in the simple random sample fall into each of the strata, sampling error may be reduced. The advantage which stratification before selection holds over this method is that with such stratification we can ensure that adequate numbers of the sample are allocated to each stratum instead of relying on chance. But with stratification after selection, we shall control in the estimation process, at least, the distribution of the sample between strata.

EXAMPLE 8.3

A simple random sample of 120 families taken at the beginning of a particular year yielded the following information about expenditure on entertainment in the previous three months—

Region	Number of Families	Mean Expenditure	Standard Deviation
		$	$
Metropolitan . .	52	61·23	9·99
Urban-Provincial .	33	38·79	10·64
Rural . . .	35	23·51	5·73
All Regions* . .	120	44·06	18·52

* The all regions values are related to the individual region values in an analogous manner to that set out in the footnote on page 178 above. Here of course the values are sample values.

The population consisted of 10,000 families, and it was known to be distributed between the three strata in the proportions 5:3:2 respectively. Estimate 95 per cent confidence limits for the population mean expenditure and for the aggregate expenditure in the given quarter.

Treating the sample as a simple random sample of 120 families, the 95 per cent confidence limits for the mean will be

$$\bar{X} \pm 1\cdot 96 \frac{s}{\sqrt{N}}$$

i.e. $44·06 ± 3·31

i.e. $40·75 to $47·37

and the 95 per cent confidence limits for the aggregate expenditure will be 10,000 times these,

i.e. $407,500 to $473,700

However, if we stratify the sample after selection on the basis of the given population distribution between strata, we get as an estimate of the population mean expenditure for the given quarter.

$$\bar{X}^* = \Sigma \alpha_s \bar{X}_s$$
$$= \$46\cdot 95$$

and the 95 per cent confidence limits will be

$$\bar{X}^* \pm 1\cdot 96 \sqrt{\Sigma\left(\alpha_s^2 \frac{s_s^2}{N_s}\right)}$$

i.e. $46·95 ± 1·78

i.e. $45·17 to $48·73

The 95 per cent confidence limits for aggregate expenditure will be 10,000 times these,

i.e. $451,700 to $487,300

Here stratification after selection narrows the range of the confidence limits. It also raises the estimate of the mean expenditure, since the metropolitan stratum was under-represented in the simple random sample.

8.7. Other Designs for Sample Surveys

The simple random sample and the stratified random sample are the simplest of the various sample designs used in practice. More complicated techniques are available, and like the simpler ones they must be based on random selection to be satisfactory. Some designs involve *multi-stratification*, i.e. stratification by two or more characteristics, e.g. a stratification by size of income within each of the three regional strata in example 8.3 above. This increases the number of distinct strata, but does not affect the design in principle. Other designs involve *multi-stage sampling*.

In multi-stage sampling the population is first divided into first-stage sampling units. A random sample of these is taken. It is customary to select the first-stage sampling units with the probabilities of selection of units proportional to their size and not equal as in the case of simple random sampling. The first-stage sampling units which have been selected in this sample are then divided into smaller second-stage sampling units which are then sampled. This process can be continued for a number of stages until the final sampling units are selected. A simple example would be a survey of nutrition of school-children, where schools might be the first-stage sampling units. A sample of these would then be taken and from the selected ones individual children sampled. This would be a two-stage sample and the children would be the final sampling units. It should be plain that when the early-stage samples contain few units, the early-stage sampling units should be stratified before sampling to ensure adequate representation of the population, unless they are fairly similar in their characteristics. If, for example, different schools cater for different classes of children, a small sample of schools may miss out important categories of the population.

Multi-stage sampling is best exemplified in what is called *area sampling*. One of the difficulties in taking a sample of a human population for any purpose is that the population may extend over a very large area. This is certainly the case in Australia. Consequently a sample drawn from the whole area will be spread out and will be costly to contact. It will also be difficult to ensure adequate supervision of the field work. Not only this, but the listing of the whole population will be a long and costly procedure. The object of area sampling is to meet these difficulties. Area sampling involves dividing the area into small areas to provide first-stage sampling units. For example, in a particular country the first-stage units might be local government areas. A random sample of these would be taken. This is the first stage of the sampling. The second stage might be to take random samples of administrative districts from the already selected first-stage sampling units; the third might be to sample blocks of dwellings from the selected administrative districts; the fourth and final stage might be to sample dwellings from the selected blocks. This procedure has the effect of concentrating the final sample into a limited number of small areas with savings in costs of collection. At the same time listing is greatly reduced, since all that is required is a list of the first-stage sampling units, then a list of the second-stage units in the selected first-stage units and so on. Multi-stage area sampling is usually combined with stratification, the early-stage sampling units being stratified before they are sampled. Area sampling is also very useful when surveys are to be taken regularly, e.g. a monthly survey of employment. The earlier-stage sampling units once selected can be used permanently and permanent survey organizations can be established in them. Examples of this type of sampling—all more or less complex—abound in the literature.[1] Multi-stage samples will have greater sampling errors than single-stage samples of the same size, but it may be less costly to take a large multi-stage sample than a small single-stage one with the same

[1] cf. Yates: *op. cit.*

error performance, so that a multi-stage sample may be the "best" design to employ *in given circumstances*. This illustrates once again that there is no *absolute* "best" sample design, but only one which is most efficient in terms of some criterion, like cost or speed.

8.8. Sample Surveys in Practice

In the past two decades sampling has been accepted as an established method for collecting economic and social data. It has been used in three main ways:

(i) *To sample a full count which has already been conducted.* The purpose of such a procedure may be to obtain preliminary information in advance of the full tabulation, to obtain more detailed information than is abstracted in the full tabulation, or to reduce costs of tabulation. An example of the first type was the 1 per cent sample of census returns of the Great Britain Census of Population taken on 8th April, 1951. This enabled preliminary results to be made available within a year of the census, instead of waiting several years. Sampling at the tabulation stage has not been considered necessary in the 1961 Census, since the use of sampling in the field and electronic data processing methods are expected to speed up the production of full tabulations sufficiently to render preliminary tabulation superfluous.

Sampling was also extensively employed in the United States Censuses of 1940, 1950 and 1960, not only for the purpose of producing preliminary results, but also to enable detailed cross-tabulations of data to be made for which full tabulations would be extremely expensive and time-consuming. Also the United States Census questionnaire included in addition to questions addressed to the whole population supplementary questions to be addressed to only a fraction of the population.[1]

When the population of a country is very large, full collection and tabulation are an enormous task and a fraction of the population will provide a sufficiently large sample in absolute numbers to ensure reasonably low sampling errors. The relative cheapness of sampling is also of considerable importance in this connexion. Thus, since July, 1952, annual income tax statistics for wage and salary earners in Australia have been estimated on a sample basis. A sample of about 300,000 returns is selected from the population of over $3\frac{1}{2}$ million returns, and the cost of processing is about 20 per cent of that of complete enumeration.

---

[1] For details of sampling used in censuses see Waksberg, J. and Hanson, R. H., *Sampling Applications in Censuses of Population and Housing* (U.S. Government Printing Office, 1965), U.S. Bureau of the Census Technical Paper No. 13.

Sampling methods are also used for checking the accuracy of the tabulation of census returns. For example, in the 1966 Australian Census the use of such methods eliminated about three-quarters of the checking which would have otherwise been necessary.

(ii) *To obtain information for a special purpose.* The activities of "The Social Survey," a unit of the United Kingdom Government, provide many examples of special purpose surveys. Over the period 1946–60, about 300 surveys were conducted. They include large-scale continuing studies and *ad hoc* single purpose inquiries. The continuing Household Budget Survey is a major activity and other subjects covered are: the organization of nursing work in general hospitals, the evaluation of the Rent Acts, the demand for post office services, statistics of tourists' movements, the staffing of children's homes, and the incomes of professional workers.[1]

(iii) *To obtain continuous information of the behaviour of certain economic quantities.* Perhaps the best-known example of a sample survey conducted regularly for this purpose is the Current Population Survey of the United States Bureau of the Census to provide up-to-date information about the labour force. The sample is an area sample consisting now of 35,000 households, and is taken monthly. As well as providing regular information on the labour force, it also provides from time to time data on special topics, e.g. income. The results are published monthly with an indication of the magnitude of the sampling errors involved.[2]

In Australia, the Commonwealth Bureau of Census and Statistics also conducts a survey of the work force each quarter involving interviews with a sample of 10 per cent of Australian households to provide comprehensive labour force estimates. The sample used is also an area sample, covering areas in the six Australian States and the two internal Territories and includes about 40,000 households. The Population Survey Sample used for the quarterly work force survey provides a vehicle for a number of supplementary surveys which are generally conducted in conjunction with the work force survey, but may involve separate field inquiries. Supplementary surveys have covered such subjects as home occupancy and rents, internal migration, travel, chronic illness and superannuation.

Regular surveys are particularly useful for getting up-to-date

---

[1] See Moser, C. A., *Survey Methods in Social Investigation* (Heinemann, 1963).
[2] See Hansen, M. H., Hurwitz, W. N., Madow, W. G.: *Sample Survey Methods and Theory* (John Wiley & Sons, 1953), Vol. I, p. 559; also U.S. Department of Commerce, Bureau of the Census, *The Current Population Survey: A Report on Methodology* (U.S. Government Printing Office, 1963), Technical Paper No. 7.

information between censuses. The periodic full count can be used to check whether the sample surveys are showing any bias over time in one direction or another (due perhaps to changes in the composition of the frame, which cannot be adequately covered in selecting the sample). A brief account of a survey of this kind (the Australian Survey of Retail Establishments) is given in Appendix B.2, pp. 478–481.

CHAPTER IX

QUALITY CONTROL

9.1. Statistical Control of Quality

An important application of the theory of sampling and significance in the industrial field is that of the statistical control of quality. It is discussed here very briefly to illustrate a different application of the principles set out in Chapter VII from that considered in Chapter VIII above. Quality control is a specialized technique and, like sample surveys, has a literature of its own. This chapter attempts to do no more than to explain what it is and to indicate how it may be used to improve the technical efficiency of production processes.[1]

The quality of industrial products can be measured in various ways, but when we are concerned with mass-produced components, quality can generally be measured by a fairly simple characteristic of the component under consideration. Thus we may be interested in the width of a screw, the hardness of a bearing, the strength of a material, etc. If we think of quality in this way, it will be immediately evident that quality will never be absolutely constant. There will always be a certain amount of variation in quality, even if the production processes are themselves constant. Variation in quality can be attributed to two main types of causes:

1. *Chance causes.* These causes are very many in number, each one exercising only a trivial effect on the quality of the product. They are inherent in the production processes, in the sense that they will operate for any given set of processes. They can be modified only if the production processes are themselves modified. Small variations in the quality of raw materials or in the skill of manual operators are examples.

2. *Assignable causes.* These are causes which can be identified and will generally be related to variations in the productive processes themselves, e.g. mechanical faults in plant, faulty raw materials, etc. These causes interfere with the efficient working of the plant, and it is usually economically worth while to attempt to eliminate them.

If the variations in quality of a particular product are such that they can be attributed to chance causes solely, then we say that the process is under *statistical control.* This means that our observations of quality could all have come from the same homogeneous population, the

---

[1] This chapter is largely based on Chapter 10 of Davies, O. L. (Ed.): *Statistical Methods in Research and Production,* published in 1958 for the Imperial Chemical Industries, Ltd., by Oliver and Boyd, Ltd., by permission of the Imperial Chemical Industries, Ltd., and the Publishers.

variations in them being due to chance alone. This being so, valid predictions about the future behaviour of the data can be made, assuming that nothing affects the nature of the population from which the data are drawn. When a process is under statistical control, the variability of the quality of the product cannot be altered unless the production process is itself altered. However, if it can be shown that a process is out of statistical control, this means that certain assignable causes must be present affecting the variability of quality. These can be sought for and removed. The object of what is known as *quality control* is to detect assignable causes of variation in quality as soon as they occur in the production process.

9.2. Control Charts

If we have a certain measurable quality characteristic X (say, the width of a pipe) and we know that its mean value when the production process is in control is μ and its standard deviation is σ, then if we take samples of size N, we know that the sample mean $\bar{X}$ will be approximately normally distributed about a mean μ with a standard error $\frac{\sigma}{\sqrt{N}}$. Approximately 1 in 20 of samples drawn from the population will have means lying outside the limits $\mu \pm 1\cdot 96 \frac{\sigma}{\sqrt{N}}$, and 1 in 500 will have them lying outside the limits $\mu \pm 3\cdot 09 \frac{\sigma}{\sqrt{N}}$. As far as the latter set of limits are concerned, we can say that only very rarely will a mean $\bar{X}$ lie further away from μ than by about $3 \frac{\sigma}{\sqrt{N}}$, just due to chance (in theory, about 27 times in 10,000). If we struck a particular $\bar{X}$ further from μ than $3 \frac{\sigma}{\sqrt{N}}$, we should suspect that there was some assignable cause present which would account for it, e.g. something had gone wrong with the production process. This is the simple essence of quality control.

The application of quality control to a process of production can be divided into two stages—

1. The estimation of μ (known as the *process average*) and σ (the *process standard deviation*), and from μ and σ the estimation of the *control limits*. In practice the limits $\mu \pm 3 \frac{\sigma}{\sqrt{N}}$, where N is the size of samples to be tested, have been found satisfactory, in that it is usually economically worth while to check the production process when an observation lies outside these limits.

QUALITY CONTROL

2. The selection of samples of the product of size N (in practice, a sample size of about 5 is satisfactory), such samples being taken from time to time as production proceeds, and the examination as to whether the means of the samples lie within the control limits. This is accomplished by means of a *control chart*.

To estimate μ and σ, it is usual to take about 100 observations. If the sample size is to be 5, these 100 observations are divided, in the order in which they were taken, into 20 samples of 5, and the 20 sample means are calculated, in order to see whether these twenty $\bar{X}$'s are themselves within the control limits. For if they are not, the process cannot be initially under control and the estimation of μ and σ cannot be relied on. Once μ and σ have been satisfactorily estimated, the control limits can easily be ascertained and drawn on a chart. The μ and σ are, of course, only estimates of the underlying population values. The principles underlying quality control can, perhaps, best be appreciated by following through an example. Below are set out twenty samples of five observations each of the width of inch steel pipes.

Table 9.1

MEASUREMENTS OF WIDTH OF STEEL PIPES
(to nearest one-thousandth of an inch)

(1)	(2)	(3)	(4)	(5)	(6)	(7)	(8)	(9)	(10)
0·981	1·005	1·029	0·979	1·012	0·989	0·987	0·975	0·991	1·000
0·993	1·013	0·994	0·996	1·011	0·972	0·999	1·005	0·994	1·007
0·997	0·997	0·996	0·998	0·982	1·005	1·001	0·995	1·000	0·984
0·999	1·003	1·003	0·998	0·993	0·995	1·013	0·995	1·020	0·988
1·004	0·987	1·010	1·004	1·000	1·000	1·011	0·991	1·002	1·007

(11)	(12)	(13)	(14)	(15)	(16)	(17)	(18)	(19)	(20)
0·984	1·001	1·003	1·005	0·990	0·986	1·006	0·996	0·993	0·993
1·023	1·007	1·003	0·996	0·999	1·003	0·997	1·014	1·009	1·006
0·996	1·006	0·985	0·997	1·004	1·009	0·994	1·015	0·994	0·994
0·996	0·989	1·008	1·001	1·017	0·991	1·019	0·986	1·010	1·002
1·004	0·997	1·012	1·000	1·015	0·990	1·000	0·992	1·013	0·999

The estimation of μ is straightforward. The mean of the above 100 observations is used as the estimate, or what amounts to the same thing, the mean of the sample means. The estimation of σ is usually carried out, not by computing it as the standard deviation of the 100 observations about their mean, but by computing the sums of squares of the deviations of the observations in each sample from the sample mean and pooling these sums. An estimate of σ^2 will then be given by the formula

$$\frac{\Sigma(X_1 - \bar{X}_1)^2 + \Sigma(X_2 - \bar{X}_2)^2 + \ldots \Sigma(X_k - \bar{X}_k)^2}{kN - k}$$

where the subscripts refer to the first, second sample, etc., and k is the number of samples. The advantage of this method of estimating σ

is that it gives an estimate of variability based on variability within samples and hence eliminates any variability which might occur between samples due to other than chance causes. The presence of $kN - k$ in the denominator rather than kN is to avoid bias. In the following table are the necessary calculations using an arbitrary origin.

Table 9.2

COMPUTATION OF PROCESS AVERAGE AND STANDARD DEVIATION

Sample Number	Mean of Deviations from Arbitrary Origin of 1 inch $\bar{x}'$	$\bar{x}'^2$	Sum of Squares of Deviations from Arbitrary Origin of 1 inch $\sum x'^2$
1	− 0·0052	0·00002704	0·000436
2	0·0010	0·00000100	0·000381
3	0·0064	0·00004096	0·001002
4	− 0·0050	0·00002500	0·000481
5	− 0·0004	0·00000016	0·000638
6	− 0·0078	0·00006084	0·000955
7	0·0022	0·00000484	0·000461
8	− 0·0078	0·00006084	0·000781
9	0·0014	0·00000196	0·000521
10	− 0·0028	0·00000784	0·000498
11	0·0006	0·00000036	0·000833
12	0·0000	0·00000000	0·000216
13	0·0022	0·00000484	0·000451
14	− 0·0002	0·00000004	0·000051
15	0·0050	0·00002500	0·000631
16	− 0·0042	0·00001764	0·000467
17	0·0032	0·00001024	0·000442
18	0·0006	0·00000036	0·000697
19	0·0038	0·00001444	0·000435
20	− 0·0012	0·00000144	0·000126
Total	− 0·0082	0·00030484	0·010503

$$\text{Estimate of } \mu = A + \frac{\bar{x}'_1 + \bar{x}'_2 + \ldots + \bar{x}'_k}{k}$$

$$= 1 - \frac{0 \cdot 0082}{20} = 0 \cdot 9996 \text{ inches}$$

$$\text{Estimate of } \sigma = \sqrt{\frac{\sum x_1'^2 - N\bar{x}_1'^2 + \sum x_2'^2 - N\bar{x}_2'^2 + \ldots \sum x_k'^2 - N\bar{x}_k'^2}{kN - k}}$$

$$= \sqrt{\frac{0 \cdot 010503 - (5 \times 0 \cdot 00030484)}{100 - 20}} = 0 \cdot 0106 \text{ inches}$$

Having estimated μ and σ, the control limits are readily established at $\mu \pm 3 \frac{\sigma}{\sqrt{N}}$, i.e. at $0 \cdot 9996 \pm 0 \cdot 0142$ for $N = 5$, i.e. the lower control

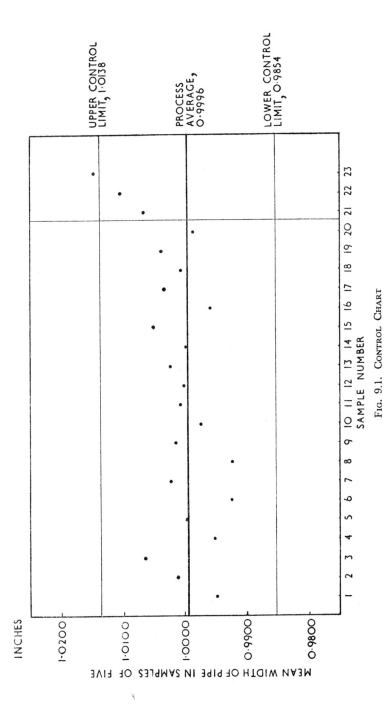

Fig. 9.1. Control Chart

limit will be 0·9854 and the upper one 1·0138. Fig. 9.1 shows these limits and also, to the left of the vertical line, the means of the 20 samples of 5 upon which the estimation of μ and σ is based.

It will be observed that the 20 sample means lie within the limits. This indicates that the process was under statistical control when these observations were made. Having established the control chart, samples of 5 are taken regularly, and their means are plotted. These are shown for samples number 21 to 23 to the right of the vertical line. It will be observed that sample 23 indicates that the process is out of control. Such a value for $\bar{X}$ would occur very rarely due to chance, and consequently there is a presumption that an assignable cause is present. It will be worth while looking for that cause immediately. However, in the case illustrated in Fig. 9.1 the trouble probably started at about the 21st or 22nd sample. This is suggested by the trend of the points at the right-hand side of the diagram. Distinct trends are unlikely to be due to chance, and a trend in the points should be just as suggestive as a point lying outside the control limits. Likewise, a number of points hovering fairly close to a control limit would be suggestive. It will also be realized that occasionally a search for trouble may be undertaken when no trouble really exists, for points will occur outside the control limits very occasionally just due to chance. This corresponds to the type I error discussed in section 7.14 above. As data are accumulated, μ and σ can be re-estimated from time to time, but only data obtained when the process is under control should be used. It will be appreciated by this stage that, in using a control chart, we are in fact testing the hypothesis that the samples come from a population with a mean and standard deviation as specified in the chart.

As well as watching the average level of quality, as in Fig. 9.1, it is customary to watch the average variability of quality. This can be done by ascertaining control limits to the value of the standard deviation of samples, s. Clearly the s's will vary from sample to sample. Methods are available for drawing up a control chart in terms of s, just as the one in Fig. 9.1 is in terms of $\bar{X}$, although in the case of s the upper control limit will alone be of practical significance, since abnormal lack of variability will not be a cause for worry. Frequently in practice the range of the values in the samples is taken as a measure of variability, and a control chart in terms of the range is used, instead of one in terms of the standard deviation. This simplifies the arithmetic.

It should be clear that it is desirable that not only the mean level of quality should be kept constant but also its variability. Hence, a control chart for s is just as important as one for $\bar{X}$. It may happen that $\bar{X}$ goes out of control, while s remains in control and vice versa. The first case would occur, for example, if the setting of a tool were wrong and the second if the bearings in a machine were becoming worn. Hence,

QUALITY CONTROL 193

the way in which the two charts behave in conjunction may be a clue to trouble in the production process. Finally, it is important to emphasize that so long as the quality characteristics remain in control, the variability in it which does occur is inevitable and cannot be modified unless the production process is itself modified.

Sometimes quality can be judged, not by taking a direct measurement, but by setting up a standard and measuring the percentage not coming up to that standard in the samples. This is called the *percentage defective, p*. Methods are available for drawing up control charts in terms of p. However, in general, these are not as satisfactory as charts in terms of $\bar{X}$, since they do not make use of as much information. The percentage defective is utilized when Go—No Go gauges are used, and, of course, in some tests of quality it alone can be utilized, e.g. in the proofing of ammunition.

Most products must be made to specifications. These specifications may state the level of quality to be aimed at, together with tolerance limits outside of which the quality may be permitted to lie only very occasionally, say 1 in 500 times. In this case, unless these tolerance limits lie outside $\mu \pm 3.09\,\sigma$, the specifications cannot be met. This is so because, even if the process is under statistical control and working at the specified level of quality, 1 in 500 items will have a value outside the limits $\mu \pm 3.09\,\sigma$ just due to chance. If the tolerance limits lie within $\mu \pm 3.09\,\sigma$, it will be necessary either to widen the tolerance limits, i.e. to relax the specifications, or else to alter the process of production to reduce σ, i.e. to reduce the variability of the product. Thus, referring back to the illustrative example, suppose that the manufacturer of the steel pipes claimed that his pipes were made to a specification of 1 inch in width within 0·015 inches either way. From the calculations made above we have an estimate of the standard deviation of 0·0106 inches. Consequently, even if the mean width of pipes were in the long run 1 inch exactly, we should expect about 15 per cent of all pipes to fail to meet the specifications. This is so because a tolerance of ± 0.015 represents a distance of $\pm \dfrac{0.015}{0.0106} = \pm 1.42$ standard deviations away from the mean. Assuming that the distribution is normal, approximately 15 per cent of the population will lie further from the mean than this. The manufacturer must either relax his specifications or modify the production processes to reduce the variability of the product.

9.3. Advantages of Quality Control

The general objective of quality control is to maintain quality. The alternative and traditional technique is *100 per cent inspection* of the output of a product. Comparing quality control with 100 per cent inspection, quality control has outstanding advantages.

1. Since quality control involves inspecting only a fraction of the output of a product, costs of inspection are greatly reduced and efficiency of inspection increased.

2. With 100 per cent inspection, unwanted variations in quality may be detected later than with the continuous sampling technique of quality control. This means that a greater volume of faulty products will have been produced, and a greater delay in the rectification of faults in the production process will occur. Quality control ensures early detection of faults and, hence, a minimum waste of reject production. The control chart provides a graphic summary of how production is proceeding.

3. Quality control enables a process to be brought into and held in a state of statistical control, i.e. a state in which variability is the result of chance causes alone. When a process is under control, the quality of the product can be accurately specified, in that limits can be specified within which, say, 99 per cent of the product will lie. So long as statistical control continues, these specifications can be accurately predicted for the future, which even 100 per cent inspection cannot guarantee to do. Consequently it is possible to assess whether the production processes are capable of turning out products which will comply with any given set of specifications.

4. Whether or not a change in the production process results in a significant change in quality can be readily detected by quality control.

5. When the test of quality is destructive, e.g. proofing of ammunition, testing breaking strength in cables, etc., 100 per cent inspection is impossible. In such cases sampling must be resorted to, and the application of the proper sampling methods of quality control ensures not only that the quality is controlled, but also that valid inferences about the total output are drawn from the samples.

CHAPTER X

REGRESSION AND CORRELATION

10.1. Descriptive Measures of Regression and Correlation for Ungrouped Data

In the preceding chapters statistical description and statistical analysis have been treated in terms of one variable only. Thus, when we had a sample of houses, we considered only one variable at a time, e.g. rent. However, if we have a sample of houses, there are a number of variables which might be observed for each unit in the sample, e.g. rent, age, size, etc. We can thus distinguish between problems involving one variable, and those involving many variables. In the former case, each unit under observation is associated with a measurement of one variable X; in the latter, each is associated with measurements of several variables X, Y, Z, etc.

For the present we restrict ourselves first to the two variable case. Suppose we have a group of N units; then for each unit we shall have a pair of values of the variables X and Y (e.g. age of house and rent). The group can be described with respect to the X- and Y-variable separately by the ordinary measures we have so far discussed, e.g. $\bar{X}$, $\bar{Y}$, s_X, s_Y, but we now need an additional measure (or measures) to describe the possible relationship between X and Y. This section and the following concentrate on the *description* of two variable data and parallels Chapter V for one variable data. The *analysis* of the data is to be found in sections 10.3 to 10.6.

Suppose, for example, we observe 11 houses and find that

Table 10.1

House	Age (years)	Monthly Rent (dollars)
1st	3	50
2nd	12	32
3rd	5	40
4th	7	33
5th	8	45
6th	19	13
7th	10	30
8th	22	14
9th	15	28
10th	8	51
11th	25	26

In this case we can call the age in years, X, and the rent in dollars,

Y. We want to find some way of measuring the relationship between X and Y. Given appropriate measures, we could say:

(i) whether there is a relationship between age and rent and, if so, whether it is weak or strong;

(ii) what rent on the average is associated with each age;

(iii) what we can expect the rent of a house of a particular age to be, i.e. we can use the relationship for the purpose of prediction.

One can always get some idea whether there is any relationship present by plotting the values on a *scatter diagram*. We measure the X-variable on the horizontal and the Y-variable on the vertical axis and plot a point for each pair of X and Y values. This is done for the data of Table 10.1 in Fig. 10.1.

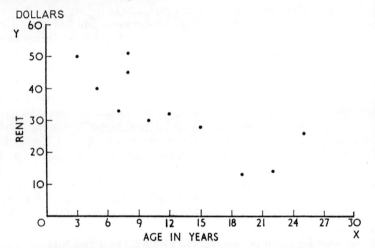

Fig. 10.1. Rent and Age of a Group of Eleven Houses

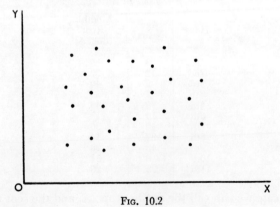

Fig. 10.2

It is evident that in this case there is some inverse relationship between rent and age, i.e. the greater the age the lower the rent.

If we had a large number of observations there would be a large number of points in the diagram. If there were no relationship we should expect the scatter to look as shown in Fig. 10.2. Here there is no evidence of a tendency for the Y values to be related to the X values in any particular way. If there were a perfect relationship between X and Y, so that Y is determined when X is given, we might get

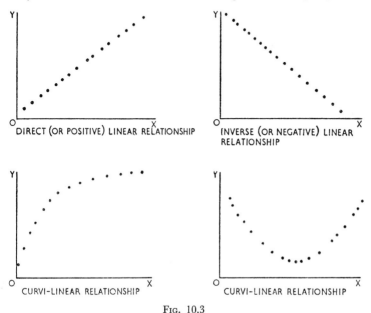

Fig. 10.3

In practice relationships are seldom perfect, and they tend to show up as in Fig. 10.4.

Linear Regression of Y on X

We consider here only the case of linear (straight line) relationships. Suppose that we have N pairs of observations, $X_1Y_1, X_2Y_2 \ldots X_nY_n$ which appear on the scatter diagram shown in Fig. 10.5.

Some relationship between X and Y is evident. If this relationship is linear, it can be written

$$Y_c = a + bX$$

this being the equation to a straight line. In this equation a and b are constants which determine the position of the line, a being the intercept of the straight line on the Y-axis and b its slope, i.e. the change in Y per unit change in X. The symbol Y_c stands for the value of Y *computed* from the relationship for a given X. Since the relationship is not perfect but expresses only a general tendency of association between X and Y, observed values of Y will not necessarily be the same as Y_c for given X. In Fig. 10.5 below a straight line relationship is drawn in. It

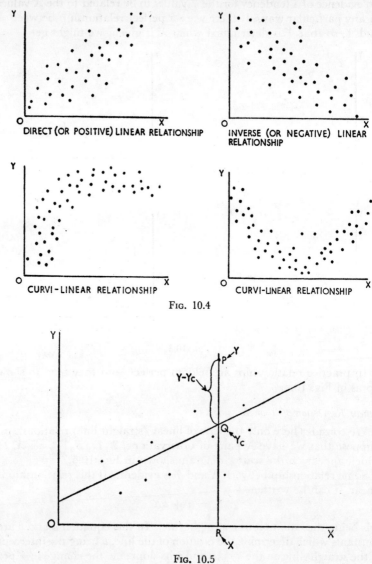

Fig. 10.4

Fig. 10.5

indicates the average relationship between X and Y. The point P is a particular observation of X and Y. The distance OR measures the value of X. With this value of X, the value of Y computed from the relationship (i.e. Y_c) is given by RQ, whereas the unit for which OR is the observed X value has an observed Y value of RP. We use the symbol Y with the subscript c to indicate Y values for given X computed from the relationship, and without the subscript to indicate observed values of Y.

The above equation expresses the average relationship between Y and given X's and is called the *linear regression of Y on X*. The constant b is known as the *regression coefficient of Y on X*. The question is how shall we fit such a line to given plot points, i.e. how shall we determine the constants a and b.

Clearly we want our line to be as close as possible on the whole to the various plot points, i.e. we want to minimize the overall discrepancy between the plot points and the line. To do this we must have some measure of this overall discrepancy. For a given X, we can measure the discrepancy between the corresponding observed Y and the corresponding point on the line by the deviation of the point from the line, i.e. by $Y - Y_c$ (the distance PQ on the above diagram). It is tempting to suggest that the sum of all such deviations would be a measure of the overall discrepancy of the points from the line, i.e. $\Sigma(Y - Y_c)$. However, since in this sum minus deviations and plus deviations from the fitted line will tend to cancel out, the sum $\Sigma(Y - Y_c)$ cannot be used as a measure of overall discrepancy, any more than the sum $\Sigma(X - \bar{X})$ could be used in Chapter V to measure overall dispersion of a variable X from its mean. It is natural, therefore, to square the deviations to get rid of the minus signs and to use $\Sigma(Y - Y_c)^2$ as a measure of overall discrepancy of the points from the line. If we now fix the line so that $\Sigma(Y - Y_c)^2$ is made as small as possible, we shall obtain a line which fits the points well; for the discrepancy between points and line, as measured by $\Sigma(Y - Y_c)^2$, will be minimized. Such a line will be representative of the relationship between X and Y. This is analogous to using $\bar{X}$ as representative of a set of values of the variable X, for it will be recalled (see section 5.4, p. 54) that the sum of the squares of the deviations of a variable X from an origin is smallest when the origin is the mean, i.e. $\Sigma(X - A)^2$ is a minimum for $A = \bar{X}$.

Let us now recapitulate. We have N pairs of observations $X_1 Y_1$, $X_2 Y_2 \ldots X_n Y_n$. We wish to fit a line of relationship to these points. The line is to be $Y_c = a + bX$, and a and b are to be selected so that $\Sigma(Y - Y_c)^2$ is a minimum, i.e. the sum of the squares of the vertical deviations of the observed points from the line is to be a minimum. This procedure is known as fitting a curve by the *method of least squares*.

We require $W = \Sigma(Y - Y_c)^2$
$\qquad\qquad = \Sigma(Y - a - bX)^2$ to be a minimum.

Differentiating W with respect to a and b, we have

$$\frac{\partial W}{\partial a} = -2\Sigma(Y - a - bX)$$

and
$$\frac{\partial W}{\partial b} = -2\Sigma X(Y - a - bX)$$
$$= -2\Sigma(XY - aX - bX^2)$$

For W to be a minimum $\dfrac{\partial W}{\partial a}$ and $\dfrac{\partial W}{\partial b}$ must both equal 0, which they will do when

$$\Sigma(Y - a - bX) = 0,$$
and $\qquad\Sigma(XY - aX - bX^2) = 0$

i.e. when
$$\Sigma Y = Na + b\Sigma X$$
$$\Sigma XY = a\Sigma X + b\Sigma X^2$$

These two equations are known as the *normal equations* for determining a and b. If we determine the numerical values of a and b such that these equations hold, the least squares equation $Y_c = a + bX$ will satisfy two algebraic properties. First, the deviations of observations about the regression line sum to zero; secondly, the sum of squared deviations from the line, i.e. $\Sigma(Y - Y_c)^2$ is a minimum. An equation fitted by the method of least squares may then be regarded as representative of the relationship between X and Y.

Solving the normal equations for a and b, we have from the first equation that
$$a = \bar{Y} - b\bar{X}$$
and from the second
$$\Sigma XY = \bar{Y}\Sigma X - b\bar{X}\Sigma X + b\Sigma X^2$$

i.e.
$$b = \frac{\Sigma XY - N\bar{X}\bar{Y}}{\Sigma X^2 - N\bar{X}^2}$$

The expression for b can be further simplified by putting $x = X - \bar{X}$ and $y = Y - \bar{Y}$, i.e. x and y are the deviations of X and Y from their respective means. Now

$$X = x + \bar{X}, \text{ and } Y = y + \bar{Y}$$

so that
$$\Sigma XY = \Sigma xy + \bar{X}\Sigma x + \bar{Y}\Sigma y + N\bar{X}\bar{Y}$$
$$= \Sigma xy + N\bar{X}\bar{Y} \quad \text{(since } \Sigma x \text{ and } \Sigma y = 0\text{)}$$

and similarly $\quad \Sigma X^2 = \Sigma x^2 + N\bar{X}^2$

REGRESSION AND CORRELATION

Hence
$$b = \frac{\Sigma xy}{\Sigma x^2}$$

Accordingly we can write the regression of Y on X as

$$Y_c = \bar{Y} - b\bar{X} + bX \quad \text{(since } a = \bar{Y} - b\bar{X}\text{)}$$

i.e.
$$(Y_c - \bar{Y}) = b(X - \bar{X})$$

where
$$b = \frac{\Sigma xy}{\Sigma x^2}$$

Detailed consideration of the interpretation of the regression equation $Y_c = a + bX$ is reserved for sections 10.3 and 10.4 below. For the present it suffices to regard the equation as a measure of the average relationship between X and Y, such that for a given X, Y_c is the value of Y which we would on the average expect to be associated with that X. The regression coefficient b measures the change in Y which occurs on the average per unit change in X. Y_c is, of course, expressed in the same units as Y. It will be noted that when $X = \bar{X}$, $Y_c = \bar{Y}$, so that the regression line passes through the means of the X's and Y's.

If there is no relationship between X and Y, the points on the scatter diagram will not show any particular pattern (see Fig. 10.2 above). Consequently, values of X above their mean $\bar{X}$ will be paired with values of Y above their mean $\bar{Y}$ as often as they are paired with values below $\bar{Y}$, and similarly for values of X below $\bar{X}$, hence $\Sigma xy = \Sigma(X - \bar{X})(Y - \bar{Y})$ will be approximately zero, since positive and negative xy's will cancel out. In this case b will be zero, and the regression line will be the horizontal line $Y_c = \bar{Y}$. Such a line indicates that the value of Y does not depend on the associated value of X, i.e. that there is no relationship between X and Y.

The actual computation of $\frac{\Sigma xy}{\Sigma x^2}$ is rendered simple by working from arbitrary origins. In machine computations it is often simplest to choose arbitrary origins of zero. In this case we can apply directly the results obtained from the normal equations, i.e.

$$b = \frac{\Sigma XY - N\bar{X}\bar{Y}}{\Sigma X^2 - N\bar{X}^2}$$

where ΣX^2, ΣXY, $\bar{X}$ and $\bar{Y}$ are readily obtainable from the observed data. For arbitrary origins other than zero, we write

$$x' = X - A, \text{ and } y' = Y - B$$

where A and B are arbitrary origins, so that

$$\bar{x}' = \bar{X} - A, \text{ and } \bar{y}' = \bar{Y} - B$$

Then
$$xy = (X - \bar{X})(Y - \bar{Y})$$
$$= (x' - \bar{x}')(y' - \bar{y}')$$
$$= x'y' - \bar{x}'y' - x'\bar{y}' + \bar{x}'\bar{y}'$$
$$\therefore \quad \Sigma xy = \Sigma x'y' - N\bar{x}'\bar{y}' - N\bar{x}'\bar{y}' + N\bar{x}'\bar{y}'$$
$$= \Sigma x'y' - N\bar{x}'\bar{y}'$$

As far as Σx^2 is concerned, we have already shown (see section 5.5, p. 71 above) that
$$\Sigma x^2 = \Sigma x'^2 - N\bar{x}'^2$$

These two results give Σxy and Σx^2 in terms of arbitrary deviations. The fitting of a regression line to actual data by this method is illustrated in the example below.

EXAMPLE 10.1

The data in the table refer to a group of eleven tenanted houses.
(a) Plot the data on a scatter diagram.
(b) Calculate the linear regression equation of rent on age of houses.
(c) Draw in this equation on the scatter diagram.

Age (years)	Rent (dollars)	Arbitrary Deviations $A = 12$ $B = 33$				
X	Y	x'	y'	$x'y'$	x'^2	y'^2
3	50	−9	17	−153	81	289
12	32	0	−1	0	0	1
5	40	−7	7	−49	49	49
7	33	−5	0	0	25	0
8	45	−4	12	−48	16	144
19	13	7	−20	−140	49	400
10	30	−2	−3	6	4	9
22	14	10	−19	−190	100	361
15	28	3	−5	−15	9	25
8	51	−4	18	−72	16	324
25	26	13	−7	−91	169	49
		2	−1	−752	518	1651

The primary calculations are as follows. The last column in the above table and the sum $\Sigma y'^2$ are required for subsequent examples.

$$\bar{x}' = \frac{\Sigma x'}{N} = 0{\cdot}1818, \quad \bar{y}' = \frac{\Sigma y'}{N} = -0{\cdot}0909$$
$$\bar{X} = A + \bar{x}' = 12{\cdot}1818 \text{ years}, \quad \bar{Y} = B + \bar{y}' = \$32{\cdot}9091$$

$\Sigma x^2 = \Sigma x'^2 - N\bar{x}'^2 = 518 - 0\cdot 3636 = 517\cdot 6364$

$\Sigma y^2 = \Sigma y'^2 - N\bar{y}'^2 = 1{,}651 - 0\cdot 0909 = 1{,}650\cdot 9091$

$\Sigma xy = \Sigma x'y' - N\bar{x}'\bar{y}' = -752 + 0\cdot 1818 = -751\cdot 8182$

The regression equation of Y on X is

$$Y_c = a + bX$$

i.e.
$$(Y_c - \bar{Y}) = b(X - \bar{X})$$

where $\bar{Y} = 32\cdot 9091, \quad \bar{X} = 12\cdot 1818$

and
$$b = \frac{\Sigma xy}{\Sigma x^2} = -\frac{751\cdot 8182}{517\cdot 6364} = -1\cdot 4524$$

i.e. $Y_c - 32\cdot 9091 = -1\cdot 4524 (X - 12\cdot 1818)$

so that $Y_c = 50\cdot 60 - 1\cdot 452 X$

is the required regression of rent on age.

This can be interpreted by saying that the data under consideration suggest that for an increase of one year in the age of a house we should expect rent on the average to fall by $1·45.

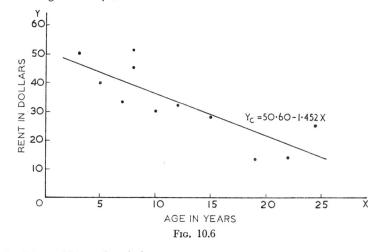

Fig. 10.6

Coefficient of Linear Correlation

The line $Y_c = a + bX$ measures the average relationship between Y and given X's. It does not measure the strength of the relationship. This depends on how well the line describes the relationship, i.e. how closely the points cluster about the line.

We may look at this matter in the following way. If we have N values of Y, they will vary amongst themselves, and a measure of this variation is $\Sigma(Y - \bar{Y})^2$. But some of this variation may be due to the fact that the Y's are associated with X's, and the X's are varying. If the relationship between the X's and Y's were quite perfect so that a particular Y is always associated with a particular X, all the variability of the Y's would be explained by the relationship. On the other hand, if there were no relationship, none of the variability could be thus explained. In Fig. 10.7 are plotted a number of observations and the regression of Y on X computed from them. A horizontal line at $\bar{Y}$ and a vertical line at $\bar{X}$ have been drawn in. As has been noted above, the regression line must pass through the intersection of these two lines. The point P is a particular observation of X and Y. The deviation of the Y value at point P from the mean of the Y's is PS, and this can be split up into two parts—one, the deviation of the regression line from $\bar{Y}$, i.e. QS, and the other, the deviation of the point from the regression line, i.e. PQ. The part QS arises because the X value of the point P is greater than $\bar{X}$ (R lies to the right of $\bar{X}$), and hence the corresponding Y_c must be greater than $\bar{Y}$, since, as has been shown above, the regression equation is $Y_c - \bar{Y} = b(X - \bar{X})$, and in this case b is positive. Consequently, this part of the deviation of PS occurs because, given the relationship between X and Y, the point P has a particular X value. It can thus be regarded as being "explained" by the regression. The other part of the deviation is not so explained. It results from the inherent variability of the Y's for a *given* X.

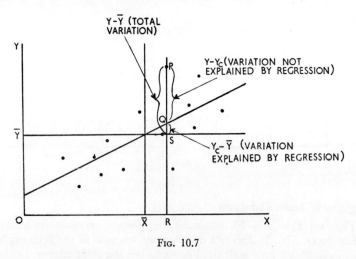

Fig. 10.7

As can be seen from the diagram, the deviation of a particular Y from the mean of the Y's, can be broken up as follows—

$$Y - \bar{Y} = (Y - Y_c) + (Y_c - \bar{Y})$$
$$\therefore \ (Y - \bar{Y})^2 = (Y - Y_c)^2 + (Y_c - \bar{Y})^2 + 2(Y - Y_c)(Y_c - \bar{Y})$$

Hence, summing for the N pairs of observations,

$$\Sigma(Y - \bar{Y})^2 = \Sigma(Y - Y_c)^2 + \Sigma(Y_c - \bar{Y})^2 + 2\Sigma(Y - Y_c)(Y_c - \bar{Y})$$

Consider the last term on the right-hand side, and bear in mind that

$$Y_c - \bar{Y} = b(X - \bar{X}), \quad \text{where} \quad b = \frac{\Sigma xy}{\Sigma x^2}$$

Then

$$\begin{aligned}\Sigma(Y - Y_c)(Y_c - \bar{Y}) &= \Sigma\Big([Y - \bar{Y} - b(X - \bar{X})][b(X - \bar{X})]\Big) \\ &= \Sigma\Big((y - bx)(bx)\Big) \\ &= b\Sigma xy - b^2\Sigma x^2 \\ &= \frac{\Sigma xy}{\Sigma x^2}\Sigma xy - \left(\frac{\Sigma xy}{\Sigma x^2}\right)^2 \Sigma x^2 \\ &= 0\end{aligned}$$

Accordingly

$$\Sigma(Y - \bar{Y})^2 = \Sigma(Y - Y_c)^2 + \Sigma(Y_c - \bar{Y})^2$$

We thus have the variability of the Y's about their mean $\bar{Y}$, as measured by $\Sigma(Y - \bar{Y})^2$, resolved into two components, one, $\Sigma(Y - Y_c)^2$, measuring the variability of the Y's about the line of regression and the other, $\Sigma(Y_c - \bar{Y})^2$, measuring the variability of the Y's computed from the line of regression about $\bar{Y}$. This second component measures the amount of variability of the Y's which is *explained* by the line of relationship in the sense that it is possible to attribute this amount to the effect of the relationship in translating variability in the X's into variability in the Y's. The first component measures the amount of *unexplained* variability.[1] The stronger the relationship, i.e. the more closely the points are clustered about the line, the larger will be the second component relative to the first.

In Table 10.2, the partition of total variation of Y into these two components is verified arithmetically with reference to our example of eleven rents. The first column records the actual observations and the second the computed values obtained by substituting the given values of X, in ascending order, into the regression equation $Y_c = 50.60 - 1.452X$. From section 5.4 we know that the sum of values from their mean is zero, and since in fitting least squares equations $\Sigma(Y - Y_c)$ must likewise be zero, we can see from Fig. 10.7 that the same must be true of the sum $\Sigma(Y_c - \bar{Y})$. Thus in Table 10.2 the sums of the three middle columns differ from zero only by small rounding errors.

[1] See footnote on page 175 above.

Table 10.2
Computation of Unexplained, Explained and Total Variation: Regression of Rent on Age of Houses

Rent		Deviations			Variation		
Observed r	Computed $\hat{r}_c$	$r - r_c$	$r_c - \bar{r}$	$r - \bar{r}$	Unexplained $(r - r_c)^2$	Explained $(r_c - \bar{r})^2$	Total $(r - \bar{r})^2$
50	46·2448	3·7552	13·3357	17·0909	14·1015	177·8409	292·0989
40	43·3400	−3·3400	10·4309	7·0909	11·1556	108·8037	50·2809
33	40·4352	−7·4352	7·5261	0·0909	55·2822	56·6422	0·0082
51	38·9828	12·0172	6·0737	18·0909	144·4131	36·8898	327·2807
45	38·9828	6·0172	6·0737	12·0909	36·2067	36·8898	146·1899
30	36·0780	−6·0780	3·1689	−2·9091	36·9421	10·0419	8·4629
32	33·1732	−1·1732	0·2641	−0·9091	1·3764	0·0697	0·8265
28	28·8160	−0·8160	−4·0931	−4·9091	0·6659	16·7535	24·0993
13	23·0064	−10·0064	−9·9027	−19·9091	100·1280	98·0635	396·3723
14	18·6492	−4·6492	−14·2599	−18·9091	21·6151	203·3447	357·5541
26	14·2920	11·7080	−18·6171	−6·9091	137·0773	346·5964	47·7357
		−0·0004	0·0003	−0·0001	558·9639	1,091·9361	1,650·9094 (a)

(a) Because of rounding errors, there is a slight discrepancy between the sum of unexplained and explained variation and the amount of total variation recorded in the last column.

Dividing both sides of the equation derived in the preceding paragraphs by $\Sigma(Y - \bar{Y})^2$, we have

$$\frac{\Sigma(Y - Y_c)^2}{\Sigma(Y - \bar{Y})^2} + \frac{\Sigma(Y_c - \bar{Y})^2}{\Sigma(Y - \bar{Y})^2} = 1 \qquad (*)$$

The first term gives the proportion of variation unexplained by the relationship, and the second the proportion explained. The stronger the relationship the larger the second component and the smaller the first. We now define

$$r = \sqrt{\frac{\Sigma(Y_c - \bar{Y})^2}{\Sigma(Y - \bar{Y})^2}}$$

as the *coefficient of linear correlation*, measuring the degree of linear relationship between X and Y. Bearing in mind that $Y_c - \bar{Y} = b(X - \bar{X})$, we have

$$r = \sqrt{\frac{\Sigma(Y_c - \bar{Y})^2}{\Sigma(Y - \bar{Y})^2}}$$

$$= \sqrt{\frac{\Sigma b^2 x^2}{\Sigma y^2}}$$

$$= \frac{\Sigma xy}{\sqrt{\Sigma x^2 \Sigma y^2}}$$

The sign of r is taken as the same as that of Σxy (i.e. as the same as that of b). We know from the expression (*) above that r^2 cannot exceed unity, and hence $-1 \leqslant r \leqslant +1$. For perfect correlation $\Sigma(Y - Y_c)^2 = 0$, since the points all lie in the regression line, so that $r^2 = 1$. Perfect negative correlation is indicated by $r = -1$ and perfect positive correlation by $r = 1$. For no relationship, the regression line is horizontal and coincides with the mean $\bar{Y}$, so that $\Sigma(Y_c - \bar{Y})^2 = 0$, and $r = 0$. The square of r, known as *the coefficient of determination*, tells us the fraction of total variability of Y explained by the relationship between Y and X.

The coefficient of determination r^2 can be obtained directly from the sums of squares $\Sigma(Y_c - \bar{Y})^2$ and $\Sigma(Y - \bar{Y})^2$ given in Table 10.2. However, this method is computationally arduous. In practice, to compute r we make use of the results

$$\Sigma xy = \Sigma x'y' - N\bar{x}'\bar{y}'$$
$$\Sigma x^2 = \Sigma x'^2 - N\bar{x}'^2$$
$$\Sigma y^2 = \Sigma y'^2 - N\bar{y}'^2$$

EXAMPLE 10.2

Calculate the coefficients of determination and linear correlation between rent and age of houses for example 10.1 above.

208   APPLIED STATISTICS FOR ECONOMISTS

We have $\quad \Sigma xy = -751 \cdot 8182, \quad \Sigma x^2 = 517 \cdot 6364, \quad \Sigma y^2 = 1650 \cdot 9091$

so that $\quad r^2 = \dfrac{(\Sigma xy)^2}{\Sigma x^2 \Sigma y^2} = 0 \cdot 6614$

Verifying this result, from Table 10.2 we have

$$r^2 = \frac{\Sigma(Y_c - \bar{Y})^2}{\Sigma(Y - \bar{Y})^2} = \frac{1{,}091 \cdot 9361}{1{,}650 \cdot 9094} = 0 \cdot 6614$$

We can say that 66 per cent of the variability of rents in this case can be explained by the relation between rent and age; $r = -0 \cdot 81$ is the coefficient of linear correlation between rents and age.

Standard Error of Estimate

As well as having a line of regression of Y on X and a measure of the degree of relationship, for certain purposes we want a measure of the absolute dispersion of the Y values about the line. Such a measure is analogous to the standard deviation calculated for a single variable measuring the dispersion of the X's about $\bar{X}$. We have already come across the value $\Sigma(Y - Y_c)^2$, which indicates the variability of the Y's about the line. We now write

$$s_Y(e) = \sqrt{\frac{\Sigma(Y - Y_c)^2}{N - 2}}$$

as a measure of dispersion about the line. This is called the *standard error of estimate of the regression of Y on X*. The divisor is $N - 2$ to avoid bias, just as the divisor in the formula for the standard deviation is $N - 1$. The measure $s_Y(e)$ can be computed as follows—

$$\begin{aligned}
\Sigma(Y - Y_c)^2 &= \Sigma[(Y - \bar{Y}) - b(X - \bar{X})]^2 \\
&= \Sigma(y - bx)^2 \\
&= \Sigma y^2 - b(2\Sigma xy - b\Sigma x^2) \\
&= \Sigma y^2 - b(2\Sigma xy - \Sigma xy) \\
&= \Sigma y^2 - b\Sigma xy
\end{aligned}$$

Dividing through by Σy^2, we may also write

$$\begin{aligned}
\Sigma(Y - Y_c)^2 &= \left[1 - \frac{(\Sigma xy)^2}{\Sigma x^2 \Sigma y^2}\right] \Sigma y^2 \\
&= (1 - r^2) \Sigma y^2
\end{aligned}$$

Hence $\quad s_Y(e) = \sqrt{\dfrac{\Sigma y^2 - b\Sigma xy}{N - 2}}$

$\qquad\qquad\qquad = \sqrt{\dfrac{(1 - r^2)\Sigma y^2}{N - 2}}$

Example 10.3

Calculate the standard error of estimate of the regression of rent on age of houses for example 10.1 above.

We have
$$s_Y(e) = \sqrt{\frac{\Sigma y^2 - b\Sigma xy}{N-2}}$$
$$= \sqrt{\frac{1{,}650{\cdot}9091 - (-1{\cdot}4524 \times -751{\cdot}8182)}{9}}$$
$$= \$7{\cdot}88$$

Using the results of Table 10.2, it can be easily checked that

$$s_Y(e) = \sqrt{\frac{\Sigma(Y - Y_c)^2}{N-2}} = \sqrt{\frac{558{\cdot}9639}{9}}$$
$$= \$7{\cdot}88$$

Linear Regression of X on Y

So far we have considered the regression of Y on X, i.e. we treated X as the independent variable and asked how Y is related to given X's. But we might look at things the other way around. We might take Y as the independent variable and consider the regression of X on Y

i.e.
$$X_c = c + dY$$

This equation is known as the *linear regression of X on Y* and d is the *regression coefficient of X on Y*. The constants c and d are determined such that $\Sigma(X - X_c)^2$ is a minimum. It can soon be appreciated that the two lines of regression will not coincide, unless the regression is perfect. This can be seen from the following simple case in Fig. 10.8, taking three points only.

The regression of Y on X is found by minimizing in the vertical direction and is given by the solid line, but the regression of X on Y is found by minimizing in the horizontal direction and is given by the broken line. In the diagram the sum of the deviations $\Sigma(X - X_c)^2$ is obviously less from the broken than the solid line.

The constants c and d can be found as before, simply by switching the X's and Y's. The regression of X on Y will also pass through the two means, so that the two regression lines intersect at $(\bar{X}, \bar{Y})$. We have

$$X_c - \bar{X} = d(Y - \bar{Y})$$

or
$$X_c = \bar{X} - d\bar{Y} + dY$$

where
$$d = \frac{\Sigma xy}{\Sigma y^2}$$

Similarly it can be shown that
$$\Sigma(X - \bar{X})^2 = \Sigma(X - X_c)^2 + \Sigma(X_c - \bar{X})^2$$
and r can be defined as $\quad r = \sqrt{\dfrac{\Sigma(X_c - \bar{X})^2}{\Sigma(X - \bar{X})^2}}$

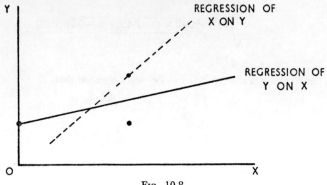

Fig. 10.8

This reduces to $\quad r = \dfrac{\Sigma xy}{\sqrt{\Sigma x^2 \Sigma y^2}}$

as before. The interpretation of r is precisely as before.

Also $\quad s_X(e) = \sqrt{\dfrac{\Sigma(X - X_c)^2}{N - 2}}$

where $\quad \Sigma(X - X_c)^2 = (1 - r^2)\Sigma x^2$
$\qquad\qquad\qquad\qquad = \Sigma x^2 - d\Sigma xy$

Further since $\quad b = \dfrac{\Sigma xy}{\Sigma x^2} \quad$ and $\quad d = \dfrac{\Sigma xy}{\Sigma y^2}$

then $\quad r = \sqrt{bd}$

The two lines will coincide when $b = \dfrac{1}{d}$. In such a situation $r = 1$, and the relationship is perfect.

We may also note that
$$b = \dfrac{\Sigma xy}{\Sigma x^2}$$
$$= \dfrac{\Sigma xy}{\sqrt{\Sigma x^2 \Sigma y^2}} \dfrac{\sqrt{\dfrac{\Sigma y^2}{N-1}}}{\sqrt{\dfrac{\Sigma x^2}{N-1}}}$$
$$= r\dfrac{s_Y}{s_X}$$

and
$$d = r\frac{s_X}{s_Y}$$

where s_X and s_Y are the standard deviations of the X and Y variables respectively. When $r = 0$, both b and d will be zero, so that the regression lines reduce to lines drawn at right angles through $\bar{X}$ and $\bar{Y}$.

EXAMPLE 10.4

Calculate for example 10.1, p. 202, the linear regression of age on rent of houses and the standard error of estimate of this regression. Draw in the regression lines of both rent on age and age on rent on the same scatter diagram.

The regression equation of X on Y is
$$X_c = c + dY$$
i.e.
$$(X_c - \bar{X}) = d(Y - \bar{Y})$$
where $\bar{X} = 12\cdot1818$ years, $\bar{Y} = \$32\cdot9091$

and
$$d = \frac{\Sigma xy}{\Sigma y^2} = -\frac{751\cdot8182}{1{,}650\cdot9091} = -0\cdot4554$$

i.e.
$$X_c - 12\cdot1818 = -0\cdot4554(Y - 32\cdot9091)$$
so that
$$X_c = 27\cdot17 - 0\cdot4554 Y$$

is the required regression of age on rent.

This can be interpreted by saying that the data under consideration suggest that for an increase of one dollar in the rent of a house we should expect the house to be on the average 0·46 years younger.

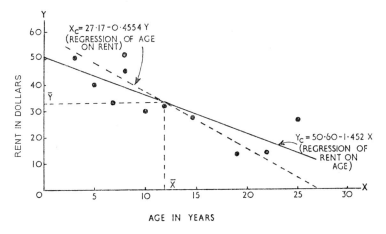

FIG. 10.9

The standard error of estimate of the regression of age on rent of houses is given by

$$s_X(e) = \sqrt{\frac{\Sigma x^2 - d\Sigma xy}{N-2}}$$

$$= \sqrt{\frac{517\cdot6364 - (-0\cdot4554 \times -751\cdot8182)}{9}}$$

$$= 4\cdot41 \text{ years}$$

10.2. Descriptive Measures of Regression and Correlation for Grouped Data

In the one variable case when we moved from ungrouped to grouped data, we introduced the frequency distribution. This frequency distribution involved a one-way classification, i.e. it consisted of items classified into groups according to one criterion of classification. In the two variable case the frequency distribution will be a two-way classification as follows—

		Y Classes		All Classes
X	Classes		f_{XY}	f_X
	All Classes		f_Y	N

With grouped data, X and Y are class mid-points. We write f_{XY} for the frequency in the cell with X and Y as mid-points, f_X for the frequency in the class with X as mid-point irrespective of Y, and f_Y for the frequency in the class with Y as mid-point irrespective of X. It follows that the f_X's and f_Y's are marginal totals and that $\Sigma f_X = \Sigma f_Y = N$, where N is the total number of observations. The effect on the formulae in section 10.1 above is slight. We merely replace Σxy with $\Sigma f_{XY}xy$,

REGRESSION AND CORRELATION 213

Σx^2 with $\Sigma f_X x^2$ and Σy^2 with $\Sigma f_Y y^2$, where x and y are the deviations of class mid-points from $\bar{X}$ and $\bar{Y}$ respectively.

It is usually convenient to work in class interval units. If h_X is the width of the X class interval and h_Y of the Y class interval, then we must multiply $\Sigma f_{XY} xy$ by $h_X h_Y$, $\Sigma f_X x^2$ by h_X^2 and $\Sigma f_Y y^2$ by h_Y^2 to obtain original units. This means that to obtain b we multiply the calculation in class interval units by $\dfrac{h_X h_Y}{h_X^2}$, i.e. by $\dfrac{h_Y}{h_X}$, and to obtain d we multiply by $\dfrac{h_X}{h_Y}$. The correlation coefficient is invariant with respect to the units in which X and Y are expressed and no adjustment is necessary.

EXAMPLE 10.5

The following data refer to the monthly rents paid for and the ages of a group of 112 houses—

Rent in Dollars

Age in Years	10 and under 20	20 and under 30	30 and under 40	40 and under 50	50 and under 60	All Rents
0 and under 3	–	2	2	1	1	6
3 ,, 6	–	2	5	8	4	19
6 ,, 9	1	4	20	10	2	37
9 ,, 12	1	4	8	7	2	22
12 ,, 15	3	3	5	4	1	16
15 ,, 18	3	5	3	–	1	12
All Ages	8	20	43	30	11	112

Calculate—
(a) The linear regression of rent on age.
(b) The linear regression of age on rent.
(c) The coefficient of linear correlation between rent and age.
(d) The standard error of estimate of the regression of rent on age.
(e) The standard error of estimate of the regression of age on rent.

We set out the table as shown on p. 214.

In the table, f stands for f_Y (horizontally) and f_X (vertically), d' stands for y' (horizontally) and x' (vertically) in class interval units, and the body of the table contains f_{XY}, in parentheses.

The sums $\Sigma f_X x'$, $\Sigma f_Y y'$, $\Sigma f_X x'^2$, $\Sigma f_Y y'^2$ are readily obtained by working in the margins of the table. To obtain $\Sigma f_{XY} x'y'$, first write the value of $x'y'$ in the top left-hand corner of each cell, and then the product $f_{XY} x'y'$ can be written in the bottom right-hand corner. Addition of these amounts for all cells gives $\Sigma f_{XY} x'y'$. The primary calculations are as follows—

Here $\bar{x}' = \dfrac{\Sigma f_X x'}{N} = 0 \cdot 5268, \quad \bar{y}' = \dfrac{\Sigma f_Y y'}{N} = 0 \cdot 1429$

Rent in Dollars (Y)
(Arbitrary Origin $B = 35$)

						10–	20–	30–	40–	50–	All Rents
	Mid-point					15	25	35	45	55	
		f				8	20	43	30	11	112
			d'			–2	–1	0	1	2	
				fd'		–16	–20	0	30	22	16
					fd'^2	32	20	0	30	44	126
Class	Mid-point	f	d'	fd'	fd'^2						
0–	1·5	6	–2	–12	24	4	2 (2) 4	0 (2) 0	–2 (1) –2	–4 (1) –4	–2
3–	4·5	19	–1	–19	19	2	1 (2) 2	0 (5) 0	–1 (8) –8	–2 (4) –8	–14
6–	7·5	37	0	0	0	0 (1) 0	0 (4) 0	0 (20) 0	0 (10) 0	0 (2) 0	0
9–	10·5	22	1	22	22	–2 (1) –2	–1 (4) –4	0 (8) 0	1 (7) 7	2 (2) 4	5
12–	13·5	16	2	32	64	–4 (3) –12	–2 (3) –6	0 (5) 0	2 (4) 8	4 (1) 4	–6
15–	16·5	12	3	36	108	–6 (3) –18	–3 (5) –15	0 (3) 0	3	6 (1) 6	–27

Age in Years (X)
Arbitrary Origin $A = 7·5$

and $\bar{X} = A + (\bar{x}' \times h_X) = 9.08$ years
and $\bar{Y} = B + (\bar{y}' \times h_Y) = \$36.43$

$\Sigma f_X x^2 = [\Sigma f_X x'^2 - N\bar{x}'^2] \times h_X^2$
$= (237 - 31.0820) \times 9 = 205.9180 \times 9$

$\Sigma f_Y y^2 = [\Sigma f_Y y'^2 - N\bar{y}'^2] \times h_Y^2$
$= (126 - 2.2871) \times 100 = 123.7129 \times 100$

$\Sigma f_{XY} xy = [\Sigma f_{XY} x'y' - N\bar{x}'\bar{y}'] \times h_X h_Y$
$= (-44 - 8.4313) \times 30 = -52.4313 \times 30$

(a) The linear regression of rent on age is given by
$$Y_c - \bar{Y} = b(X - \bar{X})$$
where
$\bar{Y} = \$36.43$, $\bar{X} = 9.08$ years, $b = \dfrac{\Sigma f_{XY} xy}{\Sigma f_X x^2} = \dfrac{-52.4313}{205.9180} \times \dfrac{30}{9} = -0.8487$

i.e. $Y_c = 44.14 - 0.8487 X$

(b) The linear regression of age on rent is given by
$$X_c - \bar{X} = d(Y - \bar{Y})$$
where
$\bar{X} = 9.08$ years, $\bar{Y} = \$36.43$, $d = \dfrac{\Sigma f_{XY} xy}{\Sigma f_Y y^2} = \dfrac{-52.4313}{123.7129} \times \dfrac{30}{100} = -0.1271$

i.e. $X_c = 13.71 - 0.1271 Y$

(c) The coefficient of linear correlation between rent and age is given by
$$r = \dfrac{\Sigma f_{XY} xy}{\sqrt{\Sigma f_X x^2 \Sigma f_Y y^2}} = -0.33$$

(d) The standard error of estimate of the regression of rent on age is given by
$$s_Y(e) = \sqrt{\dfrac{\Sigma f_Y y^2 - b\Sigma f_{XY} xy}{N-2}} = \sqrt{100.3304} = \$10.02$$

(e) The standard error of estimate of the regression of age on rent is given by
$$s_X(e) = \sqrt{\dfrac{\Sigma f_X x^2 - d\Sigma f_{XY} xy}{N-2}} = \sqrt{15.0304} = 3.88 \text{ years}$$

We have now shown how to describe a group of N pairs of observations of two variables X and Y. Each variable can be described separately by the methods outlined in Chapter V, and the relationship (if any) between X and Y can be expressed by the regression equations and the strength of the relationship by the coefficient of correlation. The use of these measures of relationship for the purpose of analysis is discussed in the following sections.

10.3. Normal Probability Distribution of Two Variables

We first consider how the above work on regression and correlation can be fitted into our general notion of samples and populations. Just as we represented the population of a single variable by a continuous probability curve in two dimensions, so we can represent one in two variables by a probability surface in three dimensions.

The volume under this surface is unity and the probability of selecting an element at random from the population with its X value between

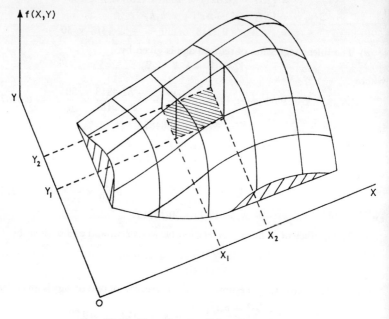

Fig. 10.10. The Normal Probability Surface

X_1 and X_2 *and* its Y value between Y_1 and Y_2 is given by the fraction of total volume standing above the area shaded in the diagram.

It will be recalled that the equation to the normal univariate (single variable) distribution is

$$f(X) = \frac{1}{\sqrt{2\pi}\,\sigma} e^{-\frac{1}{2}\left(\frac{X-\mu}{\sigma}\right)^2}$$

and that this distribution is a very important one. Similarly the normal bivariate distribution is important, and its equation is given by

$$f(X, Y) = \frac{1}{2\pi\sigma_X \sigma_Y \sqrt{1-\rho^2}} e^{-\frac{1}{2(1-\rho^2)}\left[\left(\frac{X-\mu_X}{\sigma_X}\right)^2 - 2\rho\left(\frac{X-\mu_X}{\sigma_X}\right)\left(\frac{Y-\mu_Y}{\sigma_Y}\right) + \left(\frac{Y-\mu_Y}{\sigma_Y}\right)^2\right]}$$

There are a number of points to note about this distribution. First, it depends on five parameters. These are μ_X, μ_Y, σ_X, σ_Y and ρ. The first two are the means and the second two the standard deviations of X and Y respectively. Only ρ is unfamiliar. It is the parameter of association between X and Y. It is, in fact, the *coefficient of linear correlation* of the population and corresponds to the statistic r defined in the preceding sections. It must lie within the range ± 1, and measures the extent to which X and Y are linearly associated.

If we take a particular value of X, we get a univariate normal distribution of the Y's for the given X. This can be seen by rearranging the above equation and writing it as

$$f(X, Y) = \left\{\frac{1}{\sqrt{2\pi}\sigma_X} e^{-\frac{1}{2}\left(\frac{X-\mu_X}{\sigma_X}\right)^2}\right\}$$
$$\left\{\frac{1}{\sqrt{2\pi}\sigma_Y\sqrt{1-\rho^2}} e^{-\frac{1}{2}\left[\frac{Y-\mu_Y-\rho\frac{\sigma_Y}{\sigma_X}(X-\mu_X)}{\sigma_Y\sqrt{1-\rho^2}}\right]^2}\right\}$$

For a given value of X, the first term in brackets is a constant and the second term is seen to be a univariate normal distribution in Y with mean $\mu_Y + \rho\frac{\sigma_Y}{\sigma_X}(X - \mu_X)$ and standard deviation $\sigma_Y\sqrt{1-\rho^2}$. The mean of this distribution can be written $\mu_Y(X)$, indicating that it depends on the particular value of X selected. It can be seen that the mean $\mu_Y(X)$ is linearly related to X, i.e. it can be written

$$\mu_Y(X) = \alpha + \beta X$$

where $\beta = \rho\frac{\sigma_Y}{\sigma_X}$ and $\alpha = \mu_Y - \beta\mu_X$.

This is the *regression of Y on X* in the population, giving the mean values of Y's for given X's. The standard deviation of the distribution of Y for a given X is given by

$$\sigma_Y(e) = \sigma_Y\sqrt{1-\rho^2}$$

which is the same irrespective of the value of X.

It follows that, if for example rent and age of houses are jointly distributed in a normal bivariate distribution and we take houses of a particular age, the distribution of rent of those houses will be an ordinary normal distribution, with a mean depending linearly on the particular age chosen and with a given standard deviation.

This can be illustrated diagrammatically in Fig. 10.11.

The diagram shows the regression of Y on X in the XY plane. For a given value of X, say X', there will be a distribution of Y's about a mean $\mu_Y(X')$ given by the regression line. This distribution, centred on the regression line, is shown in the diagram. It must be thought of as standing up in the third dimension, being a perpendicular cross-section in the Y-direction of the probability surface. Such cross-sections will, of course, be greater in size the nearer is X to its mean.

The standard deviation of the distribution of Y for a given X in the normal bivariate distribution is always the same irrespective of the particular value of the given X. This is illustrated in the diagram by the fact that the curves drawn there have identical absolute dispersion.

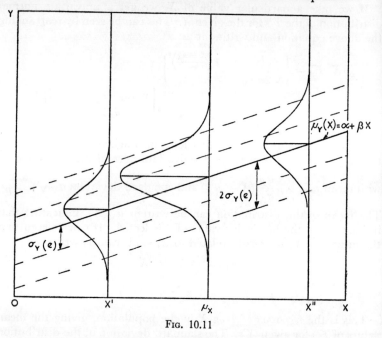

Fig. 10.11

Similarly, if we take a given value of Y, X will be normally distributed about a mean $\mu_X(Y)$ given by the line of regression of X on Y

$$\mu_X(Y) = \gamma + \delta Y$$

where $\gamma = \mu_X - \delta \mu_Y$ and $\delta = \rho \dfrac{\sigma_X}{\sigma_Y}$

with a standard error of estimate

$$\sigma_X(e) = \sigma_X \sqrt{1 - \rho^2}$$

Since for a given X, Y is normally distributed about a mean $\mu_Y(X) = \alpha + \beta X$ with a standard deviation $\sigma_Y(e) = \sigma_Y \sqrt{1 - \rho^2}$, the range $\mu_Y(X) \pm \sigma_Y(e)$ will contain about two-thirds of the Y values for the given X, and $\mu_Y(X) \pm 2\sigma_Y(e)$ will contain about 95 per cent of the values. It follows that bands running a distance $\sigma_Y(e)$ and $2\sigma_Y(e)$ from the regression of Y on X will contain approximately two-thirds and 95 per cent respectively of the population (in geometric terms two-thirds and 95 per cent of the volume) and similarly, *mutatis mutandis*, for the regression of X on Y. Such bands are shown in Fig. 10.11 above. If we pick units at random from the population, as the sample gets larger approximately two-thirds of the plot points will lie in the narrower and 95 per cent in the wider bands.

If $\rho = 0$, i.e. if there is no correlation, then

$$\mu_Y(X) = \mu_Y \quad \text{and} \quad \mu_X(Y) = \mu_X$$
and
$$\sigma_Y(e) = \sigma_Y \quad \text{and} \quad \sigma_X(e) = \sigma_X$$

In this case the value of X in no way affects the distribution of the Y's, the distributions of Y being identical for all given X's and vice versa. This is the case of complete independence of X and Y.

If we have a random sample of N pairs of observations from a bivariate normal population, we may wish to estimate the regression lines and the coefficient of linear correlation. The best way to do this is by the method of least squares, already outlined in section 10.1 above. Thus the equation $Y_c = a + bX$, calculated from the sample, is an estimate of the regression of Y on X in the population, namely $\mu_Y(X) = \alpha + \beta X$, and similarly for the regression of X on Y. Also, the correlation coefficient $r = \dfrac{\Sigma xy}{\sqrt{\Sigma x^2 \Sigma y^2}}$ is an estimate of the parameter ρ in the population. Thus a, b, c, d and r are estimates of α, β, γ, δ and ρ in the bivariate case, just as $\bar{X}$ and s are estimates of μ and σ in the univariate case. Similarly $s_Y(e)$ and $s_X(e)$, as defined in section 10.1, are estimates of the standard errors of estimates $\sigma_Y(e)$ and $\sigma_X(e)$. All these estimates, except r, are unbiased estimates of the corresponding population parameters, in the sense that with repeated sampling the means of the sample estimates will tend to the population parameters.

10.4. General Discussion on Regression and Correlation

Meaning of the Lines of Regression

In a normal bivariate population, the regression of Y on X is

$$\mu_Y(X) = \alpha + \beta X$$

This means that if we select units from the population at random all with a particular X value, then the mean value of the associated Y's will be given by the regression equation. Thus, for a given X, the equation tells us the mean value of the Y's we can expect to be associated with it. Furthermore, it tells us that for every unit increase in the X

value, the mean value of the associated Y values can be expected to increase by β units.

In practice we shall not be given the above equation, but only sample data. From these we estimate the line by the method of least squares, which gives

$$Y_c = a + bX$$

Here a and b are estimates of α and β.

Furthermore, for a given X, the Y's will be normally distributed about a mean of $\mu_Y(X) = \alpha + \beta X$ with a standard deviation, known as the standard error of estimate, of $\sigma_Y(e)$. In the population this is given by

$$\sigma_Y(e) = \sigma_Y \sqrt{1 - \rho^2}$$

and we estimate it from the sample by

$$s_Y(e) = \sqrt{\frac{(1 - r^2)\Sigma y^2}{N - 2}} = \sqrt{\frac{\Sigma y^2 - b\Sigma xy}{N - 2}}$$

This measures the standard deviation of the Y's for a given X, and it indicates the variability of the Y's about the line of regression.

Just as we have the regression of Y on X, so we have that of X on Y, and it has a similar interpretation. It tells us the mean value of X we can expect from repeated sampling in which Y is given. It is sometimes found rather puzzling to have two different regression lines or lines of relationship. Provided $\rho \neq \pm 1$ there are always two lines. These two lines are characteristics of the population and are not due to the chances of sampling. They are

$$\left. \begin{array}{l} \mu_Y(X) = \alpha + \beta X \\ \mu_X(Y) = \gamma + \delta Y \end{array} \right\} \text{ for the population}$$

and

$$\left. \begin{array}{l} Y_c = a + bX \\ X_c = c + dY \end{array} \right\} \text{ for the sample.}$$

Written in this form they are seen to be quite distinct, the first giving mean values of Y's for given X's, and the second mean values of X's for given Y's. The confusion arises because they are sometimes written in the form

$$\left. \begin{array}{l} Y = a + bX \\ X = c + dY \end{array} \right\}$$

In this form one is tempted to rewrite the second

$$Y = -\frac{c}{d} + \frac{1}{d}X$$

and to ask why a does not equal $-\dfrac{c}{d}$ and b equal $\dfrac{1}{d}$, when both lines are supposed to be "best" fits. However, the first equation is a "best" fit in the sense that it minimizes vertical deviations, and the second is a "best" fit in the sense that it minimizes horizontal deviations. The point is that the two equations are not interchangeable and that "X" in the first equation is not the same thing as the "X" in the second one, which is a mean value and should be written X_c for the sample or $\mu_X(Y)$ for the population. The two equations arise because variability in the Y's for given X's and in the X's for given Y's is possible. When $\rho = \pm 1$, this is not so, and $\beta = \dfrac{1}{\delta}$. In this case we have only one value of Y associated with every X and vice versa, and only one line of relationship.

The relationship between X and Y expressed by the lines of regression is not an exact relationship except when $\rho = \pm 1$. The regression of Y on X does not tell us that a particular value of Y is always associated with a given value of X, as the relationship

$$V = h^3$$

does, where V is the volume of a cube and h its dimension. In an exact relationship such as this, we know that if $h = 4$, say, then *the* associated value of V is 64. Rather the regression of Y on X tells us the value of Y which *on the average* will be associated with a given value of X. Thus, if the regression of monthly rent on the age of a house is

$$\mu_Y(X) = 45 - 0{\cdot}8X$$

this tells us that the *mean* rent of houses of age 10 years is $45 - (0{\cdot}8 \times 10) = \$37$ per month. Furthermore, the rents of individual houses of age 10 years may be more or less dispersed around their mean of \$37, so that the relationship between rent and age may be more or less close. Relationships of this kind expressed in regression equations are called *stochastic* relationships in contrast with *exact* relationships.

In practice we may be interested in only *one* of the two regression lines. Thus, if we know that X in a sense "determines" Y, e.g. the age of a house determines its rent rather than vice versa, we shall mainly be concerned with the Y on X regression. Similarly, if we want to predict values of Y for given values of X. Furthermore, if we are conducting an experiment in which we control the values of X so that they do not vary due to chance, the Y on X regression is alone relevant.

We have only been concerned here with the case of a normal (or in practice near normal) population, but the concept of a regression line is quite general. In general,

$$\mu_Y(X) = f(X)$$

where $f(X)$ is some function of X (not necessarily linear), defines the regression of Y on X and tells us the mean value of the Y's which are associated with given X's.

Meaning of the Coefficient of Linear Correlation[1]

The interpretation of the parameter ρ for the population is the same as that of the statistic r for a sample. Thus ρ^2 tells us the proportion of the total variability of the Y's (or the X's) explained by the relationship between the X's and the Y's.[2] Thus the Y's vary not only because they vary for given X's but also because the X's vary, and this variation in the X's is translated via the regression relation into variation in the Y's. The same holds for the X's, *mutatis mutandis*. For example, if for the population of ages of brides and ages of bridegrooms in Australia ρ were 0·9, this means that we can explain 81 per cent of the variability in the ages of brides by the relationship between age of bride and age of bridegroom. The remaining 19 per cent is unexplained by the relationship. If we were to hold the age of bridegrooms constant by regarding only the brides of men of a given age, we should reduce the variability of the age of brides (as measured by the variance) to 19 per cent of the total variability in the ages of brides of men of all ages.

The meaning of ρ may be further appreciated by considering the expression $\Sigma(X - \mu_X)(Y - \mu_Y)$ which appears as the numerator in one of the forms in which ρ may be written. A positive relationship will be indicated $(0 < \rho < 1)$ when values of X above the mean value of X tend to be associated with values of Y above the mean value of Y, and conversely for values of X below the mean value of X. Similarly a negative relationship will be indicated $(-1 < \rho < 0)$ when values of X above the mean value of X tend to be associated with values of Y below the mean value of Y, and conversely for values of X below the mean value of X. An absence of relationship will be indicated $(\rho = 0)$ when neither of the above tendencies is present. In this case X and Y can be said to be *independent*, for the particular value which Y assumes is quite unaffected by the value of X paired with it, and conversely.

The greater ρ the more reliance can we place on the line of regression as a predicting instrument. For example, suppose that X stands for age of houses and Y for rent. If we wish to predict[3] the rent of a particular house and we have no information about the house, the

---

[1] See also section 10.1, p. 207 above.
[2] This is manifest in the definition of the standard error of estimate, for

$$\sigma_Y^2(e) = \sigma_Y^2(1 - \rho^2)$$

i.e. $$\rho^2 = \frac{\sigma_Y^2 - \sigma_Y^2(e)}{\sigma_Y^2}$$

[3] On the use of the word "predict" in this connexion see section 10.6 below.

best we can do is to use μ_Y—the mean rent of houses—as a prediction. This is the best we can do in the sense that the errors in making a series of such predictions will be smaller than if we were to use any figure other than μ_Y. Actual rents will be distributed about μ_Y with a dispersion measured by σ_Y, so that σ_Y will be an indication of the errors of prediction. In fact 95 per cent of houses will have rents within $\mu_Y \pm 1 \cdot 96\, \sigma_Y$. However, if we know the regression of rent on age of house

$$\mu_Y(X) = \alpha + \beta X$$

and we know the age of the particular house under consideration, we should do better by predicting the rent of the house as $\mu_Y(X)$. For actual rents of houses of the given age will be distributed about $\mu_Y(X)$ with a dispersion measured by $\sigma_Y(e) = \sigma_Y \sqrt{1-\rho^2}$, so that $\sigma_Y(e)$ will be an indication of the errors of prediction. Again 95 per cent of houses of the given age will have rents within $\mu_Y(X) \pm 1\cdot 96\, \sigma_Y \sqrt{1-\rho^2}$. It can be seen that the accuracy of a prediction of Y can be increased if we know X and the regression of Y on X, and the greater ρ for a given σ_Y the smaller the errors in using the regression for purposes of prediction. If $\rho = \pm 1$, the relationship between X and Y is exact, and no errors of prediction are incurred in estimating Y from a knowledge of X.

Just as the mean and standard deviation of a single variable are characteristics of the population, so in the case of a bivariate population is the coefficient of correlation (and the regression lines which are defined in terms of the parameters μ_X, μ_Y, σ_X, σ_Y and ρ). Thus, if the value of ρ between ages of brides and bridegrooms were 0·9 in Australia and 0·7 in the United Kingdom, this would represent a real difference in the marriage habits of the two countries. These figures would indicate that the ages of brides and bridegrooms were more closely associated in Australia than in the United Kingdom. A bivariate normal population cannot be fully specified without the correlation coefficient.

The above discussion has been in terms of *linear* correlation only. The correlation in the bivariate normal population is linear and the calculation of r is defined in terms of linear regression lines. Although it can be shown that the ratio $\dfrac{\Sigma(Y_c - \bar{Y})^2}{\Sigma(Y - \bar{Y})^2}$ can be used as a measure of correlation where the regression is not linear, this ratio reduces to $r^2 = \dfrac{(\Sigma xy)^2}{\Sigma x^2 \Sigma y^2}$ only when $Y_c = a + bX$, i.e. when the regression is linear. Consequently, the fact that, for a sample, r is close to 0 $\left(\text{when calculated from the formula } r = \dfrac{\Sigma xy}{\sqrt{\Sigma x^2 \Sigma y^2}}\right)$ does not necessarily indicate the absence of a relationship, sampling errors apart. It indicates only the absence of a *linear* relationship.

For example, consider the following diagram—

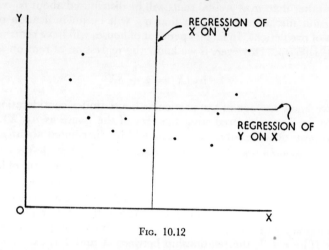

Fig. 10.12

Here r would be very small, but there is clearly a relationship (curvilinear) between X and Y. Consequently, $r = \dfrac{\Sigma xy}{\sqrt{\Sigma x^2 \Sigma y^2}}$ can be used validly as a measure of correlation only if we can suppose that the sample has been drawn from a population in which the relationship is linear or approximately linear.

Correlation and Causation

The existence of correlation does not necessarily imply causation. Correlation merely indicates association or, more strictly speaking, covariation. To infer a causal relationship between X and Y, just because a set of observations on X and Y yield a high value for r, is a very unsafe procedure. Often two variables are both related to a third one, with the result that they are highly correlated with each other, although they are not directly related in the sense that changes in the one would on the average produce changes in the other, other things being equal. This frequently happens when observations recorded over time are correlated.

As an illustration of this, Yule[1] correlated the proportion of Church of England marriages to all marriages in England and Wales for the years 1866–1911 with the standardized mortality rate for the same

[1] Yule, G. U.: "Why do we sometimes get nonsense-correlations between time series?" *Journal of the Royal Statistical Society*, Vol. LXXXIX, January, 1926, p. 1.

period and obtained a value of r of 0·95. To draw the conclusion that the greater the proportion of marriages solemnized by the Church of England the higher the death rate would be absurd. The fact is that both series of figures moved steadily downwards over the period for quite unconnected reasons, and consequently their values over the period varied together. It is easy to produce such nonsense correlations from observations recorded over time, and frequently they yield very high values for the correlation coefficient.[1] Accordingly, in such cases a high value for r, although it indicates strong covariation, has no significance whatever in indicating a *real* relationship.

Frequently correlations, while not of a nonsensical character, are brought about by the operation of a third factor. Thus we might find that fathers' I.Q.s and the sizes of their families are negatively correlated. This suggests that the less intelligent tend to have larger families. It would be unwise to infer this, however, without further investigation. It may be, for example, that there is a positive correlation between I.Q. and income and a negative one between income and family size. Consequently the relations between I.Q. and income, and family size and income would produce a negative correlation between observations on I.Q. and family size, even though for *given incomes* I.Q. and family size were quite unrelated. Such problems as these involve more than two variables and require *multivariate* analysis, an elementary introduction to which is given in sections 10.8 and 10.9 below.[2]

A high value for r indicates a high degree of covariation between the two variables under consideration. This cannot be used as evidence for association or a causal relationship unless the possibility of the covariation being due to the operation of other related variables has been carefully investigated. In any event, the best procedure is usually to postulate an hypothetical causal relation which can be tested against observed data. If r turns out to be high, the hypothesis is supported as far as that piece of evidence is concerned. But a high value of r can never prove that a relationship between X and Y exists. It can only indicate that the data are not inconsistent with one. The reasonableness of the relationship on general theoretical grounds must always be carefully examined. It follows that a detailed knowledge of the data is necessary before general conclusions can be drawn from correlations.

Correlation analysis has other uses than that of demonstrating causal relationship. The fact of association between X and Y may itself be of interest, or the object may be to determine an estimating equation (regression of Y on X), so that Y's can be predicted for given X's.

---

[1] See section 11.7, p. 269, below.
[2] In particular see pp. 249 and 250 below.

When handling sample data it must not be forgotten that a sample may throw up an appreciable value for r just due to chance, even though there is no correlation in the underlying population. In Fig. 10.13 the plot points represent a random sample drawn from a population with zero correlation. However, just due to chance, these particular points yield a regression of Y on X with $b \neq 0$, and hence $r \neq 0$. It

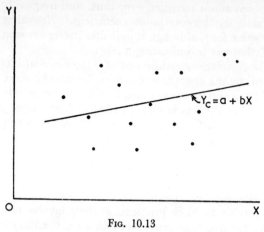

Fig. 10.13

follows that it is always necessary to test whether a sample value of r could reasonably have come from a population in which there was no linear correlation, before concluding that the two variables are correlated. This can readily be handled by devising tests for the significance of r.

10.5. Tests of Significance in Regression and Correlation Analysis

If we have a sample of N pairs of observations with a correlation coefficient of r, the question arises: could such a value of r have come from a population in which there was no linear correlation, i.e. in which $\rho = 0$? In other words, is the correlation statistically significant?

It will be readily appreciated that the distribution of sample values of r drawn from a population in which $\rho = 0$ cannot be normal, since the value of r is bounded by ± 1, and the normal distribution is asymptotic at both ends. This rather complicates matters. However, since $\beta = \rho \dfrac{\sigma_Y}{\sigma_X}$, if $\rho = 0$, β must also $= 0$ (and $\delta = 0$ also). It follows that a test of the significance of b as against the hypothesis $\beta = 0$ is equivalent to a test of the significance of r against the hypothesis $\rho = 0$. To perform such a test of significance of b, we must deduce the sampling distribution of b calculated from samples drawn from a normal bivariate population. For a sample of size N, we have

$$b = \frac{\Sigma x_i y_i}{\Sigma x_i^2} \qquad (i = 1, 2, \ldots N)$$

$$= \frac{\Sigma x_i (Y_i - \bar{Y})}{\Sigma x_i^2}$$

$$= \sum \left(\frac{x_i}{\Sigma x_i^2} Y_i \right) - \bar{Y} \sum \frac{x_i}{\Sigma x_i^2}$$

$$= \sum \left(\frac{x_i}{\Sigma x_i^2} Y_i \right)$$

since $\Sigma x_i = 0$. With repeated sampling, for the given values X_i, the observed values Y_i will be independent and normally distributed about the population regression line given by $\mu_Y(X_i) = \alpha + \beta X_i$ with a variance $\sigma_Y^2(e)$. Thus for the given X_i, b is a linear combination of the Y_i, and by virtue of section 7.2 above, we may write

$$\mu_b = \sum \left(\frac{x_i}{\Sigma x_i^2} \mu_Y(X_i) \right)$$

$$= \sum \left(\frac{x_i}{\Sigma x_i^2} \right) (\alpha + \beta X_i)$$

$$= \beta \sum \left(\frac{x_i}{\Sigma x_i^2} X_i \right)$$

$$= \beta$$

since $\sum \left(\frac{x_i}{\Sigma x_i^2} X_i \right) = \sum \left(\frac{x_i}{\Sigma x_i^2} x_i \right) + \bar{X} \sum \frac{x_i}{\Sigma x_i^2} = 1$

Thus $\mu_b = \beta$ (i.e. the mean of b's from repeated sampling equals the population β), and b is an unbiased estimate of β.

The variance of the b's obtained from repeated sampling can be similarly expressed as a linear combination of the variances of the independently distributed Y_i.

$$\sigma_b^2 = \sum \left[\left(\frac{x_i}{\Sigma x_i^2} \right)^2 \text{Var}(Y_i) \right]$$

$$= \sum \left[\frac{x_i^2}{(\Sigma x_i^2)^2} \sigma_Y^2(e) \right]$$

$$= \frac{\sigma_Y^2(e)}{\Sigma x_i^2}$$

The denominator in this formula is the variation of the given X's. Omitting the subscript i, we have

$$\sigma_b = \frac{\sigma_Y(e)}{\sqrt{\Sigma x^2}}$$

In practice we should not know $\sigma_Y(e)$, and we must estimate it from the sample. This estimate is

$$s_Y(e) = \sqrt{\frac{\Sigma(Y - Y_c)^2}{N - 2}}$$

Accordingly the estimate of σ_b is

$$s_b = \frac{s_Y(e)}{\sqrt{\Sigma x^2}}$$

When we use s_b instead of σ_b, we must turn from the normal to the t-distribution, and we have that

$$t = \frac{b - \beta}{s_b}$$

is distributed in the t-distribution with $N - 2$ degrees of freedom.

It follows that if we are drawing samples from a non-correlated population (i.e. one in which $\beta = 0$), the statistic b will be normally distributed about 0 with a standard error of $\sigma_b = \frac{\sigma_Y(e)}{\sqrt{\Sigma x^2}}$, and the statistic $t = \frac{b}{s_b}$ will be distributed in the t-distribution with $N - 2$ degrees of freedom. Hence if we wish to test whether b differs significantly from zero, i.e. whether the sample could have come from a population with $\beta = 0$, we compute $t = \frac{b}{s_b}$ and refer it to the table to see whether or not it is significant at the 5 per cent level. If the value of $|t|$ is so large that the sample could not have reasonably come from an uncorrelated population we conclude that $\beta \neq 0$.

If b differs significantly from zero, we may wish to determine within what limits we could reasonably expect the population β to lie, i.e. what are the 95 per cent confidence limits for β. These are given by

$$\pm t_{.05} s_b$$

just as

$$\bar{X} \pm t_{.05} s_{\bar{X}}$$

are the 95 per cent confidence limits for μ.

If in selecting a sample of observations of X and Y in order to estimate β, we are able to control the values of the independent variable X, as may well happen in experimental work, we can reduce the range

REGRESSION AND CORRELATION

of confidence limits based on a sample of a given size by making the variations in X as great as possible. For $s_b = \dfrac{s_Y(e)}{\sqrt{\Sigma x^2}}$ and Σx^2 will thus be made large. This procedure is quite legitimate provided the associated Y's are free to vary at random. Indeed reliable estimates of β cannot be made if there is too little variation in the X's. Diagrammatically speaking, if all the X's are about the same, there would not be sufficient scatter to discern a relationship.

The test $t = \dfrac{b}{s_b}$ can easily be put into terms of r, thus giving a direct test for r against the hypothesis that $\rho = 0$.

We have
$$t = \frac{b}{s_b}$$

$$= \frac{\Sigma xy}{\Sigma x^2} \cdot \frac{\sqrt{\Sigma x^2}}{\sqrt{\dfrac{(Y - Y_c)^2}{N - 2}}}$$

$$= \frac{\Sigma xy}{\sqrt{\Sigma x^2}} \cdot \frac{\sqrt{N - 2}}{\sqrt{(1 - r^2)\Sigma y^2}}$$

$$= \frac{r\sqrt{N - 2}}{\sqrt{1 - r^2}}$$

Hence $t = \dfrac{r\sqrt{N - 2}}{\sqrt{1 - r^2}}$ is distributed in the t-distribution with $N - 2$ degrees of freedom,[1] when r is the coefficient of linear correlation of a sample drawn from a population with $\rho = 0$. We may then test the null hypothesis H_o: $\rho = 0$ against the alternative hypothesis H_a: $\rho \neq 0$ by the procedure outlined in section 7.7 above.

EXAMPLE 10.6

A random sample of 27 factories has a coefficient of linear correlation of $r = 0.60$ between the number of industrial accidents occurring over a year and the age of the factory. Is this evidence of a linear relationship between number of accidents and age?

Hypothesis. The sample comes from a normal bivariate population in which $\rho = 0$, i.e.

$$H_o: \rho = 0$$
$$H_a: \rho \neq 0$$

---

[1] Tables giving the critical values of r direct for different levels of significance are available. See Fisher, R. A.: *Statistical Methods for Research Workers* (Oliver and Boyd, 1958), pp. 209–210.

Test—
$$|t| = \frac{|r|\sqrt{N-2}}{\sqrt{1-r^2}}$$
$$= \frac{0.60 \times 5}{0.80}$$
$$= 3.75$$

From the t-table with $\nu = 25$, $P(|t| > 3.75) < 0.01$.

Conclusion. If the null hypothesis H_o were true, repeated sampling with 25 degrees of freedom would yield a value of $|t|$ as great as or greater than 3·75 in less than one case in 100 due to chance. This means that, if H_o were true, repeated sampling would yield a value of $|r|$ as great as or greater than the one actually obtained in less than one case in 100 due to chance. We thus reject the null hypothesis H_o and accept the alternative hypothesis H_a. The sample coefficient is significant and its value is evidence of some relationship between number of accidents and age of factory. As far as this particular sample is concerned, 36 per cent of the variation in the number of accidents can be explained by the relation between number of accidents and age of factory.

EXAMPLE 10.7

Writing X for age in years and Y for weekly rent in dollars, a random sample of 11 houses yields the following results—
$$\Sigma xy = -750; \quad \Sigma x^2 = 625; \quad \Sigma y^2 = 1600$$

(a) Calculate the regression coefficient of rent on age.
(b) Test whether it is significant. If it is significant calculate the 95 per cent confidence limits for the population regression coefficient.

(a) The regression coefficient is
$$b = \frac{\Sigma xy}{\Sigma x^2}$$
$$= -1.20$$

(b) We have $b = -1.20$
$$s_Y(e) = \sqrt{\frac{\Sigma y^2 - b\Sigma xy}{N-2}} = \sqrt{77.78}$$
$$s_b = \frac{s_Y(e)}{\sqrt{\Sigma x^2}} = \sqrt{\frac{77.78}{625}} = 0.35$$

Hypothesis. The sample comes from a normal bivariate population with $\beta = 0$, i.e.
$$H_o: \beta = 0$$
$$H_a: \beta \neq 0$$

Test—
$$|t| = \frac{|b|}{s_b}$$
$$= \frac{1.20}{0.35}$$
$$= 3.43$$

From the t-table with $\nu = 9$, $P(|t| > 3.43) < 0.01$.

Conclusion. If the null hypothesis H_o were true, repeated sampling with 9 degrees of freedom would yield a value of $|t|$ as great as or greater than 3·43 in less than one case in 100 due to chance. This means that, if H_o were true, repeated sampling would yield a value of $|b|$ as great as or greater than the one actually obtained in less than one case in 100 due to chance. We thus reject the null hypothesis H_o and accept the alternative hypothesis H_a. The sample could not reasonably have come from a population in which $\beta = 0$. We say that the sample regression coefficient b is significant and conclude that the population β differs from zero. Accordingly, we are interested in determining confidence limits within which β can be expected to lie. These 95 per cent confidence limits are

$$b \pm t_{\cdot05} s_b$$

where $\quad t_{\cdot05} = 2\cdot262 \quad \text{for} \quad \nu = 9$

i.e. $\quad -1\cdot20 \pm 0\cdot79$

Hence we can be 95 per cent confident that β lies between $-0\cdot41$ and $-1\cdot99$.

In addition to significance tests concerning the coefficients r and b, it is useful for some purposes to test also whether a sample could have reasonably come from a population in which the regression parameter α is zero. The null hypothesis to be tested in such cases is that the variable Y is proportional to X, i.e. that the population regression line passes through the origin, against the alternative hypothesis that it does not:

$$H_o: \alpha = 0$$
$$H_a: \alpha \neq 0$$

The statistic a calculated from a sample of size N is defined as

$$a = \bar{Y} - b\bar{X}$$

Writing $\quad b = \sum \left(\dfrac{x_i}{\Sigma x_i^2} Y_i \right),$

we have $\quad a = \bar{Y} - \sum \left(\dfrac{x_i}{\Sigma x_i^2} Y_i \right) \bar{X}$

$$= \sum \left[\left(\frac{1}{N} - \bar{X} \frac{x_i}{\Sigma x_i^2} \right) Y_i \right]$$

Thus a is a linear combination of the independent and normally distributed Y_i, and with repeated sampling

$$\mu_a = \sum \left[\left(\frac{1}{N} - \bar{X} \frac{x_i}{\Sigma x_i^2} \right) \mu_Y(X_i) \right]$$

$$= \sum \left(\frac{1}{N} - \bar{X} \frac{x_i}{\Sigma x_i^2} \right) (\alpha + \beta X_i)$$

$$= \alpha + \frac{1}{N} \beta \Sigma X_i - \beta \bar{X} \sum \left(\frac{x_i}{\Sigma x_i^2} X_i \right)$$

$$= \alpha$$

For the variance of the a's, we have

$$\sigma_a^2 = \sum \left[\left(\frac{1}{N} - \bar{X} \frac{x_i}{\Sigma x_i^2} \right)^2 \text{Var}(Y_i) \right]$$

$$= \sum \left[\left(\frac{1}{N} - \bar{X} \frac{x_i}{\Sigma x_i^2} \right)^2 \sigma_Y^2(e) \right]$$

$$= \left(\frac{1}{N} - \frac{2\bar{X}}{N} \sum \frac{x_i}{\Sigma x_i^2} + \bar{X}^2 \sum \frac{x_i^2}{(\Sigma x_i^2)^2} \right) \sigma_Y^2(e)$$

Since $\Sigma x_i = 0$, the middle term vanishes, and

$$\sigma_a^2 = \sigma_Y^2(e) \left(\frac{\Sigma x_i^2 + N\bar{X}^2}{N \Sigma x_i^2} \right)$$

where $\Sigma x_i^2 + N\bar{X}^2 = \Sigma X_i^2$ (see section 5.5, p. 72).

Omitting the subscripts, we have

$$\sigma_a = \sigma_Y(e) \sqrt{\frac{\Sigma X^2}{N \Sigma x^2}}$$

Since $\mu_a = \alpha$ (i.e. the mean of the a's from repeated sampling equals the population α), a is an unbiased estimate of α.

Generally, $\sigma_Y(e)$ is not known and must be estimated from the sample. The estimate of σ_a is

$$s_a = s_Y(e) \sqrt{\frac{\Sigma X^2}{N \Sigma x^2}}$$

and the statistic

$$t = \frac{a}{s_a}$$

is distributed in the t-distribution with $N - 2$ degrees of freedom. For α significantly different from zero, the 95 per cent confidence limits are

$$\alpha \pm t_{.05} s_a$$

EXAMPLE 10.8

In the following regression equation, X is the amount of advertising expenditures on a certain cosmetic product and Y the amount of revenue derived from the sale of the product by a departmental store (X and Y in \$'000s).

$$Y_c = 0.6 + 4.89X$$

We are also given—

$\Sigma x^2 = 82.5 \quad \Sigma y^2 = 2,110.5 \quad \Sigma xy = 403.5$
$N = 10 \quad \bar{X} = 5.5 \quad \bar{Y} = 27.5.$

REGRESSION AND CORRELATION   233

It is claimed from these results that if advertising of the product were discontinued, the expected revenue would be zero. Can this claim be substantiated on the basis of the sample data?

First we compute

$$s_Y(e) = \sqrt{\frac{\Sigma y^2 - b\Sigma xy}{N - 2}} = 4 \cdot 14$$

The estimate of σ_a is

$$s_a = s_Y(e)\sqrt{\frac{\Sigma X^2}{N\Sigma x^2}}$$

We have $\Sigma X^2 = \Sigma x^2 + N(\bar{X})^2 = 385$, and

$$s_a = 4 \cdot 14 \sqrt{\frac{385}{825}}$$
$$= 2 \cdot 83$$

Hypothesis. The sample comes from a population with $\alpha = 0$, i.e.

$$H_o: \alpha = 0$$
$$H_a: \alpha \neq 0$$

Test—

$$t = \frac{a}{s_a} = \frac{0 \cdot 60}{2 \cdot 83} = 0 \cdot 21$$

From the t-table with $\nu = 8$, $P(|t| > 0 \cdot 21) > 0 \cdot 05$.

Conclusion. If the null hypothesis H_o were true, repeated sampling with 8 degrees of freedom would yield a value of $|t|$ as great as or greater than $0 \cdot 21$ in more than 80 cases in 100 due to chance. This means that, if H_o were true, repeated sampling would yield a value of $|a|$ as great as or greater than the one actually obtained in more than 80 cases in 100. We do not reject the null hypothesis. The sample regression coefficient a is not significant. On the basis of the sample result, the claim of zero expected revenue in the absence of advertising appears to be justified.

10.6. PREDICTION FROM A REGRESSION EQUATION

In section 10.4, p. 222 above, reference was made to the use of regression equations as predicting instruments. The discussion there was purely illustrative and in terms of the population regression. In practice we must make our predictions from sample regressions. First we should make clear what we mean by a prediction.

If from a sample of N pairs of observations we have a regression equation of Y on X, $Y_c = a + bX$ and we wish to ascertain from it the value of Y which can be expected to be associated with a particular value of X, say X', we speak of making a *prediction* of Y for the given X. The term "prediction" is used here in the sense not of a forecast but of a conditional statement, i.e. if X has a certain value, then, according to the regression, Y can be expected to have a certain value. The value of Y associated with X' will be a mean value, being the value which will on the average be associated with X'. Such a prediction can be made by putting X' in the regression equation and thus obtaining Y'. The value Y' is a sample estimate of the population value $\mu_Y(X')$, i.e. the mean value of Y when $X = X'$, and it is desirable to determine prediction limits to $\mu_Y(X')$, analogous to confidence limits for population parameters. To do this we must first derive the standard error of Y', where Y' is derived from a sample regression line by substituting $X = X'$ in it. We have

$$Y_c = a + bX$$

i.e.
$$Y_c = \bar{Y} + b(X - \bar{X})$$

For $X = X'$ this becomes

$$Y' = \bar{Y} + b(X' - \bar{X})$$

Thus, for given X's, Y' is a linear combination of independent normally distributed variables. With repeated sampling for the given values of X, the statistics $\bar{Y}$ and b will vary, and hence Y' estimated from the sample regression equation will vary. By virtue of section 7.2 above it follows that Y' will be normally distributed about a mean[1]

$$\mu_{Y'} = [\mu_Y + \beta(\bar{X} - \mu_X)] + \beta(X' - \bar{X})$$
$$= \mu_Y + \beta(X' - \mu_X)$$
$$= \mu_Y(X')$$

with a variance

$$\text{Var}(Y') = \text{Var}(\bar{Y}) + (X' - \bar{X})^2 \text{Var}(b)$$

Since the variance of Y for given X is $\sigma^2{}_Y(e)$, $\text{Var}(\bar{Y}) = \dfrac{\sigma^2{}_Y(e)}{N}$ and we have $\text{Var}(b)$ from above.

Hence
$$\text{Var}(Y') = \sigma^2{}_Y(e) \left[\frac{1}{N} + \frac{(X' - \bar{X})^2}{\Sigma x^2}\right]$$

---

[1] The mean of Y_i for a given X_i is given by the regression
$$\mu_Y(X_i) = \mu_Y + \beta(X_i - \mu_X)$$
But $\bar{Y} = \dfrac{\Sigma Y_i}{N}$ and hence the mean of $\bar{Y}$ for given X's is the mean of the linear combination $\dfrac{\Sigma Y_i}{N}$,

i.e. $\dfrac{1}{N}\sum_i [\mu_Y + \beta(X_i - \mu_X)]$

i.e. $\dfrac{1}{N}(N\mu_Y + \beta\Sigma X_i - \beta N\mu_X)$

i.e. $\mu_Y + \beta(\bar{X} - \mu_X)$

This gives the mean of $\bar{Y}$ for given X's with repeated sampling.

REGRESSION AND CORRELATION 235

Since $\mu_{Y'} = \mu_Y(X')$, i.e. the mean of Y' for given X' calculated from sample regressions arising from repeated sampling equals the population mean for given X', we can say that Y' is an unbiased prediction of $\mu_Y(X')$.

Generally we shall not know $\sigma_Y(e)$, and we must use $s_Y(e)$ as an estimate of it. Accordingly

$$t = \frac{Y' - \mu_Y(X')}{\sqrt{\text{Var}(Y')}}$$

where
$$\text{Var}(Y') = s^2{}_Y(e)\left[\frac{1}{N} + \frac{(X' - \bar{X})^2}{\Sigma x^2}\right]$$

will be distributed in the t-distribution with $N - 2$ degrees of freedom. It follows that the 95 per cent prediction limits to $\mu_Y(X')$ can be written—

$$Y' \pm t._{.05} s_Y(e) \sqrt{\frac{1}{N} + \frac{(X' - \bar{X})^2}{\Sigma x^2}}$$

The width of the prediction limits is greater the further X' is from $\bar{X}$. This is because sampling errors in b are multiplied as X' moves away from $\bar{X}$. Consequently predictions of $\mu_Y(X')$ from a sample regression equation are subject to greater sampling errors the more extreme is the value of X under consideration. The above prediction limits mean that if we calculate $Y_c = a + bX$ from repeated samples, we shall be right in 95 per cent of the cases in locating $\mu_Y(X')$ in the above range. It does not mean that 95 per cent of predictions based on *one* estimate $Y_c = a + bX$ will rightly locate $\mu_Y(X')$ in the range, for in that case the same sampling errors in the estimates a and b will be exactly repeated each time.

The above prediction limits refer to the limits for the *mean* value of Y associated with a particular value of X. The range within which a *single* value of Y associated with a particular X is likely to lie will of course be much wider, because the variation of individual Y's about the mean of the Y's must also be taken into account. For a given X', a single value of Y will be given by

$$Y' \pm d'$$

i.e. by the mean value of the Y's given by the line of regression plus or minus a deviation about the line of regression. It follows that the variance of a single Y for a given X will be given by

$$\text{Var}(Y' \pm d') = \text{Var}(Y') + \text{Var}(d')$$

But the variance of the deviations about the line of regression is given by $\sigma_Y^2(e)$, and $\mathrm{Var}(Y')$ has been derived above. Accordingly

$$\mathrm{Var}(Y' \pm d') = \sigma^2_Y(e)\left[1 + \frac{1}{N} + \frac{(X' - \bar{X})^2}{\Sigma x^2}\right]$$

Using $s_Y(e)$ as an estimate of $\sigma_Y(e)$, we shall have as 95 per cent prediction limits within which a single value of Y associated with a particular value of X is likely to fall—

$$Y' \pm t_{\cdot 05}\, s_Y(e)\sqrt{1 + \frac{1}{N} + \frac{(X' - \bar{X})^2}{\Sigma x^2}}$$

An example of the application of these formulae is given in the following section.

10.7. Regression in Empirical Economic Investigations

The empirical investigation of relationships between economic variables lies within the province of that branch of economics known as *econometrics*. The remainder of this chapter is devoted to the application of regression to measuring such relationships, but it is concerned with the statistical methods involved and not with the investigation of relationships between economic variables as such. For the latter, reference should be made to the growing literature on econometrics (see Appendix C, 3).

Relationships between economic variables are postulated by economic theory. Statistical analysis enables us to test whether the postulated relationships are consistent with the observed facts and, if so, to give empirical content to them. Suppose we have a certain variable Y (say, savings) which is postulated to depend on another variable X (say, disposable income). We do not expect the relationship between Y and X to be an exact one, but rather that the value of Y depends on X on the average, in the sense that any particular observed value of Y consists of a part determined by the corresponding value of X (a *systematic* part) plus or minus an amount due to chance (a *random* part). In the simplest case, the part determined by X may be a linear function of X (although a more complex function could easily be postulated). We can write

$$Y = \alpha + \beta X + \varepsilon$$

where ε is a random term with mean 0. Consequently repeated observations on Y for a given X will yield a mean value for Y equal to $\alpha + \beta X$, i.e. $\mu_Y(X) = \alpha + \beta X$. The above equation is called an economic *behaviour* (or *structural*) *equation*. X is called the *explanatory* variable, because we hope to explain variations in Y by the variations in X which take place. The constants α and β are the parameters of the equation. It will be appreciated that the relation between the mean value of Y for a given X and the given value of X is a regression equation of the kind discussed in the preceding sections.

Suppose we have N observations[1] on X and Y. We can use these observations to make an estimate of the behaviour equation. This can be done by calculating the line of regression of Y on X, computed by the method of least squares. This line of regression is

$$Y_c = a + bX$$

where a and b are estimates of α and β.

Two questions immediately arise—

(i) Does b differ significantly from 0? If it does not, X does not offer any explanation of Y, and our observed data cast doubt on the existence of a relation of the form

$$Y = \alpha + \beta X + \varepsilon$$

This question can be answered by testing the significance of b by the method given in section 10.5 above.

(ii) If b does differ significantly from 0, what is the degree to which the variable X offers an explanation of variation in the variable Y? This can be indicated by calculating r^2 (the square of the coefficient of linear correlation), for r^2 gives the proportion of variation of Y explained by the relation between X and Y.

To answer these questions we set up an hypothesis that

$$Y = \alpha + \beta X + \varepsilon$$

and investigate its plausibility. If the hypothesis is plausible, we may wish to use the regression equation

$$Y_c = a + bX$$

as an instrument for predicting values of Y which we can expect to be associated with particular values of X. We shall then want to attach prediction limits to our predictions. The method for doing this has been outlined in section 10.6 above.

---

[1] These observations must be thought of as a sample of observations. It may often be difficult to think of economic observations in this way, when there is no visible population from which the sample has been drawn. Thus, what is the population from which an aggregate like "savings in Australia 1967–68" has been drawn? However, once we have postulated that such an aggregate consists of a systematic part (say, $\alpha + \beta X$, where X is income) plus a random term, we have in fact created a hypothetical population of values which differ in their random terms, and the value of the random term for any particular observation—even if that observation is unique as in the above case—is a matter of chance.

238   APPLIED STATISTICS FOR ECONOMISTS

The procedures outlined in the preceding sections can be validly used for the types of problem outlined here, provided that for given X's the Y's are normally distributed about the line

$$\mu_Y(X) = \alpha + \beta X$$

with a standard deviation the same for all X's. This will be so when ε is normally distributed about a mean 0 with given standard deviation. In effect, we are applying the principles outlined in the preceding sections having regard to only *one* regression line, for we treat X as a controlled variable and are interested in explaining the variations in Y alone. This situation is in fact pictured in Fig. 10.11 above.

EXAMPLE 10.9

This example is purely for illustrative purposes, and it must not be supposed to represent an attempt to determine a genuine economic relationship.[1] We have data for the number of lambs slaughtered for human consumption in Victoria over the period 1921–22 to 1938–39. Can we determine a relationship which will "explain" the behaviour of this variable? We should expect *a priori* the number of lambs slaughtered to depend on sheep numbers, so we postulate that the number of lambs slaughtered is a linear function of sheep numbers and compute the regression of lambs slaughtered on sheep numbers.

Year	Lambs Slaughtered (millions) Y	Sheep Numbers (millions) (a) X	Arbitrary Deviations				
			$B = 2\cdot000$ y'	$A = 16\cdot000$ x'	y'^2	x'^2	$x'y'$
1921–22	1·239	12·171	− 0·761	− 3·829	0·579	14·661	2·914
1922–23	2·158	12·326	0·158	− 3·674	0·025	13·498	− 0·580
1923–24	1·242	11·766	− 0·758	− 4·234	0·575	17·927	3·209
1924–25	1·340	11·060	− 0·660	− 4·940	0·436	24·404	3·260
1925–26	1·880	12·650	− 0·120	− 3·350	0·014	11·223	0·402
1926–27	1·926	13·741	− 0·074	− 2·259	0·005	5·103	0·167
1927–28	1·554	14·920	− 0·446	− 1·080	0·199	1·166	0·482
1928–29	2·144	15·557	0·144	− 0·443	0·021	0·196	− 0·064
1929–30	2·367	16·498	0·367	0·498	0·135	0·248	0·183
1930–31	2·209	17·427	0·209	1·427	0·044	2·036	0·298
1931–32	2·541	16·478	0·541	0·478	0·293	0·228	0·259
1932–33	3·586	16·376	1·586	0·376	2·515	0·141	0·596
1933–34	3·829	17·512	1·829	1·512	3·345	2·286	2·765
1934–35	4·267	17·196	2·267	1·196	5·139	1·430	2·711
1935–36	4·583	16·784	2·583	0·784	6·672	0·615	2·025
1936–37	4·825	17·457	2·825	1·457	7·981	2·123	4·116
1937–38	4·651	17·663	2·651	1·663	7·028	2·766	4·409
1938–39	4·026	18·863	2·026	2·863	4·105	8·197	5·800
			14·367	− 11·555	39·111	108·248	32·952

(a) Sheep numbers are at March 31st in preceding financial year.
Source: Commonwealth Bureau of Census and Statistics, Australia: *Production Bulletin*, Part II, No. 43, 1948–49, pp. 123, 127

[1] For one thing, since the data used here are in the form of time series, the trends in the two series should be eliminated before the regression is computed. This procedure would complicate the illustration and for the sake of simplicity has been omitted (see section 11.7, p. 269 below).

REGRESSION AND CORRELATION

$$\bar{x}' = -0.642 \qquad \therefore \bar{X} = A + \bar{x}' = 15.358$$

$$\bar{y}' = 0.798 \qquad \therefore \bar{Y} = B + \bar{y}' = 2.798$$

$$\Sigma x^2 = \Sigma x'^2 - N\bar{x}'^2 = 100.829$$

$$\Sigma y^2 = \Sigma y'^2 - N\bar{y}'^2 = 27.649$$

$$\Sigma xy = \Sigma x'y' - N\bar{x}'\bar{y}' = 42.174$$

$$b = \frac{\Sigma xy}{\Sigma x^2} = 0.418$$

The regression of Y on X is given by

$$Y_c - \bar{Y} = b(X - \bar{X})$$

i.e. $\qquad Y_c - 2.798 = 0.418(X - 15.358)$

i.e. $\qquad Y_c = -3.622 + 0.418X$

To test the significance of this regression we set up the hypothesis that the data are a random sample from a population in which there is no regression of lamb slaughterings on sheep numbers, i.e. in which $\beta = 0$. We then perform the t-test, by calculating

$$t = \frac{b}{s_b}$$

where

$$s_b = \frac{s_Y(e)}{\sqrt{\Sigma x^2}}$$

and

$$s_Y(e) = \sqrt{\frac{\Sigma y^2 - b\Sigma xy}{N - 2}}$$

Here $s_Y(e) = 0.791$ and $s_b = 0.079$, so that $|t| = 5.29$. Referring to the t-table, we have for 16 degrees of freedom $P(|t| > 5.29) < 0.01$. If the hypothesis were true, the repeated sampling with 16 degrees of freedom would yield a value of $|t|$ as great as or greater than 5.29 in less than one case in 100 due to chance. We conclude that the regression is significant and the data are not inconsistent with the postulated relationship. The computed relationship indicates that, on the average, for every increase of 1 million in sheep numbers we can expect the number of lambs slaughtered to increase by 418,000.

The regression coefficient b is an estimate of the parameter β, and we can ascertain 95 per cent confidence limits for β as

$$b \pm t_{.05} s_b$$

i.e. $\qquad 0.418 \pm (2.120 \times 0.079)$

i.e. $\qquad 0.251$ to 0.585

so that we can be 95 per cent confident that β lies within this range.

The closeness of the fit of the regression to the data (i.e. the extent to which sheep numbers offer an adequate "explanation" of lamb slaughtering) can be seen from the coefficient of linear correlation between the two variables. Here

$$r^2 = \frac{(\Sigma xy)^2}{\Sigma x^2 \Sigma y^2} = 0{\cdot}64$$

and $r = 0{\cdot}80$

Accordingly we may say that 64 per cent of the variation in lamb slaughterings can be explained by the relationship between lamb slaughterings and sheep numbers, and 36 per cent of the variation is left unexplained.

Suppose we wish to predict the number of lambs slaughtered in a particular year for which sheep numbers are 17·523 million. From the regression equation we shall have

$$Y' = -3{\cdot}622 + 0{\cdot}418 X'$$

where $X' = 17{\cdot}523$

i.e. $Y' = 3{\cdot}703$

so that the regression equation predicts 3·703 million as the number of lambs slaughtered. This is a prediction of the mean value of Y associated with the particular X. We can ascertain 95 per cent prediction limits for the population mean value of Y for $X = 17{\cdot}523$ as

$$Y' \pm t_{\cdot 05} s_{Y(e)} \sqrt{\frac{1}{N} + \frac{(X' - \bar{X})^2}{\Sigma x^2}}$$

i.e. $3{\cdot}703 \pm \left(2{\cdot}120 \times 0{\cdot}791 \times \sqrt{\frac{1}{18} + \frac{(17{\cdot}523 - 15{\cdot}358)^2}{100{\cdot}829}} \right)$

i.e. 3·168 to 4·238 million

so that we can be 95 per cent confident that the mean value of lamb slaughterings when sheep numbers are 17·523 million lies within this range. Naturally the range within which *individual* values of lamb slaughterings can be expected to lie will be wider. It is given by

$$Y' \pm t_{\cdot 05} s_{Y(e)} \sqrt{1 + \frac{1}{N} + \frac{(X' - \bar{X})^2}{\Sigma x^2}}$$

i.e. 1·941 to 5·465 million

These prediction ranges are probably too wide to be of much practical value, but the closeness of the fit of the regression to the data as indicated by r^2 is not extremely high, and we should consider trying to improve the fit by adding additional explanatory variables (see the following section).

10.8. Multiple Regression

Simple regression deals with the case of the relation between a dependent variable (Y) and one independent or explanatory variable (X). It may well be, however, that the dependent variable is postulated to be a function of several explanatory variables. In the linear case we shall have

$$X_1 = \alpha + \beta_2 X_2 + \beta_3 X_3 + \beta_4 X_4 + \ldots + \varepsilon$$

where X_1 is the dependent variable and X_2, X_3, X_4, etc., are explanatory variables. As in the simple case, repeated observations on X_1 for given X_2, X_3, X_4, etc., will yield a mean value for X_1 equal to

$$\mu_{X_1}(X_2, X_3, X_4 \ldots) = \alpha + \beta_2 X_2 + \beta_3 X_3 + \beta_4 X_4 + \ldots$$

Here, we speak of the *multiple regression* of X_1 on X_2, X_3, X_4, etc., and β_2, β_3, β_4, etc., are known as the *partial regression coefficients*. The interpretation of these partial regression coefficients is quite simple. For example, β_2 measures the change in the mean value of X_1 that we can expect if X_2 changes by one unit, X_3, X_4, etc., remaining unchanged.

Methods similar to those used for simple regression are available for fitting multiple regression equations to observed data and for testing the significance of the partial regression coefficients. These methods require that for given X_2, X_3, etc., the variable X_1 should be normally distributed about a mean given by the multiple regression equation, with a standard deviation the same for all values of X_2, X_3, etc., i.e. that ε should be normally distributed about a mean 0 with given standard deviation. We shall illustrate these methods by considering the three variable case. When more than three variables are involved, the formulae become involved, although the methods are in principle the same. This section is purely an introduction to multiple regression methods. For advanced treatment of the subject reference should be made to the many texts available (see Appendix C, 1 and 2). In the application of multiple regression to economic data many complications occur which are beyond the scope of this text. For these complications reference should be made to texts on econometrics (see Appendix C, 3).

Suppose we have N observations of the three variables X_1, X_2 and X_3 and we postulate that X_1 is a linear function of X_2 and X_3

$$X_1 = \alpha + \beta_2 X_2 + \beta_3 X_3 + \varepsilon$$

so that the mean value of X_1 for given X_2 and X_3 is given by

$$\mu_{X_1}(X_2, X_3) = \alpha + \beta_2 X_2 + \beta_3 X_3$$

Thus X_1 might be savings, X_2 disposable income and X_3 stock of liquid assets. We wish to make an estimate of this relation from our sample data. This can be done by calculating the sample regression of X_1 on X_2 and X_3, i.e.

$$X_{1_c} = a + b_2 X_2 + b_3 X_3$$

As before we select the constants a, b_2 and b_3 so that the sum of the squares of the deviations of the observed from the computed values of X_1 is a minimum, i.e. we require

$$W = \Sigma(X_1 - X_{1_c})^2$$
$$= \Sigma(X_1 - a - b_2 X_2 - b_3 X_3)^2$$

to be a minimum. For this we must have

$$\frac{\partial W}{\partial a} = -2\Sigma(X_1 - a - b_2X_2 - b_3X_3) = 0$$

$$\frac{\partial W}{\partial b_2} = -2\Sigma(X_1X_2 - aX_2 - b_2X_2^2 - b_3X_2X_3) = 0$$

$$\frac{\partial W}{\partial b_3} = -2\Sigma(X_1X_3 - aX_3 - b_2X_2X_3 - b_3X_3^2) = 0$$

These conditions give us three normal equations for determining a, b_2 and b_3. The normal equations are

$$\left.\begin{array}{l}\Sigma X_1 = Na + b_2\Sigma X_2 + b_3\Sigma X_3 \\ \Sigma X_1X_2 = a\Sigma X_2 + b_2\Sigma X_2^2 + b_3\Sigma X_2X_3 \\ \Sigma X_1X_3 = a\Sigma X_3 + b_2\Sigma X_2X_3 + b_3\Sigma X_3^2\end{array}\right\}$$

Solving these equations in the manner set out in section 10.1 above, we get

$$X_{1_c} - \bar{X}_1 = b_2(X_2 - \bar{X}_2) + b_3(X_3 - \bar{X}_3)$$

where
$$b_2 = \frac{\Sigma x_1x_2 \Sigma x_3^2 - \Sigma x_1x_3 \Sigma x_2x_3}{D}$$

and
$$b_3 = \frac{\Sigma x_1x_3 \Sigma x_2^2 - \Sigma x_1x_2 \Sigma x_2x_3}{D}$$

and
$$D = \Sigma x_2^2 \Sigma x_3^2 - (\Sigma x_2x_3)^2$$

To compute the partial regression coefficients it is most convenient to work from arbitrary origins and to make use of the relations

$$\Sigma x_1x_2 = \Sigma x_1'x_2' - N\bar{x}_1'\bar{x}_2', \quad \text{etc.}$$

Corresponding to the coefficient of linear correlation in the bivariate case there is a measure called the *coefficient of multiple correlation*. It likewise measures the proportion of variability in the dependent variable "explained" by the explanatory variables. As before, we split up the variability of the X_1's into the variability about the regression and the variability due to the regression. We have

$$X_1 - \bar{X}_1 = (X_1 - X_{1_c}) + (X_{1_c} - \bar{X}_1)$$

so that

$$\Sigma(X_1 - \bar{X}_1)^2 = \Sigma(X_1 - X_{1_c})^2 + \Sigma(X_{1_c} - \bar{X}_1)^2 \\ + 2\Sigma(X_1 - X_{1c})(X_{1c} - \bar{X}_1)$$

REGRESSION AND CORRELATION

Consider the last term on the right-hand side. We have

$$\Sigma(X_1 - X_{1_c})(X_{1_c} - \bar{X}_1)$$
$$= \Sigma\Big([X_1 - a - b_2X_2 - b_3X_3][b_2(X_2 - \bar{X}_2) + b_3(X_3 - \bar{X}_3)]\Big)$$
$$= b_2\Sigma(X_1X_2 - aX_2 - b_2X_2^2 - b_3X_2X_3)$$
$$\quad - b_2\bar{X}_2\Sigma(X_1 - a - b_2X_2 - b_3X_3)$$
$$\quad + b_3\Sigma(X_1X_3 - aX_3 - b_2X_2X_3 - b_3X_3^2)$$
$$\quad - b_3\bar{X}_3\Sigma(X_1 - a - b_2X_2 - b_3X_3)$$

As can be seen from the normal equations, the requirement that $\Sigma(X_1 - X_{1_c})^2$ be a minimum involves that each of the above terms should be zero. So that

$$\Sigma(X_1 - X_{1_c})(X_{1_c} - \bar{X}_1) = 0$$

and hence

$$\Sigma(X_1 - \bar{X}_1)^2 = \Sigma(X_1 - X_{1_c})^2 + \Sigma(X_{1_c} - \bar{X}_1)^2$$

i.e. the total variability of the X_1's is split up into unexplained and explained variability. We now define the square of the coefficient of multiple correlation as

$$R^2 = \frac{\Sigma(X_{1_c} - \bar{X}_1)^2}{\Sigma(X_1 - \bar{X}_1)^2}$$

Clearly R^2 measures the proportion of variability in the X_1's explained by the explanatory variables in the regression. To calculate R^2 we write

$$\Sigma(X_{1_c} - \bar{X}_1)^2 = -\Sigma[(X_1 - X_{1_c} - X_1 + \bar{X}_1)(X_{1_c} - \bar{X}_1)]$$
$$= \Sigma(X_1 - \bar{X}_1)(X_{1_c} - \bar{X}_1)$$

since from above

$$\Sigma(X_1 - X_{1_c})(X_{1_c} - \bar{X}_1) = 0$$

Hence
$$\Sigma(X_{1_c} - \bar{X}_1)^2 = \Sigma[x_1(b_2x_2 + b_3x_3)]$$
$$= b_2\Sigma x_1x_2 + b_3\Sigma x_1x_3$$

From which it follows that

$$R^2 = \frac{b_2\Sigma x_1x_2 + b_3\Sigma x_1x_3}{\Sigma x_1^2}$$

9—(B.837)

As a measure of variability about the regression, i.e. of the variability of the X_1's for given X_2 and X_3, we have as before the standard error of estimate. The estimate of $\sigma_{X_1}(e)$ from the sample is

$$s_{X_1}(e) = \sqrt{\frac{\Sigma(X_1 - X_{1_c})^2}{N - 3}}$$

$$= \sqrt{\frac{\Sigma x_1^2 - b_2 \Sigma x_1 x_2 - b_3 \Sigma x_1 x_3}{N - 3}}$$

$$= \sqrt{\frac{(1 - R^2)\Sigma x_1^2}{N - 3}}$$

To test whether b_2 and b_3 are significantly different from zero, we proceed in a manner analogous to section 10.5 above. We have that

$$b_2 = \frac{\Sigma x_1 x_2 \Sigma x_3^2 - \Sigma x_1 x_3 \Sigma x_2 x_3}{D}$$

$$= \sum \left[X_1 \frac{x_2 \Sigma x_3^2 - x_3 \Sigma x_2 x_3}{D} \right]$$

i.e. b_2 is the sum of N terms like the one in square brackets, the first term corresponding to the first triad of observations, the second term to the second triad, etc. Thus, for given values of X_2 and X_3, the statistic b_2 is a linear combination of the observed values X_1, each X_1 being independently normally distributed about a mean of $\mu_{X1}(X_2, X_3) = \alpha + \beta_2 X_2 + \beta_3 X_3$ given by the regression, with a variance $\sigma_{X_1}^2(e)$. Accordingly, with repeated sampling for the given values of X_2 and X_3, b_2 will be normally distributed about a mean

$$\mu_{b_2} = \sum \left(\frac{x_2 \Sigma x_3^2 - x_3 \Sigma x_2 x_3}{D} \right) (\alpha + \beta_2 X_2 + \beta_3 X_3)$$

$$= \beta_2$$

with a variance

$$\sigma_{b_2}^2 = \sum \left[\sigma_{X_1}^2(e) \left(\frac{x_2 \Sigma x_3^2 - x_3 \Sigma x_2 x_3}{D} \right)^2 \right]$$

$$= \sigma_{X_1}^2(e) \frac{\Sigma x_2^2 (\Sigma x_3^2)^2 - 2(\Sigma x_2 x_3)^2 \Sigma x_3^2 + (\Sigma x_2 x_3)^2 \Sigma x_3^2}{D^2}$$

$$= \sigma_{X_1}^2(e) \frac{\Sigma x_3^2}{D}$$

Similarly b_3 will be normally distributed about a mean β_3 with variance $\sigma_{b_3}^2 = \sigma_{X_1}^2(e) \dfrac{\Sigma x_2^2}{D}$. Since μ_{b_2} and μ_{b_3} equal β_2 and β_3 respectively, we can say that b_2 and b_3 are unbiased estimates of β_2 and β_3.

In practice, we do not know $\sigma_{X_1}(e)$ and we must use $s_{X_1}(e)$ as an estimate, so that we have

$$s_{b_2} = s_{X_1}(e) \sqrt{\dfrac{\Sigma x_3^2}{D}}$$

$$s_{b_3} = s_{X_1}(e) \sqrt{\dfrac{\Sigma x_2^2}{D}}$$

To test whether b_2 and/or b_3 differ significantly from zero, i.e. whether the sample value of b_2 and/or b_3 could have come from a population in which β_2 and/or β_3 were zero, we use the t-test, referring

$$t = \dfrac{b_2}{s_{b_2}}$$

and

$$t = \dfrac{b_3}{s_{b_3}}$$

to the t-tables with $N - 3$ degrees of freedom. If b_2 is significantly different from zero, 95 per cent confidence limits for β_2 can be ascertained in the usual manner and will be

$$b_2 \pm t_{.05} s_{b_2}$$

and similarly for b_3. If it turns out that b_2 is not significantly different from zero, this means that doubt is cast on X_2 as an explanatory variable.

Finally, we may wish to use the equation

$$X_{1_c} = a + b_2 X_2 + b_3 X_3$$

as an instrument for prediction. Putting X_2' and X_3' for particular values of X_2 and X_3, the value of X_1 that the equation will predict will be

$$X_1' = a + b_2 X_2' + b_3 X_3'$$

The value X_1' is the estimate of the mean value $\mu_{X_1}(X_2, X_3)$ associated with the values X_2' and X_3' of X_2 and X_3. The variance of this estimate is given by

$$\mathrm{Var}(X_1') = \sigma_{X_1}^2(e) \left[\dfrac{1}{N} + \dfrac{(X_2' - \bar{X}_2)^2 \Sigma x_3^2}{D} + \dfrac{(X_3' - \bar{X}_3)^2 \Sigma x_2^2}{D} - 2 \dfrac{(X_2' - \bar{X}_2)(X_3' - \bar{X}_3) \Sigma x_2 x_3}{D} \right]$$

Using $s_{X_1}(e)$ as an estimate of $\sigma_{X_1}(e)$, we can write 95 per cent prediction limits for the mean $\mu_{X_1}(X_2', X_3')$ as

$$X_1' \pm t_{.05} s_{X_1}(e) \sqrt{\frac{1}{N} + \frac{(X_2' - \bar{X}_2)^2 \Sigma x_3^2}{D} + \frac{(X_3' - \bar{X}_3)^2 \Sigma x_2^2}{D} - 2\frac{(X_2' - \bar{X}_2)(X_3' - \bar{X}_3)\Sigma x_2 x_3}{D}}$$

where $t_{.05}$ is obtained from the t-table with $N - 3$ degrees of freedom. Also, in accordance with the argument in section 10.6 above, we can write the 95 per cent limits for a single value of X_1 as

$$X_1' \pm t_{.05} s_{X_1}(e) \sqrt{1 + \frac{1}{N} + \frac{(X_2' - \bar{X}_2)^2 \Sigma x_3^2}{D} + \frac{(X_3' - \bar{X}_3)^2 \Sigma x_2^2}{D} - 2\frac{(X_2' - \bar{X}_2)(X_3' - \bar{X}_3)\Sigma x_2 x_3}{D}}$$

EXAMPLE 10.10

We may continue here example 10.9 in the preceding section. The explanatory variable, sheep numbers, accounted for about 64 per cent of the variation in lamb slaughterings, leaving 36 per cent of the variation unexplained. We

Year	Lambs Slaughtered (millions) X_1	Sheep Numbers (a) (millions) X_2	Export price of lamb/lb / Export price of wool/lb X_3
1921–22	1·239	12·171	0·494
1922–23	2·158	12·326	0·408
1923–24	1·242	11·766	0·311
1924–25	1·340	11·060	0·286
1925–26	1·880	12·650	0·425
1926–27	1·926	13·741	0·347
1927–28	1·554	14·920	0·342
1928–29	2·144	15·557	0·398
1929–30	2·367	16·498	0·575
1930–31	2·209	17·427	0·600
1931–32	2·541	16·478	0·557
1932–33	3·586	16·376	0·469
1933–34	3·829	17·512	0·338
1934–35	4·267	17·196	0·601
1935–36	4·583	16·784	0·442
1936–37	4·825	17·457	0·373
1937–38	4·651	17·663	0·511
1938–39	4·026	18·863	0·597

(a) Sheep numbers are at 31st March in preceding financial year.

Source: Commonwealth Bureau of Census and Statistics, Australia: *Production Bulletin*, Part II, No. 43, 1948–49, pp. 93, 123, 127

now introduce a second explanatory variable. On *a priori* grounds we should expect that the relative prices of lamb and wool would influence lamb slaughterings in such a way that the higher the relative lamb-wool price the greater the number of lambs slaughtered. Accordingly, we postulate that the number of lambs slaughtered is a linear function of sheep numbers *and* the relative lamb-wool price, and compute the regression of lamb slaughterings on sheep numbers and the relative lamb-wool price. The latter is represented by the ratio of the export price of lamb per lb to the export price of wool per lb. The necessary data are given on p. 246.

A calculating machine is almost a necessity for computing multiple regressions. When using a machine, it is simplest to work from an arbitrary origin of zero, so that the arbitrary deviations are in fact the original data. The following summations have been computed on a machine—

$\Sigma X_1 = 50\cdot367$ $\Sigma X_2 = 276\cdot445$ $\Sigma X_3 = 8\cdot074$

$\Sigma X_1^2 = 168\cdot578$ $\Sigma X_2^2 = 4346\cdot489$ $\Sigma X_3^2 = 3\cdot814$

$\Sigma X_1 X_2 = 815\cdot715$ $\Sigma X_2 X_3 = 126\cdot575$ $\Sigma X_1 X_3 = 23\cdot362$

Hence $\bar{X}_1 = 2\cdot798$ $\bar{X}_2 = 15\cdot358$ $\bar{X}_3 = 0\cdot449$

and

$\Sigma x_1^2 = \Sigma X_1^2 - N\bar{X}_1^2 = 27\cdot649$

$\Sigma x_2^2 = \Sigma X_2^2 - N\bar{X}_2^2 = 100\cdot829$

$\Sigma x_3^2 = \Sigma X_3^2 - N\bar{X}_3^2 = 0\cdot198$

$\Sigma x_1 x_2 = \Sigma X_1 X_2 - N\bar{X}_1\bar{X}_2 = 42\cdot174$

$\Sigma x_1 x_3 = \Sigma X_1 X_3 - N\bar{X}_1\bar{X}_3 = 0\cdot770$

$\Sigma x_2 x_3 = \Sigma X_2 X_3 - N\bar{X}_2\bar{X}_3 = 2\cdot574$

Accordingly

$$b_2 = \frac{\Sigma x_1 x_2 \Sigma x_3^2 - \Sigma x_1 x_3 \Sigma x_2 x_3}{\Sigma x_2^2 \Sigma x_3^2 - (\Sigma x_2 x_3)^2}$$

$$= \frac{6\cdot368}{13\cdot339}$$

$$= 0\cdot477$$

and

$$b_3 = \frac{\Sigma x_1 x_3 \Sigma x_2^2 - \Sigma x_1 x_2 \Sigma x_2 x_3}{\Sigma x_2^2 \Sigma x^2 - (\Sigma x_2 x_3)^2}$$

$$= -\frac{30\cdot918}{13\cdot339}$$

$$= -2\cdot318$$

The regression of X_1 on X_2 and X_3 is given by

$$X_{1_c} - \bar{X}_1 = b_2(X_2 - \bar{X}_2) + b_3(X_3 - \bar{X}_3)$$

i.e. $X_{1_c} - 2\cdot798 = 0\cdot477(X_2 - 15\cdot358) - 2\cdot318(X_3 - 0\cdot449)$

i.e. $X_{1_c} = -3\cdot487 + 0\cdot477 X_2 - 2\cdot318 X_3$

We now test the significance of the two partial regression coefficients. First we set up the hypothesis that the data are a random sample from a population in which there is no partial regression of lamb slaughterings on sheep numbers, i.e. in which $\beta_2 = 0$. We then perform the t-test by calculating

$$t = \frac{b_2}{s_{b_2}}$$

where
$$s_{b_2} = s_{X_1}(e) \sqrt{\frac{\Sigma x_3^2}{\Sigma x_2^2 \Sigma x_3^2 - (\Sigma x_2 x_3)^2}}$$

and
$$s_{X_1}(e) = \sqrt{\frac{\Sigma x_1^2 - b_2 \Sigma x_1 x_2 - b_3 \Sigma x_1 x_3}{N - 3}}$$

Here $s_{X_1}(e) = 0.788$ and $s_{b_2} = 0.096$, so that $|t| = 4.97$.

Referring to the t-table, we have for 15 degrees of freedom $P(|t| > 4.97) < 0.01$. If the hypothesis were true, repeated sampling with 15 degrees of freedom would yield a value of $|t|$ as great as or greater than 4.97 in less than one case in 100 due to chance. We conclude that the partial regression of lamb slaughterings on sheep numbers is significant.

Secondly, we set up the hypothesis that the data are a random sample from a population in which there is no partial regression of lamb slaughterings on the relative lamb-wool price, i.e. in which $\beta_3 = 0$. Again we perform the t-test by calculating

$$t = \frac{b_3}{s_{b_3}}$$

where
$$s_{b_3} = s_{X_1}(e) \sqrt{\frac{\Sigma x_2^2}{\Sigma x_2^2 \Sigma x_3^2 - (\Sigma x_2 x_3)^2}}$$

Here $s_{b_3} = 2.166$, so that $|t| = 1.07$.

Referring to the t-table, we have for 15 degrees of freedom $P(|t| > 1.07) = 0.3$. If the hypothesis were true, repeated sampling with 15 degrees of freedom would yield a value of $|t|$ as great as or greater than 1.07 in about 30 cases in 100 due to chance. Accordingly, we do not reject the hypothesis. The observed data are not inconsistent with the absence of partial regression of lamb slaughterings on the relative lamb-wool price. This means that, as far as these data are concerned, there is no evidence that the relative lamb-wool price has any value as an explanatory variable of the behaviour of lamb slaughterings. Accordingly the multiple regression equation computed above is not an improvement on the simple regression equation computed in the example in the preceding section as an explanation of the behaviour of lamb slaughterings. This is reflected in the square of coefficient of multiple correlation. We have

$$R^2 = \frac{b_2 \Sigma x_1 x_2 + b_3 \Sigma x_1 x_3}{\Sigma x_1^2}$$
$$= 0.66$$

Hence, 66 per cent of the variation in lamb slaughterings can be explained by the multiple regression. This is a trivial increase on the 64 per cent which the simple regression explained, and is not statistically significant.

On the other hand, if it had turned out that the partial regression of lamb slaughterings on the relative lamb-wool price was significant, the multiple regression would then have provided a better explanation of the behaviour of lamb slaughterings than the simple regression, and the multiple regression could have been used for predicting lamb slaughterings for given sheep numbers and relative lamb-wool price according to the formulae given on p. 246 above. The computed multiple regression would have been interpreted by saying that, on the average for every increase of 1 million in sheep numbers, relative lamb-wool prices remaining constant, we should expect the number of lambs slaughtered to increase by 477,000, and for every increase of one unit in, say, the 1st decimal place of the ratio of lamb to wool prices, sheep numbers remaining constant, we should expect the number of lambs slaughtered to decrease by 232,000.[1]

10.9. Relation between Simple and Multiple Regression

Suppose we have a simple regression estimated from sample data, between two variables X_1 and X_2, where X_2 is the explanatory variable. The regression will be

$$(X_{1_c} - \bar{X}_1) = b(X_2 - \bar{X}_2)$$

If b is significantly different from zero, we can say that X_2 contributes to the explanation of the behaviour of X_1. If we add another explanatory variable X_3, we shall have

$$(X_{1_c} - \bar{X}_1) = b_2(X_2 - \bar{X}_2) + b_3(X_3 - \bar{X}_3)$$

If both b_2 and b_3 are significantly different from zero, we can say that both X_2 and X_3 contribute to the explanation of the behaviour of X_1, so that the addition of the variable X_3 increases the extent to which the behaviour of X_1 can be explained. In these circumstances the second equation will generally be a better predicting instrument than the first.

It is important to emphasize that the b in the first equation is not the same thing as the b_2 in the second one. The partial regression coefficient b_2 tells us the amount by which X_1 will on the average change per unit change in X_2 when X_3 is taken as constant, i.e. b_2 indicates the way in which X_1 and X_2 are related under conditions in which X_3 is always the same, whereas b measures the amount of change allowing X_3 to vary in the way it has for the particular sample of observations under consideration. Indeed it is quite possible for b_2 to be not significant, although b is significant. Consider, for example, the case referred to on p. 225 above, which was concerned with the behaviour of family size (X_1). If we compute the simple regression of family size on father's I.Q. (X_2), we may find that b is significant, but if we introduce a second explanatory variable, say, family income (X_3), we may find that b_2 in the multiple regression of family size on father's I.Q. *and* family income is not significant. This may occur if there is a relationship between

---

[1] The sign of the partial regression coefficient of lambs slaughtered on the relative lamb-wool price is the opposite to that anticipated *a priori*. If the coefficient had been significant, it would have been necessary to examine this in order to provide some explanation.

family size and family income, and one between family income and father's I.Q. As a result of these relationships a simple regression of family size on father's I.Q. may give a significant result, although there is no direct relation between family size and father's I.Q. For a given level of family income the two variables behave independently. Accordingly, if we have a simple regression of X_1 on X_2 which is significant, and we suspect that the apparent relationship may be due to the influence of a third variable (X_3), we should compute the multiple regression of X_1 on X_2 and X_3 and test the significance of the partial regression coefficient of X_1 on X_2. If this is not significant, then our suspicions are confirmed, and the data under consideration cannot be taken as evidence of a real relationship between X_1 and X_2.

CHAPTER XI

TIME SERIES

11.1. Objectives of the Analysis of Time Series

Most economic data are recorded over time. A series of observations recorded over time is called a *time series*, e.g. Australian National Income 1946–47 to 1966–67, coal production 1914–1967, etc. The objects of the analysis of time series are, first, to describe the past behaviour of time series, and, secondly, to analyse this behaviour.

Empirical economic investigations are very largely dependent on the use of data arranged in time series. Suppose, for example, we wish to investigate the relation between the price of tea and the quantity demanded. Ideally, we should take an economy, set the price of tea at 50 cents per lb, observe the quantity demanded; vary the price, observe again, and so on. This is the experimental method available to physical scientists working in the laboratory. Unfortunately, the economist can seldom utilize this method. His data at best consist of actually recorded prices and quantities at various times, i.e. a time series of prices and a parallel one of quantities purchased. The demand curve for tea, if it is to be estimated, must somehow be drawn out of these data, and an analysis of time series is an indispensable part of the process. The same sort of considerations apply if we are trying to determine the nature of the consumption function for a particular economy or the relation between changes in investment and changes in national income.

The analysis of time series has developed in the main as a result of investigations into the nature and causes of those fluctuations in economic activity called *trade cycles*. Economic theory has suggested various explanations of trade cycles. Analysis of time series has attempted to test the plausibility or otherwise of these theories. At the same time, such analysis may suggest new hypotheses for economic theorists to work on. A large number of individual time series reveal cyclical fluctuations, and the relationship between these series can be investigated. Does one series lead another or lag behind it in its fluctuations? Are the movements of one series merely the reflection of the movements in another? These are the sorts of question which the analysis of time series attempts to answer.

Some economists believe that an analysis of time series may enable them to set up a model showing how the economy works and indicating the main forces which determine whether we have boom or depression.

With such a model it might then be possible to forecast, at least for a short period ahead, what is likely to happen to the level of economic activity in the economy as a whole or even in some particular sector or industry. Thus one of the objects of the analysis of time series is to forecast how these series will behave in the future on the basis of how they have behaved in the past.

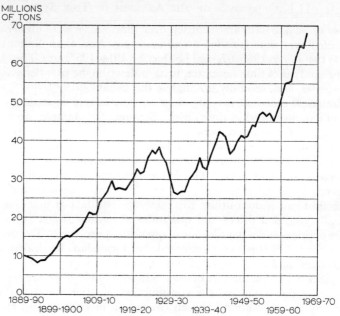

FIG. 11.1. GOODS AND LIVESTOCK CARRIED BY AUSTRALIAN RAILWAYS, 1889–90 TO 1966–67

Source: Commonwealth Bureau of Census and Statistics, Australia: *Year Book*, No. 40, 1954, p. 124; *Transport and Communication Bulletin*, No. 5, 1911, p. 12; No. 11, 1919, p. 20; No. 20, 1928, p. 10; No. 30, 1938–39, p. 10; No. 40, 1948–49, p. 6; No. 47, 1955–56, p. 9; No. 50, 1958–59, p. 7; No. 57, 1965–66, p. 7; and *Quarterly Summary of Australian Statistics*, No. 268, 1968, p. 48

The preceding paragraphs have indicated the important part which time series play in empirical economic investigations. This chapter is concerned with some of the main techniques used in the analysis of time series.

11.2. CHARACTERISTIC BEHAVIOUR OF TIME SERIES

In examining any particular time series it is essential first to graph it in order to obtain a general picture of its behaviour. As an example of time series, the tonnage of Australian railway freight over a large number of years is shown in Fig. 11.1.

If a large number of fairly long time series were examined, the following characteristics would be observed in many of them—
1. Secular trend.
2. Periodic movements.
3. Erratic movements.

The term *secular trend* is taken to mean the general long-term movement of the series. It need not have any particular shape, but the idea of trend implies a persistent movement in one direction or the other. The trend of a particular series may be due to, for example, the growth of population or a long-run increase in productivity or a steady change in economic habits. But the "trend" itself is relative to the period to which the series refers, for an upward trend over a short period may only be an upward cyclical movement contained within a longer period.

Periodic movements may be of two kinds. There is first the strictly periodic movement which is associated with *seasonal variation*. Many economic time series are subject to this type of movement. Seasonal variation is evident when the data are recorded at weekly or monthly or quarterly intervals. Although the amplitude of seasonal variations may vary, their period is fixed—being one year. As a result, seasonal fluctuations do not appear in series of annual figures. Secondly, annual figures may themselves show more or less periodic movements about the trend. These movements are called *cyclical movements*. The amplitude and the period of the cycles may not be very regular, but in many series which reflect economic activity in one way or another, a cycle with a period of some eight or nine years is not uncommon. These cycles are what are usually called *trade cycles*, but minor cycles of about three years have been discerned in United States data, and there is some evidence that very long waves of about fifty years' period exist, although these are rather mixed up with secular trends.

Erratic movements may also be of two kinds. There is first the strictly *random (chance) movements* which turn the series first one way and then another in a purely chance manner. Secondly, certain isolated or irregular, but powerful, movements crop up from time to time—the influence of a strike or a political upheaval or a war. These may be called *episodic movements*. In a very long period these might be randomly distributed, but in the sorts of periods we have to analyse these movements may occur only once or twice, and they must be regarded as distinct from the more random movements which operate all the time.

11.3. BASIC ASSUMPTIONS IN THE ANALYSIS OF TIME SERIES

The immediate objective of the analysis of time series is to break down the series into the main components which reflect the secular trend, the periodic movements and the erratic movements. In other words, the movement of the series is regarded as being compounded of these elements, and an attempt is made to reveal the magnitudes of each of these elements separately, showing how the movements of the separate components together account for the movement of the series.

The method of analysis depends very largely on the hypothesis as to how the components of the series are combined and interact. The simplest hypothesis is to assume that the separate influences have values which are additive and independent of each other. The latter assumption means that, for example, the seasonal influence will be the same irrespective of which phase of the cycle obtains. Thus we have

$$Y_t = T_t + C_t + S_t + E_t$$

where Y_t is the value of the variable at a particular time t, T_t is the trend value, C_t is the cyclical variation, S_t is the seasonal variation, E_t is the erratic variation. C_t will have positive or negative values according to whether we are in an "above normal" or "below normal" phase of the cycle; in other words, the cycle is conceived of as a fluctuation around a normal trend movement. The total positive and negative values for one cycle will be zero. Much the same considerations apply to the S_t term. S_t will be positive or negative according to the season of the year, and the total of S_t for a year will be zero. If the data are annual, the term S_t will disappear. E_t will also have positive or negative values, and in the long run ΣE_t will be zero. Occasionally E_t may take on an extreme value on account of some extraordinary occurrence, but such a movement should be regarded as an isolated occurrence.

In Fig. 11.2 there are shown separately a trend, a cyclical variation and a random term over a number of years. These are then compounded to give a time series of the form

$$Y_t = T_t + C_t + E_t$$

The data are supposed to be on an annual basis, so that no seasonal variation appears.

An alternative hypothesis to the above one is that

$$Y_t = T_t \cdot C_t \cdot S_t \cdot E_t$$

where C, S and E, instead of being positive or negative values, are indexes fluctuating above or below unity. Since, by taking logarithms, we get

$$\log Y_t = \log T_t + \log C_t + \log S_t + \log E_t$$

it will be appreciated that there is no difference in principle between the two hypotheses. Both assume that the time series components are additive, but in the second case it is the logarithms which are added. If it is believed that the cyclical, seasonal and erratic forces operate with equal absolute effect irrespective of the trend value, then the first hypothesis is appropriate. If it is believed that they operate proportionately to

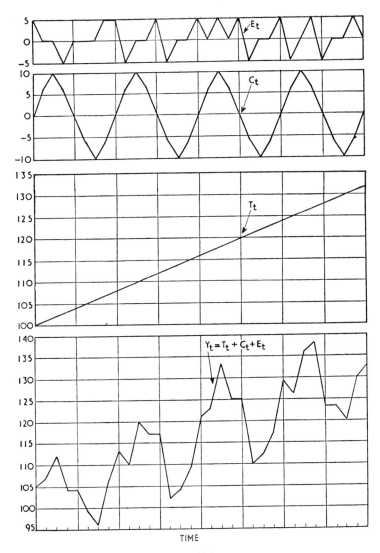

Fig. 11.2

the general level of the series, then the second is appropriate. The first regards the cyclical, seasonal and erratic variations as being plus or minus so many units, the second plus or minus such and such a percentage.

In practice a time series built up on either of the above hypotheses can be fairly readily decomposed into its elements. Fairly simple methods are available to ascertain the components T_t and S_t. If we subtract T_t and S_t from Y_t, we obtain a residual element which will be $C_t + E_t$. It may be quite difficult to separate off E_t from this, but generally one would suppose E_t to be small compared with C_t, so that this will not matter much. One of the objects of time series analysis is to obtain the residual component, so that comparison of the cyclical patterns in various series can be made.

One could cast a good deal of doubt on the plausibility of the basic hypothesis outlined above. Is it reasonable to assume that the cyclical and trend components are determined by separate forces acting independently and then simply added together? In fact it is quite likely that this year's value of the variable will depend to some extent on last year's value, so that trend and cycle will get inextricably mixed and no meaningful separation of them will be possible. In such a situation a chance variation this year may affect the whole future course of the series, and it will not be possible to say how much of the value of the variable in any given year is "due to" one or another of the components set out earlier. Thus an equation of the form

$$Y_t = [a(1 + g)^t] + [bY_{t-1} + c(Y_{t-1} - Y_{t-2})] + E_t$$

where a, g, b and c are constants and E_t is a random addition at time t, will for certain values of b and c generate a series with cycles about a trend. But, although the first square bracket contains the term which contributes the trend, being a straight-out growth function like compound interest, and E_t is the erratic component, no meaning can be attached to separating out the components, for the second square bracket which contains the mechanism which "causes" the cycle cannot stand on its own, depending as it does on the value of the variable as a whole in preceding periods. A change in the trend or in E_t will change the whole nature of the cycle in subsequent periods.[1]

It should now be clear that the analysis of time series is a matter of great complexity, depending very much on the theory held regarding the generation of time series in general and of the given time series under consideration in particular. We shall now give methods for the analysis of trend and seasonal variation, strictly applicable only on the basis of the additive hypothesis, but nevertheless of a deal of general utility.

---

[1] See, for example, Tinbergen, J., and Polak, J. J.: *The Dynamics of Business Cycles* (Routledge and Kegan Paul, 1950), Chapter 16, p. 252.

11.4. Measurement of Trend

Freehand Drawing

The immediate object of estimating the trend of a time series is to describe the general underlying movement of the series. Always the first thing to do is to graph the series. One simple way of obtaining the trend is then to draw a smooth freehand curve through the series. The main objection to this procedure is that it is necessarily rough and ready and the result will depend very much on the drawer of the trend himself, so that it is not an objective process.

Moving Averages

A second method for obtaining the trend is to attempt to smooth out the bumps in the series by a process of averaging. This can be done by what are called *moving averages*. Suppose we have annual data. Then, for example, a three-year moving average is obtained by averaging the values, first of periods 1 to 3, then of periods 2 to 4, etc. The averages are taken to represent the "smoothed-out" values relating to the period in the middle of the span of the average. Thus the three-year average of periods 1 to 3 is centred at period 2, that of 2 to 4 at period 3, etc. This is illustrated as follows—

Period t	Value Y_t	Moving Average (3-year)
1	Y_1	
2	Y_2	$\frac{1}{3}(Y_1 + Y_2 + Y_3)$
3	Y_3	$\frac{1}{3}(Y_2 + Y_3 + Y_4)$
4	Y_4	$\frac{1}{3}(Y_3 + Y_4 + Y_5)$
5	Y_5	$\frac{1}{3}(Y_4 + Y_5 + Y_6)$
6	Y_6	.
.	.	.
.	.	.

If the moving average has an even number of terms, a second averaging process must be employed to centre the moving averages at periods rather than between periods. This can be seen at the top of p. 258.

If the series consists of erratic fluctuations around a trend, a moving average will tend to reduce and smooth out these fluctuations. This must occur because the average of a number of terms must always lie between the smallest and largest of the terms. The larger the number of terms in the moving average, the smoother the resulting series; but the larger the number of terms the more information is lost at the end and the beginning of the series. A three-year moving average is often used for this kind of work.

On the other hand, if the series consists of periodic movements about

Period t	Value Y_t	Moving Average (4-year)	Moving Average (2-year) of Moving Average (4-year)
1	Y_1		
2	Y_2		
3	Y_3	$\frac{1}{4}(Y_1 + Y_2 + Y_3 + Y_4)$	$\frac{1}{2}[\frac{1}{4}(Y_1 + Y_2 + Y_3 + Y_4) + \frac{1}{4}(Y_2 + Y_3 + Y_4 + Y_5)]$
4	Y_4	$\frac{1}{4}(Y_2 + Y_3 + Y_4 + Y_5)$	$\frac{1}{2}[\frac{1}{4}(Y_2 + Y_3 + Y_4 + Y_5) + \frac{1}{4}(Y_3 + Y_4 + Y_5 + Y_6)]$
5	Y_5	$\frac{1}{4}(Y_3 + Y_4 + Y_5 + Y_6)$	$\frac{1}{2}[\frac{1}{4}(Y_3 + Y_4 + Y_5 + Y_6) + \frac{1}{4}(Y_4 + Y_5 + Y_6 + Y_7)]$
6	Y_6	$\frac{1}{4}(Y_4 + Y_5 + Y_6 + Y_7)$	$\frac{1}{2}[\frac{1}{4}(Y_4 + Y_5 + Y_6 + Y_7) + \frac{1}{4}(Y_5 + Y_6 + Y_7 + Y_8)]$
7	Y_7	$\frac{1}{4}(Y_5 + Y_6 + Y_7 + Y_8)$	.
8	Y_8	$\frac{1}{4}(Y_6 + Y_7 + Y_8 + Y_9)$	.
9	Y_9	.	.

a trend, a moving average with a span equal to the period of the cyclical movements will iron out these movements. This can be illustrated by the following example, in which an eight-year cycle has been artificially combined with a trend to produce the variable shown in the fourth column. An eight-year moving average appropriately centred eliminates the cyclical fluctuations and reduces the variable to the original trend.

Table 11.1

Period t	Trend T_t	Cycle C_t	Value $T_t + C_t$	Moving Average (8-year)	Moving Average (2-year) of Moving Average (8-year)
0	100	0	100		
1	101	6	107		
2	102	10	112		
3	103	6	109		
4	104	0	104	103·5	104
5	105	− 6	99	104·5	105
6	106	− 10	96	105·5	106
7	107	− 6	101	106·5	107
8	108	0	108	107·5	108
9	109	6	115	108·5	109
10	110	10	120	109·5	110
11	111	6	117	110·5	111
12	112	0	112	111·5	
13	113	− 6	107		
14	114	− 10	104		
15	115	− 6	109		

If the trend is not linear, moving averages will show a bias compared with the original trend so that the application of moving averages will not reveal the original trend. This can readily be seen by replacing the T_t column in the above table with a non-linear trend. Furthermore, if the period and/or the amplitude of the cycle is not constant, moving averages will not remove the whole of the cyclical element. Another difficulty in using moving averages is that, if they are calculated for a series of completely random fluctuations, they tend to produce a series with spurious periodic elements in it. Moving averages are useful in connexion with measuring seasonal variation, for the period of seasonal variation is always twelve months (see section 11.8), but we will not pursue the method in connexion with the estimation of trend.

Mathematical Curves

A third method for obtaining the trend, which we will now consider in some detail, is to fit a mathematical curve to the time series. This method has certain advantages. Once the form of the curve has been decided, the fitting of the curve, unlike the freehand drawing of one, is quite objective. Furthermore, the form of a mathematical curve can be interpreted in terms of the behaviour of the variable, so that a particular form of curve implies a certain type of trend. Thus, for example, a curve of the form $Y = a + bX$, where Y is the value of the variable and X measures time, means that the variable is changing by a constant amount per unit of time. This results in a convenient summarization of the trend, and, perhaps, a logical description of it. This is an advantage not shared by moving averages.

The main method employed to fit mathematical curves to time series data is that of least squares, which has already been discussed in section 10.1 above. We select the constants in our mathematical expression, whatever it is, such that $\Sigma(Y - Y_T)^2$ is a minimum, where the Y's are the observations over time and the Y_T's the corresponding trend values as defined by the mathematical expression.[1]

11.5. Fitting a Mathematical Trend

Linear Trend

The simplest form of trend is a *linear trend*. It is defined by
$$Y_T = a + bX$$
where Y_T is the trend value of the variable under consideration, here assumed for convenience to be on an annual basis, and X is time. Such a trend is shown in Fig. 11.3 on p. 261.

The constant a is the trend value of Y at the origin of time (i.e. where $X = 0$) and the constant b is the increase in Y_T per unit of time (i.e. per unit change in X). We can place the origin of X at any place we please, provided we remember where we have placed it, e.g. if the series extends from 1947–1967, we can put 1947 = 0 and then 1957 = 10 and 1967 = 20 or we can put 1957 = 0 and then 1947 = $-$ 10 and 1967 = + 10, etc. Since the curve $Y_T = a + bX$ is a continuous curve, the origin corresponds to a point of time. If the data refer to measurements over whole years, the origin should be at the middle of a

---
[1] The symbol Y_T used here is equivalent to the symbol T used on p. 254 above.

year, whereas if they refer to measurements at a point of time, it should be at such a point of time. For example, coal production in calendar years should have its origin at the 30th June of a year, population as at the end of calendar years should have its origin at 31st December of a year.

The normal equations to determine a and b by least squares (see section 10.1, p. 200) are

$$\Sigma Y = Na + b\Sigma X$$
$$\Sigma XY = a\Sigma X + b\Sigma X^2$$

where N is the number of years covered by the time series. If the number of years covered is odd, we can place the origin at the middle year and then $\Sigma X = 0$. This is very convenient, for then the normal equations reduce to

$$\Sigma Y = Na$$
and
$$\Sigma XY = b\Sigma X^2$$

i.e.
$$a = \bar{Y} \quad \text{and} \quad b = \frac{\Sigma XY}{\Sigma X^2}$$

When stating a trend, the units and the origin *must* be attached. An equation without them is quite meaningless. The origin should be expressed as a date, not as a year.

EXAMPLE 11.1

Fit a linear trend to the annual figures for freight carried by civil aviation overseas services over the years 1954–55 to 1966–67 given on p. 262, and compute the trend values of freight carried for the years under consideration.

Here we have $a = \dfrac{\Sigma Y}{N} = \dfrac{391 \cdot 9}{13} = 30 \cdot 15$, $b = \dfrac{\Sigma XY}{\Sigma X^2} = \dfrac{860 \cdot 6}{182} = 4 \cdot 73$

and the trend is
$$Y_T = 30 \cdot 15 + 4 \cdot 73 X$$

where Y_T is the trend value of annual freight carried in millions of ton-miles and X is financial years with origin at 31.12.60 (the middle of 1960–61).

This can be interpreted by saying that over the years under consideration the trend of freight carried has shown an increase of 4·73 million ton-miles per annum. The trend values corresponding to the observed data can easily be calculated by substitution in the trend equation and are given in the last column of the table.[1]

[1] For some purposes trend values for months (or quarters) are required. These can be obtained by substituting the appropriate value for X, expressed in years and fractions of years, in the trend equation. When the variable is such that the annual data, upon which the trend is based, are the sums of monthly (or quarterly) values, the trend value obtained upon substitution must be appropriately scaled down. For example, to obtain the trend value for September, 1964, in example 11.1, we centre this value at 15th September and substitute $X = 3\frac{11}{24}$ in the equation. This gives $Y_T = 47 \cdot 69$. This is the trend value of *annual* freight carried centred at 15.9.64. The corresponding *monthly* trend value is $47 \cdot 69 \div 12 = 3 \cdot 97$ million ton-miles.

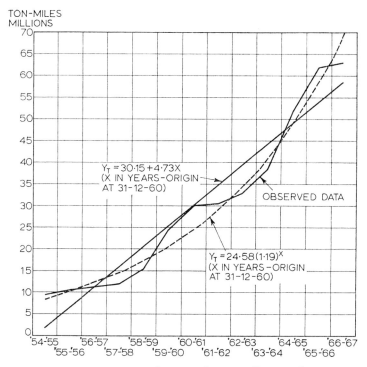

Fig. 11.3. Civil Aviation Overseas Services—Freight Carried, Australia, 1954–55 to 1966–67

Source: Commonwealth Bureau of Census and Statistics, Australia: *Year Book*, No. 42, 1956, p. 256; No. 47, 1961, p. 265; No. 51, 1965, p. 583; *Quarterly Summary of Australian Statistics*, No. 268, 1968, p. 54

Fig. 11.3 shows for the preceding example the observed data together with the linear trend line. The broken line should for the present be ignored. Although the data reveal a distinct upward trend, it can be seen that it is not really linear in character.

If the number of years happens to be even, the origin can be put between the middle two years (i.e. at the end of the $\frac{N}{2}$th year), and time expressed in terms of half-years. In this case ΣX will again equal 0. In the equation $Y_T = a + bX$, the variable X will then be expressed in half-years, and the origin will be at the end of the $\frac{N}{2}$th year. To convert the equation into one referring to full years, we shall have

$$Y_T = a + 2bX$$

262   APPLIED STATISTICS FOR ECONOMISTS

Year	Freight Carried by Civil Aviation Overseas Services (a) (millions of ton-miles) Y	X	XY	X^2	Y_T
1954–55	9·4	−6	−56·4	36	1·77
1955–56	10·5	−5	−52·5	25	6·50
1956–57	11·2	−4	−44·8	16	11·23
1957–58	12·1	−3	−36·3	9	15·96
1958–59	15·4	−2	−30·8	4	20·69
1959–60	24·6	−1	−24·6	1	25·42
1960–61	30·1	0	0	0	30·15
1961–62	30·4	1	30·4	1	34·88
1962–63	33·1	2	66·2	4	39·61
1963–64	38·6	3	115·8	9	44·34
1964–65	51·8	4	207·2	16	49·07
1965–66	61·8	5	309·0	25	53·80
1966–67	62·9	6	377·4	36	58·53
	391·9	0	860·6	182	

(a) Source: Commonwealth Bureau of Census and Statistics, Australia: *Year Book*, No. 42, 1956, p. 256; No. 47, 1961, p. 265; No. 51, 1965, p. 583; *Quarterly Summary of Australian Statistics*, No. 268, 1968, p. 54

where X is in full years and the origin is at the end of the $\frac{N}{2}$th year, since b measures the change per half-year and hence $2b$ measures it per full year. To shift the origin to the middle of a year we must add the change for a half-year to the constant a, since this constant represents the value of Y_T at the origin. We shall have

$$Y_T = a + b + 2bX$$

where X is in full years and the origin is in the middle of the $\left(\frac{N}{2} + 1\right)$th year. This is illustrated in the example below.

EXAMPLE 11.2

Fit a linear trend to the annual figures for freight carried by civil aviation overseas services over the years 1954–55 to 1965–66 given on p. 263.

Here we have $\quad a = \dfrac{\Sigma Y}{N} = \dfrac{329}{12} = 27·42, \qquad b = \dfrac{\Sigma XY}{\Sigma X^2} = \dfrac{1,295·4}{572} = 2·26$

and the trend is

$$Y_T = 27·42 + 2·26X$$

where X is in half-years with origin at 30th June, 1960. This can be readily converted to X in full years and origin at 31st December, 1960, as in the other example. The coefficient of X is the increase in Y per unit increase in X. Hence, converting X into full years doubles the coefficient, i.e.

$$Y_T = 27·42 + 4·52X$$

TIME SERIES 263

Year	Freight Carried by Civil Aviation Overseas Services (millions of ton-miles) Y	X	XY	X^2
1954–55	9·4	−11	−103·4	121
1955–56	10·5	−9	−94·5	81
1956–57	11·2	−7	−78·4	49
1957–58	12·1	−5	−60·5	25
1958–59	15·4	−3	−46·2	9
1959–60	24·6	−1	−24·6	1
1960–61	30·1	1	30·1	1
1961–62	30·4	3	91·2	9
1962–63	33·1	5	165·5	25
1963–64	38·6	7	270·2	49
1964–65	51·8	9	466·2	81
1965–66	61·8	11	679·8	121
	329·0	0	1,295·4	572

where X is in years with origin at 30th June, 1960. To shift the origin along half a year, we must add half a year's increment to the constant 27·42, in order to obtain the value of Y_T at the origin, i.e.

$$Y_T = 27·42 + 2·26 + 4·52X$$

i.e.
$$Y_T = 29·68 + 4·52X$$

where X is in financial years with origin at 31st December, 1960.

The trends in the above example and the preceding one are different, since they are based on a different number of observations.

Exponential Trend

The linear trend
$$Y_T = a + bX$$

is one in which Y_T increases by a constant absolute amount per annum. This amount is given by the coefficient b. Many other types of trends are available. A commonly used one is where Y_T increases not by a constant absolute amount per annum, but by a constant percentage per annum. Such a trend is defined by

$$Y_T = AB^X$$

where A and B are constants. For example, if $B = 1·01$, then Y_T increases by 1 per cent per annum. This type of trend is called an *exponential trend*.

It will be recalled from section 4.4 above that a function of this form when plotted on a semi-logarithmic scale gives a straight line. An exponential trend is shown on an arithmetic scale by the broken line in Fig. 11.3 on p. 261 and on a semi-logarithmic scale in Fig. 11.4 on p. 265.

An exponential trend is readily fitted by taking logarithms, for

$$\log Y_T = \log A + X \log B$$

This is now similar in form to

$$Y_T = a + bX$$

where Y_T is replaced by $\log Y_T$, a is replaced by $\log A$, and b is replaced by $\log B$.

To fit such a curve we write down the logarithms of the Y's and proceed as before. By suitable choice of the origin, we shall have

$$\log A = \frac{\Sigma \log Y}{N}$$

and
$$\log B = \frac{\Sigma X \log Y}{\Sigma X^2}$$

By looking up the anti-logarithms of $\log A$ and $\log B$, we can readily ascertain A and B. As before, when the number of years is even, X can be expressed in half-years, and the conversion of the equation into full years and appropriate shift of origin can be performed by adjusting the logarithmic form of the equation.

EXAMPLE 11.3

Fit an exponential trend to the annual figures for freight carried by civil aviation overseas services over the years 1954–55 to 1966–67, and compute the trend values of freight carried for the years under consideration.

Year	Freight Carried by Civil Aviation Overseas Services (millions of ton-miles) Y	$\log Y$	X	$X \log Y$	X^2	$\log Y_T$	Y_T
1954–55	9.4	0.9731	−6	−5.8386	36	0.9381	8.67
1955–56	10.5	1.0212	−5	−5.1060	25	1.0135	10.31
1956–57	11.2	1.0492	−4	−4.1968	16	1.0889	12.28
1957–58	12.1	1.0828	−3	−3.2484	9	1.1643	14.60
1958–59	15.4	1.1875	−2	−2.3750	4	1.2397	17.37
1959–60	24.6	1.3909	−1	−1.3909	1	1.3151	20.65
1960–61	30.1	1.4786	0	0	0	1.3905	24.58
1961–62	30.4	1.4829	1	1.4829	1	1.4659	29.23
1962–63	33.1	1.5198	2	3.0396	4	1.5413	34.77
1963–64	38.6	1.5866	3	4.7598	9	1.6167	41.37
1964–65	51.8	1.7143	4	6.8572	16	1.6921	49.21
1965–66	61.8	1.7910	5	8.9550	25	1.7675	58.55
1966–67	62.9	1.7987	6	10.7922	36	1.8429	69.65
	391.9	18.0766	0	13.7310	182		

Here we have
$$\log A = \frac{\Sigma \log Y}{N} = \frac{18 \cdot 0766}{13} = 1 \cdot 3905$$

$$\log B = \frac{\Sigma X \log Y}{\Sigma X^2} = \frac{13 \cdot 7310}{182} = 0 \cdot 0754$$

and the exponential trend is
$$\log Y_T = 1 \cdot 3905 + 0 \cdot 0754 X$$
i.e.
$$Y_T = (24 \cdot 58)(1 \cdot 19)^X$$

with X in financial years with origin at 31st December, 1960.

This can be interpreted by saying that over the years under consideration the trend of freight carried has shown an increase of 19 per cent per annum.

To obtain trend values for individual years it is easiest to work in logarithms from the equation
$$\log Y_T = 1 \cdot 3905 + 0 \cdot 0754 X$$

and then look up the anti-logarithms. This has been done in the last two columns of the above table.

Fig. 11.4 shows for the preceding example the observed data, together with the exponential trend line, on a semi-logarithmic scale. In Fig. 11.3 the exponential trend is also shown on an arithmetic scale alongside the linear trend for the same data. It can be seen that in this

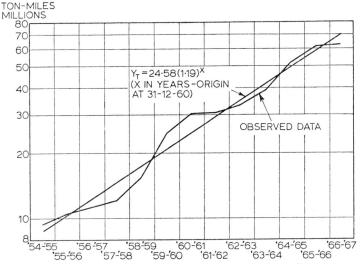

FIG. 11.4. CIVIL AVIATION OVERSEAS SERVICES—FREIGHT CARRIED, AUSTRALIA, 1954-55 TO 1966-67
(Semi-logarithmic Scale)
Source: Commonwealth Bureau of Census and Statistics, Australia: *Year Book*, No. 42, 1956, p. 256; No. 47, 1961, p. 265; No. 51, 1965, p. 583; *Quarterly Summary of Australian Statistics*, No. 268, 1968, p. 54

case the exponential trend gives a better description of the underlying movement of the series than the linear one.

Other Mathematical Trends

The linear and exponential are the simplest of trend types. Polynomials of the form

$$Y_T = a + bX + cX^2 + dX^3 + \ldots$$

may also be used, and these can be fitted by the method of least squares. For example, to fit the parabola

$$Y_T = a + bX + cX^2$$

the normal equations for determining a, b, c are derived as before from the conditions for minimizing $\Sigma(Y - Y_T)^2$ and are

$$\left.\begin{array}{l}\Sigma Y = Na + b\Sigma X + c\Sigma X^2 \\ \Sigma XY = a\Sigma X + b\Sigma X^2 + c\Sigma X^3 \\ \Sigma X^2 Y = a\Sigma X^2 + b\Sigma X^3 + c\Sigma X^4\end{array}\right\}$$

By suitable choice of the origin ΣX and ΣX^3 can be made to vanish and we have

$$\left.\begin{array}{l}\Sigma Y = Na + c\Sigma X^2 \\ \Sigma XY = b\Sigma X^2 \\ \Sigma X^2 Y = a\Sigma X^2 + c\Sigma X^4\end{array}\right\}$$

The second equation now gives b directly, and a and c can be obtained by solving simultaneously the first and third equations.

More complicated forms may be used, in which the fitting is rather more difficult. For example, a well-known curve is the *logistic* curve. Its equation is

$$Y_T = \frac{k}{1 + e^{a-bX}}$$

where k, a and b are constants (k and b positive) and $e = 2\cdot 71828$. This curve is S-shaped and is asymptotic to the X-axis for large negative X and to the abscissa $Y = k$ for large positive X. It is useful for describing a series which starts growing slowly, then grows rapidly and finally reaches a saturation point, e.g. the sales of a new product, the population of insects reproducing in a confined space, etc. It is graphed in Fig. 11.5.

In selecting a trend to fit to a given series, the choice of the curve is largely a matter of the subjective opinion of the fitter. But the idea is to use a trend which will describe the general underlying movement of the data. Furthermore, the movement in the trend should be one which is reasonable on *a priori* grounds. Thus a linear trend implies change by a constant amount per period, an exponential one change by a constant percentage, and a parabolic one implies that the amount of change per period itself changes by a constant amount per period (the first derivative of a parabola being linear), and so on. If the data when

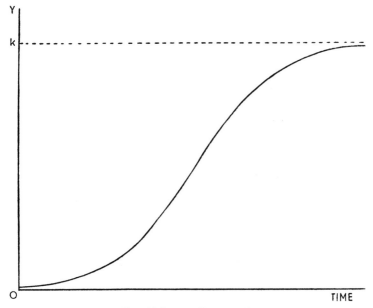

Fig. 11.5. The Logistic Curve

graphed on arithmetic paper give the impression of a straight line, then the linear trend $Y_T = a + bX$ should be used. If, when graphed on semi-logarithmic paper, they give one of a straight line, then the exponential trend $Y_T = AB^X$ should be used. For the use of more complicated trends a good deal of experience in the practice of fitting trends is required. However, higher degree polynomials should be used with caution for the very reason that the more constants in the equation the closer the fit. By selecting a function of sufficiently high degree it would always be possible to pass the curve through every point since a function with n disposable constants can be passed through n points. However, this would negate the whole idea of a trend. Moreover, it would be very difficult to justify the use of high-degree curves on theoretical grounds, since their rates of change cannot be interpreted in the simple ways in which they can for linear and exponential functions.

11.6. Interpretation of a Mathematical Trend

Mathematical curves are useful to *describe* the general movement of a time series, but it is doubtful whether any analytical significance should be attached to them, except in special cases. It is, of course, possible to interpret a mathematical trend analytically by regarding the trend value of the variable as dependent on time in some operative sense, e.g. the size of a population may be regarded as a function of the time which has elapsed since the beginning of the population. Thus we could say that a variable Y measured at any particular time is made up of a systematic component dependent on time and a random term, say

$$Y = [\alpha + \beta X] + \varepsilon$$

where X is time, α and β are parameters and ε is a random fluctuation. Here the term in the square brackets is the trend component of Y. If we have N observations of Y over time, we can fit by least squares

$$Y_T = a + bX$$

and having tested that b differs significantly from zero, use a and b as estimates of α and β respectively. This approach corresponds to that discussed in section 10.7 above, where it was shown how regression analysis can be used to test a theory about the way in which a dependent variable can be "explained" by independent variables. However, it is seldom possible to justify on theoretical grounds any *real* dependence of a variable on the passage of time. Variables do change in a more or less systematic manner over time, but this can usually be attributed to the operation of other explanatory variables. Thus many economic time series show persistent upward trends over time due to a growth of population or to a general rise in prices, e.g. national income, and the trend element can to a considerable extent be eliminated by expressing these series *per capita* or in terms of constant purchasing power. For these reasons mathematical trends are generally best regarded as tools for describing movements in time series rather than as theories of the causes of such movements. It follows that it is extremely dangerous to use trends to forecast future movements of a time series. Such forecasting, involving as it does extrapolation, can be valid only if there is theoretical justification for the particular trend as an expression of a functional relationship between the variable under consideration and time. But if the trend is purely descriptive of past behaviour, it can give few clues about future behaviour. Often the extrapolation of a trend gives ridiculous results which themselves are *prima facie* evidence that the trend could not be maintained, e.g. if the world population continued to grow at the average rate of the past ten years, there would by the year A.D. 4000 be about $2\frac{1}{2}$ million million million million people on the globe.

Moreover, the notion of trend is itself very elusive, because unless we can observe very long time series we can never be certain that what appears as a trend is not in fact the upward or downward phase of a cyclical movement and the trend is in fact quite different. Consequently, the concept of trend movements as distinct from cyclical movements is relative to the period of time observed. This does not prevent us from using mathematical curves to describe the movements of series, but it does make it difficult to sustain any contention that we have located *the* trend of a series. This point is illustrated in Fig. 11.6. If the period *AB* is considered in isolation, a clear *downward* trend is apparent. However, if a larger period is considered, it is seen to be a cyclical phase in a long-period *upward* trend.

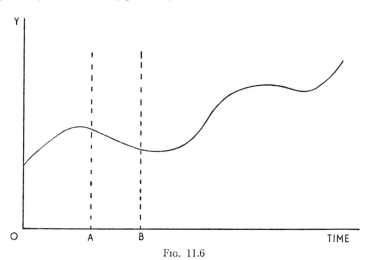

Fig. 11.6

11.7. Use of Time Series in Correlation and Regression

It was shown in section 10.7 above how correlation and regression analysis can be used to test hypotheses about economic relationships. Very often the only data available are in the form of time series, and special care is needed in correlating data which are in this form. It may well happen that two variables exhibit a high degree of correlation over time not because they are related in any way but because other factors have produced persistent trends causing both series to rise together or the one to rise and the other to fall steadily.[1] Such circumstances as these may produce nonsense correlations. Consequently, in order to render a correlation between time series meaningful we

[1] See examples on pp. 5–6 and 225.

must try to eliminate the influence of these other factors on the two series. But it may be very difficult to determine precisely what these factors are. One way out is to regard the trends in the series as measures of their influence. We then remove the trends from the series and correlate the *trend-free* series. The resulting correlation will indicate whether the position of the one variable relative to its trend is related to that of the other variable relative to *its* trend.

The removal of trend from a series is quite straightforward. If we regard the variable as being compounded as follows—

$$Y_t = T_t + C_t + E_t$$

the trend-free series is given by

$$Y_t - T_t$$

however T_t is arrived at. If T_t has been derived as a mathematical trend, it will correspond to Y_T in section 11.5 above. On the other hand, if we regard the variable as

$$Y_t = T_t \cdot C_t \cdot E_t$$

the trend-free series is given by

$$\frac{Y_t}{T_t}$$

Once the trends (if any) are removed, the two series can be correlated in the usual manner.

As an illustration of the need for using trend-free series in certain cases, we may quote an example given by Tippett.[1] Consider the annual marriage rate per 1,000 of population and an index of real wages for England and Wales, 1850–1890. The former shows a downward trend and the latter an upward one, both trends being approximately linear. Hence a correlation of the two original series will give substantial negative correlation, i.e. the higher the level of real wages the lower the marriage rate. But is this correlation of any value? We cannot conclude from it that higher real wages "cause" a lower marriage rate, because the trends in the two series were caused by independent unrelated factors—the trend in the marriage rate being due to later marriage and an ageing population and that in real wages to increasing industrial efficiency. What we want to know is whether, when the index of real wages is above its trend, the marriage rate tends to be above or below its trend or whether the two variables are independent. By correlating trend-free series we may be able to throw some light on this, and in this particular case such a correlation gives an appreciable degree of positive correlation, i.e. the marriage rate and the level of real wages tend to move above or below their respective trends

[1] Tippett, L. H. C.: *Statistics* (O.U.P., 1943), 1st edition, pp. 37, 56.

together. This result would support the hypothesis that economic conditions influence the marriage rate.

The above refers to the correlation of two variables the data for which are in the form of time series. Similarly, if we are interested in the regression of a dependent variable Z on an independent variable Y, where the data are in the form of time series, the regression should be performed in terms of trend-free data, so that it will be[1]

$$Z - Z_T = h(Y - Y_T)$$

where h is the regression coefficient of trend-free Z on trend-free Y. Moreover, if both the trends are linear, so that Z_T and Y_T can be written

$$Z_T = l + mX \quad \text{and} \quad Y_T = p + qX$$

where l, m, p and q are constants and X is time, the above regression equation can be written

$$Z = (l - hp) + hY + (m - hq)X$$

This is an ordinary multiple regression equation of Z on Y and X. The regression coefficient of Y is h, as in the first formulation, so that h can be determined by the multiple regression methods of section 10.8 above. It measures the change in Z per unit change in Y, with the influence of time (i.e. the trends) eliminated.

11.8. Measurement of Seasonal Variation

When data are expressed annually there is no seasonal variation. But monthly or quarterly data frequently exhibit strong seasonal movements (see Fig. 11.7, p. 275) and considerable interest attaches to devising a pattern of average seasonal variation. It may be desired to compare the seasonal patterns of different series, but more often we may want to know the extent to which we should discount the most recently available statistics for seasonal factors. For example, average weekly earnings in Australia were $61·90 per week in the March quarter 1967–68 and then rose to $65·90 in the June quarter. Was this due to an underlying upward tendency or simply because the June quarter is usually seasonally higher than the March quarter? If we knew by how much the June quarter is usually above or below the March quarter for seasonal reasons, we could answer this question. In what follows we shall deal with quarterly movements, but monthly movements can be treated in identical fashion.

In order to analyse seasonal variation it is necessary to assume that the seasonal pattern is superimposed on a series of values and is

---

[1] Since $\Sigma(Z - Z_T) = \Sigma(Y - Y_T) = 0$, there will be no constant term in the regression,

independent of these in the sense that the same pattern is superimposed irrespective of the level of the series, e.g. the September quarter always contributes so much more or so much less to the series. For this purpose we can use the hypothesis about the composition of a time series set out in section 11.3 above

$$Y_{it} = T_{it} \cdot C_{it} \cdot S_i \cdot E_{it}$$

where the subscripts refer to the ith quarter and the tth year. Here however, it is only necessary to assume the independence of S_i and of E_{it} which is much more reasonable than assuming that all the terms are mutually independent. Consequently, the use of this hypothesis for the estimation of seasonal variation is of much greater generality than its use for analysing trend and cyclical movements. The component S_i will be above or below unity according to the quarter of the year, and is assumed to be the same from year to year. Its average value for the year will be unity,[1] e.g. S_i might have the values 1·00, 1·04, 0·95, 1·01, indicating that the March quarter has no seasonal influence, the June quarter adds 4 per cent for seasonal reasons, the September quarter subtracts 5 per cent and the December quarter adds 1 per cent. The component E_{it} is a random term with a mean value tending to unity for any given i.

The problem is: given a series of values for Y_{it}, to determine the unknown component S_i. There are several ways of doing this but the method of moving averages given below is perhaps the most satisfactory in practice and is relatively simple to apply. The *first step* is to take a four-quarter moving average of the series. This will give values centred between quarters, and a further two-quarter average of the four-quarter averages must be taken to centre the moving averages at quarters. Since the period of the average is the same as the period of the seasonal variations, such an average will tend to iron out the seasonal variations, and it will also smooth out the random fluctuations represented by the E_{it} term. As a result the moving averages will be an estimate of $(T_{it} \cdot C_{it})$, i.e. of the trend-cyclical components. This is illustrated in Fig. 11.7 below. If we now express the observed data (Y_{it}) as a ratio to the moving averages, we shall have a series consisting of seasonal and random components, i.e.

$$R_{it} = \frac{Y_{it}}{T_{it} \cdot C_{it}} = S_i \cdot E_{it}$$

where R_{it} is the ratio of the observed data to the moving average in the ith quarter of year t. Ratios such as these are illustrated in Fig. 11.8

---

[1] Strictly speaking, the average used here should be a geometric average, since the hypothesis is multiplicative, i.e. additive in the logarithms of the variables.

below. If we take a particular quarter and average the ratios for this quarter, we shall get for the ith quarter

$$\bar{R}_i = S_i \cdot \bar{E}_{it}$$

where the average is taken over the various years. But in the long run $\bar{E}_{it} = 1$, so this procedure will average out the random fluctuations, leaving us with the seasonal component $\bar{R}_i = S_i$. This is the *second step*. This would complete the process were it not for the fact that the moving averages are often biased estimates of $T_{it} \cdot C_{it}$. As has been pointed out on p. 258 above, a moving average may not only smooth out periodic and erratic fluctuations, but in doing so may produce a series which is biased with respect to the original series on which the periodic and erratic fluctuations have been superimposed. Consequently, $\bar{R}_i$, the mean of the ratios for a particular quarter, may not in fact give S_i, since R_{it} may not equal $(S_i \cdot E_{it})$. In this case we can write

$$\bar{R}_i = S_i \cdot B$$

where B stands for the bias resulting from the moving averages being biased estimates of $(T_{it} \cdot C_{it})$, assumed here (not unreasonably) to be the same for all four quarters. This bias can be detected by averaging $\bar{R}_i$ for the four quarters. Since the average of S_i equals unity we shall have

$$\bar{R} = B$$

where $\bar{R}$ is the mean of the four $\bar{R}_i$. Where $\bar{R}$ differs from unity, bias exists. The *third step* is then to adjust S_i for bias by writing

$$S_i = \bar{R}_i / \bar{R}$$

so that the mean of S_i equals unity. The values of S_i are called an *index of seasonal variation*.

Since the pattern of seasonal variation is itself likely to change, it is usual to base an index of seasonal variation on a moderately short period. It must be emphasized that any seasonal pattern is valid only for the period on which it is based, and it can be used beyond that period only on the assumption that it remains unchanged. It is hardly necessary to point out that it is not worth while computing an index of seasonal variation unless there is clear evidence of a regular seasonal pattern. This evidence can best be obtained by graphing the data.

The above treatment is rather complicated, and perhaps the simplest way to explain the estimation of an index of seasonal variation is to do so by working through an example. We take as data the average weekly earnings of wage- and salary-earners in Australia 1958–59 to 1967–68. The observed values are given in Table 11.2 below and shown as the bold line in Fig. 11.7. The presence of seasonal variation is evident from the diagram. The computation is carried through in three steps.

Table 11.2

Year and Quarter		Average Weekly Earnings. Observed Values (a) $	4-quarter Moving Totals $	Centred Moving Averages $	Ratios of Observed Values to Moving Averages
1958–59	S	40·40			
	D	42·80			
	M	38·40	162·80	40·99	0·937
	J	41·20	165·10	41·54	0·992
1959–60	S	42·70	167·20	42·28	1·010
	D	44·90	171·00	43·30	1·037
	M	42·20	175·40	44·21	0·955
	J	45·60	178·30	44·95	1·014
1960–61	S	45·60	181·30	45·53	1·002
	D	47·90	182·90	45·84	1·045
	M	43·80	183·80	46·05	0·951
	J	46·50	184·60	46·29	1·005
1961–62	S	46·40	185·70	46·61	0·995
	D	49·00	187·20	47·03	1·042
	M	45·30	189·00	47·39	0·956
	J	48·30	190·10	47·68	1·013
1962–63	S	47·50	191·30	47·96	0·990
	D	50·20	192·40	48·28	1·040
	M	46·40	193·80	48·68	0·953
	J	49·70	195·60	49·34	1·007
1963–64	S	49·30	199·10	50·03	0·985
	D	53·70	201·10	50·58	1·062
	M	48·40	203·50	51·40	0·942
	J	52·10	207·70	52·34	0·995
1964–65	S	53·50	211·00	53·23	1·005
	D	57·00	214·80	54·16	1·052
	M	52·20	218·50	55·04	0·948
	J	55·80	221·80	55·70	1·002
1965–66	S	56·80	223·80	56·25	1·010
	D	59·00	226·20	56·79	1·039
	M	54·60	228·10	57·43	0·951
	J	57·70	231·30	58·24	0·991
1966–67	S	60·00	234·60	59·11	1·015
	D	62·30	238·30	60·16	1·036
	M	58·30	243·00	61·18	0·953
	J	62·40	246·40	62·09	1·005
1967–68	S	63·40	250·30	63·03	1·006
	D	66·20	253·90	63·91	1·036
	M	61·90	257·40		
	J	65·90			

(a) Source: Commonwealth Bureau of Census and Statistics, Australia: *Wage Rates and Earnings*, November 1966, p. 14 and June 1968, p. 36

1. Calculating the Moving Averages and the Ratios of Observed Data to Moving Averages

First we calculate four-quarter moving averages of the data and then two-quarter moving averages of the four-quarter moving averages to centre them properly. This is best done by calculating four-quarter moving totals, adding adjacent totals and dividing by 8. These moving averages will represent the series with the seasonal and random components largely smoothed out. They are graphed in Fig. 11.7 as the

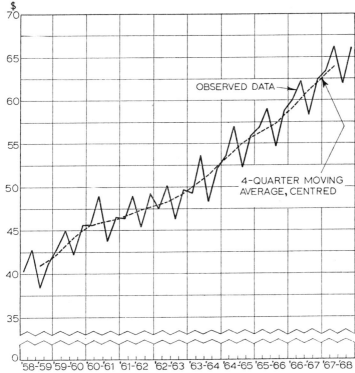

Fig. 11.7. Average Weekly Earnings by Quarters—Australia, 1958–59 to 1967–68
Source: Commonwealth Bureau of Census and Statistics, Australia: *Wage Rates and Earnings*, November 1966, p. 14 and June 1968, p. 36

dotted line. Next we calculate the ratios of observed values to moving averages. These ratios will contain the seasonal and random components together with any bias resulting from the application of the moving average procedure. They are shown in Fig. 11.8 and indicate a clear seasonal pattern, the December quarter being high and the March quarter low.

2. Averaging the Ratios

We arrange the ratios in columns for each quarter and average the values in each column. This will eliminate the random fluctuations in the ratios. In averaging these ratios the straight-out arithmetic mean is not always to be recommended. This is because occasionally an irregular large fluctuation may occur, which must be regarded as of

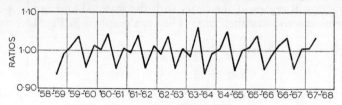

Fig. 11.8. Average Weekly Earnings by Quarters—Australia, 1958-59 to 1967-68
Ratio of Observed Data to Moving Averages

an *episodic* character and should be completely omitted from the calculations. Hence the median is often used or else a modified mean of the middle so many items. When the number of observations per quarter is small, the latter is probably to be preferred. However, in this case there seems to be no reason for not taking an arithmetic mean of all the items. It is convenient at this stage to omit the decimal point, using 1,000 as a base instead of unity.

Table 11.3

Ratios of Observed Values to Moving Averages

	September Quarter	December Quarter	March Quarter	June Quarter
1958–59			937	992
1959–60	1,010	1,037	955	1,014
1960–61	1,002	1,045	951	1,005
1961–62	995	1,042	956	1,013
1962–63	990	1,040	953	1,007
1963–64	985	1,062	942	995
1964–65	1,005	1,052	948	1,002
1965–66	1,010	1,039	951	991
1966–67	1,015	1,036	953	1,005
1967–68	1,006	1,036		
Average	1,002	1,043	950	1,003

3. Adjusting the Averages to Obtain the Index

As explained above, the average ratios just obtained may contain bias. With a base of 1,000, these averages should average 1,000, i.e. add to 4,000. In this particular case they add to 3,998, indicating negligible bias. Even so they should be adjusted to average 1,000. This is done by multiplying each by the factor 4,000/3,998, so that we have the index, as in the following table—

Table 11.4

	Average Ratios	Index of Seasonal Variation
September Quarter	1,002	1,002
December Quarter	1,043	1,044
March Quarter	950	950
June Quarter	1,003	1,004
Mean	999·5	1,000

The index can be interpreted by saying that due to purely seasonal influences average weekly earnings in the December quarter, for example, tend to be about 4·4 per cent above what they otherwise would be. The seasonal pattern is shown diagrammatically in Fig. 11.9. It must be thought of as being superimposed on what the series would be without seasonal influences.

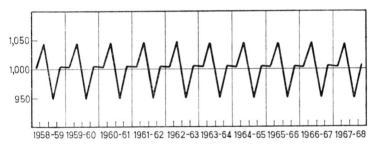

Fig. 11.9. Average Weekly Earnings—Australia, 1958–59 to 1967–68
Index of Seasonal Variation

To *deseasonalize* a series, for which we have an index of seasonal variation, we divide the observed data by the index. The resulting figures then give a series indicating movements abstracted from seasonal influences. Consider the years 1966–67 and 1967–68 in Table 11.5.

It will be noted that the fluctuations over these two years can be wholly accounted for by seasonal influences. We should not conclude, for example, that because average earnings rose by 4 dollars per week between the March and June quarters 1967–68, average earnings are on a strong upward trend, This rise was very largely seasonal, and the same applies to the substantial rise between the September and December quarters.

Whenever up-to-date information on current economic trends is required, monthly or quarterly data are necessary. If these are subject to seasonal influences, they must be deseasonalized before they can be used as indicators of current trends. Nevertheless, unless the seasonal

Table 11.5

AVERAGE WEEKLY EARNINGS—AUSTRALIA,
1966–67 AND 1967–68

Year and Quarter		Observed Data $	Seasonal Index	Deseasonalized Data $
1966–67	S	60·00	1,002	59·88
	D	62·30	1,044	59·67
	M	58·30	950	61·37
	J	62·40	1,004	62·15
1967–68	S	63·40	1,002	63·27
	D	66·20	1,044	63·41
	M	61·90	950	65·16
	J	65·90	1,004	65·63

pattern is reasonably constant, little reliance can be placed on an index purporting to measure seasonal variation. If a seasonal index is to be applied continuously as fresh data become available, a careful watch should be kept for any changes in the seasonal pattern, and in any case the index should be kept up to date by frequent revisions. By a careful examination of the ratios of observed values to the corresponding moving averages changes in seasonal pattern can be detected, and no index should be computed without such examination.

Reference is made to the official publication of deseasonalized economic time series in Australia in Appendix B.1, p. 474 below. The method used, although sophisticated in its technique, is based on the principles set out above.

CHAPTER XII

SOCIAL ACCOUNTS AND THE MEASUREMENT OF NATIONAL INCOME

12.1. The Nature of Social Accounts

BUSINESSES keep records of their transactions in the form of accounts. Amongst other things these accounts provide information upon which the business enterprises can base policy decisions. They provide a continuous commentary on the affairs of business concerns and are essential to management in controlling efficiency, in ensuring financial stability and a satisfactory rate of profit, and in facilitating balanced growth. The content of the accounts differs from business to business and is dictated, to a large extent, by the type of business and by the information which it is desired to extract from the accounts. In effect, the accounts may be regarded as a system through which the many individual transactions of the business are classified. In the course of a year a business enters into many transactions—it buys, it sells, it pays wages, it distributes profits, it borrows money, etc. A list of all these transactions, in chronological order, could be kept, but it would not be very useful. It would be very long and would present a mass of unorganized data. By entering the transactions into appropriate accounts, the transactions can be classified into like categories, can be aggregated and summarized, and presented in a form which leads to much more interesting and relevant information than can be obtained from the mass of unorganized transactions. Looked at from this point of view, the accounts of a business are essentially a system of classification.

Just as there are many reasons which make it necessary for individual businesses to keep records of their transactions, so it is most important to have records of the transactions which take place in the national economy as a whole. These records are called *social accounts*. If appropriate classifications are made, information can be derived about the annual income of the nation, how it is produced, distributed and spent, how the wealth of the nation is being built up, and so on. Such information provides a basis for national economic policy; it helps governments in their attempts to maintain economic stability and prosperity, and to ensure an efficient distribution of economic resources and a balanced growth of the economy. Moreover, empirical investigation of the working of an economy depends on the availability of data about aggregates of transactions of the kind recorded in the social accounts.

In drawing up social accounts for an economy there are two main problems. First, the system of classification of transactions must be selected, i.e. criteria of classification must be set up; and secondly, the transactions (or their aggregates) must be measured. As far as the former is concerned, there is no unique system of classification. The system to be used will depend on the purposes for which the social accounts are intended and on certain postulated theoretical relationships between different classes of transactions. Generally speaking, social accounts are designed to reveal the significant interrelationships between those aggregates of transactions which play important roles in the theory of the determination of the level of economic activity,[1] e.g. consumption, investment, etc. Consequently, the criteria of classification are chosen so that the social accounting system will provide the desired information in a relevant form. These criteria are discussed in detail below. Moreover, given the criteria, there is no unique form of the resulting accounts. The system described in the following sections is one which is very useful for purposes of exposition, although it does not correspond exactly with the systems used by various countries in their official publications.[2]

12.2. CLASSIFICATION OF ECONOMIC TRANSACTIONS

We may start by defining an *economic transaction* as a transaction between two entities which can be measured in money, e.g. the sale of boots by a shop to an individual, the payment of wages by a firm to an employee, the payment of tax by an individual to the Government, the loan of money by a bank to a firm, etc. Since two entities are involved in a transaction, it must have a two-sided character, for what one entity gives, the other must receive. Consequently, any individual transaction may be recorded from the point of view of the giver and from that of the receiver. When both the giving and receiving aspects are recorded in the one set of accounts, the system is described as *double-entry accounting*. The double-entry method is used both in ordinary commercial accounting and in the kind of social accounting system with which we are mainly concerned in this chapter, which we may describe more explicitly as national income accounting.

Economic transactions are measured in money terms. In some cases the *quid pro quo* for the money which passes is a good or service, e.g. the purchase of cigarettes or the employment of a tradesman. In these cases the money value of the transaction can be taken

---

[1] See, for example, Harcourt, G. C., Karmel, P. H., and Wallace, R. H.: *Economic Activity* (Cambridge University Press, 1967), Chapters 2, 3 and 4.

[2] These are influenced to a considerable extent by the availability of data. The Statistical Office of the United Nations has published a standard system of accounts in an attempt to standardize the practice of individual countries and organizations. (See United Nations, Studies in Methods No. 2: *A System of National Accounts and Supporting Tables.*) The accounts set out in this chapter differ in detail, although not generally in principle, from this standard form. It is believed that the system described here has certain advantages for purposes of exposition.

also as a measure of a "real" transaction. In other cases the *quid pro quo* is a financial claim, e.g. borrowing by a business from a bank. Often there may be no *quid pro quo* at all, e.g. an allowance by a father to his son. Certain other events, which may not appear to come under the definition of economic transactions, nevertheless need to be recorded in the accounting system. Such events may be internal operations affecting only one entity, e.g. depreciation of fixed assets or produce consumed by a farmer himself. Although these events do not involve the passing of money, they may be valued in terms of money and can be included with the economic transactions already described.

Over a period, say a year, there are myriads of transactions in an economy. To obtain any useful information about the working of the economy, it is necessary to classify the transactions in some way so that like transactions can be aggregated and summarized. Each transaction can be classified in two basic respects—

(i) according to the entities affected, i.e. according to the types of transactors; and

(ii) according to the nature of the economic activity involved in the transaction.

1. *Types of Transactors*

Reduced to the simplest terms, the economic process consists of the production of goods and services. The rationale of this process is the use of these goods and services. Except in the simplest type of economy the producer of a good is not the consumer of the good, and the good must be transferred from producers to consumers. This is usually done through the market, although some services are not provided through the market, e.g. government services, the services of housewives, etc. Generally, goods are sold by producer to consumer for money, which the consumer has acquired by assisting in the production of other goods. We can, therefore, immediately distinguish those entities which are responsible for organizing the production of goods and services. These we call *trading enterprises*, and we speak of the *trading enterprises sector* of the economy. The term "trading enterprise" is used in the widest sense. It includes all forms of productive activity—manufacturing businesses and trading organizations, as well as entities selling services to the community. It covers incorporated businesses such as public and private companies, state-owned business undertakings and unincorporated businesses such as partnerships and sole proprietorships. Under the latter would be such diverse enterprises as the master plumber and the lawyer, the landlord and the farmer. The distinguishing characteristic of the trading enterprise is that it organizes the factors of production (labour, capital equipment, natural resources) and thereby produces goods or services for sale.

Just as the producing entities form a sector of the economy, so do the consuming entities. These are the individuals who enjoy the fruits of the production organized by the trading enterprises. These we call *persons* and the sector is called the *personal sector*. The trading enterprises produce goods and services. This production results in incomes being paid to the owners of the factors of production in the form of profits and wages, etc. For the most part, the owners of the factors of production are persons (exclusively so in the case of labour), and the incomes so earned enable the persons to consume or use up the production. The distinction between trading enterprises and persons is clear. The former organize production, usually with the object of making profits—the latter enjoy consumption with the object of satisfying their wants. Of course, a particular individual may be both a trading enterprise and a person, as is the case with the proprietor of an unincorporated business. This may seem confusing, but the individual's dual capacity is clear, for when he buys raw material for his business he is acting in a capacity different from that when he is buying chops for his dinner. The *personal sector* is sometimes called the *households sector*, and non-profit making institutions, which can be regarded as aggregations of persons, e.g. clubs and associations, are included in it.

The logical distinction between the trading enterprises sector and the personal sector can be made in even the simplest type of economy, though the separation and measurement of transactions may be difficult in the case of an economy of subsistence farmers. However, in the modern complex economy other sectors must also be distinguished, for there are entities involved in economic transactions which do not fit into the mould of either of the two sectors already delineated. First are governments, so we must consider a third sector, namely the *public authority* or *government sector*. Included here are all public authorities, e.g. Commonwealth, State and Local Authorities in Australia. The economic activities of governments are mainly concerned with the provision of *collective goods and services* for the collective consumption or use by the community. These are provided to the community as a whole—law and order, defence, health, education, roads, etc. They are financed by levying taxes or raising loans. In addition, by means of taxation and cash social service payments, governments are able to effect a redistribution of income. However, government-owned business undertakings, e.g. Post Office, Railways, Water and Sewerage, etc., are separated off from the government sector and included in the trading enterprises sector, because they, like private trading enterprises, produce goods and services for sale.

A fourth sector is the *financial enterprises sector*. In this are included banks, life assurance offices, superannuation funds and various lending agencies. These institutions are not included with trading enterprises because their function is essentially different. Their activities lie mainly in borrowing and lending. They are concerned with the provision of financial facilities to the economy as a whole—they "oil the wheels" of industry. They are thus more analogous to governments than to trading enterprises, although most of them do in fact operate in order to make profits. Their peculiarities make it desirable to treat them as a separate sector.

Since we are classifying the economic transactions of a particular economy, the four sectors already mentioned will cover the trading enterprises, persons, governments and financial enterprises of the particular economy under consideration only. However, there will be many transactions involving entities resident overseas. Consequently, we must have a fifth sector entitled *rest of world* to take care of those entities which are not resident in the economy, e.g. the entities to which we export and from which we import goods and services, the entities from which we raise overseas loans and so on.

We may now summarize the above by saying that we can classify the entities which undertake economic transactions into the following five sectors—

Trading Enterprises	(T)
Personal	(P)
Government	(G)
Financial Enterprises	(F)
Rest of World	(W)

We could make a more detailed classification—for example, by classifying trading enterprises by type of industry (e.g. primary, secondary or tertiary), or persons by occupational status (employer, worker-on-own-account, employee). But the above is adequate for most purposes.

2. *Nature of Transaction*

Goods and services can be used to satisfy current wants or to add to the accumulated wealth of the economy. Thus we can distinguish between *current* and *capital* transactions. On the current side, further distinctions can be made. The production of goods and services can be termed the production of income—the "income" being the money measure of the physical goods and services produced. Once income has been produced, it must be distributed (e.g. distributed from trading enterprises to persons) and once distributed, it must be disposed of (e.g. consumed). The three processes of *production, distribution* and *disposal* can be distinguished. Consequently, we can classify economic transactions according to whether they are concerned with

$$
\begin{aligned}
&\text{Production} &&(1)\\
&\text{Distribution} &&(2)\\
&\text{Disposal} &&(3)\\
&\text{Accumulation} &&(4)
\end{aligned}
$$

It is customary to distinguish sub-categories of transactions within the above four categories, e.g. transactions concerned with distribution may be subdivided into dividends, wages, interest, etc.

We may now combine our classifications according to type of transactor and nature of transaction. If we take a transaction involving a trading enterprise, it can fall into any one of the four classes 1, 2, 3, 4. If it falls into class 1, we can call it a T1 transaction. But from T's point of view it may be either a payment or a receipt. Consequently, in any class of transactions (such as T1) we must distinguish between payments and receipts. Thus we can regard the class T1 as a pigeonhole, subdivided into two compartments for the purpose of recording payments on the left-hand side and receipts on the right-hand side. For each sector we shall have four pigeon-holes subdivided in this way, so that altogether we shall have twenty pigeon-holes. Since economic transactions are necessarily two-sided,[1] any particular transaction can be recorded twice, once in the payments compartment of one pigeonhole and once in the receipts compartment of another pigeon-hole. For example, if I buy cigarettes from a tobacconist, the transaction will be sorted into the payments compartment of P3 (because it is a disposal of income by a person) and the receipts compartment of T1 (because it is connected with the production of income by a trading enterprise). What we have done in effect is to establish a system of double-entry accounts. For each sector we have set up a production account, a distribution account, a disposal account and a capital account. These are the *national income accounts*. An accounting form for the classification is convenient, but it is the system of classification rather than the form in which it is set down which is the important thing.

If we had a record of all transactions taking place in an economy over a year, we could sort them into their appropriate pigeon-holes and find out the total value of all the main types of transactions. We should then have a summary of the transactions for the economy. But before we proceed with this line of thought, it is necessary to consider just what sort of items we shall include in the four accounts of each sector.

---

[1] Some transactions take place between two accounts within the one entity, e.g. depreciation allowances. However, they should still be regarded as two-sided.

12.3. Content of National Income Accounts

A hypothetical set of national income accounts drawn up along the lines of this section is given on pp. 293–8. Reference to it may assist in the understanding of what follows. The production of goods and services is for the most part organized by the trading enterprises sector, although the government sector organizes the production of collective goods and services. The form of the production account (T1 Account) can best be understood by considering the contribution of an individual trading enterprise to the production of goods and services over a particular year. Clearly this contribution is not simply the value of the physical output of the enterprise because this includes the value of goods and services used in the production process which have been produced by other enterprises. The contribution to production of a particular enterprise is not great just because it uses expensive raw materials, the value of which is reflected in the value of its output. Rather, the productive contribution of an enterprise reflects *the work done within* the enterprise itself. Accordingly, we define the contribution to production of an enterprise as the value of its output over the course of the year under consideration, less the value of the inputs purchased from other enterprises and used up in the course of production. These latter are known as *intermediate* products, e.g. raw materials, legal advice, etc. They enter into the *final* products of the enterprise under consideration. Products which are intermediate to one enterprise are of course the final products of the enterprise producing them. The contribution to production as defined here is often referred to as the *value added*, for it is the value added to the raw materials, etc., by the process of production applied by the particular enterprise. It is also referred to as the *value of production* and should not be confused with the *value of output*, which is physical output valued at appropriate market prices.

The value of the annual output of the enterprise is made up of the value of its annual sales, plus (or minus) the value of any addition to (or subtraction from) its inventories of finished or semi-finished goods over the year. The value of its annual inputs purchased from other enterprises and used up in the course of production is its annual purchases from other enterprises for current productive activity (i.e. purchases of intermediate products), plus (or minus) the value of any net subtraction from (or addition to) inventories of inputs over the year. It follows that the contribution to production of an enterprise, which we now call its *gross product*, is

{ Sales of output during year
{ + Additions to inventories of finished or semi-finished goods during year
minus
{ Purchases of intermediate goods and services during year
{ + Subtractions from inventories of intermediate goods during year.

Stated in summary form, *gross product* is

{ Sales of output during year
+ Net additions to inventories of all kinds during year
minus
Purchases of intermediate goods during year.

This can be set out in account form as follows—

Production Account of a Trading Enterprise

Purchases of intermediate goods and services . . x	Sales of output . . x
Gross product . . . x	Additions to inventories . x

So far we have ignored the capital equipment (e.g. buildings, machinery, etc.) which the enterprise uses in conjunction with labour and intermediate goods to produce its final output. Purchases of such equipment are capital transactions and are not recorded in the production account. However, capital equipment is itself used up in the process of production (e.g. machinery wears out and becomes obsolete, etc.). This using up is called *depreciation*. It is as much an input as are intermediate products. Consequently, to obtain the net contribution of an enterprise to production we must deduct from the gross product an allowance for depreciation, so that we shall have

Net product = Gross product − Depreciation allowances

However, it is convenient to leave the production account as above and charge depreciation to the distribution account. This is because the estimation of depreciation is, to a considerable extent, arbitrary, since there can be no exact measure of how much of the value of a piece of capital equipment has passed into the output of the enterprise in a particular period. Accordingly, it can be regarded as an appropriation from the gross product.

The quantities referred to above are all expressed in money terms, and should be valued at current market prices. So long as price levels are stable this presents no difficulties, but if price levels are changing the valuation of the items "additions to inventories" and "depreciation allowances" presents great problems. Additions to inventories are usually measured by subtracting the value of inventories at the beginning of the year from that at the end. Accordingly during a period of changing prices some of the difference in the values may be due to the price changes themselves—for example, physical stocks might remain unchanged, but the closing value might be higher than the opening one. Similarly, if depreciation allowances are calculated on the basis of the original cost of the equipment divided by its expected life (a customary procedure in accounting), they will not reflect the current value of the equipment used up if prices have been changing. Since gross and net product are intended to be measures of the contribution of an enterprise to the production of goods and services measured in current money values, the effect of changing price levels is distorting. For certain purposes, e.g. the measurement of production, the item "additions to inventories" should be the value at current market prices of the physical additions to inventories (and not the additions to the value of inventories); and the item "depreciation allowances" should be the value at current market prices of the estimated portion of capital equipment "used up" over the course of the year. However, in general, the proper adjustments to make to "additions to inventories" and "depreciation allowances" depend on the purposes for which the social accounts are being used. The topic is extremely complicated, and it is not proposed to discuss it further.

We may now aggregate the production accounts of all trading enterprises to obtain the trading enterprises production account. The sales of outputs of trading enterprises can be split up as follows—

Sales to persons,
Sales of intermediate goods[1] to other trading enterprises,
Sales of capital equipment to other trading enterprises,
Sales to financial enterprises,
Sales to governments,
Sales to rest of world,

and the purchases of intermediate products will be—

Purchases from other trading enterprises,
Purchases from rest of world.

Since sales by all enterprises of intermediate goods to other enterprises must equal purchases by all enterprises of intermediate goods from other enterprises, these two items will cancel out in the aggregation and we shall have—

TRADING ENTERPRISES—PRODUCTION ACCOUNT (T1 ACCOUNT)

Purchases of intermediate goods and services from rest of world x	Sales to persons . . . x
	Sales of capital equipment to enterprises . . . x
	Sales to government . . x
	Sales to financial enterprises . x
	Sales to rest of world . . x
Gross product . . . x	Additions to inventories . . x

This explains the content of the Trading Enterprises Production Account (T1 Account). It will be appreciated that it is an account in a highly summarized form, involving the aggregation of myriads of individual transactions of a like nature. The balancing item in this account is the gross product of the trading enterprises sector.

[1] The goods are "final" from the point of view of the seller.

The Government Production Account (G1 Account) is more notional than real. Governments purchase certain goods and services from trading enterprises and the rest of world, employ public servants and, as a result, provide collective goods and services to the community. They do this without making direct charges (although the goods and services provided by governments are paid for indirectly through taxation). By analogy with the T1 Account we should define the "gross product" of the government sector as the value of the collective goods and services provided, *minus* the value of the goods purchased from trading enterprises and rest of world which "enter into" the collective goods and services. However, since these collective goods and services are not sold, they have no market price. We value them, conventionally, at cost, i.e. at the cost of the purchases entering into them, *plus* the wages of the public servants. We can call this amount the *value of government output*. Thus we have

GOVERNMENT—PRODUCTION ACCOUNT (G1 ACCOUNT)

Purchases of goods and services from trading enterprises	x	Value of government output	x
Purchases of goods and services from rest of world	x		
Gross product	x		

It should be evident that the gross product of the government sector is in fact equal to the wages paid to the public servants.

The financial enterprises sector is treated in precisely the same way as the government sector, so that its gross product is in fact equal by definition to the wages the sector pays out. The services which financial enterprises provide to the community are not fully charged for directly, but their cost is for the most part covered by the difference between interest receipts and interest payments. Interest is treated as a transfer payment passing through the distribution accounts of sectors, and it does not appear in the production accounts.[1]

The personal sector is by definition not concerned with production and hence has no production account.[2] The rest of world sector does not contribute to the production of the economy under consideration, although rest of world entities do, of course, contribute to the production of their own economies. Accordingly, there is no production account for the rest of world sector.

[1] This is the method of treatment used in the Australian social accounts and differs somewhat from the more complicated method suggested by the Statistical Office of the United Nations (*op. cit.*, p. 32). There is a good deal to be said for the Australian method. See Brown, H. P.: "Some Aspects of Social Accounting—Interest and Banks," *Economic Record*, Supplement, August, 1949, p. 73 and Harcourt, G. C., Karmel, P. H., and Wallace, R. H.: *op. cit.*, p. 14.

[2] Sometimes a production account is set up for persons, to include the contribution to production of domestic servants. However, there is no reason why domestic servants should not be regarded as trading enterprises.

So much for the production accounts of the various sectors. The gross products recorded in these accounts are carried down to the distribution accounts, where they are distributed to the owners of factors of production in the form of factor payments (wages, profits, etc.) in return for services rendered by the factors in the process of production. In addition, all current payments between sectors not in direct return for the rendering of services are recorded in the distribution accounts. Such payments are called *transfer payments*, e.g. personal taxes paid to the government, interest paid to and by financial enterprises, etc.

The balance of the distribution account of a sector is called the *disposable income* of the sector and is carried down to the disposal account. The disposal of income takes the form of the purchase of current goods and services for *final use* by the sector under consideration, i.e. consumption by persons and expenditure on current goods and services by governments or financial enterprises. That part of the disposable income of a sector not expended for this purpose is called the *savings* of the sector, saving being interpreted simply as not spending for current purposes. The savings item, which is carried down to the capital account of the sector, together with borrowing from other sectors (if any) finances the accumulation of fixed capital equipment and inventories and also any lending to other sectors. All these items are recorded in the capital accounts.[1]

Certain special features of the treatment of the accounts of the individual sectors should be noted.

1. *Trading Enterprises*

The gross product is carried down to the distribution account, where it is distributed to factors of production. Depreciation allowances are also charged to this account—their double-entry appears in the capital account, since they form part of the *gross savings* of enterprises from which the purchase of capital equipment is partly financed. Taxation is similarly charged to the distribution account. The profits of all unincorporated businesses (which include farms, professions, etc.) are deemed to be wholly withdrawn to the persons' distribution account, because it is not usually possible to distinguish between the personal income of the proprietor and the income of the business. In so far as proprietorship funds in the businesses are increased, they are deemed to be increased by lending (i.e. by the provision of capital) from the proprietors as persons to the enterprises they control. This process involves

---

[1] In some systems of social accounts, for example, in Australia (see footnote 2 on p. 280 above) the distribution and disposal accounts are combined to form one account, known as the appropriation account. For purposes of exposition it is better to keep the two accounts separate as they refer to essentially different activities.

entries in the capital accounts of the personal and trading enterprises sectors. The surpluses of government business undertakings are transferred to the distribution account of the government sector. It follows that the balance of the distribution account, i.e. the disposable income of the trading enterprises sector, must be the disposable income of companies, since the remainder of the revenue recorded in the distribution account will be paid out in one form or another. Furthermore, since companies cannot by their nature indulge in the final use of current goods and services, their disposable income is wholly saved and is known as *undistributed company profits*. These undistributed company profits are in fact *enterprise savings*. This item, enterprise savings, is carried down to the right-hand side of the capital account which also includes depreciation allowances of private trading enterprises and net borrowing from other sectors. (Depreciation allowances of government trading enterprises are credited to the capital account of the government sector.) The left-hand side of the capital account records purchases of capital equipment by private trading enterprises from other enterprises and additions to inventories, the double entries for which items appear on the right-hand side of the production account. Purchases of capital equipment and additions to inventories constitute the bulk of *gross private investment*, sometimes called *gross private capital formation*.[1] Purchases of capital equipment by government business undertakings and additions to their inventories are recorded in the capital account of the government sector.

2. *Government*

The gross product is carried down to the distribution account where it is paid out as wages to public servants (the gross product and these wages being identically equal by definition). The revenue side of the distribution account records taxation receipts, and the expenditure side records subsidies, social service cash payments, interest, etc. The balance is government disposable income, which, in the disposal account, is available for expenditure on current collective goods and services (e.g. current expenditure on health and education services), known as *government expenditure on current goods and services*. The surplus of disposable income over current expenditure is *government savings*. These savings are carried down to the capital account, where they are available to finance *government expenditure on capital works* (e.g. roads and buildings). This capital expenditure includes the capital formation of government business undertakings, and is sometimes called *public capital formation*. To the extent to which these undertakings make provision for depreciation, the depreciation allowances are credited to

---

[1] The two other items usually included are capital formation by financial enterprises and by persons in the form of houses for owner-occupation.

the capital account. When capital expenditure exceeds savings, the government sector must borrow from other sectors. The amount of borrowing is the *net increase in indebtedness of the government sector*. This will be negative if the government is a net lender. The net increase in indebtedness, sometimes called the *government deficiency*, must not be confused with the *published budget deficit*, which usually refers to the overall cash deficit or to the deficit in the main current account of a single government, whereas the government deficiency in the social accounts embraces current and capital activities of all public authorities.

The sum of government expenditure on current goods and services and on capital works is known simply as *government expenditure on goods and services*. On the assumption that governments do not make direct charges for administrative services rendered, this sum is identically the same as the value of government output as defined on page 288 above. In so far as governments do make such charges (e.g. for scientific advice, etc.) government expenditure on goods and services will be less than the value of government output by the amount of these fees, such fees being charged to the appropriate sector as expenditure items.

3. *Financial Enterprises*[1]

The treatment of this sector is analogous to that of the government sector. Corresponding to government expenditure on current goods and services is an item *financial enterprises expenditure on current goods and services*. This is equal by definition to the value of financial enterprises output *less* fees charged to other sectors; but, unlike governments, financial enterprises are deemed to purchase their capital equipment from trading enterprises rather than create it themselves, so that the double entry of the capital account item "purchases of capital equipment" is to be found in the trading enterprises production account and not in the financial enterprises production account.

The main items in the distribution account are interest receipts and payments for banks, and premiums and claims for life assurance companies. Because many of these enterprises are privately owned, dividends and taxes appear as debit items in the distribution account. Naturally the principal items in the capital account refer to borrowing and lending, since these are the main activities of financial enterprises. These items correspond to the changes which occur over the year in the liabilities (e.g. bank deposits) and assets (e.g. bank advances, bank holdings of government securities) of the enterprises.

4. *Personal*

Personal incomes are recorded in the distribution account and include not only incomes received for factor services rendered (e.g.

[1] See footnote 1 on p. 288 above.

wages and salaries) but also government social services cash payments (e.g. old-age pensions). The main debit in the distribution account is personal income taxation. The balance of the account is *personal disposable income*, which is carried to the disposal account to be expended on goods and services or saved. The former item is called *consumers' expenditure* and the latter *personal savings*. Personal savings is carried down to the capital account. It is customary to exclude from consumers' expenditure the purchase of houses for owner-occupation, since this expenditure is seldom financed from personal current disposable income. All other expenditure by persons is regarded as current consumers' expenditure, which therefore includes expenditure by consumers on durable goods such as cars, furniture, etc. Since houses are excluded from consumers' expenditure they must either be charged to persons capital account or house-ownership must be treated as a trading enterprise.[1] The balance of the personal capital account is lent to other sectors.

5. *Rest of World*

The accounts of this sector are drawn up to express the economic transactions of the home economy with the rest of the world, although the entries are made from the point of view of rest of world entities. Income received by rest of world entities from the home economy (e.g. dividends, interest, etc.) are credited to the distribution account, whereas similar payments by rest of world entities to the home economy are debited to that account. The balance is the disposable income of the rest of world sector. It may be positive or negative. It is the difference between the income produced by rest of world factors in the home economy and payable to the rest of world, and the income produced in the rest of world by home factors and receivable from the rest of world. This balance is carried down to the disposal account where it can be regarded as available for financing the excess of the rest of world sector's purchases of goods and services from the home economy (the home economy's *exports*) over the rest of world sector's sales of goods and services to the home economy (the home economy's *imports*). The balance in this account is *rest of world savings*. It is savings from the point of view of the rest of world sector and may be positive or negative.

The item "rest of world savings (or dissavings)" is carried down to the capital account, where it is balanced by lending by the rest of world sector (or borrowing by the rest of world sector). Lending by the rest

---

[1] The latter is the customary practice and is explained on pp. 306–308 below. However, for the sake of simplicity the former is used in the hypothetical example in the next section.

of world sector is, of course, equivalent to *borrowing from the rest of world* by the home economy, sometimes called *net increase in indebtedness of the home economy to the rest of world*. This occurs when income receivable from abroad and the current export proceeds of the home economy are insufficient to cover income payable abroad and the cost of imports. Contrariwise, borrowing by the rest of world sector is equivalent to *lending to the rest of world* by the home economy, and is sometimes called *net increase in indebtedness of the rest of world to the home economy*. The distribution and disposal accounts, viewed from the point of view of the home country, make up what is generally called the *balance of international payments on current account*, and rest of world savings (dissavings) correspond to a deficiency (surplus) in that account. The capital account is the *balance of international payments on capital account*.

Below is set out a hypothetical set of social accounts constructed in accordance with the principles outlined above. For simplification, the financial enterprises sector has been restricted to banks. Only major items have been included. More detail could be shown, if desired. Each entry contains a reference to the other account involved in the double entry. The entries are for the most part recorded on a net basis, e.g. some persons borrow from banks and some persons lend to banks, but if the amount lent exceeds that borrowed, the net amount is recorded as lending by persons to financial enterprises. These social accounts record the flows of money, goods and services between the different sectors of the economy classified according to the economic significance of the flows. They are a complete set of double-entry accounts and for this reason are said to be *articulated*. They give a record for the economy, just as the accounts of a private firm give a record for the firm. Basically they are a system of classification of economic transactions. As such they are not sacrosanct. Other systems are possible, but for many purposes the system given here is very useful.

TRADING ENTERPRISES
$million
PRODUCTION ACCOUNT (T1 ACCOUNT)

9. Purchases of intermediate goods from rest of world (W3)	330	1. Sales to persons (P3)		1,500
10. Gross product (T2)	2,255	2. Sales to persons (houses) (P4)		40
		3. Sales of capital equipment to trading enterprises (T4)		430
		4. Sales of capital equipment to financial enterprises (F4)		5
		5. Sales to governments (G1)		80
		6. Sales of current goods and services to financial enterprises (F1)		5
		7. Sales to rest of world (W3)		500
		8. Additions to inventories (T4)		25

Distribution Account (T2 Account)

13. Depreciation allowances (T4) . 100	11. Gross product (T1) . . 2,255
14. Indirect taxes (G2) . . 230	12. Government subsidies (G2) . 50
15. Direct taxes (G2) . . . 80	
16. Surplus of government business undertakings (G2) . . 20	
17. Wages (P2) 900	
18. Withdrawals, etc. (P2) . . 730	
19. Dividends payable to rest of world (W2) . . . 30	
20. Dividends payable at home (P2) 40	
21. Interest (F2) 100	
22. Disposable income (T3) . . 75	

Disposal Account (T3 Account)

24. Savings (T4) 75	23. Disposable income (T2) . 75

Capital Account (T4 Account)

30. Purchases of capital equipment from trading enterprises (T1) 430	25. Savings (T3) . . . 75
31. Additions to inventories (T1) . 25	26. Depreciation allowances (T2) . 100
	27. Borrowing (F4) . . . 130
	28. Borrowing (P4) . . . 120
	29. Borrowing (W4) . . . 30

Notes

1. Consumers' expenditure.
7. Exports.
9. Part of imports. Imports are purchased by trading enterprises and by governments. All imports for consumption (as well as for capital) purposes are the intermediate goods of importers, and enter into item 1 above.
14. Taxes not levied on income or wealth.
15. Income taxes on companies. All incomes of unincorporated businesses and persons are paid to P2 account and taxed there.
16. Excess of revenue over working expenses of business undertakings owned by governments.
18. Profits of unincorporated enterprises deemed to be wholly withdrawn by the proprietors.
19. Dividends paid by domestic companies to rest of world shareholders.
21. Interest on bank overdrafts, etc.
22. In effect, undistributed company profits.
27. Borrowing from banks, e.g. by overdrafts.
28. Borrowing from persons, e.g. by share issues or loans, or, in the case of unincorporated businesses, by the provision of capital by the proprietors.
29. Borrowing from rest of world by issue of shares and other securities to foreigners.
30 and 31. These items, together with item 2 above debited to the P4 account and item 4 debited to the F4 account, make up gross private investment.

GOVERNMENT
$million

Production Account (G1 Account)

3. Purchases of goods and services from trading enterprises (T1) .	80		1. Value of government output (current items) (G3) .	. 175
4. Purchases of goods and services from rest of world (W3) . .	70		2. Value of government output (capital items) (G4) .	. 125
5. Gross product (G2) . . .	150			

Distribution Account (G2 Account)

12. Wages (P2)	150		6. Gross product (G1) . .	. 150
13. Subsidies (T2) . . .	50		7. Direct taxes (P2) . .	. 250
14. Interest (F2)	55		8. Direct taxes (T2) . .	. 80
15. Interest (P2)	40		9. Direct taxes (F2) . .	. 5
16. Interest (W2) . . .	60		10. Indirect taxes (T2) . .	. 230
17. Cash benefits (P2) . .	100		11. Surplus of government business undertakings (T2) . .	. 20
18. Disposable income (G3) .	280			

Disposal Account (G3 Account)

20. Government expenditure on current goods and services (G1)	175		19. Disposable income (G2) .	. 280
21. Savings (G4) . . .	105			

Capital Account (G4 Account)

25. Government expenditure on capital works (G1) . .	125		22. Savings (G3) . . .	. 105
			23. Borrowing (P4) . .	. 10
			24. Borrowing (F4) . .	. 10

Notes

1 and 2. See items 20 and 25 below.
4. Part of imports.
7. Mainly personal income tax.
8 and 9. Company income tax.
12. Identically equal to gross product, by definition of gross product.
14 and 15. Interest on public debt domiciled at home.
16. Interest on public debt domiciled abroad.
17. Cash transfers not in return for goods or services rendered, e.g. old-age pensions, child endowment, etc.
20 and 25. These expenditures are deemed to be made from the disposal and capital accounts to the production account and constitute the value of government output, valued at cost.
21. Government surplus on current account.
23 and 24. These items represent the net increase in government indebtedness, financed by borrowing from persons and banks.

FINANCIAL ENTERPRISES
$million

Production Account (F1 Account)

2. Purchases of current goods and services from trading enterprises (T1) 5	1. Value of financial enterprise output (F3) 40
3. Gross product (F2) . . . 35	

Distribution Account (F2 Account)

7. Wages (P2) 35	4. Gross product (F1) . . . 35
8. Direct taxes (G2) . . . 5	5. Interest (T2) 100
9. Interest (P2) 100	6. Interest (G2) 55
10. Dividends (P2) . . . 5	
11. Disposable income (F3) . . 45	

Disposal Account (F3 Account)

13. Financial enterprise expenditure on current goods and services (F1) 40	12. Disposable income (F2) . . 45
14. Savings (F4) . . . 5	

Capital Account (F4 Account)

17. Purchases of capital equipment from trading enterprises (T1) 5	15. Savings (F3) 5
18. Lending (G4) . . . 10	16. Borrowing (P4) . . . 185
19. Lending (T4) . . . 130	
20. Lending (W4) . . . 45	

Notes

1. See item 13 below.
5. Interests on overdrafts. It is assumed in this example that no overdrafts are granted to persons.
6. Interest on government securities held by banks.
7. Identically equal to gross product, by definition of gross product.
9. Interest on deposits.
13. This expenditure is deemed to be made from the disposal account to the production account and constitutes the value of financial enterprise output, valued at cost.
16. Net increase in personal deposits over the year.
17. Mainly expenditure on buildings for office accommodation. Part of gross private investment.
18. Net increase in government securities held by banks.
19. Net increase in net debit balances (overdrafts) of trading enterprises over the year.
20. Net increase in loans to rest of world entities over the year.

PERSONAL
$million

Distribution Account (P2 Account)

11. Direct taxes (G2) . . . 250	1. Wages (T2) 900
12. Disposable income (P3) . 1,855	2. Wages (F2) 35
	3. Wages (G2) 150
	4. Interest (F2) 100
	5. Interest (G2) . . . 40
	6. Dividends (F2) . . . 5
	7. Dividends (W2) . . . 5
	8. Dividends (T2) . . . 40
	9. Withdrawals, etc. (T2) . . 730
	10. Cash benefits (G2) . . . 100

Disposal Account (P3 Account)

14. Purchases of goods and services (T1) 1,500	13. Disposable income (P2) . 1,855
15. Savings (P4) . . . 355	

Capital Account (P4 Account)

17. Lending (F4) . . . 185	16. Savings (P3) 355
18. Lending (G4) . . . 10	
19. Lending (T4) . . . 120	
20. Purchases of capital equipment from trading enterprises (T1) 40	

NOTES
14. Consumers' expenditure.
15. Personal savings.
17. Net increase in personal deposits in banks over the year.
18. Net increase in holdings of government securities.
19. Net increase in equity in and loans to trading enterprises.
20. Expenditure on houses for owner-occupation. Part of gross private investment.

REST OF WORLD
$million

Distribution Account (W2 Account)

3. Dividends (P2) . . . 5	1. Dividends (T2) 30
4. Disposable income (W3) . . 85	2. Interest (G2) 60

Disposal Account (W3 Account)

6. Purchases of goods and services (T1)	500	5. Disposable income (W2)	.	. 85
7. *less* Sales of intermediate goods and services to trading enterprises (T1)	− 330			
8. *less* Sales of goods and services to government (G1)	− 70			
9. Savings (W4)	− 15			

Capital Account (W4 Account)

12. Lending (T4) . . . 30	10. Savings (W3) . . . − 15
	11. Borrowing (F4) . . . 45

Notes

1 and 2. These items make up income received by rest of world from home country.
3. Dividends paid by rest of world companies to persons at home.
6. Exports of home country.
7 and 8. These items make up imports of home country.
9. In this case rest of world savings is negative. This deficiency of the rest of world sector is financed by net borrowing from the home country (item 11 less item 12). This represents a net decrease in the indebtedness of the home country to the rest of the world.

12.4. Consolidation of Sector Accounts

Given a set of accounts such as that above, it is natural to go one step further by consolidating the accounts of the five sectors into one set of accounts for the whole economy. Indeed, this step is of the greatest importance and leads to fruitful results. The consolidation of the social accounts is effected by pooling the five sectors for each account, and cancelling out items which appear on both sides of the resulting national consolidated account. As far as the consolidated national production account is concerned, items 5 and 6 of the T1 Account cancel with item 3 of the G1 Account and item 2 of the F1 Account respectively. On the left-hand side this leaves, apart from the gross products, purchases of rest of world goods and services by trading enterprises and governments. These make up total imports. Hence the left-hand side can be cleared (except for the gross products) by subtracting imports from both sides. The final result is shown in the consolidated accounts below. The accounts are numbered 1, 2, 3 and 4, and, as before, the figures in brackets indicate the double entry.

SOCIAL ACCOUNTS

CONSOLIDATED NATIONAL ACCOUNTS

$million

Production Account (1)

Gross product of T (2)	2,255	Consumers' expenditure (3)		1,500
Gross product of G (2)	150	Gross private investment (4)		500
Gross product of F (2)	35	Government expenditure (current) (3)		175
		Government expenditure (capital) (4)		125
		Financial enterprise expenditure (current) (3)		40
		Exports (3)		500
		less Imports (3)		− 400
	2,440			2,440

Distribution Account (2)

Depreciation allowances (4)	100	Gross product of T (1)	2,255
Disposable income of T (3)	75	Gross product of G (1)	150
Disposable income of G (3)	280	Gross product of F (1)	35
Disposable income of F (3)	45		
Disposable income of P (3)	1,855		
Disposable income of W (3)	85		
	2,440		2,440

Disposal Account (3)

Consumers' expenditure (1)	1,500	Disposable income of T (2)	75
Government expenditure (current) (1)	175	Disposable income of G (2)	280
		Disposable income of F (2)	45
Financial enterprises expenditure (current) (1)	40	Disposable income of P (2)	1,855
Exports (1)	500	Disposable income of W (2)	85
less Imports (1)	− 400		
Enterprise savings (4)	75		
Government savings (4)	105		
Financial enterprise savings (4)	5		
Personal savings (4)	355		
Rest of world savings (4)	− 15		
	2,340		2,340

CAPITAL ACCOUNT (4)

Gross private investment (1)	500	Savings by T (3)	75
Government expenditure on capital works (1)	125	Savings by G (3)	105
		Savings by F (3)	5
		Savings by P (3)	355
		Savings by W (3)	− 15
		Depreciation allowances (2)	100
	625		625

The consolidation of the production accounts records on the left-hand side the gross products of the trading enterprises, government and financial enterprises sectors. Since all production takes place within these three sectors, and since it is all valued at market prices, the resulting gross product is termed *gross national product at market prices*. This follows the Australian terminology, which differs in the following respect from what is now standard international terminology. Gross national product at market prices is defined in Australia as the market value of the product attributable to the factors of production located in the territory of the given economy. The standard international terminology refers to this aggregate as "gross domestic product at market prices" and uses the term "gross national product at market prices" to describe the market value of the product attributable to the factors of production supplied by the normal residents of the economy. This latter is, then, the gross domestic product, *plus* the excess of the income produced in the rest of world by factors of the home economy over the income produced by rest of world factors in the home economy.[1] In the Australian terminology, this would be referred to as the *gross resident product at market prices*. The Australian terminology is adhered to for the rest of this chapter.

It will be appreciated that the gross national product at market prices measures the total value of production of goods and services in the economy over the year, i.e. the total output of *all* goods and services, *less* that part of this total used up during the year in producing other parts of it. Gross national product at market prices is given by the total of the left-hand side of the consolidated national production account (in our example, $2,440 million).[2]

In the process of producing the goods and services which make up the gross national product, trading enterprises, governments and

[1] This is equivalent to gross domestic product, *minus* the disposable income of the rest of world sector.

[2] If we add imports of goods and services to gross national product, we have *gross market supplies*, or to use the terminology of the Australian social accounts, *national turnover of goods and services*. These market supplies are matched by the five categories of final expenditure: consumption expenditure, gross private investment expenditure, government expenditure, financial enterprises expenditure and exports.

financial enterprises create incomes. These are distributed as factor incomes, which after transfers between sectors become the disposable incomes of the sectors. This is shown in the consolidated national distribution account. Not quite all of the gross national product represents disposable income, for some of it is set aside to make good the wear and tear on capital equipment. If we deduct these depreciation allowances from the gross national product, we get *net national product at market prices*, which is identically equal to the disposable income of the sectors. This disposable income is either spent for current purposes or saved (see consolidated national disposal account). In so far as it is spent, it absorbs part of the gross national product. The savings, together with the depreciation allowances, are used to finance the absorption of the other part, i.e. public and private capital formation, and this is shown in the consolidated national capital account.

Each of the four consolidated accounts results in an important economic identity. The production account shows how national production is absorbed by different forms of expenditure. We have

$$\text{Gross national product} = \begin{cases} \text{consumers' expenditure} \\ + \text{ gross private investment} \\ + \text{ government expenditure} \\ + \text{ financial enterprise expenditure} \\ + \text{ exports} \\ - \text{ imports} \end{cases}$$

The total of the right-hand side of the consolidated national production account is sometimes called *gross domestic expenditure at market prices*.[1] It is, of course, identically equal to gross national product.

The distribution account shows how the net national product becomes the disposable income of one sector or another. The disposal account shows how disposable income is either spent or saved. The capital account shows the balance between investment and savings items. The identity relating these items can be written in various ways. One way in which it can be written is—

$$\begin{matrix} \text{Gross private investment} \\ + \text{ government expendi-} \\ \text{ture on capital works} \end{matrix} = \begin{cases} \text{personal savings} \\ + \text{ undistributed company profits} \\ + \text{ financial enterprise savings} \\ + \text{ depreciation allowances} \\ + \text{ government savings} \\ + \text{ net increase in indebtedness to rest of world} \end{cases}$$

The above identities which emerge from the accounts do not represent anything more than *ex post* accounting identities. We have not

[1] This term does not appear in the Australian social accounts, but is sometimes used in textbooks to distinguish the aggregate that it represents from *gross national expenditure*, which is the sum of consumers' expenditure, gross private investment, government expenditure and financial enterprise expenditure; that is to say, gross national expenditure is equal to gross domestic expenditure *minus* exports *plus* imports.

"proved" anything from the accounts. The identities emerge because the accounts have been designed so that they do emerge. They are the result of the structure of the accounts and the definitions of the items in the accounts. This structure and these definitions have been adopted because they are appropriate for modern use in a wide field of economic investigation. The identities are of the same character as the accounting identity—

$$\text{Revenue} = \text{Expenses} + \text{profit}$$

So long as profit is defined as the excess of revenue over expenses, this identity must always be true.

12.5. Gross National Product and Allied Aggregates

The gross national product at market prices can be looked at from two distinct viewpoints.

1. *Production Viewpoint*

Gross national product represents a measure of the volume of goods and services produced over the year, valued at current market prices, such goods and services being in a *final* form as far as the year under consideration is concerned. Final goods and services are those which are produced during the year and are subject to no further productive processes during the year. It is necessary to include the phrase "in a final form" for clearly we do not include separately any goods and services which during the year enter into other goods and services, e.g. wheat which has been used to produce bread is included within the bread, and only wheat which has been exported or which forms an addition to stocks is included separately. This can be seen by examining the right-hand side of the consolidated national production account, where only goods and services which enter into final expenditure are included. Similarly, the left-hand side of that account is the sum of values added, so that only the value added by the bread manufacturer is included, the value of the flour he uses being included in the values added by the flour-miller and the wheat grower at previous stages of production.

Although the gross national product at market prices does measure the total production of goods and services, it makes no allowance for the depreciation of capital equipment which occurs during the year. Consequently, if the whole of gross national product were used up for current purposes (as it was more or less during the latter years of the Second World War), the nation would be impoverishing itself through not maintaining its capital intact. This could continue for only a few years without damage to the productive capacity of the economy. The allowance for depreciation is an estimate of fixed capital depletion over the year, and by deducting it from gross national production we get *net national product at market prices*. This measures the total *net* volume of goods and services produced whilst maintaining capital intact. It indicates the volume of production which can be used for current purposes without diminishing the economy's capital equipment and hence is the true figure for the annual income produced by an economy. However, it must be remembered that the depreciation allowances included in the social accounts are often based on estimates made by trading enterprises and accordingly may be in error. Moreover, they are no more than financial provisions for depreciation and do not correspond to expenditure on replacement of worn-out equipment. Parallel with net national product, we define *net private investment* as gross private investment, *less* depreciation allowances. Subject to the above qualifications net private investment measures the net increase in privately-owned real capital (equipment and inventories) over the year.

2. *Income Viewpoint*

The gross products of the government and financial enterprises sectors are simply the wages paid to the employees of these sectors. Except for the items of depreciation allowances and net indirect taxes,[1] the gross product of the trading enterprises sector is wholly made up of income earned by factors of production.[2] Consequently, if depreciation allowances and net indirect taxes are deducted from gross national product, we obtain a measure of the aggregate of money income earned by factors of production. This aggregate is called *net national product at factor cost*[3]. If from net national product at factor cost we deduct net income payable overseas, we have *national income*, which is the sum of the incomes earned by factors of production resident in the domestic economy. The relation between gross national product at market prices and net national product at factor cost can best be appreciated by considering in isolation a home-produced final good entering gross national product. If an individual good is considered, its whole market price, except for those elements in it representing depreciation and indirect taxes, can be seen to be distributed to the factors of production as income. Thus, suppose we exclude these two

---

[1] The excess of indirect taxes over subsidies.
[2] See left-hand side of the distribution account, some of the components of which represent transfers of original factor earnings. Company income taxation, dividends and undistributed company profits make up company income, the income earned by factors of production owned by companies. Surplus of government business undertakings is income earned by factors owned by governments.
[3] This treatment differs slightly from that used in the Australian social accounts, where net indirect taxes are deducted first from gross national product to give *gross national product at factor cost*, from which depreciation allowances are then deducted to give *net national product*.

elements and consider a radio sold for $40. Of this, say $16 represents the wages paid by the retailer and his profit, and the other $24 is the cost of the radio paid to the manufacturer. Of this, say $14 represents profit and wages and $10 cost of raw materials, and so on. Ultimately, the whole $40 accrues as income to one entity or another. But if the $40 includes depreciation allowances and indirect taxes, these elements are paid into the capital account and to the government respectively, and only the remainder becomes incomes of factors of production.

The income of the personal sector, before the payment of direct taxes, is known as *personal income*. It is the disposable income of the personal sector, plus the direct taxes paid by persons, and is in fact the total of the right-hand side of the personal distribution account. It is the total income of all private individuals in the economy in the everyday sense of the word "income." But, because not all of net national product at factor cost is distributed to persons and because the incomes of some persons are derived as transfers of income from other sectors (e.g. pensions) or as incomes payable from overseas, personal income is not the same as net national product at factor cost.

The relationships between the various aggregates referred to above are illustrated below by using the figures from the hypothetical set of accounts.

		$m
Gross national product at market prices		2,440
less depreciation allowances		− 100
Net national product at market prices		2,340
less indirect taxes *minus* subsidies		− 180
Net national product at factor cost		2,160
less Income accruing to sectors other than persons:		
company income excluding dividends to persons	185	
surplus of government business undertakings	20	
interest paid by trading enterprises to financial enterprises	100	
		− 305
plus Income transferred from other sectors to persons:		
dividends from rest of world	5	
dividends from financial enterprises	5	
cash benefits from governments	100	
interest from governments	40	
interest from financial enterprises	100	
		+ 250
Personal income		2,105

The official published social accounts of an economy are usually not set out in the same systematic manner as has been employed above. This is largely due to the limitations of the statistics available to estimate the various items. But we should regard the systematic treatment as

SOCIAL ACCOUNTS 305

lying behind the published figures. As an illustration of the published presentation of social accounts (and this presentation differs considerably from economy to economy) the Australian national income accounts are set out in Appendix B.3, pp. 481–487, with comments.

12.6. THE PROBLEM OF IMPUTATION

As a measure of the volume of goods and services produced over a year the concept of gross national product as developed above suffers from at least one important limitation. Included in it are only those goods and services which pass through or are valued in the market. In any year many goods and services are produced which do not pass through the market, but which belong to the total volume of production just as surely as those which do pass through the market. Many examples could be quoted but a few will suffice. If a man owns a dwelling and lets it, the provision of dwelling space is a service which is paid for by rent. This service can be measured by the net rent earned by the dwelling, i.e. the excess of the rent paid over costs of collection and maintenance expenditure, and net rent enters into gross national product as the value added by the enterprise of house-letting. On the other hand, if a man owns a dwelling and lives in it himself, no rent is paid, and the value of the service rendered to the man is not included. It follows that the greater the proportion of owner-occupied dwellings, other things being equal, the lower gross national product— clearly an anomalous situation. Again, a certain amount of farm produce is consumed on the farm. This does not pass through the market and does not directly enter into gross national product. The same holds for fruit and vegetables produced by the home-gardener. Similarly, the services of housewives are not counted in gross national product, whereas those of hired domestic servants are, so that when a man marries his housekeeper gross national product falls!

The question arises whether in drawing up the social accounts, and in estimating gross national product, we should include an estimate of the value which these items would have if they passed through the market, i.e. whether we should *impute* values to these items. It has been customary in national income computation to include the imputed value of the services of owner-occupied dwellings and of farm produce consumed on the farm, but of no other items. The reason for including these particular items is two-fold. In the first place, they are readily calculated, for data on the number of owner-occupied dwellings and on the volume of farm produce consumed on the farm are usually available; and since there are active markets in comparable dwelling accommodation and farm produce, an imputation of the value of these items can be made. In the second place, the proportion of owner-occupied dwellings and the proportion of farm produce consumed on the farm is likely to vary from period to period and from economy to economy. Since, for example, a significantly larger proportion of people own their own homes in Australia today than some years ago, a comparison of gross national product for Australia today with that for Australia some years ago would lead to a relative understatement of the current figure, unless the annual value of owner-occupied dwellings were included. Similarly, since in India a far greater proportion of farm produce is consumed on the farm than in Australia, a comparison of the gross national products of India and Australia would lead to a relative understatement of gross national product for India, unless the annual value of farm produce consumed on the farm were included. Other items, like housewives' services, which might be imputed, but which are not in fact so imputed, are either extremely difficult to estimate or else of no great significance from a comparative point of view.

The effect on the social accounts and gross national product of imputing values to items such as those mentioned above can be traced through by taking owner-occupied dwellings as an example. Owner-occupiers must now be treated as trading enterprises. Consequently, the purchase and financing of houses for owner-occupation will appear entirely in the T4 account. Any personal savings devoted to the finance of these houses must be conceived of as lent from the P4 to the T4 account. The effects on the hypothetical social accounts are shown below. All figures are changes to be applied to the accounts as set out on pp. 293–8 above. We assume the imputed net rental value of owner-occupied dwellings (i.e. imputed gross rents *less* maintenance expenditure) to be $60 million, annual depreciation $10 million and new construction $40 million.

T1 Account

4. Gross product (T2) . . + 60	1. Sales to persons (P3) . . + 60
	2. *eliminate* Sales to persons (houses) (P4) . . . − 40
	3. Sales of capital equipment to trading enterprises (T4) . + 40

T2 Account

6. Depreciation allowances (T4) + 10	5. Gross product (T1) . . + 60
7. Withdrawals, etc. (P2) . . + 50	

T4 Account

10. Purchases of capital equipment from trading enterprises (T1) . . . +40		8. Depreciation allowances (T2) . +10	
		9. Borrowing (P4) . . . +30	

P2 Account

12. Disposable income (P3) . +50	11. Withdrawals, etc. (T2) . +50

P3 Account

14. Purchases of goods and services (T1) . . . +60	13. Disposable income (P2) . +50
15. Savings (P4) . . . −10	

P4 Account

17. *eliminate* Purchases of capital equipment (T1) . . −40	16. Savings (P3) . . . −10
18. Lending (T4) . . . +30	

NOTES

1. Imputed net rents paid by owner-occupiers as persons to themselves as trading enterprises for the use of house-room. This equals imputed gross rent *less* maintenance expenditure (already charged to consumers' expenditure) and becomes a part of consumers' expenditure (item 14).

2 and 3. Item 2 is eliminated, and item 3 replaces it. The double entry of item 2 is item 17, and of item 3 is item 10.

7. Imputed income earned by owner-occupiers through their renting house-room to themselves.

9 and 18. Imputed lending by owner-occupiers as persons to owner-occupiers as trading enterprises.

The effect on the consolidated national accounts will be as follows—

1. Account

Gross product of T (2) . . +60	Consumers' expenditure (3) . +60

2. Account

Depreciation allowances (4) . +10	Gross product of T (1) . . +60
Disposable income of P (3) . +50	

3. Account

| Consumers' expenditure (1) . + 60 | Disposable income of P (2) . + 50 |
| Personal savings (4) . . . − 10 | |

4. Account

| | Depreciation allowances (2) . + 10 |
| | Personal savings (3) . . . − 10 |

It can be seen from the above accounts that gross national product will be $60 million more, and net national product $50 million more, than without the imputation of rents for owner-occupied dwellings. The latter figure represents the net annual value of the dwellings to their occupiers. This increases personal disposable income. But owner-occupiers must now charge themselves rent, and personal savings in fact fall by the amount of depreciation on these dwellings. This occurs because previously no depreciation was specifically allowed on these dwellings. Total gross savings are, of course, unchanged, the fall in personal savings being offset by the rise in depreciation allowances. This process of imputation does not affect the *cash* position of any entity, it only affects the classification of certain items and the measures of *national* aggregates. The above discussion incidentally illustrates the usefulness of the social accounting framework for analysing the effects of alternative treatments for specific items.

12.7. Estimation of Items in Social Accounts

The compilation of the social accounts of an economy would be quite a simple matter if records were kept of every single economic transaction. These transactions could then be classified into the appropriate pigeon-holes of the social accounts and added through to yield the various national aggregates. Unfortunately, records of only a small portion of all the economic transactions are kept, and of these few are in a suitable form. In practice, it is necessary to make direct estimates of aggregates such as gross national product and consumers' expenditure, and to use the double-entry structure of the social accounts to fill in certain other items. The main sources for estimation of the items in the social accounts are statistical collections of data on income, production and expenditure, usually collected for purposes other than that of social accounting. Sometimes additional information is collected by sample surveys. The sampling of the accounts of individual businesses can also be used to provide detailed information. Reference is made to the sources used for the Australian social accounts in Appendix B.2.

Historically the estimation of gross national product came well

SOCIAL ACCOUNTS 309

before the development of social accounting. We shall therefore first concern ourselves with the estimation of gross national product. In this matter an understanding of the structure of the social accounts is very useful. The consolidated national accounts show that gross national product can be regarded as the sum of one of the following—

1. Gross products of the trading enterprises, government and financial enterprises sectors (left-hand side of production account),

2*a*. Consumers' expenditure, government expenditure, financial enterprises expenditure, gross private investment and the excess of exports over imports (right-hand side of production account),

2*b*. Consumers' expenditure, government current expenditure, financial enterprises current expenditure, the excess of exports over imports, depreciation allowances and savings of the five sectors (left-hand side of disposal account, together with depreciation allowances),

3*a*. Disposable incomes of the five sectors and depreciation allowances (left-hand side of distribution account),

3*b*. Incomes received by the five sectors and depreciation allowances.

Under number 3*a*, the disposable incomes are reckoned as the final incomes at the various sectors' disposal after the payment of taxes and after transfers between sectors, but these incomes could be readily reckoned before the payment of taxes and transfers between sectors. We designate incomes reckoned in this latter way as "incomes received," because they represent the income received by the various sectors in return for services rendered by the sectors in the production process. Broadly speaking, the term "income received" corresponds to what we ordinarily mean by income received before the payment of taxes. The totals of incomes received and of disposable incomes are the same, although for some sectors disposable income will be higher than income received at the expense of the other sectors, on account of transfer payments. Number 3*b* therefore also corresponds to gross national product. "Incomes received" can be found among the items in the left-hand side of the distribution accounts of the three producing sectors.

It follows that we can estimate gross national product via any of the five sums enumerated above. In practice numbers 1, 2*a* and 3*b* have proved most tractable. These are termed respectively the *production*, the *expenditure* and the *incomes-received* methods of estimating national income.

Ideally, estimates should be made by the three methods, and they would provide checks on each other. In practice the availability of suitable data dictates the method. As far as the production method is concerned, it is necessary to estimate the gross products of trading enterprises and the wages paid to employees of governments and financial enterprises. If annual censuses of production are conducted, the gross products of primary and secondary industries can be fairly readily ascertained because they correspond to the values added by enterprises in the process of production. However, tertiary industries (distributive industries, service industries, professions, etc.) usually present some difficulty, and it may be necessary to estimate their gross product as the incomes received in those industries, thus resorting to the incomes-received method.

The expenditure method can be applied only if data are available on the various items of final expenditure—consumers' expenditure, government and financial enterprises expenditure, gross private investment, and exports and imports. Here, the items of consumers' expenditure and gross private investment usually present the greatest difficulty.

The incomes-received method naturally must place great reliance on statistics collected as a result of income taxation. The gross national product consists of the gross products of the trading enterprises, government and financial enterprises sectors. Referring back to our hypothetical set of accounts and having regard to the distribution account of trading enterprises, we shall have—

Gross national product =

$$\left.\begin{array}{l}\text{Depreciation allowances} \\ +\text{ Indirect taxes} \\ -\text{ Subsidies} \\ +\text{ Direct company taxes} \\ +\text{ Dividends} \\ +\text{ Undistributed company profits} \\ +\text{ Surplus of government business undertakings} \\ +\text{ Withdrawals, etc.} \\ +\text{ Interest paid by trading enterprises} \\ +\text{ Wages paid by trading enterprises}\end{array}\right\} \text{(gross product of trading enterprises)}$$

+ Wages paid by governments (gross product of governments)
+ Wages paid by financial enterprises (gross product of financial enterprises)

Bracketing certain items together and putting in the figures from our example this becomes—

Gross national product =

	$m
Depreciation allowances	100
+ Net indirect taxes	180
+ Company income	225
+ Surplus of government business undertakings	20
+ Withdrawals, etc.	730
+ Interest paid by trading enterprises	100
+ Wages	1,085
	2,440

This latter sum gives the gross national product broken up into the major components of incomes received, and indicates how gross national product may be estimated by the incomes-received method.

It is not proposed to discuss the estimation of the individual items which enter into the social accounts. This is a practical problem, usually fraught with great difficulty on account of gaps in the available data. However, some comment on the *balancing properties* of the accounts is desirable in this connexion.

If every economic transaction were recorded and classified into the appropriate accounts, the accounts would automatically balance (apart from clerical errors). This must be so on account of the formal definition of the accounts. However, in practice the various items are estimated from various statistical sources and not from complete records and are subject to quite wide margins of error. Since each account must balance, then provided that all the items are directly estimated a comparison of the sums of the items on the left-hand side and on the right-hand side will provide a very valuable check on accuracy. If these sums are not identical, then errors of estimation must have been made and the items must be adjusted to bring the two sides of the account into equality.

It frequently happens that one particular item in an account is extremely difficult to estimate. In that case, it can be estimated as a balancing item (i.e. the property of the account that the two sides must be equal is utilized), the item obtained as a difference being the amount necessary to ensure the balance. This method of estimation of a particular item is an indirect one and, although convenient, is inferior to direct estimation. In the first place, all errors in the other items of the account are absorbed into the balancing item. Consequently, small errors in a number of items directly estimated may result in a larger error in the balancing item, if they do not by chance cancel out. In the second place the use of a balancing item removes any possibility of utilizing the balancing property of the account as a check. Nevertheless, when data are scarce, balancing items may have to be used.

The number of items which can be estimated in this way is, of course, limited. If there are N distinct accounts in a system of social accounts, at most only $N-1$ items can be estimated as balancing items. For, given any $N-1$ of the accounts, the other one can be immediately filled in by using the missing double entries of the items already included in the $N-1$ accounts. Consequently, if $N-1$ accounts are made to balance by estimating one item in each as a balancing item, the other account must necessarily balance, and no balancing item can be derived from it.

12.8. Uses of the Social Accounting Framework

Economists have always been interested in aggregates such as national income, but the systematic treatment of aggregated transactions in a social accounting framework dates only from about 1940. A social accounting system enables the structure of economic transactions to be set out in a consistent way and makes clear the dependence of the definition of any given aggregate on the particular system chosen. It helps to elucidate the relations between associated concepts, e.g. the distinction between gross national product at market prices and net national product at factor cost (see p. 303 above). It readily reveals the effects of any change in the treatment of particular items on the various aggregates of transactions (see pp. 306–308 above). An examination of the systems employed by various countries makes international comparisons possible, since the comparability or otherwise of concepts is made clear.

Social accounts provide a framework for the classification of transactions and hence suggest the form in which data should be collected. The aggregates recorded in the social accounts constitute a major source of data for empirical economic investigation. Research into the behaviour of the economy as a whole requires estimates of such aggregates as gross national product, consumers' expenditure, gross private investment, etc. In addition, movements in gross national product, valued at constant prices (see section 14.2, p. 370 below) and expressed per head of the population, give an indication of movements in the standard of living. Similarly, movements in gross national product, valued at constant prices and expressed per head of the working population, give an indication of movements in the level of productivity. Changes in the components of national income throw light on questions of income distribution.

Social accounts also facilitate the actual estimation of the aggregates of transactions. They indicate alternative routes to the estimation of a given concept, e.g. the estimation of gross national product by the production, expenditure and incomes-received method (see pp. 308–310 above); and make clear the precise nature of the routes necessary to obtain the same final result. Moreover, if all transactions can be measured directly, the classification of transactions into a social accounting framework reveals any discrepancies and errors and provides a basis for making adjustments of discrepancies. By the same token, if all transactions cannot be measured directly, the nature of the accounts permits some items to be estimated by double-entry or as balancing items (see p. 311 above).

The social accounts are a meeting place for economic theory and practical measurement. The general structure of the accounts must be designed to encompass the categories of economic theory (e.g. income, consumption, investment) in such a way that these categories are capable of actual measurement. The accounts show the formal relationships between aggregates (e.g. personal disposable income = consumers' expenditure + personal savings). They form the basis for economic models for the purposes of analysing the behaviour of the economy as a whole, of economic forecasting and of illuminating problems of economic policy.

The social accounts give a picture *ex post* of the outcome of economic activity. They can also be used as a framework for drawing up an *ex ante* forecast of the likely outcome of the economy in the future.[1] In this connexion the social accounts ensure consistency of forecasts, both internally and in relation to other, external, known facts. Moreover they enable one to judge whether the expected outcome on the basis of known circumstances is likely to be consistent with over-riding policy objectives; and, if not, to formulate appropriate policy. An example from what is known as *national budgeting* will help to illustrate this important use of social accounts.

We assume a relatively simple closed economy, in which all firms are companies that distribute all profits after tax as dividends. We amalgamate the production, distribution and disposal accounts of the sectors to obtain the current accounts of the trading enterprises, persons and government sectors. From these we can obtain the national current account. We imagine that we are standing at 30th June, 1968. The *ex post* accounts for the actual outcome of 1967–68 are set out

$ million

TRADING ENTERPRISES

Wages	6,100	Sales to consumers	6,400
Dividends	2,400	Investment (equipment)	1,500
Company tax	1,200	Investment (stocks)	500
		Sales to government	1,300

GOVERNMENT

Government expenditure*—		Company tax	1,200
Purchases from enterprises	1,300	Personal income tax	2,000
Wages	1,500		
Surplus	400		

* Both current and capital expenditure.

[1] The distinction between *ex post* and *ex ante* measurement is a common one in economic theory. *Ex post* measurements relate to events which have actually taken place. They are a record of what has happened. Ordinary commercial book-keeping is an *ex post* record. *Ex ante* measurements relate to expectations, intentions, plans, forecasts, etc. Budgets consist of essentially *ex ante* measurements.

Personal

Personal income tax . . . 2,000	Wages—	
Consumers' expenditure . . 6,400	From government . . 1,500	
Personal savings . . . 1,600	From enterprises . . . 6,100	
	Dividends 2,400	

National Current Account

Wages 7,600	Consumers' expenditure . . 6,400
Company profits . . . 3,600	Investment 2,000
	Government expenditure . . 2,800
Gross national product . . 11,200	11,200

above. No distinction is drawn between current and capital government expenditure. The national capital account is omitted. However, it can be used as a check on the other accounts, for investment must equal personal savings plus government surplus.

Now, suppose that we are attempting to make an estimate of what we expect to happen in 1968–69, on the following basis—

(i) The plans of the government sector are known and are specified in the government current account as follows—

Government expenditure—	Company tax 2,000
Purchases from enterprises . 2,000	Personal income tax . . . 3,200
Wages 1,500	
Surplus 1,700	

(ii) We believe, on the basis of market forecasts, that consumers' expenditure and investment will be $7,500 million and $1,500 million respectively.

This information is sufficient to fill in the social accounts completely. This is done in the following tables. Unnumbered items are "known" ones. Numbered items are derived as balancing items or by double entry, and the numbers indicate the order in which they can be derived.

Trading Enterprises

(1) {Wages / Dividends} . . . 9,000	Sales to consumers . . 7,500
Company tax . . . 2,000	Investment (equipment) } . 1,500
	Investment (stocks)	
	Sales to government . . 2,000

PERSONAL

Personal income tax	.	. 3,200	Wages—		
Consumers' expenditure		. 7,500	From government	.	. 1,500
(3) Personal savings	.	. − 200	(2) { From enterprises Dividends }	.	. 9,000

NATIONAL CURRENT ACCOUNT

Wages Company profits }	.	. 12,500	Consumers' expenditure Investment . . Government expenditure	. 7,500 . 1,500 . 3,500
Gross national product.		. 12,500		12,500

From these tables, we see immediately that the "known" factors imply negative personal savings of $200 million. We may ask: is this reasonable compared with the previous year's actual experience? Is it reasonable to suppose that consumers' expenditure will be $7,500 million, when personal disposable income is only $7,300 million? We also see that a gross national product of $12,500 million is implied in these forecasts. Again we may ask: is this possible with our given resources? If the answers to these questions are negative, the "known" factors are not likely to eventuate as predicted.

This illustrates the inter-relatedness of the various aggregates. It also illustrates that the specification of some of the aggregates may imply nonsense values in others unless the functional dependence of some of the aggregates on others is taken into account. Aggregates must be estimated in a logical order. Thus consumers' expenditure clearly depends on personal disposable income and cannot be forecast prior to it or independently of it.

In order to make the social accounting framework an effective guide for economic policy, it is necessary to specify certain functional relationships between some of the aggregates, as distinct from the formal relationships inherent in the system of accounting adopted. The distinction between functional and formal relationships can be seen by considering personal savings. Personal savings equals personal disposable income *minus* consumers' expenditure. This is a formal relationship; it always holds by definition. However, the manner in which personal savings varies with changes in personal disposable income is a functional relationship and can take many forms. The formal relationships are revealed by the social accounting framework; the functional relationships are not. They must be derived from theoretical reasoning and by empirical observation.

These points can be illustrated by extending the above analysis to investigate whether certain known plans about the future are consistent with some overall objective of economic policy. Continuing with our example, suppose—

(1) In 1967–68 the economy was operating at full employment. We believe that capacity will expand by about seven per cent in 1968–69, so that gross national product at full employment will be about $12,000 million.

(2) We know the government intends to spend $3,000 million on goods and services.

(3) From our knowledge of private investment plans we believe that investment in equipment will be $1,300 million, and that this will take place independently of the level of activity actually achieved.

We now ask: is full employment likely to be achieved, or will there be a tendency towards unemployment or inflation? Before proceeding we must assume certain functional relationships. For simplicity we assume that the relationships between certain aggregates will be the same percentagewise in 1968–69 as in 1967–68, namely:

(a) 63 per cent of the gross product of trading enterprises is paid out in wages;

(b) the rate of company tax is $33\frac{1}{3}$ per cent;

(c) the rate of personal income tax is 20 per cent;

(d) 80 per cent of personal disposable income is devoted to consumers' expenditure;

(e) 54 per cent of government expenditure is spent on the wages of public servants.

Forecasts of the accounts for 1968–69 are given below. In these accounts we set out the magnitudes of the various items on the assumption that full employment is achieved and that the other assumptions listed above hold. As before, the numbers indicate the order in which the items are filled in. The letters indicate where use has been made of the functional relationships (a) to (e) above.

TRADING ENTERPRISES

(a)	(4)	Wages	. . . 6,540	(15)	Sales to consumers	.	. 6,860
		Dividends	. . . 2,560		Investment (equipment)	.	1,300
(b)	(5)	Company tax	. . . 1,280	(16)	Investment (stocks)	.	. 840
				(2)	Sales to government	.	. 1,380
	(3)	Gross product*	. . . 10,380				10,380

* $12,000 million *minus* government expenditure on wages.

Government

		Government expenditure—		(12) Company tax	. .	. 1,280
		Purchases from enter-		(13) Personal income tax	.	. 2,140
(e)	(1)	prises .	. . 1,380			
		Wages .	. . 1,620			
	(14) Surplus	.	. . 420			

Personal

(c)	(9) Personal income tax	. 2,140		Wages—		
(d)	(10) Consumers' expenditure	6,860	(6)	From governments		. 1,620
	(11) Personal savings .	. 1,720	(7)	From enterprises	.	. 6,540
			(8)	Dividends	. .	. 2,560

National Current Account

Wages	. 8,160	Consumers' expenditure .	. 6,860
Company profits . .	. 3,840	Investment . . .	. 2,140
		Government expenditure .	. 3,000

It will be noted that the final item to be filled in, namely "investment in stocks," is a balancing item. Since this item emerges as the difference between the components of expenditure and gross national product at full employment, it can be interpreted as indicating whether expenditures are likely to be too low to produce full employment or are likely to be so high that inflation will emerge. In this particular case, if it is judged that trading enterprises would not willingly accumulate stocks to the extent of $840 million, then it can be said that full employment is unlikely to emerge in 1968–69 unless there is some change in economic policy. Other things being equal, gross national product would be lower than $12,000 million, as enterprises would not in fact produce at a rate resulting in an unwanted accumulation of stocks. To avoid this situation, the government would have to modify its policy, for example by spending more itself or reducing taxes to induce consumers to spend more.

This analysis is known as *gap analysis*. The extent to which the implied changes in stocks are not planned by trading enterprises measures the gap between full employment supplies and planned expenditures. In the example the gap indicated a deficiency of expenditure (i.e. a *deflationary gap*). If the gap had been negative, it would have indicated an excess of expenditure (i.e. an *inflationary gap*).

There are many ways of using a social accounting system for the purposes of national budgeting. The preceding example is only one of these and is given here purely by way of illustration. No special significance should be attached to the detailed treatment in this example. Nor is it appropriate here to pursue further this analysis, which relates essentially to a branch of applied economics.

In the following two sections, two important extensions of social accounting methods are discussed in some detail.

12.9. Flow-of-funds Accounting

Social accounts of the kind usually published are oriented to the analysis of the production, distribution and disposal of income and are mainly concerned with the recording of flows of goods and services. With the exception of certain borrowing and lending items, purely financial flows are ignored. This may be illustrated by referring back to the example (p. 284 above) of my buying cigarettes from a tobacconist. Such a transaction is recorded twice in the social accounts —once as a sale of goods in the trading enterprises production account and once as a purchase of goods in the persons disposal account. This concentrates on the movement of the goods. However, whereas the cigarettes will pass from the tobacconist to me, cash will pass from me to the tobacconist. If one takes into account the financial flows (including flows of financial claims as well as cash), it will be necessary to record each transaction four times. Such recording results in *flow-of-funds accounting*. In the case of the above example, we shall have—

	Trading Enterprises	*Personal*
Non-financial flow	Sale of goods	Purchase of goods
Financial flow	Receipt of cash	Payment of cash

Reading horizontally, the first line indicates a flow of goods, the second a flow of cash. Reading vertically, the columns indicate re-arrangements of assets.

Flow-of-funds accounting involves the quadruple recording of all transactions between separate economic entities (i.e. purely internal transactions, like depreciation, are not recorded) in which financial transfers occur. In particular it encompasses purely financial flows in which no real transaction is involved (e.g. the sale of securities), and the transfer of existing real assets through which no production takes place (e.g. the sale of second-hand equipment). The flow-of-funds accounts show, for sectors of the economy, how funds have been derived (through the earning of revenue or the incurring of financial obligations) and used (through the incurring of costs or the creation of assets). Such information is useful in analysing the financial structure of the economy and the impact of monetary and fiscal policy on that structure.

Flow-of-funds accounting can be used to analyse changes in the structure of financial assets and liabilities of each sector, reconciling these changes with the net increase in wealth (i.e. the savings) of the particular sector under consideration. Indeed the information to do this has already in principle been included in the capital accounts of the social accounting system set out above, although it is not usually included in published statements of social accounts. The capital accounts show on the right-hand side savings and on the left-hand side capital formation. The balances are shown as borrowing and lending items, with double-entries for the borrowers and lenders. If these items were aggregated to read simply "net borrowing" or "net lending" and these balances carried down to a fifth account, entitled "changes in financial claims account," these fifth accounts could be used to analyse the changes in the financial structure of each sector.[1]

This is illustrated for the hypothetical accounts on pp. 293–298 above. The capital accounts would read—

T4 ACCOUNT

Purchases of capital equipment from trading enterprises (T1)	. 430	Savings (T3) . . .	. 75
Additions to inventories (T1) .	. 25	Depreciation allowances (T2)	. 100
		Net borrowing (T5) . .	. 280

G4 ACCOUNT

Government expenditure on capital works (G1)	. 125	Savings (G3) . . .	. 105
		Net borrowing (G5) . .	. 20

F4 ACCOUNT

Purchases of capital equipment from trading enterprises (T1)	. 5	Savings (F3)	. 5
Net lending (F5)	. 0		

P4 ACCOUNT

Purchases of capital equipment from trading enterprises (T1) .	. 40	Savings (P3)	. 355
Net lending (P5)	. 315		

W4 ACCOUNT

	Savings (W3) . . .	− 15
	Net borrowing (W5) . .	. 15

All borrowing and lending transactions (i.e. all financial capital transactions) will now be recorded in the "changes in financial claims

[1] The authors are indebted to Professor R. L. Mathews, for this suggestion. See also, Mathews, R. L.: *Accounting for Economists* (Cheshire, 1965), Ch. 19.

accounts." These are set out below. They include rather greater detail than shown on pp. 293–298 above, but are consistent with the accounts on those pages. It will be noted that changes in assets are shown on the left-hand side and changes in financial obligations on the right-hand side. In this particular example all changes happen to be increases. Decreases would be recorded as negative items.

T5 Account

Net borrowing (T4) . . . 280	Changes in—		
Changes in—	Bank advances (F5) . . . 160		
Bank balances (F5) . . . 25	Equity capital held by persons		
Cash (F5) 5	(P5) 90		
	Debentures held by persons (P5) 30		
	Equity capital held by foreigners		
	(W5) 30		

G5 Account

Net borrowing (G4) . . . 20	Changes in—
Changes in—	Government securities held by
Bank balances (F5) . . . 5	banks (F5) 15
Cash (F5) 0	Government securities held by
	persons (P5) 10

F5 Account

Changes in—	Net lending (F4) 0
Advances to trading enterprises	Changes in—
(T5) 160	Deposits of trading enterprises
Advances to persons (P5) . . 20	(T5) 25
Government securities (G5) . 15	Deposits of governments (G5) . 5
International reserves (W5) . 45	Deposits of persons (P5) . . 200
	Cash*
	trading enterprises (T5) . . 5
	governments (G5) . . . 0
	persons (P5) 5

*Increased liabilities of central bank.

P5 Account

Changes in—	Net lending (P4) 315
Bank balances (F5) . . . 200	Changes in—
Cash (F5) 5	Bank advances (F5) . . . 20
Government securities (G5) . 10	
Equity capital (T5) . . . 90	
Debentures (T5) . . . 30	

W5 Account

Net borrowing (W4) . . . 15	Changes in—	
Changes in—	International reserves held by	
Equity capital in trading enterprises (T5) 30	banks (F5) 45	

If the changes in financial claims accounts are consolidated for the five sectors, the consolidated account will show the net borrowing and lending positions of the sectors, as follows—

Consolidated Changes in Financial Claims Account

Net borrowing of T . . . 280	Net lending of F 0
Net borrowing of G . . . 20	Net lending of P 315
Net borrowing of W . . . 15	

If the capital and changes in financial claims accounts are aggregated for each sector without further netting out, the resulting accounts will set out the uses (left-hand sides) and sources (right-hand sides) of capital funds for each sector.

The above example uses the same sector classification as for the national income accounts. In practice a more detailed classification is employed, especially in respect of financial institutions.

12.10. Inter-industry Analysis

In the preceding section we were concerned with an extension of social accounting, developed mainly through a more detailed recording of financial transactions. In this section we shall deal with an extension arising from a more detailed classification of production activities.

If the trading enterprises sector is split up into a number of sub-sectors according to industries, the production accounts of the new sectors will have to record the sales and purchases of intermediate goods and services between sub-sectors as well as the sales of goods and services to final purchasers. These intermediate transactions are netted out in the aggregated trading enterprises production account, but once the production accounts are shown on an industry basis *inter-industry* transactions become explicit. This is illustrated in the tables below. Three industrial sub-sectors are assumed—primary, secondary and tertiary—and production accounts for each sub-sector are set up. The transactions shown below are consistent with those in the tables on pp. 293–298 above. However, to simplify the presentation private investment in equipment, houses, inventories and financial enterprises' equipment is amalgamated; and the current activities of financial enterprises have been amalgamated with those of governments.

$ million

PRODUCTION ACCOUNT—PRIMARY INDUSTRIES

Purchases from secondary industries	130	Sales to secondary industries	550
Purchases from tertiary industries	80	Sales to tertiary industries	30
Purchases from rest of world	30	Sales to government	10
Gross product	1,100	Sales to persons	300
		Sales to enterprises (capital equipment and inventories)	50
		Sales to rest of world	400
Total purchases	1,340	Total sales	1,340

PRODUCTION ACCOUNT—SECONDARY INDUSTRIES

Purchases from primary industries	550	Sales to primary industries	130
Purchases from tertiary industries	80	Sales to tertiary industries	40
Purchases from rest of world	300	Sales to government	50
Gross product	540	Sales to persons	700
		Sales to enterprises (capital equipment and inventories)	450
		Sales to rest of world	100
Total purchases	1,470	Total sales	1,470

PRODUCTION ACCOUNT—TERTIARY INDUSTRIES

Purchases from primary industries	30	Sales to primary industries	80
Purchases from secondary industries	40	Sales to secondary industries	80
Purchases from rest of world	0	Sales to government	25
Gross product	615	Sales to persons	500
		Sales to enterprises (capital equipment and inventories)	0
		Sales to rest of world	0
Total purchases	685	Total sales	685

PRODUCTION ACCOUNT—GOVERNMENT

Purchases from primary industries	10	Value of government output	340
Purchases from secondary industries	50		
Purchases from tertiary industries	25		
Purchases from rest of world	70		
Gross product	185		
Total purchases	340	Total value	340

The division of the trading enterprises sector into three sub-sectors has greatly increased the detail and complexity of the accounts. If a fine industrial classification were used, the accounts would become very complicated indeed. One way of simplifying the presentation, whilst underlining the essential inter-dependence of the economic structure, is to present the data in the form of a *matrix*, that is, a table

INTER-INDUSTRY TABLE
$ million

SOLD TO: SOLD BY:	PURCHASING INDUSTRIES			PURCHASES FOR FINAL USE				TOTAL SALES
	Primary	Secondary	Tertiary	Consumers' Expenditure	Gross Private Investment	Government Expenditure	Exports	Aggregate Final Market Expenditure	
Primary Industries	—	550	30	300	50	10	400	760	1,340
Secondary Industries	130	—	40	700	450	50	100	1,300	1,470
Tertiary Industries	80	80	—	500	—	25	—	525	685
Rest of World	30	300	—	—	—	70	—	70	400
Factors of production	1,100	540	615	—	—	185	—	185	2,440
TOTAL PURCHASES	1,340	1,470	685	1,500	500	340	500	2,840	6,335

in which the entries run both horizontally and vertically, as, for example, in a contingency table (see p. 144 above). This is done for the above data in the table on p. 323 Any system of inter-related accounts can be expressed in this form. A finer industrial classification would increase the numbers of columns and rows.

Reading horizontally, the table tells us, for example, how the *output* of primary industries valued at $1,340 million was absorbed, partly as intermediate goods by other industries and partly as final goods by final purchasers. Reading vertically, the table tells us, for example, the *inputs* which were used up to produce the output of primary industries, including the inputs from other industries, from the rest of the world and from the factors of production who earned incomes ($1,100 million in this case, including depreciation) in producing this output. Such a table is often called an *input–output table*. There are many forms in which these tables can be drawn up, but the table shown indicates the general pattern. It should be noted that the value of gross national product can be obtained from this table either by adding the "factors of production" row, or by adding the "aggregate final market expenditure" column and subtracting imports.

The input-output table reveals the inter-dependence of the various sectors of the economy. Thus, if an element in final market expenditure is changed, practically all other elements in the table must be affected. Suppose secondary exports are increased. Other things being equal, this will increase secondary output and hence inputs. Consequently primary and tertiary outputs and imports will increase. But the increases in primary and tertiary outputs will themselves involve further increases in the inputs from the three industrial sectors, and hence in their outputs, and so on. At each "round" the adjustments become smaller, so that ultimately there will be a set of values for the various elements consistent with the increase in secondary exports.

The table itself will not enable us to work out the effects of changes in one element on the others. To do this one must know the technical input-output relationships. However, by making relatively simple assumptions, one can estimate these relationships from the table; for example, one could postulate that values of inputs are proportional to values of outputs. This assumption implies that relative prices and relative combinations of inputs do not vary with output. Accordingly it is somewhat unrealistic, although for small shifts in output it may do as a first approximation. The table below sets out for the previous table the input-output coefficients on the basis of this postulate.

Given the input-output coefficients, the determination of the effects of a change in a component of final market expenditure on the other aggregates in the economy requires mathematical manipulation. This is not the place for an exposition of what has become an important branch of economics in its own right,[1] but it is possible to indicate the line of argument in the case of a simple table such as the one below Suppose we require an increase of $1 worth of output of secondary

VALUE OF INPUTS PER UNIT VALUE OF OUTPUTS

Inputs from:	Primary	Secondary	Tertiary
Primary	—	0·38	0·04
Secondary	0·10	—	0·06
Tertiary	0·06	0·05	—
Imports	0·02	0·20	—
Gross products	0·82	0·37	0·90
Total	1·00	1·00	1·00

industries for final expenditure, and we wish to ascertain the total increase in primary, secondary and tertiary output which this will imply. This can best be done in terms of a series of "rounds."

Round 1. An increase of $1 in secondary output will require increases of $0·38 and $0·05 in primary and tertiary inputs, respectively, and hence in outputs.

Round 2. The $0·38 increase in primary output will require increases of $0·38 × 0·10 = $0·04 and $0·38 × 0·06 = $0·02 in secondary and tertiary inputs, respectively, and hence in outputs.

The $0·05 increase in tertiary output will require negligible increases in primary and secondary inputs ($0·05 × 0·04 and $0·05 × 0·06).

Round 3. The $0·04 increase in secondary output will require $0·04 × 0·38 = $0·02 increase in primary input and a negligible increase in tertiary input ($0·04 × 0·05). The $0·02 increase in tertiary output will require negligible increases in primary and secondary inputs ($0·02 × 0·04 and $0·02 × 0·06).

To the order of accuracy here employed it is unnecessary to go further. We shall require for an increase of $1 of secondary output for final expenditure an increase in:

Primary output of:
$0·38 (*Round* 1) + $0·02 (*Round* 3) = $0·40
Secondary output of:
$1·00 (*Round* 1) + $0·04 (*Round* 2) = $1·04
Tertiary output of:
$0·05 (*Round* 1) + $0·02 (*Round* 2) = $0·07

The application of inter-industry analysis to the problems of economic policy is limited by the difficulty of defining and deriving accurately the technical input-output coefficients and the need to assume their constancy at least in the short period. Nevertheless, inter-industry analysis is a more refined tool than analysis based on the simple

[1] For a simple account, see Edey, H. C. and Peacock, A. T.: *National Income and Social Accounting* (Hutchinson's University Library, 1954), Ch. VIII. For further references see Appendix C.7.

national accounting aggregates. Thus, suppose a government decides, under conditions of full employment, to increase defence expenditure by $100 million. Simple aggregative analysis would indicate that, say, consumers' expenditure be cut by $100 million to release resources for defence. This is in the right direction; but resources are not homogeneous. The resources required for defence works and services will probably not coincide with resources released by cutting back consumers' expenditure, and some industries will have capacities which are not readily enlarged. Inter-industry analysis offers some hope of ascertaining what increased defence expenditure means in terms of increased outputs in different industries, and hence whether these outputs are physically possible, and how one should go about releasing resources to achieve the desired increase in defence expenditure. Again, suppose a country is committed to a programme of rapid economic growth, what industries should be expanded and in what order? And what are the likely changes in the demand for imports and in the balance of payments? Inter-industry analysis throws light on questions such as these.

CHAPTER XIII

PRICE INDEX NUMBERS

13.1. The Concept of an Index Number

An *index number* is a device for comparing the general level of magnitude of a group of distinct, but related, variables in two or more situations. If we want to compare the output of, say, consumer-durable goods in Australia in 1967 with what it was in 1939, we shall have to consider a group of variables, such as the outputs of refrigerators, radios, carpets, etc., which have the common attribute of being consumer-durable goods. If all these variables change in exactly the same ratio, there will be no difficulty in speaking of the change in the output of consumer-durable goods as a whole. But in practice the outputs of individual items change in different ratios. Consequently, we shall have to examine the movements in a large number of distinct variables. The significance of such a host of diverse movements cannot be readily comprehended. What we want is one figure as an indicator or *index* of the change in the magnitude of the output of consumer-durable goods as a whole, so that we can say that the general level of output of these goods in 1967 is, say, 50 per cent or 100 per cent higher than in 1939. Thus, an index number performs a function similar to that of an average. An average is useful as a figure for representing the general level of magnitude of a particular variable. Similarly, an index number represents the general level of magnitude of the *changes* between two or more situations, of a number of variables taken as a whole.

Index numbers can be used for many different purposes. The example given in the preceding paragraph is one of an index of physical volume, the variables under consideration being the physical outputs of consumer-durable goods. A quite different type of index would be one of the relative wealth of the Australian states. This would have to combine those variables reflecting wealth. Or we might have one designed to reflect the relative social status of individuals, combining those variables reflecting social status. But the best-known index numbers are those of prices, and we shall be concerned for the rest of this chapter with price index numbers.

13.2. Price Index Numbers

A price index number is used for comparing changes in the general level of prices of a group of commodities. It may be an index of wholesale prices, of retail prices, of the prices of building materials, of the prices of agricultural products, etc. Generally the index number refers to changes in the prices obtaining in a particular area over time, and it is expressed by putting a particular period (called the *base*) equal to 100 or 1,000 and expressing the other periods under consideration relatively to 100 or 1,000. The selection of a period for the base depends upon the purpose of the index number. For example, if we wished to compare current price levels with the pre-war price level, we should use, say, 1938 or 1939 as base, or if we wished to make the comparison with the immediate post-war period we should use 1946 or 1947. The base is quite flexible and generally can be easily shifted (see section 13.8, p. 350). The table below gives an example of an Australian price index number. According to this index the general level of wholesale prices in Australia was 10 per cent higher in 1966–67 than it had been in 1959–60.

Table 13.1

WHOLESALE PRICE INDEX
(Basic Materials and Foodstuffs)
AUSTRALIA, 1959–60 TO 1966–67
(Base 1959–60 = 100)

Period	Index
1959–60	100
1960–61	103
1961–62	97
1962–63	98
1963–64	99
1964–65	102
1965–66	107
1966–67	110

Source: Derived from Commonwealth Bureau of Census and Statistics, Australia: *Monthly Review of Business Statistics*, No. 370, 1968, p. 25

Although price index numbers are most frequently used to make comparisons over time, they can also be used for making comparisons between the general level of prices obtaining in different areas at particular times.

If the prices of all the individual goods which constitute the group under consideration moved between two situations in exactly the same ratio, that ratio would measure the general movement in prices, and the construction of the required index number would be simple. However, in practice the prices of individual goods move differently. This is tantamount to saying that *relative* prices change, i.e. the prices of goods relative to each other. Relative prices are affected by changes in both the supply of and the demand for particular goods. Changes in the technical conditions of production and changes in tastes can be cited as the most important factors. At the same time changes in the price of one particular commodity relative to other prices react back on the other prices. Thus the prices of competitive goods (substitutes) tend to move in the same direction (e.g. if the price of beef falls, the price of mutton will tend to fall), whereas the prices of complementary goods tend to move in opposite directions (e.g. if an excise duty on petrol raises its price, the number of cars demanded, and hence the price of cars, will tend to fall). Apart from these influences which operate on particular relative prices, there are the broader influences which operate on prices in general. In particular, there is the level of economic activity obtaining in the economy as a whole. All prices tend to move upwards in times of full employment and tend to move downwards in times of unemployment. However, this influence itself affects different groups of prices differently, some being more sensitive to changes in the economic tempo than others. It will be readily appreciated that the movements of prices of individual goods will be diverse and that some device for measuring the general level of magnitude of changes in prices is necessary. Price index numbers furnish such a device.

13.3. Index Number Formulae—Aggregative Type

In this and the following sections a number of price index number formulae will be developed in terms of a comparison between two periods of time. These two periods will be designated "period 0" and "period 1." The price of an individual good in period 0 will be written p_0 and its price in period 1 will be written p_1. The index number which is designed to measure the change in the general level of prices of a group of goods between periods 0 and 1 will be designated P_{01}. The

Table 13.2

Commodity	Unit	Prices		Quantities		Aggregate Costs			
		Period 0 p_0	Period 1 p_1	q_0	q_1	$p_0 q_0$	$p_1 q_0$	$p_0 q_1$	$p_1 q_1$
(1)	(2)	(3)	(4)	(5)	(6)	(7)	(8)	(9)	(10)
		cents	cents			cents	cents	cents	cents
Bread	2 lb	18	21	100	90	1,800	2,100	1,620	1,890
Tea	lb	80	90	10	8	800	900	640	720
Potatoes	7 lb	49	35	20	30	980	700	1,470	1,050
Butter	lb	56	56	30	30	1,680	1,680	1,680	1,680
Milk	qt	21	18	7	10	147	126	210	180
Beef	lb	42	63	80	60	3,360	5,040	2,520	3,780
Mutton	lb	30	40	70	60	2,100	2,800	1,800	2,400
		296	323			10,867	13,346	9,940	11,700

figure P_{01} is a ratio between periods 0 and 1, such that if we take period 0 as a base and put its level of prices equal to 100, the index number for period 1 will be $100 \times P_{01}$.

The easiest way to outline the various index-formulae is to follow through a simple arithmetical example. In the first four columns of Table 13.2 are set out the necessary data concerning the prices of seven foodstuffs in two periods. Our task is to devise an index number for measuring the change in the general level of prices of this group of seven goods.

We have now to find a way of combining the movements in the prices of distinct commodities to form a single index number. One way of doing this is to ask the question: "What would be the change from one period to another in the aggregate cost of purchasing a certain collection of the commodities under consideration?" This gives rise to index numbers of the *aggregative* type. We can imagine that we have a basket of goods and find the cost of the contents in the two periods. The relative movement in the cost will be our index number. Clearly, the main consideration will be the composition of the basket.

Simple Aggregate of Prices

A simple suggestion is to take the ratio of the total of the prices in the two periods. Such an index is called a *simple aggregate of prices*. The formula is

$$P_{01} = \frac{\Sigma p_1}{\Sigma p_0}$$

In our example, we shall have as our index number for period 1

$$(323 \div 296) \times 100 = 109$$

with base, period $0 = 100$. This indicates a 9 per cent rise in the general level of prices of the group under consideration.

It is not difficult to see what this index measures. It measures the change in the aggregate cost of purchasing a collection of goods consisting of one unit each of the units in which the prices are quoted. In our example this collection is 2 lb bread, 1 lb tea, 7 lb potatoes, 1 lb butter, 1 quart milk, 1 lb beef and 1 lb mutton. Such a collection cost 296 cents in period 0 and 323 cents in period 1, an increase of 9 per cent.

The simple aggregate of prices depends for its value very largely on the units in which the prices are quoted. If these are changed, the collection being priced itself changes. Thus, if potatoes and milk were quoted in hundredweights and gallons respectively, our data would appear as in Table 13.3 below.

Our index number would be

$$(902 \div 1{,}094) \times 100 = 82$$

PRICE INDEX NUMBERS 331

Table 13.3

Commodity	Unit	Prices	
		Period 0 p_0	Period 1 p_1
		cents	cents
Bread	2 lb	18	21
Tea	lb	80	90
Potatoes	cwt	784	560
Butter	lb	56	56
Milk	gal	84	72
Beef	lb	42	63
Mutton	lb	30	40
		1,094	902

showing a fall in prices of 18 per cent as against a rise of 9 per cent when the original units were used. The second index is different from the first because the collection now contains more potatoes and milk—the two commodities whose prices have fallen.

Weighted Aggregate of Prices

The simple aggregate of prices is an unsatisfactory index because it covers a collection of goods determined arbitrarily by the units in which the prices happen to be quoted. What we must do is to pick a collection in which the various goods are given weight according to their importance in some sense or other. This leads to indexes known as *weighted aggregates of prices*. We write q for the quantity of an individual good to be included in the market basket to be priced. The q's are known as *quantity weights*. Our weighted aggregate index then measures the change in the cost of a collection consisting of q units of each item, q being different from item to item. The formula is

$$P_{01} = \frac{\Sigma p_1 q}{\Sigma p_0 q}$$

where $\Sigma p_0 q$ is the cost of the collection in period 0 and $\Sigma p_1 q$ the cost in period 1. Incidentally, the simple aggregate of prices is a special case of the above formula, where each q is one unit.

Clearly the q's ought to be fixed according to the relative importance of the commodities in question. The relative importance can be assessed on the basis of the quantities of the goods purchased relative to each other. If the quantities of the various goods purchased never changed or if they changed only in the same ratio, then the quantities purchased in any year could be used as weights, for the *relative* composition of the basket would always be the same, and hence the change in its cost would be the same irrespective of which period's purchases were used to determine the weights. However, quantities of various goods purchased do not all change in the same ratio, i.e. relative quantities purchased change from time to time. Consequently, there is no unique set of weights available. It was pointed out above that if prices all changed in the same proportion the construction of an index number would present no difficulties. We now see that the difficulties in index number construction arise because both relative prices and relative quantities change. Indeed these changes are related because the quantities purchased of those goods whose prices have risen more than the average will, in general, tend to rise less (or fall more) than the average.

The question now is: "How shall we determine the numerical values of the q's?" The most obvious suggestion is to use the quantities of the goods purchased in period 0, which we designate q_0, or to use the quantities purchased in period 1, which we designate q_1. The first alternative gives rise to what is known as Laspeyres's price index number (P_{01}^{La}) and the second to Paasche's (P_{01}^{Pa}). The formulae[1] are

$$P_{01}^{La} = \frac{\Sigma p_1 q_0}{\Sigma p_0 q_0} \quad \text{and} \quad P_{01}^{Pa} = \frac{\Sigma p_1 q_1}{\Sigma p_0 q_1}$$

The precise meaning of the four money aggregates in the above formulae should be noted. $\Sigma p_0 q_0$ measures the cost of the base-year purchases at base-year prices, i.e. it is the value expended in the base year; $\Sigma p_1 q_1$ measures the cost of the given-year purchases at given-year prices, i.e. it is the value expended in the given year; $\Sigma p_1 q_0$ measures the cost of the base-year purchases at given-year prices, and $\Sigma p_0 q_1$ the cost of the given-year purchases at base-year prices. P_{01}^{La} is said to be a *weighted aggregate of prices with base-period weights* and P_{01}^{Pa} a *weighted aggregate of prices with given-period weights*. Both indexes measure the change in the cost of a certain collection of goods—in Laspeyres's the collection is the quantities purchased in period 0, while in Paasche's it is the quantities purchased in period 1. There is no reason why the change in the cost of two distinct collections should be the same, and P_{01}^{La} and P_{01}^{Pa} will in general be different.

In our example above (see Table 13.2), columns (5) and (6) give the q_0 and q_1 quantities expressed in the units in which the prices are quoted. The aggregate costs are computed in the final four columns. For Laspeyres's index we have

$$(13{,}346 \div 10{,}867) \times 100 = 123$$

and for Paasche's

$$(11{,}700 \div 9{,}940) \times 100 = 118$$

[1] These two formulae were named by C. M. Walsh in *The Measurement of Exchange Value* (Macmillan, 1901) after their originators—E. Laspeyres (*Jahrbücher für Nationalökonomie und Statistik*, Vol. III, 1864, p. 81 and Vol. XVI, 1871, p. 296) and H. Paasche (*ibid.*, Vol. XXIII, 1874, p. 168).

PRICE INDEX NUMBERS 333

Both of these index numbers seem to be based on reasonable principles and *prima facie* there is no reason to prefer one formula to the other. Consequently, suggestions have been made for taking a mean between the two indexes. Thus Marshall and Edgeworth have suggested the formula[1]

$$P_{01}^{ME} = \frac{\Sigma p_1(q_0 + q_1)}{\Sigma p_0(q_0 + q_1)}$$
$$= \frac{\Sigma(p_1 q_0 + p_1 q_1)}{\Sigma(p_0 q_0 + p_0 q_1)}$$

This is equivalent to using the mean of the q_0's and q_1's as weights. Fisher has also suggested a formula, which he called the Ideal index number[2]

$$P_{01}^{Id} = \sqrt{P_{01}^{La} \cdot P_{01}^{Pa}} = \sqrt{\frac{\Sigma p_1 q_0}{\Sigma p_0 q_0} \cdot \frac{\Sigma p_1 q_1}{\Sigma p_0 q_1}}$$

This is the geometric mean of Laspeyres's and Paasche's. In our example, for the Marshall-Edgeworth index we have

$$[(13{,}346 + 11{,}700) \div (10{,}867 + 9{,}940)] \times 100 = 120$$

and for Fisher's Ideal index

$$\sqrt{123 \times 118} = 120$$

If we had worked to another figure we should have found the former to be 120·4 and the latter 120·2. These two indexes are always very close.

The above formulae involve weights which change when the periods of comparison change. Thus the weights in Paasche's index change whenever the given period changes, and the weight in Laspeyres's whenever the base period changes. In contrast to these indexes with *changing weights* is the aggregative index with *fixed weights*, i.e.

$$P_{01} = \frac{\Sigma p_1 q}{\Sigma p_0 q}$$

where the q's are fixed weights irrespective of the periods of comparison. This type of index is the most commonly used form in practice (see sections 13.8 and 13.10).

13.4. Relation between Laspeyres's and Paasche's Index Numbers

Laspeyres's and Paasche's index numbers are probably the two most important index numbers in the statistical and economic

---

[1] See Edgeworth, F. Y.: *Papers Relating to Political Economy* (Macmillan, 1925), Vol. I, p. 213.
[2] See Fisher, I.: *The Making of Index Numbers* (Houghton Mifflin, 1922), p. 220.

theory of index numbers. As has been pointed out above, these two indexes will in general give different results when applied to the same data. However, if the prices of all the goods change in the same ratio, the two indexes will be equal, for then the weighting system is irrelevant; or, if the quantities of all the goods change in the same ratio, they will be equal, for then the two weighting systems are the same relatively. When, as occurs in practice, neither all prices nor all quantities move in the same ratio, the relation between the two indexes depends on the correlation between price and quantity movements. Normally, we shall expect negative correlation, i.e. people will buy relatively less of those commodities which have become relatively dearer.[1] In this case, those goods whose prices have risen more than the average when prices in general are rising (or whose prices have fallen less than the average when prices in general are falling) will tend to have q_1's relatively smaller than the corresponding q_0's, and consequently will have relatively less weight in Paasche's than in Laspeyres's index. Accordingly under these circumstances Paasche's index will tend to be less than Laspeyres's. This means that Paasche's index will show a smaller rise when prices are rising and a greater fall when prices are falling. It will be noted that in our examples Paasche's is less than Laspeyres's index (see also pp. 346–7 below).

The relation between the two indexes can be derived analytically by means of the formula for the coefficient of linear correlation. If we have a series of pairs of observations of X and Y, each pair being weighted by a frequency f, the sum of the frequencies being N, the formula for the coefficient of linear correlation between X and Y will be

$$r_{XY} = \frac{\Sigma fxy}{\sqrt{\Sigma fx^2 \Sigma fy^2}}$$

$$= \frac{\frac{\Sigma fxy}{N}}{s_X s_Y}$$

$$= \frac{\frac{\Sigma fXY}{N} - \frac{\Sigma fX}{N} \frac{\Sigma fY}{N}}{s_X s_Y}$$

---

[1] This assumes that changes in prices and quantities over time are mainly the result of changes in supply conditions. In terms of the geometry of economic theory, it assumes that demand curves are stable while supply curves are shifted up and down, so that actual prices and quantities are located along the downward-sloping demand curves. If the reverse held and supply curves were upward-sloping (which they certainly are in the short period), the correlation between prices and quantities would be positive.

where s_X and s_Y are the standard deviations of the X's and Y's respectively, with the f's included as weights.[1]

We now write p_1/p_0 for X, q_1/q_0 for Y and $p_0 q_0$ for f in the above formula; X then stands for relative price movements and Y for relative quantity movements of individual commodities. The statistics s_X and s_Y measure the dispersion of price and quantity movements respectively and r_{XY} measures the correlation between price and quantity movements. We shall have

$$r_{XY} s_X s_Y = \frac{\Sigma\left(\frac{p_1}{p_0} \cdot \frac{q_1}{q_0} \cdot p_0 q_0\right)}{\Sigma p_0 q_0} - \frac{\Sigma\left(\frac{p_1}{p_0} p_0 q_0\right)}{\Sigma p_0 q_0} \frac{\Sigma\left(\frac{q_1}{q_0} p_0 q_0\right)}{\Sigma p_0 q_0}$$

i.e.

$$\frac{\Sigma p_1 q_0}{\Sigma p_0 q_0} \frac{\Sigma p_0 q_1}{\Sigma p_0 q_0} = \frac{\Sigma p_1 q_1}{\Sigma p_0 q_0} - r_{XY} s_X s_Y$$

and writing

$$V_{01} = \frac{\Sigma p_1 q_1}{\Sigma p_0 q_0}$$

this becomes

$$\frac{\Sigma p_1 q_0}{\Sigma p_0 q_0} \frac{\Sigma p_0 q_1}{\Sigma p_1 q_1} = 1 - \frac{r_{XY} s_X s_Y}{V_{01}} \qquad (*)$$

where V_{01} is an index of the values expended in the two periods.

It can be seen that the expression on the left-hand side of the relation (*) is the ratio $P_{01}^{\text{La}}/P_{01}^{\text{Pa}}$. This ratio will be equal to unity only if one of r_{XY}, s_X or s_Y is equal to zero, i.e. either if there is no correlation between price and quantity movements, or if all prices or all quantities move in the same ratio so that there is no dispersion in one or other of price or quantity movements. Normally we shall have $-1 < r_{XY} < 0$, $s_X \neq 0$ and $s_Y \neq 0$, so that the right-hand side of the relation will be greater than unity and we can conclude that normally

$$P_{01}^{\text{La}} > P_{01}^{\text{Pa}}$$

Furthermore the relative discrepancy between P_{01}^{La} and P_{01}^{Pa} will be greater the greater the degree of correlation between price and quantity movements and the greater the dispersion in these two sets of movements, relative to the overall movement in values expended. In practice, Laspeyres's and Paasche's indexes are usually fairly close so long as the periods being compared are not too far distant. But the further apart are the periods being compared the greater the opportunity for dispersion in price and quantity movements. Similarly, the

---

[1] See section 10.1, p. 207 above. Here the standard deviations have N as their denominators.

force of habit in consumption patterns will tend to break down, and as a result the correlation between price and quantity movements is likely to be greater in the long period than in the short period.

13.5. Tests of Adequacy of Index Number Formulae

In section 13.3 two important formulae (Laspeyres's and Paasche's) were presented together with two other formulae based on them. Are there any technical reasons why we should prefer one of these?

Fisher has suggested two tests which he believes should be met by index number formulae.[1] These tests are derived by analogy with the behaviour of individual prices. If we take a single commodity, using period 0 as the point of reference, its price movement between period 0 and period 1 is measured by the ratio p_1/p_0. On the other hand, using period 1 as the point of reference, its price movement between period 1 and period 0 is measured by p_0/p_1. Clearly

$$\frac{p_0}{p_1} = \frac{1}{\frac{p_1}{p_0}}$$

By analogy, the index number P_{01} measures price movements between period 0 and period 1, and the index number P_{10} measures price movements between period 1 and period 0. Accordingly we ought to have

$$P_{01} = \frac{1}{P_{10}}$$

i.e. $\qquad P_{01} P_{10} = 1$

This is called the *time reversal test*. In P_{10} the times are reversed. Thus, whereas

$$P_{01}^{\text{La}} = \frac{\Sigma p_1 q_0}{\Sigma p_0 q_0}$$

we have $\qquad P_{10}^{\text{La}} = \dfrac{\Sigma p_0 q_1}{\Sigma p_1 q_1}$

This test is tantamount to saying that a formula which shows a rise in prices of, say, 25 per cent between period 0 and period 1, should show a fall in price of 20 per cent between period 1 and 0, $(1 \cdot 25 \times 0 \cdot 80 = 1)$, for this would certainly be true for one commodity.

Referring back to the relation (*) in the preceding section, it can be seen that the left-hand side can be expressed either as $P_{01}^{\text{La}} P_{10}^{\text{La}}$ or as

---

[1] Fisher, I: *op. cit.*, Chapter IV, p. 62.

$\frac{1}{P_{01}^{Pa} P_{10}^{Pa}}$, so that neither Laspeyres's nor Paasche's formula will meet the time reversal test except in the unlikely circumstances of either r_{XY} or s_X or s_Y equalling zero. Furthermore, normally Laspeyres's will show an upward bias and Paasche's a downward one of the same relative magnitude *in relation to the test*.

In our example

$$P_{01}^{La} = 123 \text{ and } P_{10}^{La} = (9{,}940 \div 11{,}700) \times 100 = 85$$

Hence

$$P_{01}^{La} \times P_{10}^{La} = 1{\cdot}23 \times 0{\cdot}85 = 1{\cdot}05,$$

which is greater than unity ($0{\cdot}85 \neq 1/1{\cdot}23$).

Similarly

$$P_{01}^{Pa} = 118 \text{ and } P_{10}^{Pa} = (10{,}867 \div 13{,}346) \times 100 = 81$$

Hence

$$P_{01}^{Pa} \times P_{10}^{Pa} = 1{\cdot}18 \times 0{\cdot}81 = 0{\cdot}96,$$

which is less than unity ($0{\cdot}81 \neq 1/1{\cdot}18$).

The second test is known as the *factor reversal test*. Again, for an individual good the price movement between periods 0 and 1 is measured by p_1/p_0. Similarly, the quantity movement is measured by q_1/q_0. Since value expended is the product of price and quantity, the movement in values is measured by $p_1 q_1 / p_0 q_0$. Clearly

$$\frac{p_1 q_1}{p_0 q_0} = \frac{p_1}{p_0} \times \frac{q_1}{q_0}$$

By analogy, if P_{01} is an index number measuring price movements between periods 0 and 1 and Q_{01} is an index number measuring quantity movements, we ought to have

$$V_{01} = P_{01} \cdot Q_{01}$$

i.e.

$$\frac{P_{01} Q_{01}}{V_{01}} = 1$$

where

$$V_{01} = \frac{\Sigma p_1 q_1}{\Sigma p_0 q_0}$$

is an index measuring the movement in aggregate values expended. Thus, whereas the Laspeyres's price index number is defined by

$$P_{01}^{La} = \frac{\Sigma p_1 q_0}{\Sigma p_0 q_0}$$

i.e. the prices weighted by base-year quantities, the Laspeyres's quantity number will be defined by

$$Q_{01}^{\text{La}} = \frac{\Sigma q_1 p_0}{\Sigma q_0 p_0}$$

i.e. the quantities weighted by base-year prices.

This test is tantamount to saying that if values have increased by 80 per cent, then a formula which shows a price rise of 20 per cent ought, in its quantity form, to show a quantity rise of 50 per cent ($1\cdot 20 \times 1\cdot 50 = 1\cdot 80$), for this would certainly be the case for one commodity.

Referring back to the relation (*) in the preceding section, it can be seen that the left-hand side can be expressed either as

$$\frac{P_{01}^{\text{La}} Q_{01}^{\text{La}}}{V_{01}}, \text{ or as } \frac{V_{01}}{P_{01}^{\text{Pa}} Q_{01}^{\text{Pa}}}$$

so that neither Laspeyres's nor Paasche's formula will meet the factor reversal test, except in the unlikely circumstances of either r_{XY} or s_X or s_Y equalling zero. Furthermore, normally Laspeyres's will show an upward bias and Paasche's a downward one of the same relative magnitude *in relation to the test*.

In our example

$$P_{01}^{\text{La}} = 123 \text{ and } Q_{01}^{\text{La}} = (9{,}940 \div 10{,}867) \times 100 = 91$$

and

$$V_{01} = (11{,}700 \div 10{,}867) \times 100 = 107 \cdot 7$$

Hence,

$$(P_{01}^{\text{La}} \times Q_{01}^{\text{La}}) \div V_{01} = (1 \cdot 23 \times 0 \cdot 91) \div 1 \cdot 077 = 1 \cdot 04,$$

which is greater than unity ($1 \cdot 23 \times 0 \cdot 91 \neq 1 \cdot 077$).

Similarly

$$P_{01}^{\text{Pa}} = 118 \text{ and } Q_{01}^{\text{Pa}} = (11{,}700 \div 13{,}346) \times 100 = 88$$

Hence

$$(P_{01}^{\text{Pa}} \times Q_{01}^{\text{Pa}}) \div V_{01} = (1 \cdot 18 \times 0 \cdot 88) \div 1 \cdot 077 = 0 \cdot 96,$$

which is less than unity ($1 \cdot 18 \times 0 \cdot 88 \neq 1 \cdot 077$).

Turning to the Marshall-Edgeworth and Fisher indexes, substitution of the formulae in the two tests reveals that the Marshall-Edgeworth index meets the time reversal but not the factor reversal test, whereas Fisher's meets both. It is largely for this reason that Fisher termed his formula the "ideal" index.

However, it is not at all certain that there are good logical reasons for claiming that an index number ought to meet these tests. As far as the time reversal test is concerned, one could hardly hope for consistent results. This test requires that $P_{01}^{La} P_{10}^{La} = 1$, but the collection of goods included in P_{01}^{La} is different from that included in P_{10}^{La} (q_0 as against q_1). As far as the factor reversal test is concerned there may be a quantity index which in conjunction with the price index will satisfy the test. Thus, although

yet
$$P_{01}^{La} Q_{01}^{La} \neq V_{01}$$
$$P_{01}^{La} Q_{01}^{Pa} = V_{01}$$

and, although
$$P_{01}^{Pa} Q_{01}^{Pa} \neq V_{01}$$

yet
$$P_{01}^{Pa} Q_{01}^{La} = V_{01}$$

Furthermore, although the Ideal index does meet these tests, its meaning is not at all clear. Laspeyres's index measures the change in the cost of the collection being purchased in period 0 and Paasche's the change in the cost of that being purchased in period 1, but what does the geometric mean of the two measure? Clarity of meaning is always desirable in index number construction.

The application of the above two tests gives rise to the notion of "bias" in index number formulae. Laspeyres's is sometimes spoken of as being upwardly biased and Paasche's as being downwardly biased. But to say a measure is biased implies that there is some "true" value from which it is biased. No such "true" value, however, can be postulated in this connexion, consequently talk of bias, except in relation to the two tests, is meaningless. The question of what it is we are "ideally" trying to measure is discussed in section 13.9 below. But we can conclude at this stage that the fact that Laspeyres's and Paasche's formulae do not conform to the two tests is not very important. Nevertheless, a knowledge of how these two formulae fit in with the tests is important for a proper understanding of the mechanics of index number construction.

13.6. Index Number Formulae—Average Type

An alternative to constructing formulae in terms of aggregate costs, is to concentrate on the movements of the prices of individual commodities. If we take period 0 as base, the movement in the price of an individual commodity is given by p_1/p_0. Ratios such as this are known as *price relatives*. They refer to individual price movements. The price relatives for the seven commodities in our example are shown in column (5) of Table 13.4 below.

We now have a number of individual price movements, and a process of averaging will produce an index number which can be used to indicate the general movement in prices.

Table 13.4

Commodity	Unit	Prices		Price Relatives p_1/p_0	$\log p_1/p_0$	Value Expended $v = p_0 q_0$	$\frac{p_1}{p_0} \times v$
		Period 0 p_0	Period 1 p_1				
(1)	(2)	(3)	(4)	(5)	(6)	(7)	(8)
		cents	cents			cents	cents
Bread	2 lb	18	21	1·167	0·06707	1,800	2,100
Tea	lb	80	90	1·125	0·05115	800	900
Potatoes	7 lb	49	35	0·714	$\bar{1}$·85370	980	700
Butter	lb	56	56	1·000	0·00000	1,680	1,680
Milk	qt	21	18	0·857	$\bar{1}$·93298	147	126
Beef	lb	42	63	1·500	0·17609	3,360	5,040
Mutton	lb	30	40	1·333	0·12483	2,100	2,800
				7·696	0·20582	10,867	13,346

Simple Arithmetic Mean of Price Relatives

The simplest approach is to take a straight-out arithmetic mean of the price relatives. The formula will be

$$P_{01} = \frac{1}{N} \Sigma \frac{p_1}{p_0}$$

where N is the number of commodities. In our example this will be

$$(7·696 \div 7) \times 100 = 110$$

The arithmetic mean, however, is not a particularly suitable type of average for averaging ratios. It seems reasonable that if one ratio indicates a doubling and another a halving, the average of the two should indicate no change. But an arithmetic average will show a change in this situation of + 25 per cent, for $\frac{1}{2}(2·00 + 0·50) = 1·25$. A geometric mean does not suffer from this defect, for $\sqrt{2·00 \times 0·50} = 1·00$. For this reason, when a straight-out average of ratios is desired, the geometric mean is the appropriate average.

Simple Geometric Mean of Price Relatives

The formula for this index will be

$$P_{01} = \sqrt[N]{\Pi \left(\frac{p_1}{p_0}\right)}$$

PRICE INDEX NUMBERS 341

where the symbol Π means "the product of." For purposes of calculation it is best to work in logarithms. We have

$$\log P_{01} = \frac{1}{N} \Sigma \log \frac{p_1}{p_0}$$

In our example the logarithms of the price relatives are given in column (6), and we have

$$\log P_{01} = 0 \cdot 20582 \div 7 = 0 \cdot 02940$$

The antilogarithm of 0·02940 is 1·07, so that our index number is 107. It should be noted that the geometric mean is always less than the arithmetic mean.

If we want an average of price relatives, in which we regard each commodity as of equal importance, then the simple geometric mean of price relatives is appropriate. However, generally we do not wish to regard each commodity as having equal importance—a large price change in an unimportant commodity may be of considerably less significance than a small change in an important one. This leads us to weighted averages of price relatives.

Weighted Arithmetic Mean of Price Relatives

In weighting price relatives the weights must be *values*. If we write v for value expended on a particular commodity, then the total value expended on all the commodities under consideration will be Σv. How will this sum change between periods 0 and 1 due to price changes? To obtain the same quantity of a particular commodity as we should have obtained by spending v in period 0, we must now spend $v \times p_1/p_0$ in period 1. Consequently, total value expended would have to change to $\Sigma \frac{p_1}{p_0} v$. Hence, our index number will be

$$P_{01} = \frac{\Sigma \frac{p_1}{p_0} v}{\Sigma v}$$

This is a weighted arithmetic mean of price relatives, where the v's are the weights.

How should we determine the numerical values of the v's? One suggestion would be to use the values expended in period 0, i.e. put $v = p_0 q_0$. We should then have

$$P_{01} = \frac{\Sigma \left(\frac{p_1}{p_0} p_0 q_0 \right)}{\Sigma p_0 q_0}$$

$$= P_{01}^{\text{La}}$$

Thus an arithmetic mean of price relatives weighted by base-year values is in fact Laspeyres's index number.

The calculation of a weighted arithmetic mean of price relatives of this type is shown in columns (5), (7) and (8) of Table 13.4. The index is

$$(13{,}346 \div 10{,}867) \times 100 = 123$$

and is, of course, identically equal to Laspeyres's.

Similarly, if we put $v = p_0 q_1$, we should get

$$P_{01} = \frac{\Sigma \left(\frac{p_1}{p_0} p_0 q_1\right)}{\Sigma p_0 q_1}$$
$$= P_{01}^{\text{Pa}}$$

Alternatively, if we use given-year values and take an *harmonic* mean, we also get

$$P_{01} = \frac{1}{\dfrac{\Sigma \left(\dfrac{p_0}{p_1} p_1 q_1\right)}{\Sigma p_1 q_1}}$$
$$= P_{01}^{\text{Pa}}$$

13.7. Relation between Aggregative and Average Types

It has been shown above that both Laspeyres's and Paasche's index numbers can be expressed as weighted arithmetic means of price relatives. In fact, any aggregative type can be expressed in average terms, and any arithmetic or harmonic average type can be expressed in aggregative terms.

Thus we can always express an aggregative type as follows—

$$P_{01} = \frac{\Sigma p_1 q}{\Sigma p_0 q} = \frac{\Sigma \left(\dfrac{p_1}{p_0} p_0 q\right)}{\Sigma p_0 q}$$

where $p_0 q$ is the value expended by pricing the q's at base-year prices. Conversely, we can always express a weighted arithmetic mean type as follows—

$$P_{01} = \frac{\Sigma \dfrac{p_1}{p_0} v}{\Sigma v} = \frac{\Sigma \left(p_1 \dfrac{v}{p_0}\right)}{\Sigma \left(p_0 \dfrac{v}{p_0}\right)}$$

where v/p_0 is the quantity which can be purchased by spending v at base-year prices. In interpreting these formulae it is important to remember that q is in physical units, and v is in money terms.

PRICE INDEX NUMBERS

From the practical point of view the aggregative form is not only simpler to calculate but simpler to comprehend. The idea of a measure of the change in cost of a given collection of goods is straightforward. Consequently, the aggregative form is to be preferred when questions of interpretation are important However, the average form has its uses. We may be interested in the price relatives themselves, for the purpose of analysing the presence or absence of homogeneity in price movements or for distinguishing between the characteristic behaviour of sub-groups of commodities within our group. Furthermore, if we wish to analyse the relative importance of the items in an index, it is essential to consider it in its price relative form. An aggregative type index can be written

$$P_{01} = \frac{\Sigma p_1 q}{\Sigma p_0 q} = \frac{\Sigma \left(\frac{p_1}{p_0} p_0 q\right)}{\Sigma p_0 q} = \Sigma \left[\frac{p_1}{p_0} \cdot \frac{p_0 q}{\Sigma p_0 q}\right]$$

Since the sum of the fractions $\frac{p_0 q}{\Sigma p_0 q}$ equals unity, these fractions give the *relative* (or percentage) weights attaching to the price relative of each individual item. They are the relative expenditures in the index on individual items. Accordingly, the fraction $\frac{p_0 q}{\Sigma p_0 q}$ for an individual item gives its relative importance in the index in the sense that it is the weight attached to any change in the price of that item between periods 0 and 1. Some important applications of the price-relative form are given below.

Effect on an Index of a Change in the Price of One Item

The concept of relative weights can be used to ascertain the effect on the index, of the change in the price of one item, other things being equal. Suppose we have an index measuring price changes between periods 0 and t and we wish to consider the effect on the index of a given proportionate change in the price of one item. Let the index be

$$P_{0t} = \frac{\Sigma p_t q}{\Sigma p_0 q}$$

and the price of the particular item under consideration p_t' in period t and its weight q'. Let α be the assumed proportionate increase in price of that item, so that the assumed new price for the item is $p_t' + \alpha p_t'$. Then this increase in price will add $\alpha p_t' q'$ to the aggregate cost in period t, so that the index will become

$$\frac{\alpha p_t' q' + \Sigma p_t q}{\Sigma p_0 q}$$

which is a proportionate increase of

$$\left(\frac{\alpha p_t' q' + \Sigma p_t q}{\Sigma p_0 q} - \frac{\Sigma p_t q}{\Sigma p_0 q}\right) \bigg/ \frac{\Sigma p_t q}{\Sigma p_0 q}$$

on the original index. This reduces to

$$\alpha \left(\frac{p_t' q'}{\Sigma p_t q}\right)$$

The term in brackets is the relative weight attached to the particular item in the index in period t, so that the proportionate change in the index is given by the product of the proportionate change in the price of the item and the relative expenditure in the index on the item in the period from which the change is measured. This should be self-evident, for if I spend 20 per cent of my income on food and the price of food goes up by 10 per cent, I shall have to increase my expenditure by 10 per cent of 20 per cent, i.e. by 2 per cent, to purchase the same collection as before. For example, Laspeyres's index gives an index number of 123 in the example set out in Table 13.2 on page 329. What would be the effect on the index of an increase in the price of beef from 63 cents to 70 cents? This is an increase of one-ninth. The relative importance of beef in period 1 is $\frac{5{,}040}{13{,}346}$, so that the rise in the price of beef will increase the index by $\frac{1}{9} \times \frac{5{,}040}{13{,}346}$, i.e. by 4 per cent. The index will increase from 123 to $123 \times 1{\cdot}04 = 128$.

The relative expenditure on a particular item in the index in period t $\left(\text{i.e. } \frac{p_t q}{\Sigma p_t q}\right)$ will in general vary from period to period as relative prices change, and consequently the effect of a given percentage change in the price of an item will vary according to the period from which the change is supposed to take place. It might be noted that in fixed-weight indexes the relative expenditure *in the index* of those items whose prices are rising consistently more than the average will increase over time (see section 13.11, p. 360 below).

Combination of Indexes

Sometimes the items in an index are divided into groups for which separate indexes are given. The relation between these group indexes and the full index depends on the relative weights of the groups within the full index in the base period. This can be seen as follows. If

C_t^G is the aggregate cost (i.e. $\Sigma p_t q$) for a particular group of the regimen in period t, a group index is given by

$$P_{0t}^G = \frac{C_t^G}{C_0^G}$$

The index for the whole regimen will be obtained by summing the numerators and denominators respectively of the group indexes to get

$$P_{0t} = \frac{\Sigma C_t^G}{\Sigma C_0^G}$$

where the summation is over the different groups. This is equivalent to

$$P_{0t} = \Sigma \left[P_{0t}^G \cdot \frac{C_0^G}{\Sigma C_0^G} \right]$$

i.e. the complete index is the weighted average of the group indexes, the weights being the relative aggregate costs in the base year. A shift in the base year will, of course, change the relative weights to be applied. For example, if the base is changed from period 0 to period 1, we shall have

$$P_{1t} = \Sigma \left[P_{1t}^G \cdot \frac{C_1^G}{\Sigma C_1^G} \right]$$

As an example we may take the Australian Consumer Price Index. This index is divided into five groups for which separate group indexes are published. The table below gives these indexes for June Quarter, 1967, based on December Quarter, 1963.

Table 13.5

	June Quarter, 1967 Base: December Quarter, 1963 = 1,000	Percentage of Aggregate Expenditure in Index, December Quarter, 1963
Food	1,158	32·1
Clothing and drapery	1,063	16·9
Housing	1,148	12·6
Household supplies and equipment	1,044	14·5
Miscellaneous	1,161	23·9
Total	1,125	100·0

Source: Derived from Commonwealth Bureau of Census and Statistics, Australia: *Labour Report*, No. 51, 1964, p. 9, and *Monthly Review of Business Statistics*, No. 370, 1968, p. 27

The index number for the whole group (1,125) is in fact the average of the five group indexes weighted by the relative expenditure given in the right-hand column.

Comparison of Different Index Number Formulae

The price relative form is also useful in analysing why two different aggregative type indexes give different results when applied to the same data. Suppose we have two such indexes, one with quantity weights q and the other with weights $\hat{q}$, namely

$$\frac{\Sigma p_1 q}{\Sigma p_0 q} \text{ and } \frac{\Sigma p_1 \hat{q}}{\Sigma p_0 \hat{q}}$$

These can be rewritten

$$\Sigma \left[\frac{p_1}{p_0} \cdot \frac{p_0 q}{\Sigma p_0 q} \right] \text{ and } \Sigma \left[\frac{p_1}{p_0} \cdot \frac{p_0 \hat{q}}{\Sigma p_0 \hat{q}} \right]$$

The fractions $\frac{p_0 q}{\Sigma p_0 q}$ and $\frac{p_0 \hat{q}}{\Sigma p_0 \hat{q}}$ are the *relative* weights attaching to each individual price relative under the two systems of weighting. Consequently, by comparing these relative weights we can see whether the one system of weighting attaches more or less weight to particular price relatives than the other. To illustrate this we can consider the discrepancy between P_{01}^{La} and P_{01}^{Pa} in our example in section 13.3 above. In this example P_{01}^{La} was 123 and P_{01}^{Pa} was 118.

Table 13.6

Commodity	Price Relative	Relative Weights	
		Laspeyres's	Paasche's
	$\dfrac{p_1}{p_0}$	$\dfrac{p_0 q_0}{\Sigma p_0 q_0}$	$\dfrac{p_0 q_1}{\Sigma p_0 q_1}$
(1)	(2)	(3)	(4)
Bread	1·167	0·165	0·163
Tea	1·125	0·074	0·064
Potatoes	0·714	0·090	0·148
Butter	1·000	0·155	0·169
Milk	0·857	0·014	0·021
Beef	1·500	0·309	0·254
Mutton	1·333	0·193	0·181
		1·000	1·000

In the above table columns (3) and (4) are the relative distributions of columns (7) and (9) respectively in Table 13.2 above. The sum of the products of columns (2) and (3) will give Laspeyres's index and that of columns (2) and (4) Paasche's. It can be seen that less relative weight is given in Paasche's index to beef, which had the greatest price rise, and more to butter, which had no price rise, and potatoes, which fell in price. This helps to explain why Paasche's index was lower than Laspeyres's. Both indexes are weighted averages of the *same* price movements (the p_1/p_0's), but the relative weighting is different. The difference in relative weights in this example reflects normal negative correlation between prices and quantities. The order of magnitude of the price relatives is beef, mutton, bread, tea, butter, milk, potatoes, whereas that of the quantity relatives (see columns (5) and (6) of Table 13.2) is, with one exception, the reverse.

13.8. Comparisons between more than Two Points of Time

The above discussion has been exclusively concerned with making comparisons between two periods only. These are called *point-to-point* or *binary* comparisons. However, we are usually interested in making comparisons between a number of periods, say periods 0, 1, 2, 3, etc., and we will have a series of index numbers P_{01}, P_{02}, P_{03}, etc., each one expressing the level of prices relatively to the base period, in this case period 0. Each of the indexes P_{01}, P_{02}, P_{03}, etc., is itself a binary comparison between period 0 and the period under consideration.

However, if we wish to compare prices in period 2 relatively to period 1, we can make a direct binary comparison and calculate P_{12} or we can obtain a figure by dividing P_{02} by P_{01}. Thus, if prices rise to 120 in period 1 on base period $0 = 100$ and rise further to 180 in period 2 on the same base, period 2 prices based on period 1 should be $(180 \div 120) \times 100 = 150$. Thus it would appear that we could use either P_{12} or P_{02}/P_{01}.

Similarly, we could compare prices in period 2 with those in period 0 by a direct binary comparison P_{02} or by the product $P_{01} \times P_{12}$. Thus, if prices are up to 110 in period 1 on base period $0 = 100$ and are up to 120 in period 2 on base period $1 = 100$, period 2 prices based on period 0 should be $1 \cdot 10 \times 1 \cdot 20 \times 100 = 132$. It should be plain that these two cases are the same in principle.

Circular Test

At first sight it would appear desirable that

$$P_{02} = P_{01} \cdot P_{12}$$

i.e.
$$\frac{P_{01} P_{12}}{P_{02}} = 1$$

This requirement is called the *circular test*. This test is met by practically none of the index number formulae outlined above (exceptions are the simple geometric mean of price relatives and the weighted aggregate with fixed weights). It is not met by Laspeyres's and Paasche's. Thus, for Laspeyres's, we have

$$P_{02}^{\text{La}} = \frac{\Sigma p_2 q_0}{\Sigma p_0 q_0}, \quad P_{01}^{\text{La}} = \frac{\Sigma p_1 q_0}{\Sigma p_0 q_0} \quad \text{and} \quad P_{12}^{\text{La}} = \frac{\Sigma p_2 q_1}{\Sigma p_1 q_1}$$

Consequently, we have

$$\frac{P_{01} P_{12}}{P_{02}} = \frac{\Sigma p_2 q_1}{\Sigma p_1 q_1} \div \frac{\Sigma p_2 q_0}{\Sigma p_1 q_0}$$

which is the quotient of two differently weighted aggregative indexes comparing periods 1 and 2 and hence not necessarily equal to unity.[1]

While it may seem reasonable to argue that if a price index between periods 0 and 1 has risen to M and between periods 1 and 2 to N, then between periods 0 and 2 it should have risen to MN, a moment's reflection will show that this requirement is not reasonable. An index number has meaning only in terms of the system of weighting adopted, and one may produce many numerically different but quite valid indexes for comparing two periods. The weighting system used in P_{02}^{La} is the same as that in P_{01}^{La}, but different from that in P_{12}^{La}; consequently the increase M is an increase in something different from that in which N is the increase. The product MN is therefore a mixture, the exact meaning of which is not clear and which could not be expected to equal a direct comparison between periods 0 and 2.

Chained Indexes

The argument in the preceding paragraph may be taken as suggesting that direct binary comparisons are to be preferred to roundabout

---

[1] The relation between P_{02} and $(P_{01}P_{12})$ can be developed by the method used in section 13.4 above. Writing $X = p_2/p_1$, $Y = q_1/q_0$ and $f = p_1 q_0$ in the formula on page 334, we get a formula equivalent to the starred one on page 335, namely

$$\frac{\Sigma p_2 q_0}{\Sigma p_1 q_0} \frac{\Sigma p_1 q_1}{\Sigma p_2 q_1} = 1 - \frac{r_{XY} s_X s_Y}{\frac{\Sigma p_2 q_1}{\Sigma p_1 q_0}}$$

The left-hand side is the reciprocal of $\frac{P_{01}P_{12}}{P_{02}}$, and this will be equal to unity only if one of r_{XY}, s_X, s_Y equals zero. Here s_X and s_Y measure the dispersion in price movements between periods 1 and 2 and in quantity movements between periods 0 and 1 respectively, and they will not in general be zero. The statistic r_{XY} measures the correlation between quantity movements over periods 0 to 1 and price movements over periods 1 to 2. It is not possible to say anything very precise about this correlation other than there is no *a priori* reason for it to be zero. Indeed, since q_1/q_0 and p_1/p_0 are likely to be negatively correlated the correlation between q_1/q_0 and p_2/p_1 is likely to be considerably influenced by the correlation between p_1/p_0 and p_2/p_1, i.e. by the way in which the price movements of particular goods over one period are related to those in the subsequent period. The above result is for a comparison between three consecutive points, but it can easily be generalized for any kind of circular comparison.

ones, because the meaning which can be attached to the numerical results is clearer. Nevertheless, if we are making comparisons between two distant periods, another consideration enters. Suppose the periods are 0 and t. If we are using Laspeyres's index, our quantity weights will be q_0, if Paasche's, they will be q_t. If period t is distant from period 0, it is likely that the q_0's and the q_t's will be rather different. The q_0 weights are relevant to period 0 and the q_t ones to period t but neither set of weights is relevant to both periods. This is the problem of increasing out-of-dateness of weights as the periods being compared become further and further apart. For this reason it is sometimes considered desirable to make comparisons between closely situated periods in which the q's are not likely to have changed much and to obtain longer-term comparisons by a process of chaining binary comparisons. Such an index is called a *chained* index and the formula is

$$P_{0t}^{Ch} = P_{01} \cdot P_{12} \cdot P_{23} \cdots P_{t-1,t}$$

where the separate links in the chain are binary comparisons between adjacent periods made according to some index-number formula, e.g. Laspeyres's or Paasche's. For example, if we have $P_{01} = 110$, $P_{12} = 120$, $P_{23} = 90$, then the chained index

$$P_{03} = 100 \times 1 \cdot 10 \times 1 \cdot 20 \times 0 \cdot 90 = 119$$

Although the precise meaning of P_{0t}^{Ch} is not simple in character, because it is based on a changing collection of goods, nevertheless there is a sense in which the weights are kept up to date in the chained index, because the q's are unlikely to change radically as between adjacent periods. As explained above, the value of P_{0t}^{Ch} will not be the same as that of P_{0t} where a direct point-to-point comparison is involved. In any case, not a great deal of meaning can be attached to an index number which compares distant periods for which the relative quantities are very different.

The reason why Laspeyres's and Paasche's index numbers (and their derivatives, the Marshall-Edgeworth and the Ideal indexes) do not meet the circular test is because the weights in these index numbers depend on the periods between which comparisons are being made. If these periods change, the weights change; for example, if the base period is taken as period 2 rather than period 0, the weights in Laspeyres's index are no longer q_0 but q_2. However, an aggregative index with fixed weights of the form

$$P_{0t} = \frac{\Sigma p_t q}{\Sigma p_0 q}$$

does meet the circular test. Thus

$$P_{01} P_{12} = \frac{\Sigma p_1 q}{\Sigma p_0 q} \cdot \frac{\Sigma p_2 q}{\Sigma p_1 q} = \frac{\Sigma p_2 q}{\Sigma p_0 q} = P_{02}$$

A fixed-weight index is much simpler to calculate from period to period, since the weights do not have to be changed. Consequently, less information needs to be collected in order to calculate it. On the other hand, the weights being fixed, they become increasingly out of date as time passes. Sometimes it is convenient to use fixed weights for a number of periods and then revise the weights. The new series can then be chained to the old one, in a manner similar to that used in the chained index set out above.

Changing the Base

Sometimes we have an index for a number of periods with a certain base and wish to change the period used as a basis for comparison. Suppose the original base is period $0 = 100$ and we wish to change the base to period 2. The usual practice is to divide through the whole series by the original index number for the new base, e.g., in our case, P_{02}. It can be seen that, in the case of a chained index or an aggregative index with fixed weights, this correctly accomplishes the change in base. In the case of the chained index we have, for example,

$$P_{2t}^{Ch} = \frac{P_{0t}^{Ch}}{P_{02}^{Ch}}$$

$$= \frac{P_{01}P_{12}P_{23}P_{34}\ldots P_{t-1,t}}{P_{01}P_{12}}$$

$$= P_{23}P_{34}\ldots P_{t-1,t}$$

and in the case of an aggregative index with fixed weights

$$P_{2t} = \frac{P_{0t}}{P_{02}}$$

$$= \frac{\Sigma p_t q}{\Sigma p_0 q} \div \frac{\Sigma p_2 q}{\Sigma p_0 q}$$

$$= \frac{\Sigma p_t q}{\Sigma p_2 q}$$

Strictly speaking, however, such a procedure is not valid where indexes with changing weights are being used, for a change in the period of reference then requires a change in weights. In practice, this is usually ignored. Thus, if we have index numbers 100, 170, 200, 230 for the years 1964, 1965, 1966 and 1967 respectively, with base 1964 = 100, to change the base to 1966 we simply divide through by 200 and multiply by 100, giving the index numbers 50, 85, 100, 115 for the four years respectively, with base 1966 = 100.

13.9. Choice of an Index Number—A Cost of Living Index

It is evident from the preceding sections that there are a large number of index number formulae available. Applied to the same situation, these formulae generally give different results, although in practice weighted index numbers do not differ greatly unless the dispersion of individual price movements is substantial, a factor which is likely to be important only when the comparisons are over long periods of time. However, it should be plain that there is no *a priori* "best" index number formula. Each formula has a precise meaning—most of them measure the change in the aggregate cost of a certain collection of goods. The important question is: *Does a particular formula measure what we want to measure?* Each formula measures something, but is it the thing we want? This very important consideration is now illustrated by a discussion of the measurement of changes in the cost of living.

If we want to measure changes in the cost of living, we have first to determine what we mean by "changes in the cost of living" and secondly to devise a formula to make the measurements. Various formulae for index numbers are available. These formulae all have precise and exact meanings. Consequently, if we have defined the concept to be measured and selected a formula, the following fundamental question arises. "Is the precise thing which our formula measures, the thing which our concept requires us to measure?" The whole justification for the use of particular index numbers depends on the closeness of what we are in fact measuring to what we want to measure. Thus, we may have a concept for changes in the cost of living and an index number of retail prices. We know what we mean by the former, and we know what the latter means. Can the retail price index be used to measure changes in the cost of living?

Unfortunately, while we can usually understand the precise meaning of, say, a retail price index number, it is by no means easy to define what we mean by changes in the cost of living. It is plain that particular price changes will affect the cost of living of different individual persons in different ways, e.g. if the price of beer rises while the prices of other goods fall, the cost of living of the heavy beer drinker may rise while that of other people may fall. Consequently, we must consider the cost of living of an individual person as a starting point. It is reasonable to measure the change in the cost of living of an individual between two periods as the change in his money income which will be necessary for him to maintain his *original* standard of living—no more or no less. We now consider where this definition leads. First, it must be pointed out that, if all prices change in the same proportion, that proportion will measure the change in the cost of living and there will be no problem of measurement. However, all prices do not change in the same proportion. Suppose in period 0 an individual spends his income by purchasing q_0 of various commodities at prices p_0. Assuming he saves nothing, his total income equals his total expenditure, i.e. $\Sigma p_0 q_0$. In period 1 prices change to p_1. How much income does he need in period 1 to make him as well off as he was in period 0? If the quantities of goods which he would need to buy to leave him exactly as well off as he was in period 0 are $\bar{q}_1$, then with an income of $\Sigma p_1 \bar{q}_1$ he would be exactly as well off. It follows that the change in the cost of maintaining his original (period 0) standard of living will be given by

$$C_{01}^0 = \frac{\Sigma p_1 \bar{q}_1}{\Sigma p_0 q_0}$$

where the superscript (0) indicates that the change in his cost of living is being measured in terms of his period 0 standard of living.

Unfortunately we do not know the $\bar{q}_1$'s because they are not the quantities he actually buys in period 1, but only what he would need to buy to be as well off as before. Consequently, we do not know the aggregate $\Sigma p_1 \bar{q}_1$. But we do know the aggregate $\Sigma p_1 q_0$. This is the amount of income he would require in period 1 to enable him to purchase in period 1 the *quantities* he purchased in period 0. It can be shown that $\Sigma p_1 q_0$ will be greater than $\Sigma p_1 \bar{q}_1$, provided his tastes have remained unchanged. If he did have the income $\Sigma p_1 q_0$ in period 1, he would not buy the same quantities as he bought in period 0 (the q_0 quantities) because he would take advantage of the changes in relative prices and would alter his allocation of income, buying relatively more of those goods whose prices had fallen relatively more or risen relatively less. The fact that he would do this in preference to buying the q_0 quantities indicates that an income of $\Sigma p_1 q_0$ would make him better off than the original income of $\Sigma p_0 q_0$ and, hence, better off than an income of $\Sigma p_1 \bar{q}_1$ which is its equivalent. Accordingly we have

$$\Sigma p_1 q_0 > \Sigma p_1 \bar{q}_1$$

The proviso that his tastes must have remained unchanged is vital because, if they have changed, we cannot conclude that the quantities he would buy in period 1 with an income of $\Sigma p_1 q_0$ are preferred to those he actually did buy in period 0.

This can be illustrated, by means of indifference curves, in the case of an individual spending his income on two goods. We call the two goods A and B. Let the price of A in period 0 be p_0^A and of B be p_0^B. If the individual's money income in period 0 is M_0, we shall have

$$M_0 = p_0^A q^A + p_0^B q^B$$

where q^A and q^B are the quantities of A and B respectively which he can purchase under these circumstances (the greater q^A, the smaller q^B). On the diagram below, q^A is measured on the X-axis and q^B on the Y-axis. The line RS corresponds to the equation

$$M_0 = p_0^A q^A + p_0^B q^B$$

i.e.
$$q^B = \frac{M_0}{p_0^B} - \frac{p_0^A}{p_0^B} q^A$$

The individual can take up any position on the line, i.e. his income will buy him any combination of A and B which lies on the line. RS can be

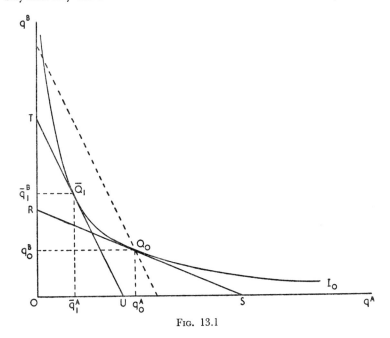

Fig. 13.1

called the *budget line*. Its slope is given by the price of A relative to that of B, i.e. by p_0^A/p_0^B. Given this price ratio the budget line will be more to the right the greater M_0, i.e. the greater the money income.

In fact, the individual will move to the point Q_0 on the budget line, at which point he reaches his highest *indifference curve*.[1] This curve

---

[1] An indifference curve shows the combinations of A and B between which the individual is indifferent. The individual will prefer a point on a higher indifference curve to one on a lower curve, since more of a good is preferred to less. See Stigler, G. J.: *The Theory of Price*, 3rd edn. (Macmillan, 1966), Chap. 4, p. 48ff, or the section on the theory of consumers' choice in any modern economics textbook.

is labelled I_0 in the diagram. Consequently in period 0 the individual will purchase q_0^A of A and q_0^B of B. Suppose that in period 1 prices change to p_1^A and p_1^B. If relative prices change, the slope of RS will change. How much income will the individual now need to be as well off as he was before? He will need sufficient to enable him to take up a position just on the indifference curve I_0. We now draw the line TU with a slope given by p_1^A/p_1^B such that it just touches the indifference curve. With an income corresponding to this line, he would take up the position $\bar{Q}_1$, purchasing $\bar{q}_1^A$ of A and $\bar{q}_1^B$ of B. This would require an income of $\bar{M}_1 = p_1^A \bar{q}_1^A + p_1^B \bar{q}_1^B$ as against his old income $M_0 = p_0^A q_0^A + p_0^B q_0^B$. But he is indifferent between the positions Q_0 and $\bar{Q}_1$, so that the ratio $\bar{M}_1/M_0$ measures the change in his old income necessary to make him as well off in period 1 as he was in period 0.

Suppose that the individual is given income in period 1 sufficient to enable him to purchase the same quantities as he did in period 0. This income must be sufficient to make the budget line for the new prices pass through Q_0 and must be equal to $p_1^A q_0^A + p_1^B q_0^B$. This is shown as the broken line in the diagram, and it must lie parallel to but to the right of TU, indicating that it represents a larger income than TU, so that

$$p_1^A q_0^A + p_1^B q_0^B > p_1^A \bar{q}_1^A + p_1^B \bar{q}_1^B$$

This inequality can be extended to the case of more than two commodities, and summing for all the purchases of the individual, we have

$$\Sigma p_1 q_0 > \Sigma p_1 \bar{q}_1$$

as above. Clearly with the income represented by the broken line the individual will not purchase q_0^A of A and q_0^B of B, but will move along the broken line to the left until contact is reached with the highest indifference curve possible. Such a curve will be higher than I_0. He will purchase more of B which has become relatively cheaper.

From the inequality

$$\Sigma p_1 q_0 > \Sigma p_1 \bar{q}_1$$

we can write[1]

$$\frac{\Sigma p_1 q_0}{\Sigma p_0 q_0} > \frac{\Sigma p_1 \bar{q}_1}{\Sigma p_0 q_0}$$

i.e. $$P_{01}^{\text{La}} > C_{01}^0$$

Thus, if we define the change in the cost of living as being the change in income necessary to make the individual as well off in period 1 as he was in period 0, Laspeyres's index number overstates rises and understates falls in the cost of living in this sense.

---

[1] Strictly speaking, the inequality should read $P_{01}^{\text{La}} \geqslant C_{01}^0$. The equality occurs when all prices have changed in the same proportion.

However, we might just as easily have defined the change in the cost of living as the ratio of the individual's income in period 1 to what he would have needed in period 0 (i.e. at period 0 prices) to have been as well off as he is in period 1. By analogy with the above this would give

$$C_{01}^1 = \frac{\Sigma p_1 q_1}{\Sigma p_0 \bar{q}_0}$$

where the superscript (1) indicates that the change in the individual's cost of living is being measured in terms of his period 1 standard of living, and we should have

$$\Sigma p_0 q_1 > \Sigma p_0 \bar{q}_0$$

The term on the left-hand side gives the income necessary to enable him to purchase the q_1 quantities at the original prices, and this will be

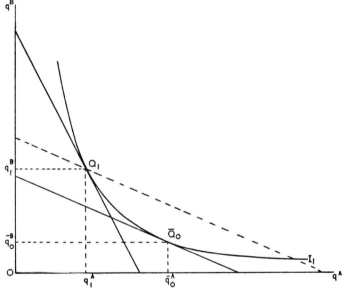

Fig. 13.2

greater than that necessary to make him just as well off. This is illustrated for the two goods case in Fig. 13.2.

The broken line lies to the right of the line passing through $\bar{Q}_0$, so that

$$p_0^A q_1^A + p_0^B q_1^B > p_0^A \bar{q}_0^A + p_0^B \bar{q}_0^B$$

It follows from the inequality[1]

$$\Sigma p_0 q_1 > \Sigma p_0 \bar{q}_0$$

that

$$\frac{\Sigma p_0 q_1}{\Sigma p_1 q_1} > \frac{\Sigma p_0 \bar{q}_0}{\Sigma p_1 q_1}$$

Inverting the inequality, this becomes

$$P_{01}^{\text{Pa}} < C_{01}^1$$

Thus Paasche's index understates rises and overstates falls in the cost of living, when the latter is defined in this sense.

It is important to realize that C_{01}^0 and C_{01}^1 are not the same thing.[2] Both measure changes in the cost of maintaining a given standard of living, but in C_{01}^0 the standard maintained is the period 0 standard and in C_{01}^1 it is the period 1 standard. Thus, for example, a particular individual may consume more whisky and less beer as his standard of living rises. If his period 1 standard is higher than his period 0 standard (because his money income has risen relatively to the price level), the change in the cost of maintaining his period 1 standard will be greater than that of maintaining his period 0 standard in a situation in which whisky prices rise more than beer prices. Clearly, in this case C_{01}^0 will differ from C_{01}^1. Consequently, we cannot use P_{01}^{La} and P_{01}^{Pa} as limits to changes in the cost of living. P_{01}^{La} is an upper limit when we are using C_{01}^0, and P_{01}^{Pa} a lower limit when we are using C_{01}^1, but they cannot be used in conjunction.

We may sum up the situation as follows. In the simplest possible case, when we are considering one individual with constant tastes in two different price situations, we can define *two* conceptually satisfactory measures for measuring the change in his cost of living, namely C_{01}^0 and C_{01}^1. However, in practice we should probably pay attention to C_{01}^0 only, for this measures the cost of maintaining a given base period standard of living, and that accords with our usual ideas about changes in the cost of living. This being so, we may take C_{01}^0 as our concept of an index number for measuring changes in the cost of living, i.e. the movement in money income necessary for an individual to maintain his original standard of living. But in practice, however clear C_{01}^0 may be conceptually, we cannot measure it precisely, because it contains hypothetical quantities of what would be necessary to make the individual as well off as he was originally. Nevertheless we can compute an index number P_{01}^{La}, which will overestimate increases and underestimate decreases in the cost of living, although by how much we

---
[1] See footnote on page 354.
[2] This is illustrated in Figs. 13.1 and 13.2. The indifference curve in the former, labelled I_0 (representing the period 0 standard), is quite distinct from that in the latter, labelled I_1 (representing the period 1 standard).

do not know. But, if an individual starting off with a given income is compensated from time to time for changes in the cost of living by varying his income in accordance with P_{01}^{La}, he will never be worse off but will in general be better off, so long as his tastes do not change.

If we had a community of individuals, all with exactly the same tastes and exactly the same incomes, so that they each purchased the same collections of goods when faced with the same prices, and if their tastes remained constant, then C_{01}^0 would measure changes in the cost of living of each individual, and P_{01}^{La} could be used as an upwardly biased estimate of it, where

$$P_{01}^{La} = \frac{\Sigma p_1 q_0}{\Sigma p_0 q_0}$$

and the q_0's represent the total purchases of each commodity by the community in period 0. In these circumstances, if we were to measure changes in the cost of living by P_{01}^{La}, we should know precisely what P_{01}^{La} measured and its relation to C_{01}^0, which is what we want to measure. Such a state of affairs would not be as satisfactory as if we could measure C_{01}^0, itself, but it would be fairly satisfactory.

However, individuals have different incomes and different tastes, so that if we use P_{01}^{La} as an index of the change in the cost of living, it will only be relevant to those particular individuals who purchase the commodities in the same relative proportions as the community as a whole. Clearly the change in P_{01}^{La} will not be relevant to an individual with a markedly different consumption pattern from the average. Furthermore, if tastes change over time, it becomes impossible to define even the concept of a change in the cost of living.

Accordingly, if for a base period we ascertain the quantities of retail goods and services which a community purchases, q_0, at their then prices, p_0, and next calculate an index of the prices of the goods and services, $P_{01}^{La} = \frac{\Sigma p_1 q_0}{\Sigma p_0 q_0}$, this retail price index can be taken as an upwardly biased estimate of the change in the cost of living of those individuals in the community whose consumption patterns correspond to the consumption pattern of the community as a whole and whose tastes do not change. For other individuals, P_{01}^{La} bears no necessary relation to changes in their cost of living. This is all that can be said.

The above analysis illustrates a number of important points, namely:

1. The difficulty of defining changes in the cost of living as a concept.

2. The need to know (*a*) precisely what we want to measure, (*b*) what the index number measures, and (*c*) the relation between them.

3. The relation between the usual type of fixed-weight aggregative retail price index and the measure of changes in the cost of living which the former often purports to be but seldom is.

13.10. Constructing an Index Number

In constructing a price index number for any purpose, four questions immediately arise:

1. What formula is to be used?
2. What commodities are to be included?
3. What are the weights to be?
4. What price quotations are to be used?

The answers to these questions depend, of course, on the purpose for which the index is being constructed. However, attention may well be drawn to some general considerations.

1. *Formula*

No formula should be used which does not allow for a logical system of weights. This immediately rules out formulae like the simple aggregate of prices or the simple arithmetic mean of price relatives. Apart from these, formulae can be classified according to whether they involve changing or fixed weights. Formulae with changing weights involve a greater collection of data, since information on quantities must be obtained continuously as well as information on prices, and consequently, they are not always practicable. In addition, if the weights are fixed, the meaning of the index is clearer, since it will refer to changes in the cost of a *given* collection. Furthermore, comparisons between two periods will be the same whether made directly or by chaining. Consequently, the fixed-weight type of index is the most common found in practice, although it is customary to revise the fixed weights from time to time (see section 13.11). Nevertheless, occasionally one of the more complex formulae is used; for example, over the period 1928 to 1962 one of the Export Price Indexes published in Australia was a Fisher Ideal index. The Fisher index is practicable in this case because information on quantities of exports is collected continuously. However, in general the formulae used are of the form

$$P_{0t} = \frac{\Sigma p_t q}{\Sigma p_0 q}$$

where P_{0t} is the price index for the period t, with period 0 as base, and the q's are fixed quantity weights. This formula can also be written as a weighted arithmetic mean of price relatives, and is sometimes found in this form:

$$P_{0t} = \frac{\Sigma \frac{p_t}{p_0} v}{\Sigma v}$$

where the v's are fixed value weights.

2. Regimen

The list of commodities included in an index is sometimes called the *regimen*. It is clear that, if we are constructing, say, an index of retail prices, we shall not be able to include every commodity sold retail. To do so would be a task of gigantic proportions, not only from the point of view of computation but also from that of collection. However, we should include all the more important commodities, and we should try to make our regimen as representative as possible. This is a problem in sampling, but, because the regimen is usually largely governed by the practicability of getting the data, the ordinary procedure of statistical random sampling can seldom be applied. Nevertheless, it is most important to emphasize that if we are regarding our index of, say, retail prices as one of retail prices *generally* (and not simply as an index of the prices of those goods *included in the regimen*), we are making the tacit assumption that the prices of goods excluded from the regimen move on the average in the same way as the index calculated from those goods in the regimen.

3. Weights

Naturally the weights depend on the purpose of the index, but they ought to reflect the relative importance of the commodities in the regimen in the relevant sense. Thus the weight attached to a commodity in a retail price index should not be the total production of the commodity but rather its total consumption. Sometimes the prices of certain items omitted from the regimen are believed to move more in accordance with particular items in the regimen than with the regimen as a whole. In these cases, the weights of the omitted items are added to those of the particular "representative" items.

In constructing an index number there is naturally a tendency to include all the major items and omit many of the less important ones. If the weights of the items are based on the actual consumption or production of the goods, the major items as a group will be relatively more important in the index number than they are in the whole field to which the index refers, due to the omission of many items which, although minor individually, are in total quite substantial. It is then said that the major items *dominate* the index. Such dominance will not be of great moment if the price behaviour of the items in the dominant group is similar to that of other items, for then the dominant items may be as representative as other items. However, in practice the price behaviour of the dominant items may be quite atypical. This will especially be the case in economies in which government controls (e.g. price control) are customary, for it is the important items which are generally those subject to control. This can be illustrated from the Australian Wholesale Price (Basic Materials and Foodstuffs) Index, which contains seventy-six items, but is dominated by fourteen (see Appendix B. 4). These fourteen accounted for 80 per cent of the aggregate expenditure in the regimen in 1960, and two of them accounted for 25 per cent. Many of these items are subject to controls of various sorts, and the markets for others have special characteristics which make them unlikely to be representative of the many minor items omitted from the index.

4. *Prices*

In the first place we have to decide at what point in the production process we wish to ascertain our prices, e.g. whether retail, wholesale, factory, etc. This depends on the purpose of the index. Apart from this, we must decide how to get our price quotations. Here we have a straight-out problem in sampling. We cannot obtain all the prices charged for particular commodities by different selling agencies, so we must take a sample. But we must see that our sample is representative and that the price quotations we use closely correspond to the prices actually charged. Thus, in constructing an index of retail prices under conditions of price control, it would be wrong to include only price quotations from the more reputable firms which adhere to the control, if we knew that there existed a substantial black market. Furthermore, when we are getting the prices for a particular commodity, we must be certain that the particular commodity is the same commodity for all the times at which we are taking the price quotations. That is, we must fix a standard of quality for each commodity and stick to it. For example, if we are pricing flour, we must not price sometimes loose flour and sometimes packaged flour. The problems associated with changes in the quality of goods are discussed in the section below.

13.11. Changes in the Regimen and Quality Changes

As has already been pointed out, the weights in the fixed-weight aggregative index will get out of date if comparisons are being made over long periods of time. This is due to three main factors. First, technical innovation alters the range of goods and services available, e.g. the replacement of kerosene by electricity as a form of lighting. Secondly, rising incomes bring a greater range of goods within the possibility of popular consumption, e.g. the growth of a mass market for consumer durables. Thirdly, whenever there is any substantial dispersion of movements in the prices of different goods, substitution takes place against those goods whose prices have risen relatively the most; as a result, a fixed weight index inevitably overweights those items whose prices have risen more than the average.

This latter point can be illustrated by comparing, for the now obsolete Australian "C" Series Retail Price Index, the weighting of the period, at about when the weights were formulated, with that of nearly thirty years later. Between 1923–27 and 1952 clothing prices rose much more than other prices in the "C" Series and, consequently, the percentage of aggregate expenditure on clothing in the regimen

Table 13.7

Group	Percentage of Aggregate Cost in "C" Series Regimen	
	1923–27	December Quarter, 1952
I. Food and Groceries	38·7	40·8
II. Housing	21·3	11·3
III. Clothing	23·0	31·1
IV. Miscellaneous	17·0	16·8
	100·0	100·0

Source: Commonwealth Bureau of Census and Statistics, Australia: *Labour Report*, No. 41, 1952, p. 16

rose considerably. In reality, people did not allocate their expenditure as in the last column, but devoted less to clothing and more to housing. It is to a large extent *diverse* price changes themselves which make the weights of an index out of date.

It follows from the argument in the preceding two paragraphs that the regimen and weights of an index number must be revised from time to time. We might have, for example, an index for periods 0 to 5 on the basis of a certain regimen and a certain set of weights, q, with base period $0 = 100$, and then change the regimen and weights to $\hat{q}$, with base period $5 = 100$. In effect, we shall have two quite distinct indexes and, strictly speaking, no valid comparisons can be made between periods before period 5 and those after. In practice, however, the two indexes will often be chained together so that periods after period 5 can be compared with the original base. We might have, for example

$$P_{06} = P_{05} \times \hat{P}_{56}$$

where $P_{05} = \dfrac{\Sigma p_5 q}{\Sigma p_0 q}$ (the old regimen and weights)

and $\hat{P}_{56} = \dfrac{\Sigma p_6 \hat{q}}{\Sigma p_5 \hat{q}}$ (the new regimen and weights)

It is important to realize that the ratio $\Sigma p_6 \hat{q} / \Sigma p_0 q$, which measures the change in the aggregate cost of two different collections, is quite meaningless. It will be affected by the changes from q to $\hat{q}$ as well as by price changes.

Indeed, the periodic revision of regimen and weights is a customary practice in index number construction although in many cases the revisions occur at rather lengthy intervals, perhaps every twenty years or so. In the case of the Australian Consumer Price Index, for example, the technique has been adopted of revising the regimen and weights at relatively short intervals (three or four years) and computing the index as a chain of short-term links.

In computing a price index it is essential that the price quotations for an item be for a standard quality of that item. If this is not so, some of the movement in the index will be due to changes in quality, whereas the index aims at measuring pure price changes. Consequently, the exact quality of each item in the index must be specified, and the specifications must be rigidly adhered to when securing price quotations. In practice, however, it is inevitable that the quality of items will change. Usually there is more than one quality of each item, and it frequently happens that over a period of time the quality originally in predominant use declines in popularity and is replaced by another one; the replacing of silk stockings by nylons is an example. When this happens, it would be preferable to measure the price changes of the item under consideration by the price movement of the new rather than the old quality, because the new quality is now more representative of the item. In fact, the old quality may sooner or later go right out of use, forcing a change in procedure.

The procedure for handling quality change is as follows. Suppose we have an item whose price is represented by that of variety A. There is also a variety B, which has been ousting A from its original popularity, and it is decided in, say, period 4 to base the variations in the price of the item in future on variety B rather than on A. The weight attached to this item is q. The contributions of this item to aggregate cost in the various periods up to and beyond period 4 are given by

$$p_0^A q, \ldots p_4^A q, \ p_5^B q, \ p_6^B q \ldots$$

Some of the change in this contribution between periods 4 and 5 may be due to the change in quality (e.g. if B is of a higher quality than A, p_5^B may be higher than p_4^A for this reason alone). Since we are interested in pure price changes this quality change must be eliminated. Assuming that variety A and B are both freely available in period 4, we can take the ratio p_4^A/p_4^B as an index of the relative valuation which consumers place on the two varieties. We can then argue[1] that the appropriate weight to be attached to the price of B is

$$\frac{q \, p_4^A}{p_4^B}$$

[1] This argument is based on practical rather than theoretical considerations. If perfect competition obtains on the buying side of the market (as it usually does in markets for consumption goods) the ratio p_4^A/p_4^B equals the *marginal* relative valuation which consumers place on the two varieties. It does not follow that q of A will be valued as the equivalent of $\frac{q \, p_4^A}{p_4^B}$ of B.

PRICE INDEX NUMBERS

This is the quantity of B which can be purchased in period 4 by an expenditure equal to the expenditure involved in purchasing q units of A in the same period. We can use this weight from period 5 onwards, so that the contributions to aggregate cost will be

$$p_0^A q, \ldots p_4^A q, \quad p_5^B \frac{q\, p_4^A}{p_4^B}, \quad p_6^B \frac{q\, p_4^A}{p_4^B} \ldots$$

It will be observed that the new weight attached to the price of B in period 4 gives the same contribution as the old weight attached to the price of A in period 4. In effect, the weight has been modified so that the contribution to aggregate cost of the item in period 4 is the same irrespective of the variety priced.

The above contributions can be rewritten

$$p_0^A q, \ldots p_0^A q \cdot \frac{p_4^A}{p_0^A}, \quad p_4^A q \cdot \frac{p_5^B}{p_4^B}, \quad p_4^A q \cdot \frac{p_6^B}{p_4^B} \ldots$$

from which it can be seen that from periods 0 to 4 the expenditure on the particular item was being varied by the change in the price of variety A, but after that period it is being varied by the change in the price of variety B. The reason for the change is that the price movements of variety B are now regarded as more representative of the price movements of the item under consideration than those of variety A. If the prices of the two varieties moved in a similar fashion there would, of course, be no point in making the change.

This procedure which aims at eliminating price changes due to quality changes is called *splicing*, and it is achieved in any period t by replacing the old weight q by a new weight $q\dfrac{p_t^A}{p_t^B}$. For example if, in an index of retail prices, it is decided in period 4 to replace variety of A of tea with variety B, the procedure would be as follows. The necessary data are given in the table below.

Table 13.8

Commodity	Unit	Period 0			Period 4			Period 5		
		Price	Quantity	Aggregate Cost	Price	Quantity	Aggregate Cost	Price	Quantity	Aggregate Cost
Tea A	lb	cents 36	10	cents 360	cents 48	10	cents	cents 54	10	cents
Tea B.	lb				60	8*	480	72	8	576
Other items.				3,240			3,600			3,960
				3,600			4,080			4,536

* The new weight after splicing is given by $10 \times \dfrac{48}{60} = 8$. Note that in the period in which the splicing takes place, both variety A and variety B make the same contribution to aggregate cost, i.e. $48 \times 10 = 60 \times 8$.

With base period $0 = 100$,
$$P_{04} = (4{,}080 \div 3{,}600) \times 100 = 113$$
If variety A had been adhered to, P_{05} would have been
$$(4{,}500 \div 3{,}600) \times 100 = 125$$
but switching to B makes
$$P_{05} = (4{,}536 \div 3{,}600) \times 100 = 126$$
This is a higher figure because the price of B has risen more than that of A between periods 4 and 5. Had variety B been introduced without splicing (a completely inadmissible procedure), P_{05} would have been $(4{,}680 \div 3{,}600) \times 100 = 130$. This would have been quite wrong as it would include an increase of price due to a quality change (the price of variety B was 60 cents as against 48 cents for variety A in period 4).

13.12. Index Numbers in Practice

The statistical offices of most countries publish various price index numbers referring to the movements in the price levels of certain groups of commodities. These indexes are used by economists and others for many purposes. The greatest care should be exercised in their use to ensure that they are appropriate for the purpose in hand. Those who use index numbers can judge their appropriateness or otherwise only if they know the details of their regimens and understand the difficulties which must be overcome in their construction.

An understanding of the difficulties and intricacies of index number construction can be appreciated by studying the anatomy of particular index numbers. Such a study is set out (although in limited detail) in reference to the Australian Consumer Price Index in Appendix B.4, pp. 487–496 below. This section is restricted to discussion of the interpretation of index numbers in practice, referring to a retail price index by way of illustration.

A retail price index, like the Australian one referred to above, is designed to measure variations in the retail prices of goods and services which enter into consumers' expenditure. This is done by selecting a regimen of goods and services which is believed to be representative (or as representative as is practicable, given the difficulties of collection, specifications, etc.) and attaching to the goods and services weights reflecting the pattern of consumers' expenditure in some sense. At the very least, the index can then be said to measure changes in the prices of the collection of goods and services specified by quality and quantity in the regimen.

In practice retail price indexes are usually interpreted much more broadly than this. They are used both as measures of changes in retail prices *generally* and as measures of changes in the cost of living.

In such uses they have their limitations. But they can hardly be criticized for not being entirely suitable for purposes for which they were not designed.

A consumer price index can be used as a measure of changes in retail prices generally only if the prices of items omitted from the regimen change in approximately the same way on the average as those in the regimen. The fact that a consumer price index *omits* certain items is itself unimportant unless the price behaviour of these items differs appreciably from the included items. But even if the coverage of a consumer price index is good, care must be taken not to strain its meaning, by, for example, treating it as if it measured changes in the price component of gross national product at current market prices; for, in this case, the price index could scarcely be relevant to the investment and government expenditure categories of gross national product unless the prices of those categories moved on the average in the same way as the index (see pp. 372–3, below).

We have already adverted to the difficulties of measuring changes in the cost of living (see section 13.9 above) and pointed out the conditions under which a retail price index can be used as an approximate measure of such changes. A particular retail price index can in no way be a measure of changes in *the* cost of living. Indeed the concept of "changes in the cost of living" is only fully meaningful in relation to a particular household. Households may range in type from the unmarried man to the married couple with children and elderly relatives. Their consumption patterns vary enormously. But in so far as given persons or groups of persons have consumption patterns approximating to that included in the particular index number's regimen and weights, that index number can be used to indicate changes in the cost of living of such persons or groups.[1]

When a retail price index is used as a measure of changes in the cost of living, misconceptions about the nature of index numbers often occur. These misconceptions are associated with the notion that, if

[1] Until 1953 the Australian "C" Series Retail Price Index (see App. B.4, p. 487 below) was for many years used by the Commonwealth Court of Conciliation and Arbitration to vary the basic wage. The Court periodically fixed the level of the basic wage and then varied it quarterly in accordance with movements in the "C" Series. The "C" Series measured accurately the cost of a certain regimen. That is all it was meant to do. The use of it by the Court as a measure of changes in the cost of living due to variations in prices was the Court's own responsibility. Between 1953 and 1961 cost of living variations to the basic wage were abandoned. However, in its judgment on the 1961 Basic Wage Case, the Commonwealth Conciliation and Arbitration Commission (as the arbitral wing of the Court is now called) indicated that it would in future vary the basic wage annually with the Consumer Price Index unless cause was shown why it should not. This policy was strongly challenged by employers' counsel in the 1964 and subsequent National Wage Cases. In 1965 it was disapproved in the majority judgment of the Commission. However, in later cases the practice of adjusting award wages for price increases appears to have commanded the Commission's general support.

the regimen omits certain items, changes in the cost of living will be understated. This is, of course, erroneous; but it is a common error and arises from a confusion about the function of price index numbers. It arises particularly in cases where wages are adjusted from time to time according to changes in a price index number. It is argued that the regimen of the index number in some way dictates the standard of living of the wage-earner whose wages are adjusted by it. If the index number omits fresh fruit and vegetables, for example, it is said that the wage-earner's standard of living cannot be meant to include these items. This confuses the whole function of price index numbers, which is to measure *changes* in prices. A price index number measures *changes* in prices of a *given* regimen; or, when it is being used in relation to the cost of living, *changes* in the cost of a *given* standard of living. The adjustment of a wage by a price index number only attempts to preserve the real value of the wage in terms of purchasing power at the level *originally* determined. It is this level which determines the wage-earners' standard of living. If the index number does not contain certain items this can only affect the issue in so far as the prices of these items move differently from the index number. Indeed, if when prices are rising generally, the prices of omitted items are rising less quickly than the average, the inclusion of such items would reduce the rate of increase of the index, and hence the measure of the change of the cost of living and any cost of living adjustment to wages. It does not matter how many or how few are the items in the regimen so long as they are properly representative of price changes generally. But, of course, if there is any considerable dispersion of price movements, a single index cannot be expected to measure changes in the cost of living of different households, which enjoy standards of living of markedly different quality and quantity.

CHAPTER XIV

REAL NATIONAL PRODUCT AND INDEXES OF PRODUCTION

14.1. Money National Product and Real National Product

THE transactions which are recorded in the social accounts and the aggregates which make up gross national product, consumption, investment, etc., are all expressed in money terms and reflect current prices and factor payments. A figure for money national product for a single year is not very meaningful unless we have some idea of the value (i.e. the purchasing power) of money itself. A series of figures for money national product over a number of years has more meaning. It tells us how money national income has changed. For some purposes money aggregates are all we want, e.g. for the estimation of tax yields, for assessing the extent to which the balance of payments is "in balance," etc. However, the value of money will, in general, be changing from year to year as the general level of prices moves up and down, and this vitiates the use of money national product figures for examining trends in productivity or the standard of living or for examining changes in the allocation of resources between different sectors in the economy or between different purposes. Clearly, what we need is to go behind the façade of prices and to measure the various relevant aggregates in "real" terms.

Consider gross national product. Gross national product can be regarded either as the value of goods and services produced for final purposes during the year or else as the sum of values added in the process of production. In the former sense, the "real" element in gross national product can be clearly discerned, for gross national product consists of a great list of quantities of goods and services produced (the "real" physical things) multiplied by the appropriate current market prices. Thus, in year 0, we can represent it by $\Sigma P_0 Q_0$, and in year 1 by $\Sigma P_1 Q_1$, and so on, where the Q's are quantities and the P's the corresponding prices. The difference (if any) between the money aggregates $\Sigma P_1 Q_1$ and $\Sigma P_0 Q_0$ will be due to changes both in the Q's and in the P's. If we are to measure real gross national product, we must eliminate the changes in the P's, so that our measure refers only to quantity or "real" changes, i.e. *quantum* changes. If all the Q's changed in exactly the same proportion, that proportion would represent the change in real gross national product, and our task would be simple. However, this is never the case. In practice we have to find ways of adding the Q_0's, and then the Q_1's, and comparing the totals. But the various Q's refer to different goods and services, and the question of how they can be added arises. How can we add yards of cloth and pounds of butter? One way would be to work out a table of equivalences of the form: 1 yard of cloth $=$ 3 lb of butter. These equivalences would have to be in terms of how the different goods and services are valued relatively to one another. Such a valuation is provided by the market. An equivalence between cloth and butter, for example, can be derived from their market prices. Thus if the price of one yard of cloth is $1·20 and the price of one pound of butter is 40 cents, these prices imply an equivalence between the two goods of 1 yard of cloth $=$ 3 lb of butter.

It follows that to measure real gross national product, we must weight the quantities in some way so as to reflect their relative values before adding them. Thus, although real gross national product is a quantity measure, it measures quantity not simply in the physical sense, but rather in the economic sense of quantity of "real value." For example, if two grades of refrigerators are produced, an increase in the proportion of the more valuable grade, total numbers remaining constant, implies an increase in "real" production in the economic sense. It will be readily appreciated that the main problem is in finding appropriate weights to reflect relative economic values. These weights can be derived from market prices, and if relative market prices never changed there would be no difficulty. In this case all prices would move up or down by the same proportion, and the real value of $\Sigma P_1 Q_1$ relative to $\Sigma P_0 Q_0$ would be given by dividing $\Sigma P_1 Q_1$ by the change in prices. In practice, all prices do not move up and down in the same ratio, and thus relative prices change.

The difficulty which arises as a result of this can be seen as follows. Suppose we wish to compare year 0 with year 1. We can use the prices of either year 0 or year 1 as weights. Thus we can compare

$$\Sigma P_0 Q_0 \text{ and } \Sigma P_0 Q_1$$

or

$$\Sigma P_1 Q_0 \text{ and } \Sigma P_1 Q_1$$

The first compares the gross national products of year 0 and year 1 valued at constant year 0 prices, and the second compares gross national product of year 0 and year 1 valued at constant year 1 prices. Both comparisons give us figures for real gross national product, valued at constant prices. Real gross national product, as here defined, is, of course, expressed in terms of money units, but it is "real" in the sense that it is expressed in terms of money units of constant purchasing power. Both comparisons are equally valid, but they will in general differ. The first comparison will generally show a greater rise or a smaller fall than the second, so long as the usual negative correlation between prices and quantities obtains. This can be proved in a manner analogous to that used in section 13.4 above. It is clear that there can be no single unambiguous measure of changes in real gross national product, just as there can be no single unambiguous measure of changes in the price level.

It would be possible to make the comparison between year 0 and year 1 with a set of prices other than those of year 0 or year 1—for example, by comparing

$$\Sigma P_n Q_0 \quad \text{and} \quad \Sigma P_n Q_1$$

This comparison, however, would rely on relative valuations which are relevant to neither year 0 nor year 1.

The statistical meaning of a series of figures for real gross national product valued at a set of constant prices is clear enough, but the greatest caution is needed in interpreting such a series in economic terms. In practice, such series expressed per head of population are frequently used to indicate movements in the standard of living of an economy. But we here come up against the type of problem discussed in section 13.9 above, namely the relation between a statistical measure and the theoretical concept we wish to measure. As yet no way of measuring the standard of living of even an individual has been satisfactorily defined, for a person's standard of living must be conceived of not in terms of the goods and services which are available to him but in terms of the satisfactions he derives from those goods and services, and these cannot be measured.[1] Moreover, if we are considering a group of individuals, we have no way of comparing the satisfactions of the individuals within the group. Thus, in two periods total production might be the same, but its distribution between the individuals might differ. Can we say that people are on the whole as well off as they were before?

In spite of these conceptual difficulties, however, we do frequently assume that changes in real national product per head can be taken to indicate changes in the standard of living of the economy, provided that there have been no substantial changes in income distribution. We should not, of course, attach any importance to a small change in real product, but a large increase would suggest that the economy was, at least, in a position to enjoy a higher standard of living.

One serious difficulty in using real national product figures is that there is no unique measure of real national product. Real national product is measured in terms of a set of constant prices, and its movements over time depend on the set used. Thus gross national product

---

[1] There is a substantial body of literature on this problem. See, for example, Allen, R. G. D.: "The Economic Theory of Index Numbers," *Economica*, N.S., Vol. XVI, August, 1949, p. 197; Hicks, J. R.: "The Measurement of Real Income," *Oxford Economic Papers*, Vol. 10, June, 1958, p. 125.

for the years 1938–39 to 1967–68, valued at constant 1938–39 prices, will move differently and possibly very differently from the same gross national product valued at constant 1967–68 prices, since the weights attached to the various quantities will be different. There is no reason to prefer one set of constant prices to the other, so that if the results are very divergent their interpretation becomes virtually impossible. This difficulty increases the more distant are the periods of comparison.

14.2. Measurement of Real National Product

The two comparisons set out on p. 368 can be put into index number form. Using year 0 as a base, we shall have—

$$Q_{01}^{\text{La}} = \frac{\Sigma P_0 Q_1}{\Sigma P_0 Q_0}$$

and

$$Q_{01}^{\text{Pa}} = \frac{\Sigma P_1 Q_1}{\Sigma P_1 Q_0}$$

The first of these indexes is Laspeyres's quantity index, and the second Paasche's. These indexes are indexes of quantity, sometimes called indexes of *quantum* or *physical volume*. They measure the change in the volume of a collection of goods and services valued at constant prices.

Although the exposition in this section is in terms of measuring the quantum of gross national product, these indexes of quantum can be applied quite generally, e.g. to measure changes in the volume of exports, the output of primary products, etc. Their central feature is that they involve the weighting of quantities of items by constant prices, which are taken to reflect the relative values of the items.

If we knew all the P's and Q's it would be quite a simple matter to work out Q_{01}^{La} or Q_{01}^{Pa} for gross national product. In practice, of course, we do not have all this information, and we have to use the fact that

$$\text{value} = \text{quantity} \times \text{price}$$

Thus, for an individual good, we can write

$$\frac{V_1}{V_0} = \frac{P_1}{P_0} \times \frac{Q_1}{Q_0}$$

i.e. relative value change equals relative price change multiplied by relative quantity change. By analogy we can write, in index number form, for an aggregate of goods and services

$$V_{01} = P_{01} \cdot Q_{01} \qquad (*)$$

where $V_{01} = \dfrac{\Sigma P_1 Q_1}{\Sigma P_0 Q_0}$ is an index of values, P_{01} is an index of prices, and Q_{01} is an index of quantum. If the relation (*) is true we can obtain Q_{01} from

REAL NATIONAL PRODUCT

$$Q_{01} = V_{01} \div P_{01}$$

As far as gross national product is concerned, we know V_{01} (i.e. the change in the money gross national product) and there are available various possible indexes of prices, calculated on a sample basis (i.e. on a sample of goods and services). The process of dividing value aggregates by price aggregates to obtain quantity aggregates is called the process of *deflation*, and the price index used is known as the *deflator*.

The relation (*) is only true when the indexes P_{01} and Q_{01} are correlative in a certain sense. Thus we know from the factor reversal test (see section 13.5, p. 339) that

$$V_{01} \neq P_{01}^{\mathrm{La}} \cdot Q_{01}^{\mathrm{La}}$$

and
$$V_{01} \neq P_{01}^{\mathrm{Pa}} \cdot Q_{01}^{\mathrm{Pa}}$$

However, we have
$$V_{01} = P_{01}^{\mathrm{La}} \cdot Q_{01}^{\mathrm{Pa}}$$

and
$$V_{01} = P_{01}^{\mathrm{Pa}} \cdot Q_{01}^{\mathrm{La}}$$

Consequently, if we desire to estimate Q_{01}^{La}, we must divide V_{01} by the appropriate Paasche price index. Then

$$Q_{01}^{\mathrm{La}} = V_{01} \div P_{01}^{\mathrm{Pa}}$$

and similarly,
$$Q_{01}^{\mathrm{Pa}} = V_{01} \div P_{01}^{\mathrm{La}}$$

On the other hand, if we use the quantities of some third year as weights in the price index, we get

$$Q_{01} = \frac{\Sigma P_1 Q_1}{\Sigma P_0 Q_0} \div \frac{\Sigma P_1 Q_n}{\Sigma P_0 Q_n}$$
$$= \frac{\Sigma P_0 Q_n}{\Sigma P_0 Q_0} \div \frac{\Sigma P_1 Q_n}{\Sigma P_1 Q_1}$$
$$= Q_{0n}^{\mathrm{La}} \div Q_{1n}^{\mathrm{La}}$$

and it is not clear what Q_{01} means. This is in contrast with Q_{01}^{La} and Q_{01}^{Pa} which have quite clear meanings.

The process of deflation can be accomplished either in the index number form, as above, or by directly deflating gross national product itself. To obtain gross national product at constant period 0 prices, we have

period 0: $\Sigma P_0 Q_0$

period 1: $\Sigma P_1 Q_1 \div P_{01}^{\mathrm{Pa}} = \Sigma P_0 Q_1$

To obtain gross national product at constant period 1 prices, we have

period 0: $\Sigma P_0 Q_0 \times P_{01}^{\text{La}} = \Sigma P_1 Q_0$
period 1: $\Sigma P_1 Q_1$

It is important to emphasize that the process of deflation gives a strictly valid result only if the correct formula for the price index is used and if the coverage of the price index is coextensive with the gross national product. In deflating an aggregate like gross national product, it is of fundamental importance to see that the deflator has a coverage coextensive with the quantity aggregate which is to be obtained from the process of deflation. In so far as the deflator does omit the prices of certain items, its use must imply the assumption that their prices have moved in the same degree as the movement of the deflator as a whole. Frequently this assumption is obviously invalid. In practice price indexes have to be used which are not of the correct form and are only partial in character. Thus, although we often divide $\Sigma P_1 Q_1$ by a price index P_{01} and designate the result as "gross national product of year 1 valued at constant period 0 prices," this is seldom strictly correct.

Consider, as an example, a comparison of real gross national product for Australia for the two years 1963–64 and 1967–68. Valued at current market prices gross national product for the two years was $18,008 million and $23,911 million respectively. We wish to deflate these figures to obtain gross national product at constant 1963–64 prices. Hence, we must find a suitable price index number. Suppose, for argument's sake, we use the Consumer Price Index. On a base of 1952-53 = 100·0, this index rose from 125·7 to 143·4 over the period under consideration, i.e. from 1,000 to 1,141. Deflating by the Consumer Price Index we have $23,911 \div 1\cdot141 = \$20,956$ million as an estimate of the gross national product of 1967–68 valued at 1963–64 prices, giving a rise in real product of some 16 per cent over the four-year period. Apart from the fact that the price index is, strictly speaking, technically unsuitable, because its quantity weights should relate to production in 1967–68 (i.e. the deflator should be a Paasche type), this estimate of the movement in real gross national product is unreliable because of the inadequate coverage of the deflator. The gross national product is a collection of final goods and services of all kinds, but the Consumer Price Index represents price movements in a restricted sector. Thus—

1. The index does not include goods and services falling into the categories of gross private investment (other than private motor vehicles) and government expenditure.

2. Gross national product covers goods and services produced in Australia, whereas the Consumer Price Index represents consumption. Thus gross national product includes exports but excludes imports. The Consumer Price Index includes some part of imports and covers

export prices only to the extent that certain exportable goods enter home consumption. There is no weight attaching to exports as such in the Consumer Price Index.

3. At best the Consumer Price Index might be said to cover the elements of gross national product which enter into consumers' expenditure. But even here the coverage is not complete.

This inadequate coverage would not matter if the prices of items included in gross national product, but not represented in the Consumer Price Index regimen, moved on the average in a similar way to the prices of those items included in the regimen. In fact, this has not been the case, so that the Consumer Price Index is an unsuitable deflator.

An alternative might be to use the Wholesale Price Index. On the base of 1936–37 to 1938–9 = 100 this index rose from 346 to 388 over this period, i.e. from 1,000 to 1,121. Deflating by this index, we have $23{,}911 \div 1{\cdot}121 = \$21{,}330$ million as an estimate of the gross national product of 1967–68 valued at 1963–64 prices, giving a rise in real product of 18 per cent over the four-year period. But the Wholesale Price Index, like the Consumer Price Index, is unsuitable. Its coverage is unsatisfactory for the following reasons—

1. The prices included in the index are wholesale, but the market prices used to value the goods and services entering gross national product are for the most part retail prices. Changes in the ratio of retail to wholesale prices, i.e. changes in the relative costs of distribution, would immediately invalidate the use of this index.

2. The index is of basic materials and excludes manufactured articles and services.

3. The index relates to consumption of basic materials and foodstuffs, whereas gross national product covers production.

The main trouble with using index numbers like either the Consumer Price Index or the Wholesale Price Index as deflators is that they are sectional indexes and only partially cover the content of gross national product. This consideration suggests that we should deflate gross national product in sections, using appropriate sectional price indexes for the various sections. This is done by way of illustration in Table 14.1. The sectionalization of gross national product has been to some extent dictated by considerations as to whether or not appropriate indexes are available, and could in fact be made more detailed. A more accurate and more detailed computation could be made if sufficient trouble were taken, and the table should be understood to be purely an example. It certainly does not contain an estimate of *the* gross national product of 1967–68 valued at 1963–64 prices, and no great reliance should be placed on the results of the table. Criticisms similar to those levelled above against the use of the Consumer Price Index and the Wholesale Price Index could easily be levelled against some of the

Table 14.1

ESTIMATION OF AUSTRALIAN GROSS NATIONAL PRODUCT 1967–68 VALUED AT 1963–64 PRICES

Item and Deflator	Deflators as Published (a)		Deflators 1967–68 (Base 1963–64 = 1,000)	Gross National Product at Current Prices		Gross National Product at 1963–64 Prices
	1963–64	1967–68	(b)	1963–64 $m	1967–68 $m	1967–68 $m (c)
	(1)	(2)	(3)	(4)	(5)	(6)
Personal Consumption—						
Food				2,510	3,230	2,746
(Food Group, Consumer Price Index)	126·0	148·2	1,176			
Clothing, footwear, drapery, etc.				1,197	1,460	1,363
(Clothing and Drapery Group, Consumer Price Index)	114·0	122·1	1,071			
Gross rent of dwellings				1,225	1,695	1,445
(Housing Group, Consumer Price Index)	159·6	187·2	1,173			
Electrical and other durable goods				872	1,052	1,002
(Household Supplies and Equipment Group, Consumer Price Index)	111·0	116·5	1,050			
Other				5,272	7,146	6,061
(Miscellaneous Group, Consumer Price Index)	129·9	153·1	1,179			
Gross Private Investment—						
Housing and construction				1,326	1,938	1,690
(Building Materials Group, Wholesale Price Index)	473·0	514·0	1,147 (d)			
(Minimum Weekly Wage Rates—Adult Males)	133·0	160·7				
Other fixed equipment				1,558	2,030	1,908
(Metal manufactures, machinery groups, etc.—Import Price Index) (e)	101·6	108·1	1,064			
Stocks (f)				174	311	277
(Wholesale Price Index)	346·0	388·0	1,121			
Government and Financial Enterprise Expenditure—				3,576	5,667	4,864
(Wholesale Price Index)	346·0	388·0	1,165 (d)			
(Minimum Weekly Wage Rates—Adult Males)	133·0	160·7				
Exports				3,162	3,541	4,038
(Export Price Index including gold—	114·0	100·0	877			
Less Imports				−2,864	−4,159	−4,069
(Import Price Index)	109·8	112·2	1,022			
Gross National Product				18,008 (g)	23,911 (g)	21,325

Source: For columns (1) and (2): Commonwealth Bureau of Census and Statistics, Australia: *Monthly Review of Business Statistics*, No. 334, 1965, p. 27; No. 370, 1968, pp. 23, 25, 27, 30; and for the Import Price Index: Reserve Bank of Australia: *Statistical Bulletin*, January 1967, p. 162 and September 1968, p. 91.
For columns (4) and (5): Commonwealth of Australia: *National Income and Expenditure*, 1967–68, pp. 7, 11, 15 (these figures are subject to revision in subsequent publications).

(a) Base years:
　Consumer Price Index, 1952–53 = 100·0.
　Wholesale Price Index, 1936–37 to 1938–39 = 100·0.
　Export Price Index, 1959–60 = 100·0.
　Import Price Index, 1952–53 = 100·0.
　Import Price Index—sectional indexes, 1962–63 = 100·0.
(b) (Col. (2) ÷ col. (1)) × 1,000.
(c) (Col. (5) ÷ col. (3)) × 1,000.

Table 14.1 (contd.)

(*d*) The index used here is a weighted average of the two indexes shown giving both indexes equal weight.

(*e*) The metal manufactures, motor vehicles, electrical machinery and other machinery groups of the Import Price Index weighted by value of imports in 1962–63 in these groups. See Reserve Bank of Australia: *Statistical Bulletin*, Economic Supplement, Dec. 1967, pp. 41–43.

(*f*) The figures of $174 m. and $311 m. for stocks for the two years respectively represent the changes in the book values of stocks over the course of the two years. In so far as prices were rising during each of the two years these figures will contain the increased value of a given volume of stocks. This increased value should be removed from the figures, to make them represent the value of the change in physical stocks valued at current market prices. When this is done, the figures can then be deflated. However, since prices remained relatively stable within 1963–64 and 1967–68, and due to the difficulty of estimating the effect of changing prices within individual years on stock values, this step has been omitted from the above table.

(*g*) Excluding statistical discrepancy as an element of expenditure; see Commonwealth of Australia, *National Income and Expenditure* 1967–68, p. 23.

sectional deflators. According to the computation in the table, gross national product measured in constant 1963–64 prices rose by 18 per cent from $18,008 million to $21,325 million as between 1963–64 and 1967–68. The sectional deflators show a substantial degree of divergence of price movements, but, by coincidence, the change in real gross national product obtained by sectional deflation turns out to be the same as that obtained by deflation by the Wholesale Price Index. A brief discussion of the estimation of Australian gross national product at constant prices is contained in Appendix B.5, pp. 500–503 below.

The process of deflation has been illustrated above in reference to deflating gross national product to obtain an index of the movement in aggregate real production. Aggregate real production is a useful concept if one is interested in indicating movements in physical productivity. But frequently one is interested not so much in productivity as in the standard of living. These two concepts do not necessarily go hand in hand. One important reason for divergence lies in foreign trade. Gross national product consists of *home-produced* goods and services for consumption, investment, government purposes and exports. These are the goods and services actually produced in the economy, and consequently, are relevant to a study of the quantum of production. But from the point of view of the standard of living, what is relevant is the aggregate of home-produced goods and services for consumption, investment and government purposes, together with the purchasing power over *imports* which the economy's exports give it. This means that instead of deflating exports by export prices and imports by import prices as was done in the example above, we should deflate them both by import prices, i.e. deflate net exports by import prices. If this is done for the data in Table 14.1 above, gross national product in 1967–68 valued at 1963–64 prices is $20,752 million, showing a rise not of 18 per cent but of only 15 per cent over the 1963–64 gross national product. The reduction in the increase in real gross national product measured in this way is due to the marked deterioration in the terms of trade which has reduced the real value to Australians of export production in terms of its purchasing power over imports, quite apart from any change in the physical production of exports. Similarly, the question arises as to the extent to which current investment adds to the current standard of living. Its effect on productivity will be in the future, but it does add to the potential current standard of living in the sense that the economy could have devoted the resources bound up in the investment goods to current consumption. Consequently, if we are interested in the standard of living, the aggregate of investment in gross national product should perhaps be deflated by an index of consumer good prices.

The idea of deflating gross national product to obtain a measure of the quantum of production is relatively simple, for gross national product is an aggregate which has a price component and a quantity component. But frequently money aggregates are deflated which cannot conceptually be split into these two components. The question then is: What does the deflated aggregate measure? Consider the case of personal income. Personal income is not an aggregate of prices multiplied by quantities produced. A deflated personal income can only measure the purchasing power of the income over the regimen of the deflator. Thus the statistical interpretation of the aggregate of personal incomes divided by a consumer price index is clear, but its meaningfulness may be limited. If the consumer price index covered all items on which income is normally spent, then the meaningfulness of the deflated aggregate would undoubtedly be increased. The same considerations apply *a fortiori* to a deflation of, for example, bank clearings by a price index.

14.3. MEASUREMENT OF MOVEMENTS IN THE QUANTUM OF INDUSTRIAL PRODUCTION

In the preceding section we discussed the measurement of movements in the quantum of national production in the aggregate. We now consider how to measure movements in the quantum of production of the individual firms and industries which contribute to the national aggregate. The indexes used in this connexion are usually called *indexes of industrial production*. Such indexes are generally limited to production taking place in secondary industries. Their compilation enables us to compare the rates of change of production in the various industries of an economy and to compare these rates with changes in employment, etc.

It is convenient to start by considering the measurement of movements in the quantum of production of an individual firm. For the sake of simplicity, we consider first a firm producing only one product (i.e. one output). Clearly we cannot measure the quantum of production of the firm by reference to the quantity of its output, for the output contains raw materials, etc. (i.e. inputs) which have been produced by other firms. If, for example, output were to remain constant while inputs were reduced, the quantum of production of the firm should be regarded as having increased. It follows that the quantum of production of a firm should be defined as the quantity component of what we have previously called the gross product of or value added by the firm (see section 12.3, p. 285). This is equivalent to relating the quantum of production to the amount of "work done" by the firm. Thus, suppose that our firm in producing its one output uses only one input. Clearly, it is not possible to subtract the input from the output, for how can one subtract, for example, pounds of flour from loaves of bread. In order to make the subtraction, it is necessary to express the units of output or input in terms of the other, and this can be done by using market prices as an indication of the equivalence between the output and the input. Thus, if we write P and Q for the price and quantity respectively of the output and p and q for the price and quantity of the input, the value of the production of the firm will be given by

$$PQ - pq$$

This is, in fact, the gross product of or value added by the firm. It is a money figure, and statistically it is impossible to resolve it into a price and a quantity component in the manner in which we can resolve the value of output (PQ) into a price component (P) and a quantity component (Q). However, as will be seen below, we can attempt to resolve changes in the value of production into price changes and quantity changes.

If we wish to measure the change in the value of production of the firm as between period 0 and period 1, we can do so by comparing $P_1Q_1 - p_1q_1$ with $P_0Q_0 - p_0q_0$. This comparison yields the ratio

$$\frac{P_1Q_1 - p_1q_1}{P_0Q_0 - p_0q_0}$$

as an index of the movement in the value of production. This ratio can be written

$$\frac{P_1Q_1 - p_1q_1}{P_0Q_0 - p_0q_0} = \frac{P_0Q_1 - p_0q_1}{P_0Q_0 - p_0q_0} \times \frac{P_1Q_1 - p_1q_1}{P_0Q_1 - p_0q_1}$$

The first term on the right-hand side measures the movement in value added when the quantities are valued at constant period 0 prices, and the second term the movement in prices when these are weighted by period 1 quantities. We can regard the former as a quantity and the latter as a price component of the movement in the value of production.

The quantity component tells us how value added would have changed had prices remained constant at the period 0 level.

However, we can also write

$$\frac{P_1Q_1 - p_1q_1}{P_0Q_0 - p_0q_0} = \frac{P_1Q_1 - p_1q_1}{P_1Q_0 - p_1q_0} \times \frac{P_1Q_0 - p_1q_0}{P_0Q_0 - p_0q_0}$$

giving similar interpretations to the two terms on the right-hand side. Thus the first term tells us how value added would have changed had prices remained constant at the period 1 level, and can be regarded as a quantity component in the movement of the value of production.

It follows that, even under the highly simplified conditions assumed here, we shall have two measures of movement of the quantum of production—

$$\frac{P_0Q_1 - p_0q_1}{P_0Q_0 - p_0q_0} \quad \text{and} \quad \frac{P_1Q_1 - p_1q_1}{P_1Q_0 - p_1q_0}$$

each comparing values added in which outputs and inputs have been valued at constant prices, and hence abstracting from the effects of price changes. These two measures are in fact quantity indexes with different price weights. They are the Laspeyres and Paasche types respectively, and they will in general give different results unless either relative quantities (i.e. here the technical input-output ratio) or relative prices (i.e. here the price of input relative to that of output) remain the same for the two periods under comparison. It follows that the quantum of production of a firm producing only one output with one input cannot be unambiguously defined. This is in sharp contrast with the situation in which one is measuring the quantum of output. For, although we cannot unambiguously define changes in the quantum of output when it refers to more than one product (thus, we have the two measures $\Sigma P_0Q_1/\Sigma P_0Q_0$ and $\Sigma P_1Q_1/\Sigma P_1Q_0$), nevertheless we can unambiguously define changes in the quantum of one output as Q_1/Q_0. However, there is no simple ratio corresponding to Q_1/Q_0 in the elementary case, when one is measuring quantum of production instead of quantum of output. This emphasizes the basic conceptual difficulty in trying to measure quantum of production. The reason for this basic difficulty is that in measuring the quantum of production at least two goods must always be involved—one output and one input—so that one is immediately faced with a problem of weighting.

In practice, an individual firm will have usually more than one output and always more than one input, and the same applies *a fortiori* to an industry. However, in measuring the quantum of production, the same principle which was used in the elementary case can be applied, namely the comparison of values added in which outputs and inputs have been valued at constant prices. This immediately leads us to two possible formulae—

$$N_{01}^{\text{La}} = \frac{\Sigma P_0 Q_1 - \Sigma p_0 q_1}{\Sigma P_0 Q_0 - \Sigma p_0 q_0}$$

and

$$N_{01}^{\text{Pa}} = \frac{\Sigma P_1 Q_1 - \Sigma p_1 q_1}{\Sigma P_1 Q_0 - \Sigma p_1 q_0}$$

where N_{01}^{La} and N_{01}^{Pa} are indexes of the quantum of production of the Laspeyres and Paasche type respectively, and the summations extend over all outputs and inputs of the industrial sector under consideration.[1] Like the simpler Laspeyres's and Paasche's index numbers of prices and outputs they do not in general yield the same numerical results when applied to the same data. Their divergence will be greater, the greater the dispersion in the movements of prices of outputs, quantities of outputs, prices of inputs and quantities of inputs. However, in practice, the two formulae should lead to fairly close results, provided the periods of comparison are not too far apart. These formulae can be applied to the measurement of the quantum of production of a single firm, a single industry, a group of industries or industry as a whole.

Let us suppose we have computed N_{01}^{La} indexes for individual industries and we wish to combine them into an index of industrial production as a whole. This can be done by adding the numerators and the denominators and obtaining

$$\frac{\sum[\Sigma P_0 Q_1 - \Sigma p_0 q_1]}{\sum[\Sigma P_0 Q_0 - \Sigma p_0 q_0]}$$

where the small Σ's are summations over outputs and inputs within the individual industries and the large $\sum$'s are summations of industries. Precisely the same result can be achieved by taking the arithmetic mean of the individual N_{01}^{La} indexes weighting them by base-year values added,[2] i.e.

$$\frac{\sum[N_{01}^{\text{La}}(\Sigma P_0 Q_0 - \Sigma p_0 q_0)]}{\sum[\Sigma P_0 Q_0 - \Sigma p_0 q_0]}$$

This latter approach is useful in connexion with the method of indicators discussed below.

If we have the N_{01}^{La}'s computed for all industries in the economy (including primary and tertiary industries as well as secondary

---

[1] This type of formula is known as Geary's formula, having been proposed by R. C. Geary in a paper "The Concept of Net Volume of Output with Special Reference to Irish Data," *Journal of the Royal Statistical Society*, Vol. CVII, Parts III, IV, 1944, page 251. However, it was suggested earlier by R. Wilson in a paper "Prices, Quantities and Values," read to the Victorian Branch of the Economic Society of Australia and New Zealand in September, 1937, and published as a pamphlet.

[2] Compare section 13.7, p. 345 above, on the combination of group indexes.

industries), their combination in the manner outlined in the preceding paragraph will lead to the Laspeyres's index of movements in real gross national product. The outputs of one industry which are inputs of another will subtract out when the denominators and numerators are summed, leaving only final outputs. This shows that indexes of the quantum of production for particular industries can be interpreted as indexes of the contributions of the particular industries to real national product. It also shows how real national product can be measured in terms of the gross products of individual enterprises rather than in terms of final expenditure, which was the approach used in section 14.2 above.

There are two important considerations to bear in mind in using the above formulae. First, account should be taken, as far as possible, of any changes in the quality of outputs and inputs, of any diversification of qualities of outputs and inputs, and of the introduction of new outputs and inputs. This can be done, in theory at least, by introducing new items into the index, i.e. by treating different qualities as different goods. However, if one is using the N_{01}^{La} index, there may be some difficulty in ascertaining appropriate prices of the new items in the base year, when perhaps the items were not being produced in any appreciable volume, although it would usually be possible to estimate notional prices based on the price movements of allied items. Secondly, the indexes must not be allowed to be distorted by changes in the volume of work in progress. Suppose the q's represent actual intake of raw materials, etc., into the factory, then if the Q's represent actual deliveries of outputs from the factory, the expression $\Sigma PQ - \Sigma pq$ might bear little relation to the quantum of work done. The deliveries might have been made from stock, or the inputs might be going into work-in-progress without any immediate effect on outputs. It is conceivable that the expression might even become negative. This difficulty can be readily overcome, in principle, by including in the Q's not deliveries from the factory, but completed production together with any uncompleted intermediate products (valued at appropriate prices) held by the factory at the end of the period under consideration, and by including in the q's not only intake of inputs, but also intermediate products held at the beginning of the period under consideration. This is consistent with the original definition of the gross product of a firm or industry.[1]

In computing an index of industrial production as a whole the area which the index is to cover must first be defined. Generally indexes of this kind are limited to the fields of secondary production. Building and construction is sometimes regarded as a doubtful item for inclusion. A more difficult problem arises in the field of repair work. Some repair work is clearly not carried out under industrial conditions, e.g. repair work connected with a service trade such as boot and shoe

[1] Compare p. 285 above.

repairs, automobile maintenance, etc., whereas other repair work such as ship and locomotive repair is. It is clear that there will always be a certain amount of blurring around the edges of any definition of secondary production.

At this point it is well to emphasize the limitations of indexes of industrial production. Indexes of industrial production suffer from the usual limitations of index numbers in that different formulae give different numerical results and the results of any given formula depend on the weighting system adopted. Moreover, the statistical material upon which they are based is usually imperfect and incomplete. Furthermore, apart from these factors, the concept of "production" is surrounded by the same complexities as those referred to in section 14·1 above in connexion with the measurement of real national product (see pp. 368–9). "Production" in the economic sense cannot be interpreted as merely physical output, but rather it must be conceived in terms of the satisfactions which are produced by physical output. Accordingly, indexes of production should be interpreted with caution. This applies particularly to comparisons over long periods of time.

Having decided on the scope of the index, the main task is to find means of filling in the terms in the formula. Suppose we direct our attention to the formula

$$\mathcal{N}_{01}^{\text{La}} = \frac{\Sigma P_0 Q_1 - \Sigma p_0 q_1}{\Sigma P_0 Q_0 - \Sigma p_0 q_0}$$

If we knew the individual P's, p's, Q's and q's, our task would be simple and straightforward, but in practice there is seldom such a wealth of statistical material available, and we must use approximate methods. There are two broad types of methods available. They are known as the *method of deflation* and the *method of indicators*.

14.4. Computation of an Index of Industrial Production by the Method of Deflation

We may here restrict ourselves to the consideration of an index for a particular industry, since the extension to industrial production as a whole is straightforward. We consider the formula

$$\mathcal{N}_{01}^{\text{La}} = \frac{\Sigma P_0 Q_1 - \Sigma p_0 q_1}{\Sigma P_0 Q_0 - \Sigma p_0 q_0}$$

The object of the method of deflation is to estimate the terms in the numerator by deflating the current value added ($\Sigma P_1 Q_1 - \Sigma p_1 q_1$) by appropriate price indexes. Thus we estimate $\Sigma P_0 Q_1$ by $\Sigma P_1 Q_1 \div P_{01}$, where $\Sigma P_1 Q_1$ is the current value of output and P_{01} is an index of the relevant output prices, and we estimate $\Sigma p_0 q_1$ by $\Sigma p_1 q_1 \div p_{01}$, where

$\Sigma p_1 q_1$ is the current value of input and p_{01} is an index of the relevant input prices. If

$$P_{01} = \frac{\Sigma P_1 Q_1}{\Sigma P_0 Q_1} \quad \text{and} \quad p_{01} = \frac{\Sigma p_1 q_1}{\Sigma p_0 q_1}$$

i.e. if the price indexes are of the Paasche type, and if they are coextensive with the fields of production under consideration, the estimates will give the correct values. But if sufficient information were available to construct such indexes, $\Sigma P_0 Q_1 - \Sigma p_0 q_1$ could be computed directly and there would be no point in using the method of deflation. In practice, the price index numbers used will be based on only a portion of the field and will seldom be of the Paasche type, so that only an approximation to $\mathcal{N}_{01}^{\text{La}}$ can be obtained. Clearly, the method requires at least a knowledge of $\Sigma P_1 Q_1 - \Sigma p_1 q_1$ as well as of $\Sigma P_0 Q_0 - \Sigma p_0 q_0$ and will be feasible only if regular statistics of current value added are collected, and if appropriate price indexes can be constructed. An example of an application of this method now follows.

Our object is to estimate the movement in the quantum of production of the Australian Tobacco, Cigarettes and Cigars Industry between 1958–59 and 1963–64. The data were obtained from the *Secondary Industries Bulletins* and *Manufacturing Industries Summaries*, published by the Commonwealth Bureau of Census and Statistics.[1] We wish to obtain an estimate of

$$\mathcal{N}_{01}^{\text{La}} = \frac{\Sigma P_0 Q_1 - \Sigma p_0 q_1}{\Sigma P_0 Q_0 - \Sigma p_0 q_0}$$

The data for the denominator, being 1958–59 values, are directly available. The value of output for 1958–59 was $98,772,000 and of input was $71,870,000. Our main task is to estimate the numerator, and we do this by considering output and input separately. We have 1963–64 values, and we must calculate a price index number with which to deflate these values, i.e. we must calculate P_{01}, so that we can use $\Sigma P_1 Q_1 \div P_{01}$ as an estimate of $\Sigma P_0 Q_1$.

We have quantity and value data of the main five items of output. We do not have price data, but by dividing the value of output by the quantity, we get what are called *unit values*. If an item of output is homogeneous, its unit value and its price are the same thing, but if the item contains different qualities, then unit value is the price of a composite unit, and movements in unit value may not correctly reflect price movements if the relative importance of the different qualities is

---

[1] Values and quantities in Tables 14.2 to 14.5 are from Commonwealth Bureau of Census and Statistics, Australia: *Secondary Industries Bulletin*, No. 53, 1958–59, p. 63; *Manufacturing Industry Bulletin*, No. 1, 1963–64, p. 57; *Manufacturing Industries Summary —Tobacco, Cigars and Cigarettes*, 1959–60, No. 26, pp. 3, 4, 7; 1963–64, No. 26, pp. 4, 7.

changing. However, since we have no price data, we must make do with unit values. Since the index number formula should be of the Paasche type, it is necessary to compare 1963–64 unit values weighted by 1963–64 quantities (i.e. 1963–64 total values) with 1958–59 unit

Table 14.2
COMPUTATION OF OUTPUT PRICE INDEX

Output	Unit	1958–59			1963–64		1958–59 Unit Values Weighted by 1963–64 Quantities (c)
		Quantity	Value	Unit Value (a)	Quantity	Value (b)	
		'000	$'000	$	'000	$'000	$'000
		(1)	(2)	(3)	(4)	(5)	(6)
Flake	lb	10,040	14,917	1·4858	5,753	9,084	8,548
Ready Rubbed	lb	6,682	10,198	1·5262	4,145	6,660	6,326
Cigars and Cigarettes	lb	35,273	72,755	2·0626	46,040	107,344	94,962
						123,088	109,836

(a) Col. (3) = col. (2) ÷ col. (1).
(b) Col. (5) is identically 1963–64 unit values weighted by 1963–64 quantities.
(c) Col. (6) = col. (3) × col. (4).

values weighted by 1963–64 quantities. The calculations are performed in Table 14.2.

From the above table we have as our index of output prices for 1963–64, with base 1958–59 = 100

$$(123{,}088 \div 109{,}836) \times 100 = 112\cdot 07$$

The coverage of this index is good, for whereas the total value of output in 1958–59 was $98,772,000, the sum of column (2) indicates that the value of the items in the index is $97,870,000, and thus there are few items omitted from the index. The method used assumes that the prices of these omitted items move in the same way as the average of the included items. To estimate the value of 1963–64 output valued at 1958–59 prices we divide the 1963–64 value by our index number. This latter value is $124,122,000, so that the required figure is $124,122,000 ÷ 1·1207 = $110,754,000. The ratio of this figure to the value of output in 1958–59 (i.e. $\Sigma P_0 Q_1 / \Sigma P_0 Q_0$) gives an index of quantum of output (sometimes called *gross output*). In this case we have

$$(110{,}754{,}000 \div 98{,}772{,}000) \times 100 = 112\cdot 13$$

as the 1963–64 index for quantum of output, with base 1958–59 = 100.

On the input side, we require a price index number of inputs, p_{01}, so that we can use $\Sigma p_1 q_1 \div p_{01}$ as an estimate of $\Sigma p_0 q_1$. It is convenient to divide the inputs into four groups and deflate each separately. These groups are materials, fuels, containers, tool repairs. The first two groups are treated in a manner similar to that for output above. The quantum of containers is assumed to move in the same way as the index of gross output computed in the preceding paragraph, and the

Table 14.3
COMPUTATION OF MATERIALS INPUT PRICE INDEX

Materials Input	Unit	1958–59			1963–64		1958–59 Unit Values Weighted by 1963–64 Quantities $'000
		Quantity '000	Value $'000	Unit Value $	Quantity '000	Value $'000	
		(1)	(2)	(3)	(4)	(5)	(6)
Tobacco—							
Australian	lb	2,807	2,655	0·9458	3,342	3,156	3,161
Imported	lb	11,494	13,422	1·1677	4,998	5,774	5,836
Tobacco for Cigarettes—							
Australian	lb	5,986	7,488	1·2509	16,814	20,934	21,033
Imported	lb	28,528	37,376	1·3102	25,641	32,310	33,595
						62,174	63,625

The materials input price index number is $(62{,}174 \div 63{,}625) \times 100 = 97{\cdot}72$.

Table 14.4
COMPUTATION OF FUELS INPUT PRICE INDEX

Fuels Input	Unit	1958–59			1963–64		1958–59 Unit Values Weighted by 1963–64 Quantities $'000
		Quantity '000	Value $'000	Unit Value $	Quantity '000	Value $'000	
		(1)	(2)	(3)	(4)	(5)	(6)
Black coal	ton	7·47	62·90	8·4204	5·76	49·00	48·50
Fuel oil	gal	388·77	51·08	0·1314	744·24	71·34	97·79
Electricity	—	—	152·26	—	—	333·00	384·57 (a)
						453·34	530·86

(a) This figure is obtained by deflating 1963–64 value by an index of unit values calculated from the electricity generating industry. This index stood at 86·59 for 1963–64, with base 1958–59 = 100, and we have $333{\cdot}00 \div 0{\cdot}8659 = 384{\cdot}57$.

The fuels input price index number is $(453{\cdot}34 \div 530{\cdot}86) \times 100 = 85{\cdot}40$.

tool repairs are deflated by the Engineering, Metals, Vehicles, etc., group of the Minimum Weekly Wage Rate Index (Adult Males). The calculation of the price indexes for materials and fuels are set out in the two tables above. The coverage of both indexes is good, for whereas the total values of materials and fuels in 1958–59 were $61,524,000 and $318,000 respectively, the sums of columns (2) in the tables above indicate that the values of the items in the indexes are $60,941,000 and $266,000 respectively.

The calculation of the index of quantum of production (i.e. of quantum of value added, sometimes called *net output*) is given below.

Table 14.5

COMPUTATION OF INDEXES OF QUANTUM OF OUTPUT, INPUT AND PRODUCTION BY METHOD OF DEFLATION

	Values		Price Indexes (Base: 1958–59 = 100)	Estimate of 1963–64 Quantities Valued at 1958–59 Prices $'000 (a)	Indexes of Quantum (Base: 1958–59 = 100) (b)
	1958–59 $'000	1963–64 $'000			
	(1)	(2)	(3)	(4)	(5)
Output	98,772	124,122	112·07 (d)	110,754	112·13
Materials	61,524	63,250	97·72 (e)	64,726	
Fuels	318	508	85·40 (f)	595	
Containers	9,324	14,132	—	10,455 (c)	
Tool repairs	704	998	115·44 (g)	865	
Input	71,870 (h)	78,888 (h)		76,641 (h)	106·64
Value added	26,902 (i)	45,234 (i)		34,113 (i)	126·80

(a) Col. (4) = (col. (2) ÷ col. (3)) × 100.
(b) Col. (5) = (col. (4) ÷ col. (1)) × 100.
(c) Quantum of containers is assumed to move with gross output, so that we have 9,324 × 1·1213 = 10,455.
(d) from Table 14.2 above.
(e) from Table 14.3 above.
(f) from Table 14.4 above.
(g) This index is based on the Engineering, Metals, Vehicles, etc., Group of the Minimum Weekly Wage Rate Index (Adult Males).
(h) The sum of the preceding four lines.
(i) The difference between output and input.

From the above table we see that the quantum of production (value added) in the Tobacco, Cigarettes and Cigars Industry rose by nearly 27 per cent between 1958–59 and 1963–64. This is against a rise of 68 per cent in current values added ($26,902,000 to $45,234,000), some of which was a price rise. Output rose in quantum by 12 per cent, but input rose by only about 7 per cent; consequently value added or net output rose by appreciably more than the rise in gross output, and the different movements in quanta of output and production are thus explained. The relatively large difference between the indexes of quantum of gross output and value added in this example illustrates the danger of using an index of quantum of gross output as an approximation to one of quantum of value added.

14.5. Computation of an Index of Industrial Production by the Method of Indicators

The basis of the method of indicators is the breaking up of the industrial area under consideration into small sectors for which the values added in the base period are known and the estimation of what such values added would be in subsequent periods (if they were valued at constant base period prices) by marking them up by indicators of the movements in the quanta of production under consideration. The indicators may, for example, be movements in physical outputs, or in physical inputs or in employment. Thus, if V_0 is the value added in period 0 in a small sector (i.e. $V_0 = \Sigma P_0 Q_0 - \Sigma p_0 q_0$, where the summations are over the items contained in the small sector), and I_{01} is an indicator of the movement of the quantum of production from period 0 to period 1 in the small sector, we shall have $I_{01} V_0$ as an estimate of the value added in period 1 at constant period 0 prices (i.e. as an estimate of $\Sigma P_0 Q_1 - \Sigma p_0 q_1$). Aggregating small sectors, we obtain as an index of the quantum of production for the larger field

$$\mathcal{N}_{01} = \frac{\Sigma I_{01} V_0}{\Sigma V_0}$$

This index, which is a weighted arithmetic mean of indicators, is an estimate of

$$\frac{\Sigma[\Sigma P_0 Q_1 - \Sigma p_0 q_1]}{\Sigma[\Sigma P_0 Q_0 - \Sigma p_0 q_0]}$$

$$= \frac{\Sigma \left[\frac{\Sigma P_0 Q_1 - \Sigma p_0 q_1}{\Sigma P_0 Q_0 - \Sigma p_0 q_0} (\Sigma P_0 Q_0 - \Sigma p_0 q_0) \right]}{\Sigma[\Sigma P_0 Q_0 - \Sigma p_0 q_0]}$$

where the small Σ's are summations over outputs and inputs within small sectors and the large Σ's are summations of small sectors.[1] It will be appreciated that the I_{01}'s are used as approximations to

$$\frac{\Sigma P_0 Q_1 - \Sigma p_0 q_1}{\Sigma P_0 Q_0 - \Sigma p_0 q_0}$$

for small sectors in the above formula.

[1] Compare p. 379 above.

Accordingly, the principal difference between the method of deflation and the method of indicators is that in the former method

$$\frac{\Sigma P_0 Q_1 - \Sigma p_0 q_1}{\Sigma P_0 Q_0 - \Sigma p_0 q_0}$$

is estimated directly in detail, whereas in the latter it is approximated by an indicator series. The method of indicators can be used when the values added in some base period are known and indicators of quanta of production in small sectors of industry can be constructed. It is, in general, a more approximate method than the method of deflation, but is usually more practicable and is the commonest method found in practice.

There are many possible indicators. Some will be more satisfactory than others, but generally the actual choice of an indicator depends on the availability of suitable statistics. The main indicators are: output series, input series and employment series. These will be discussed in turn.

1. *Output Indicators*

If we are concerned with a small sector which produces only one product and we have data on the volume of output of that product, then clearly we can use this series as an indicator. Such an indicator has one big advantage—the series is itself in physical terms, and there is no price problem involved. Here the indicator will be Q_1/Q_0, where the Q's refer to the volume of output of the commodity under consideration. However, even in this case there are two limitations to be noted. First, we are using the series as an indicator of work done or quantum of production and not as a measure of quantum of output, which is what the series really is. Any change in the technical relations between input and output will upset its validity. Moreover, if the indicator relates to deliveries or sales of output, any accumulation of work-in-progress will make it understate changes in the quantum of production. Finally, changes in quality will tend to be ignored.

Most small sectors will in fact be producing more than one product. Sometimes it will be possible to add these products directly. This will be the case where their unit of measurement is homogeneous. Thus we might have a series of "square yards of wool textiles." This yardage would consist of textiles of very different types. These different types may have very different value added contents. Clearly, an increase in yardage of 10 per cent will involve a quite different increase in quantum of production in the case where all types of textiles increase by 10 per cent from that case where the increase in yardage is mainly in the coarser forms of cloth. Strictly speaking, such an indicator series would be valid only if the average value added per square yard (measured in constant base-year prices) remained constant. Frequently, however, there will be many quite distinct products produced by a "small sector," which cannot be added. Data may not be available for many of them, and in that case we may be forced to assume that movements in the output of these items can be represented by movements in the output of items for which we have data. If we do have data for several items, the question of adding the quite distinct items arises. This can be done by using prices to determine the equivalences of different physical units and using an index of physical output, e.g. $\Sigma P_0 Q_1 / \Sigma P_0 Q_0$, as an indicator. The use of such an indicator will not be completely valid if the movements in the Q's do not properly reflect movements in the quantum of production because of changes in the technical relations between the various inputs and outputs.

2. *Input Indicators*

In some sectors movements in the quantum of production can be indicated by movements in inputs of materials, etc. The main advantage of input indicators, when available, is that for many industries the number of principal materials used is quite small. However, it can be readily appreciated that input indicators suffer from the same sort of limitations as do output indicators.

3. *Employment Indicators*

The relative movement in the number of persons employed in the small sector under consideration can be used as an indicator. Such employment indicators have the great advantage of being directly related to work done and hence avoid the problems associated with changes in work-in-progress. However, it is necessary to adjust employment series to take into account such factors as holidays and changes in the length of the working week. Even so, the possibility of overtime and slack time may upset an employment series as an indicator of the quantum of production. The principal disadvantage of using employment indicators is that they assume that labour productivity (i.e. value added measured in constant base-year prices per unit of employment[1]) is constant. For short-period comparisons this may be unimportant, but it becomes increasingly important in the long run.

An example of an application of the method of indicators now follows. This example makes use of a variety of types of indicators for illustrative purposes. Our object is to estimate the movement in the quantum of production of the Australian Beverage Industry as between 1958–59 and 1963–64. The data were obtained from *Secondary Industries*

[1] See section 14.7, p. 392 below.

REAL NATIONAL PRODUCT 389

Bulletins.[1] The industry can be divided into four small sectors, namely aerated waters and cordials, malting, brewing, distilling and wine-making. The formula to be used is

$$\mathcal{N}_{01} = \frac{\Sigma I_{01} V_0}{\Sigma V_0}$$

For indicator series we use employment for aerated waters and cordials; physical input of barley in bushels for malting; physical output of ale, stout, and beer in gallons for brewing. For distilling and wine-making we use an index of physical output obtained by combining the outputs of fortified wines and potable spirits, using base-period unit values as weights. The calculation of the first three indicators is shown in Table 14.6 below, and that of the fourth in Table 14.7 on the next page.

Table 14.6

COMPUTATION OF INDICATORS

Small Sector	Indicator Series	Units	Value of Indicator Series		Indicator for 1963–64 on 1958–59 base (a)
			1958–59	1963–64	
			(1)	(2)	(3)
Aerated waters and cordials	Employment	persons	4,877	5,193	1·065
Malting	Input of barley	'000 bus	8,198	11,886	1·450
Brewing	Output of ale, stout and beer	'000 gal	223,597	262,344	1·173

(a) Col. (3) = col. (2) ÷ col. (1).

We must now combine the four indicators by weighting them with 1958–59 values added. This is done in Table 14.8 on the next page. The index of the quantum of production for the Beverage Industry for 1963–64, with base 1958–59 = 100, is then

$$(75{,}170 \div 65{,}572) \times 100 = 114\cdot64$$

Before concluding this section some comments are necessary on the treatment of small sectors within the industrial field for which no indicators or their equivalents exist. Three alternative treatments are:

1. The coverage of the index of industrial production under consideration can be defined to exclude these particular sectors. This is unsatisfactory, for although it results in a more accurate description of

[1] Values and quantities in Tables 14.6 to 14.8 are from Commonwealth Bureau of Census and Statistics, Australia: *Manufacturing Industries*, 1958–59, No. 25, pp. 4, 5; 1959–60, No. 22, p. 3; No. 24, pp. 3, 5; No. 25, pp. 3–5; 1963–64, No. 22, p. 3; No. 24, p. 4 and No. 25, p. 6; *Secondary Industries Bulletin*, No. 53, 1958–59, pp. 63, 85 and *Year Book*, No. 53, 1967, p. 919.

Table 14.7
COMPUTATION OF INDICATOR

Output	Unit	1958–59			1963–64	
		Quantity '000	Value $'000	Unit Value (a) $	Quantity '000	Quantity Valued at 1958–59 Unit Values (b) $'000
		(1)	(2)	(3)	(4)	(5)
Fortified wines	gal	9,699	8,478	0·8742	10,390	9,083
Potable spirits	gal	2,731	4,052	1·4838	3,093	4,589
			12,530			13,672

(a) Col. (3) = col. (2) ÷ col. (1).
(b) Col. (5) = col. (4) × col. (3).

The distilling and wine-making indicator is $13,672 \div 12,530 = 1 \cdot 091$.

Table 14.8
COMPUTATION OF INDEX OF QUANTUM OF PRODUCTION BY METHOD OF INDICATORS

Small Sector	Value Added in 1958–59 $'000	Indicator of Movement in Quantum of Production 1958–59 to 1963–64	Estimate of Value Added in 1963–64 at 1958–59 Prices (a) $'000
	(1)	(2)	(3)
Aerated waters and cordials	20,048	1·065	21,351
Malting	3,816	1·450	5,533
Brewing	33,928	1·173	39,798
Distilling and wine-making	7,780	1·091	8,488
Beverage industry	65,572		75,170

(a) Col. (3) = col. (1) × col. (2).

the index actually computed, it may mean that the field covered is ragged and does not cover industrial production in the usual sense.

2. The index can be calculated leaving the particular sectors right out of account, but the coverage can be defined, formally, to include all industrial production. This is tantamount to assuming that the excluded sectors move in the same way as the average movement of the included sectors. Under (1) and (2) the index has precisely the same value but the interpretation of it differs.

3. The sectors for which data are lacking can be included both formally and statistically by assuming that their quanta of production behave in the same fashion as those of some other sector or sectors or part of a sector. This can be done by adding to the weight attached to the representing indicator an amount for the weight of the represented sector. This may be a much more reasonable procedure than that under (2), but it is necessary to make certain that there is some rational basis for selecting the representing indicator.

References to Australian indexes of quantum of production are contained in Appendix B.5, pp. 502–503 below.

14.6. COMPARISONS OVER SHORT AND LONG PERIODS

Indexes of industrial production are usually required for the purpose of measuring both short-period and long-period movements in the quantum of production. As far as the former are concerned, monthly indexes are the ideal. The question then arises whether corrections ought to be made for the varying number of working days per month. This largely depends on the purpose of the index. If we wish to measure the actual course of production, an unadjusted series will be required, but if we are trying to discern trends in the monthly figures, then it will be necessary to correct the series for the varying number of working days per month (and particularly for Easter, which comes in different months in different years). Further, it may be useful to calculate an index of seasonal variation on the basis of past experience and deseasonalize the index of production. But the limitations of indexes of seasonal variation must be borne in mind (see p. 278 above).

As far as long-period comparisons are concerned, the main difficulty is, as usual, one of weighting. If the index is calculated with fixed base-year weights, these weights will sooner or later get out of date, i.e. the base-year relative prices at which outputs and inputs are being valued will cease to be relevant. The influence of changes in relative prices can be seen by computing both the Laspeyres and Paasche forms of the index and comparing the results. The greater the changes in relative prices, the more likely that the two indexes will be far apart.[1]

In practice, it is customary to compute indexes of industrial production by the method of indicators with fixed weights, the weights being base-year values added. Indicators can often be obtained in monthly

[1] For example, Carter, C. F., Reddaway, W. B., and Stone, J. R. N., in *The Measurement of Production Movements* (Cambridge University Press, 1948), p. 68, calculated indexes of industrial production for the United Kingdom, comparing years 1935 and 1946 on the basis of 1935 weights and 1946 weights. The former index revealed a *rise* in quantum of production of about 3 per cent and the latter revealed a *fall* of about 2 per cent. The difference in *direction* of change of the two indexes is striking, but it must be remembered that the indexes were in fact quite close, being only some 5 per cent apart.

series, in which cases a monthly index can be computed. For comparisons extending over a few years the use of fixed weights in this way may be quite satisfactory. But the weights should be revised from time to time as data on values added become available. Thus, if, for example, Censuses of Production are taken every five years, the weights can be revised every five years. This will mean a new index every five years, but quasi-continuity of the indexes can be achieved by a chaining process (see section 13.11, p. 361 above). Not only should the weights be revised from time to time, but the possibility of using new and better indicators should always be borne in mind.

14.7. Measurement of Labour Productivity

Indexes of industrial production, whether of production as a whole or of particular industries, are often used in conjunction with employment series to measure labour productivity. According to the purpose for which the measurement of productivity is required, the employment figures used may refer to the available labour supply or the numbers actually in work or the number of man-hours worked per period under consideration.

We first consider productivity in reference to industrial production as a whole. If we write L_0 for the employment figure (however defined) in period 0, average value of production per employment unit in period 0 at period 0 prices will be

$$\frac{\sum[\Sigma P_0 Q_0 - \Sigma p_0 q_0]}{L_0}$$

where the small Σ's are summations over outputs and inputs within individual industries and the large $\sum$'s are summations of industries. Similarly average value of production per employment unit in period 1 at period 0 prices will be

$$\frac{\sum[\Sigma P_0 Q_1 - \Sigma p_0 q_1]}{L_1}$$

An index of productivity can be obtained by dividing the latter measure by the former. This reduces to

$$\frac{N_{01}}{L_{01}}$$

where N_{01} is an index of the quantum of production for industry as a whole and $L_{01} = \frac{L_1}{L_0}$ is an index of employment.

This index of productivity measures *overall productivity* and is affected both by changes in productivity in individual industries and by changes in the relative importance of industries. Thus, even though productivity in all individual industries rose, the index could fall if

REAL NATIONAL PRODUCT 393

activity shifted from industries with higher than average to ones with lower than average productivity. Often, however, we may be concerned to see how productivity has moved quite apart from shifts in activity, i.e. we may wish to measure the first of the two factors mentioned above. For want of a better term we may call this *productivity proper*.

An index of productivity proper must be based on the movements in productivity of individual industries. For a single industry, productivity in periods 0 and 1 respectively, at period 0 prices, will be given by

$$\frac{\Sigma P_0 Q_0 - \Sigma p_0 q_0}{l_0} \quad \text{and} \quad \frac{\Sigma P_0 Q_1 - \Sigma p_0 q_1}{l_1}$$

where l_0 and l_1 are the employment figures for the industry in periods 0 and 1 respectively. If we combine these productivities for individual industries, weighting them by their base-period employments, i.e. by l_0, we shall obtain the index

$$\frac{\Sigma \left[\frac{\Sigma P_0 Q_1 - \Sigma p_0 q_1}{l_1} \times l_0 \right]}{\Sigma [\Sigma P_0 Q_0 - \Sigma p_0 q_0]}$$

This is an index of productivity proper. The numerator is the value of production in period 1 valued at period 0 prices which the period 0 employment units would have produced if they had exhibited period 1 productivity in each industry. Consequently, this index eliminates the effects of changes in the distribution of employment.[1]

Table 14.9 gives hypothetical data to illustrate the points made above.

Table 14.9

Industry	Value Added in Period 0 at Period 0 Prices $'000	Employment in Period 0	Productivity in Period 0 (a) $'000	Value Added in Period 1 at Period 0 Prices $'000	Employment in Period 1	Productivity in Period 1 (b) $'000
	(1)	(2)	(3)	(4)	(5)	(6)
I	2,000	1,000	2·00	1,435	700	2·05
II	3,000	1,000	3·00	4,960	1,600	3·10
All	5,000	2,000	2·50	6,395	2,300	2·78

(a) Col. (3) = col. (1) ÷ col. (2).
(b) Col. (6) = col. (4) ÷ col. (5).

[1] The same index can be obtained by weighting the movements in productivity of individual industries (i.e. the ratios of the two expressions on the first half of this page) by their base-period values added (i.e. by $\Sigma P_0 Q_0 - \Sigma p_0 q_0$).

An index of overall productivity for period 1, with base-period 0 = 100 is given by

$$(2 \cdot 78 \div 2 \cdot 50) \times 100 = 111$$

This index contains not only the effect of changes in productivity in the two industries, but also the effect on overall productivity of the change in the distribution of employment. An index of productivity proper can be calculated by weighting the productivities for the two industries by their base-period employments. The resulting index is 103. Thus productivity proper has increased by 3 per cent, whereas overall productivity has increased by 11 per cent. The excess of the latter over the former is due to the shift in relative employment from industry I to industry II, i.e. from an industry with lower than average to one with a higher than average productivity.

It should be emphasized that both changes in productivity proper and shifts in the distribution of employment are important in affecting the overall productivity of an economy. Indeed, the finer the classification of the industries of an economy, the greater will be the contribution of shifts in employment to any given change in overall productivity.

This section has been confined to a technical consideration of the formulae for measuring productivity. It should be hardly necessary to add that indexes of productivity, being based on indexes of quantum of production, are subject to the same limitations as those indexes (see pp. 369–370 and 381 above) and should be interpreted with caution.

CHAPTER XV

DEMOGRAPHY

15.1. Demography and Demographic Data

DEMOGRAPHY is the study of *the measurement of human populations.* It studies the size of a group of people, how the number in the group has changed in the past and how it is likely to change in the future. Demography is mainly concerned with the growth of populations, but it also covers the distribution of the population by industry, occupation, geographical area and so on. In fact the interests of demography are at least coextensive with the population census and in some respects go beyond it. This chapter is devoted to the techniques by which measurements of population and, in particular, measurements of population growth can be made.

It is hardly necessary to emphasize the important part which a study of population movements plays in the social sciences. The balance between population and resources is a problem which has been of great interest to economists since the days of Malthus, and interest in it has been revived today in the study of the economics of the so-called underdeveloped areas. This balance exerts a great influence on the standard of living enjoyed in any given area, and a knowledge of how population growth has behaved in the past and how it is likely to behave in the future is of the first importance. Populations in different parts of the world grow at different rates, and these differential rates determine very largely the distribution of world population, with its political and strategic implications. A knowledge of the statistical techniques of population measurements is basic to any general analysis of population questions.

Data on population are acquired periodically through censuses and continuously through birth and death registrations and marriage and migration records. These data enable records to be compiled of the numbers of population, births, deaths, marriages and migrants. In addition, censuses usually provide information on the distribution of the population by sex, age, marital status, duration of marriage and number of previous issue. Likewise, birth registration usually provides information on place of birth, sex, age of parents, legitimacy, number of previous issue and their sexes and ages, father's occupation and birthplace of parents; and death registration provides information on place of death, sex, age, marital status, number of previous issue, birthplace, occupation and cause of death. Similar information is also usually collected with respect to marriages and migrants.

As far as data on births are concerned, several points should be noted. Most published data on births refer to *live-births* only, *still-births* being shown separately, if at all. The ratio of still-births to live-births varies from country to country, but in the last ten years in Australia, the United Kingdom and the United States it has been from about 12 to 22 still-births per 1,000 live-births. Births are usually classified by *date of registration* and not by *date of birth*. Thus in Australia in 1967 there were 229,296 births. This was the number of births registered in 1967 and not the number actually occurring. However, as long as the lag between birth and registration is small, the number registered in a year will be very close to the number actually taking place. The distinction between number of *confinements* and number of *births* should be noted, the difference being due to confinements resulting in *multiple live-births*. Likewise the division of births into *nuptial* and *ex-nuptial* (i.e. legitimate and illegitimate) should be noted.

The following table illustrates the relation between the numbers of confinements and births—

Table 15.1

CONFINEMENTS AND BIRTHS—AUSTRALIA, 1967 (a)

	Confinements	Births
Nuptial—		
Single	207,265	207,265
Multiple	2,164	4,297
Ex-nuptial—		
Single	17,368	17,368
Multiple	186	366
Total	226,983	229,296

(a) Excludes confinements where the births were of still-born children only.

Source: Commonwealth Bureau of Census and Statistics, Australia:
Australian Demographic Review, No. 245, p. 9

Births are frequently classified by geographical area. Such a classification is one by *place of birth*. In border areas this may differ considerably from a classification by *usual residence of mother* and the latter is often the more relevant classification. The case of the Australian Capital Territory provides an interesting illustration as shown on the next page.

Before there were extensive hospital facilities in the Australian Capital Territory, many Australian Capital Territory mothers had their confinements in Queanbeyan in New South Wales, just over the border. With improved hospitalization in 1939 this practice was reversed, and many New South Wales mothers living near the border

DEMOGRAPHY

Table 15.2
BIRTHS PER 1,000 PER ANNUM OF POPULATION IN THE
AUSTRALIAN CAPITAL TERRITORY

	By Place of Birth	By Usual Residence of Mother
1931–35	15·8	19·5
1941–45	26·8	23·1
1946–50	37·9	29·1
1951–55	32·3	28·4
1956–60	30·3	30·3

Source: Commonwealth Bureau of Census and Statistics, Australia: *Year Book*, No. 47, 1961, p. 339

had their confinements in Canberra. The situation was again reversed in 1952, when improved maternity accommodation was provided in Queanbeyan. Clearly in this case, classification of births by place of birth is misleading.

As far as deaths are concerned, the terms *infant deaths* and *neo-natal deaths* should be noted. They refer respectively to deaths of children aged under one year and under one month. Neither includes stillbirths. The term *maternal deaths* refers to deaths of females from causes arising in childbirth.

15.2. MEASUREMENT OF TOTAL POPULATION

Population at a Date

Total population is usually expressed as at a date, e.g. the Australian population, excluding full-blood Aborigines, as at the census taken on 30th June, 1961, was 10,508,186. The total population measured at a census is usually very accurate in advanced countries. In any case, it is the most accurate information we can get, and we take it as being correct. However, censuses are taken only every so often—usually every five or ten years, and consequently it is necessary to make *intercensal estimates* of the total population.

Since populations increase by:

(a) natural increase, i.e. the excess of births over deaths; and

(b) net migration, i.e. the excess of immigration over emigration,

we need only add the figures of this year's natural increase and net migration to last year's total population to get an estimate of the total for this year. Table 15.3 on the next page illustrates this.

How accurate are intercensal estimates? We have an opportunity of checking them whenever a census is taken. For example, censuses were taken in Australia at 30th June, 1961 and 1966. At the 1961 Census the total population was 10,508,186. Casting this forward by adding natural increase and net migration year by year, the intercensal

Table 15.3

ESTIMATE OF TOTAL POPULATION OF AUSTRALIA
AT 30TH JUNE, 1966 (a)

Population at 30/6/65 (Intercensal estimate)		11,359,510
Births registered 1/7/65–30/6/66	222,553	
Deaths registered 1/7/65–30/6/66	100,702	
Natural increase		121,851
Overseas arrivals 1/7/65–30/6/66	540,463	
Overseas departures 1/7/65–30/6/66	448,074	
Net migration		92,389
Estimated population at 30/6/66		11,573,750

(a) Excludes full-blood Aborigines.

Source: Commonwealth Bureau of Census and Statistics, Australia: *Quarterly Summary of Australian Statistics*, No. 260, 1966, p. 1; No. 262, pp. 2, 5

estimate at 30th June, 1966, was 11,573,750. This is as shown in Table 15.3. The 1966 Census revealed the true figure to be 11,550,462, excluding full-blood Aborigines. The intercensal estimate was thus shown to be 23,288 too high—a surplus of just over $\frac{1}{5}$ per cent or an average of about $\frac{1}{25}$ per cent per annum over the five years between the two censuses. This surplus can possibly be attributed to inaccuracies in migration records. Given the results of the 1966 Census, we know that our intercensal estimates between 1961 and 1966 were too high, so we must adjust them. This can be done by marking down the recorded annual increases in the intercensal years by the proportion in which the census total increase over the five years fell short of the recorded total increase.

Mean Population

In a census a population is counted on a particular day. Similarly, intercensal estimates are estimates of the population as at particular dates. If we have an annual figure (e.g. annual births or annual primary production) which we wish to relate to the population (e.g. to obtain annual births per 1,000 of the population or annual primary production *per capita*), we must consider whether it is appropriate to relate an annual figure which has been built up over the course of a year to a population figure measured at a particular date. In fact, it is not appropriate to do this. We must always relate things to what is relevant. If we have an annual figure built up over the course of the year, we must relate it not to the population at a date but to the *mean population* existing over the course of the year.

The calculation of the mean population for a year is in principle quite simple. If we knew the population on each day of the year, we could add up all the figures and divide by 365 in order to get the mean population.[1] In practice, we do not have all this information. Usually the most we have are quarterly intercensal estimates as at the end of quarters. Suppose we know that

Date	Size of Population
31st December—	a
31st March—	b
30th June—	c
30th September—	d
31st December—	e

and we wish to estimate the mean population for the calendar year. Our data are illustrated graphically in Fig. 15.1.

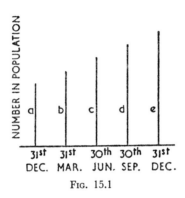

Fig. 15.1

If we knew how the population had changed within quarters, so that we could draw a curve through the tops of the five ordinates in the diagram to represent its track, the mean population would be given by the average ordinate under the curve. We do not have this information, so we must make reasonable assumptions. If we assume that within each quarter the population has increased steadily, so that a straight line joining the tops of a and b would represent the track of population during the first quarter, etc., then the following formula will give an estimate of mean population—

$$\text{Mean population} = \frac{a + 2b + 2c + 2d + e}{8}$$

However, a further refinement leading to a smoother movement over the year is to assume that the tops of a, b and c are joined by a parabola and the tops of c, d and e by another parabola, and these parabolas

[1] In theory we should need to know the population continuously at every point of time within the year.

represent the track of population during the year. This gives the following formula—

$$\text{Mean population} = \frac{a + 4b + 2c + 4d + e}{12}$$

This formula is used by the Commonwealth Statistician in calculating mean annual population for Australia. It is the most satisfactory of the formulae given here. Frequently, however, quarterly data are not available. If we know the population only at the end of years, we may have to take the simple arithmetic mean of end-of-year populations, i.e.

$$\text{Mean population} = \frac{a + e}{2}$$

Alternatively we may be forced to take the population as at 30th June as an estimate of the mean population for the calendar year. These various estimates of the mean population will all be close unless the rate of change of population over the year is very uneven. Table 15.4 illustrates the calculation of mean annual population.

Table 15.4
CALCULATION OF MEAN POPULATION
AUSTRALIA, 1967

Date	Population (a) P	Linear Method		Parabolic Method		End-of-year Data Only	
		W (b)	W × P	W (b)	W × P	W (b)	W × P
31/12/66	11,710,387	1	11,710,387	1	11,710,387	1	11,710,387
31/3/67	11,772,927	2	23,545,854	4	47,091,708		
30/6/67	11,810,165	2	23,620,330	2	23,620,330		
30/9/67	11,864,410	2	23,728,820	4	47,457,640		
31/12/67	11,928,889	1	11,928,889	1	11,928,889	1	11,928,889
		8	94,534,280	12	141,808,954	2	23,639,276

(a) Population estimates used in this table include full-blood Aborigines.
(b) Weights in formula.

Method	Mean Population
Linear	94,534,280 ÷ 8 = 11,816,785
Parabolic	141,808,954 ÷ 12 = 11,817,413
End-of-years	23,639,276 ÷ 2 = 11,819,638
Mid-year estimate	11,810,165

Source: Commonwealth Bureau of Census and Statistics, Australia: *Australian Demographic Review*, No. 243, pp. 3, 5

It should be noted that semi-logarithmic charts are very useful for illustrating the movements of total population over time. Such a chart is shown in Fig. 4.17 on p. 38 above.

15.3. Sex and Age Distribution

Census Results

A classification of the population by sex and by age is usually obtained from censuses. Thus we have—

Table 15.5

SEX AND AGE DISTRIBUTION OF AUSTRALIAN POPULATION
CENSUS 30TH JUNE, 1966

Age Last Birthday	Males	Females	Persons
Under 1 year . .	110,334	105,233	215,567
1 year . .	112,291	106,891	219,182
2 years . .	118,810	112,489	231,299
3 years . .	121,269	115,256	236,525
4 years . .	123,245	117,326	240,571
5 years . .	123,322	117,626	240,948
6 years . .	120,396	114,297	234,693
.	.	.	.
.	.	.	.
.	.	.	.
All Ages . . .	5,816,359	5,734,103	11,550,462

Source: Commonwealth Bureau of Census and Statistics, Australia: *1966 Census Bulletin*, No. 9.3, pp. 6, 8, 10

This is a frequency distribution of the population classified by sex and by single age groups. It tells us that at 30th June, 1966, there were 110,334 males aged 0 and under 1; 112,291 aged 1 and under 2, etc.

The age distribution as it actually emerges from a census is called the *recorded* age distribution. It records the number of persons in each age group who stated themselves to be a certain age (or, in the case o children, whose parents stated them to be a certain age). But some people fail to state their ages in answering the census questionnaire. Thus the recorded age distribution would contain a group entitled "age not stated." In the Australian 1966 Census whenever a missing age was encountered on a census schedule the person was first allocated to a range of ages by using the other information on the schedule and was then allocated at random to a single age group within that range. Such a procedure gives what is called the *adjusted* age distribution. It contains no "age not stated" group and is the one given in Tables 15.5 and 15.6.

The adjusted age distribution can be expected to portray more nearly the true age distribution of the population than the recorded distribution. However, it is by no means absolutely accurate because of misstatement of their ages by people at the census. These misstatements are of two kinds:

1. There is a persistent bias on the part of some groups of people towards overstating or understating their ages.
2. There is a tendency to state ages to the nearest ten, and to a lesser extent to the nearest five or nearest even number.

It is difficult to detect misstatements of the first kind, but those of the second kind are revealed by an examination of the adjusted age distribution and can be seen graphically by drawing a histogram of that distribution. For example—

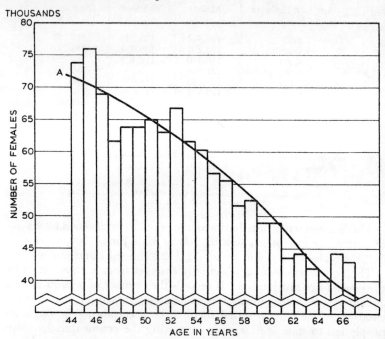

Fig. 15.2. Adjusted Distribution of Australian Females in Certain Age Groups, Census 30th June, 1966

Source: Commonwealth Bureau of Census and Statistics, Australia: *1966 Census Bulletin*, No. 9.3, pp. 7, 8

In the foregoing diagram there are several concentrations to notice namely those at ages 45, 50, 52, 65 and 66. Those at ages 45, 50 and 65 can be readily explained by the tendency for people to round their age to the nearest ten or five. Those at ages 52, and 66, have a more subtle explanation. Some people calculate their ages from their years of birth. The two ages above correspond to years of birth of 1914 and 1900; apparently at the 1966 Census many people took the commencement of World War I and the turn of the century as being near enough to their year of birth.

DEMOGRAPHY

It should now be clear that the adjusted-age distribution is somewhat inaccurate. Attempts are usually made to improve on it. The artificial peaks may be smoothed out, by allocating the excess numbers at, say, 60 years to the age groups on both sides. Various methods are available —some mathematical ones being quite complicated. The simplest procedure is to draw a smooth line through the diagram in such a way that the area under the curve is the same as that under the histogram. This is done by the curve A in the histogram above. The new distribution thus obtained is called the *graduated age distribution*. Thus, reading off the histogram, whereas the adjusted distribution gives 75,615 at age 45, the graduated gives approximately 70,000, and whereas the adjusted gives 68,894 at age 46, the graduated gives approximately 69,500. The graduated distribution attempts to smooth out the artificial bumps and to present more accurately the correct age distribution. The smoothing process is very arbitrary, however, and there is a danger that genuine bumps due to fluctuations in past births will be smoothed out. It is doubtful whether in fact a graduated distribution is much superior to an adjusted one.

Table 15.6 sets out the adjusted age distribution for Australia at the 1966 Census, in *quinquennial* age groups. The age group 0–4 means

Table 15.6

ADJUSTED SEX AND AGE DISTRIBUTION OF AUSTRALIA
CENSUS 30TH JUNE, 1966(a)

Age Last Birthday (Years)	Males	Females	Persons
0–4	585,949	557,195	1,143,144
5–9	595,538	567,358	1,162,896
10–14	556,251	530,197	1,086,448
15–19	536,848	511,378	1,048,226
20–24	436,709	417,232	853,941
25–29	384,336	361,729	746,065
30–34	355,654	331,700	687,354
35–39	397,463	367,099	764,562
40–44	396,536	377,215	773,751
45–49	343,033	334,639	677,672
50–54	323,810	317,824	641,634
55–59	276,100	266,916	543,016
60–64	215,590	219,759	435,349
65–69	161,376	195,020	356,396
70–74	115,084	160,887	275,971
75 and over	136,082	217,955	354,037
All ages	5,816,359	5,734,103	11,550,462

(a) Excludes full-blood Aborigines.

Source: Commonwealth Bureau of Census and Statistics, Australia: *1966 Census Bulletins*, No. 9.3, p. 3; No. 9.7, pp. 57–59

aged 0 and under 5 years, the age group 5–9 means aged 5 and under 10 years and so on.

Intercensal Estimates

We obtain direct information about the sex and age distribution only at censuses. These have to be kept up to date by making intercensal estimates. The techniques for making intercensal estimates vary according to the data available; but they are all based on the obvious proposition that the people in a particular age group this year must be those who were a year younger last year, depleted by death and augmented by net migration.

Suppose we wish to estimate the male age distribution at 30th June, 1967, given

> The age distribution at 30th June, 1966.
> Births over the year 1st July, 1966 to 30th June, 1967.
> Deaths by age at death over the year 1st July, 1966 to 30th June, 1967.
> Migration by age at entry or exit over the year 1st July, 1966 to 30th June, 1967.

Let us ignore migration for the time being and consider as an example the age group 15 years and under 16. Clearly, we shall have—

> Number aged 15 years and under 16 at 30th June, 1967
>
> *equals*
>
> (*a*) Number aged 14 years and under 15 at 30th June, 1966
>
> *minus*
>
> (*b*) Number who were aged 14 years and under 15 at 30th June, 1966, but who died during 1966–67

We know (*a*), but we do not know (*b*). Instead, we know the number of those who died when aged 14 years and under 15 years in 1966–67, which is not the same thing. Of those who were aged 14 years and under 15 at 30th June, 1966, but who died during 1966–67, some will have died in the age group 14 years and under 15 and some in the age group 15 years and under 16. But we know only the total deaths in the age groups 14 years and under 15, and 15 years and under 16, and we do not know how many of these would have been aged 14 years and under 15 at 30th June, 1966. This latter group of deaths, however, will consist of a fraction of those who died in 1966–67 aged 14 years and under 15 at death and a fraction of those who died in 1966–67 aged 15 years and under 16 at death. How can we estimate these fractions?[1]

These fractions will be evidently close to one-half. If deaths occur evenly through the year and if the ages of those who die are evenly distributed within each age group, exactly one-half of those aged 14 years and under 15 who die in 1966–67 will have had their fourteenth birthday

---

[1] This problem would not arise if deaths were classified by date of birth. In such a case the formula in the preceding paragraph in the text could be applied directly.

prior to 1st July, 1966, and hence will have been aged 14 years and under 15 at 30th June, 1966. Similarly, exactly one-half of those who die aged 15 years and under 16 years in 1966–67 will have had their fifteenth birthday on 1st July, 1966, or after, and hence will have been aged 14 years and under 15 at 30th June, 1966. This can be illustrated diagrammatically. The square in Fig. 15.3 represents the number of deaths aged 14 and under 15 during 1966–67. The date of death is shown along the horizontal axis and the age at death along the vertical axis. We assume that the number of deaths are distributed evenly over

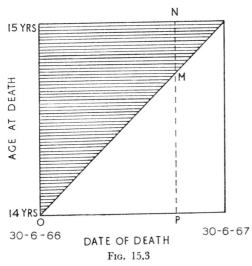

Fig. 15.3

the year, and that the ages of those who die at any date are distributed evenly over the year of age.

Draw in the ascending diagonal. Consider the date shown by point P, i.e. OP days from 30th June, 1966. Then NP is the number of deaths occurring at that date. Since the angle MOP is 45°, $MP = OP$, and persons dying at an age less than (14 years + MP days) must have had their fourteenth birthday subsequently to 30th June, 1966. Thus, of the NP deaths occurring at date P, MP will have had their fourteenth birthday after 30th June, 1966, and NM before 30th June, 1966. It follows that, if we let P move along over the whole year, the hatched area will represent the deaths of those persons who were in the age group 14 and under 15 at 30th June, 1966. This area is evidently half of the whole.

Similarly, if we consider the number of deaths aged 15 and under 16 during 1966–67, we obtain Fig. 15.4 below, and the hatched area (again half of the whole) represents the deaths of those persons who were in age group 14 and under 15 at 30th June, 1966, but who died aged 15 and under 16 during 1966–67.

In fact, deaths and ages are very nearly evenly distributed within single years, so that

Number aged 15 years and under 16 at 30th June, 1967
equals
Number aged 14 years and under 15 at 30th June, 1966
minus
{One-half of deaths during 1966–67 aged 14 years and under 15}
{One-half of deaths during 1966–67 aged 15 years and under 16}

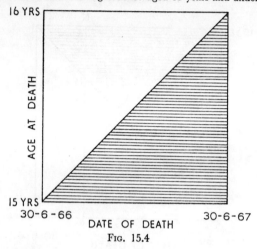

Fig. 15.4

This method can be used for all age groups except the first two. To obtain those aged 0 and under 1 at 30th June, 1967, we must take all the births over the year 1966–67 and subtract the deaths of those babies born during that year. We know how many infants under the age of 1 year died in 1966–67, but some of these will have been born in 1965–66. Deaths are not distributed evenly over the first year of life. In Australia, for example, the mean age at which infants die is about two months. In such a case[1] we can attribute about 85 per cent of infant deaths in 1966–67 to births of 1966–67 and the other 15 per cent to births of 1965–66. Thus 15 per cent of infant deaths in 1966–67 will already have been aged 0 years and under 1 at 30th June, 1966. Hence we have

Number aged 0 years and under 1 at 30th June, 1967
equals
Births during 1966–67
minus
85 per cent of deaths during 1966–67 aged 0 years and under 1

[1] The method given here requires modification according to the conditions of infant mortality in the country under consideration.

and

Number aged 1 year and under 2 at 30th June, 1967

equals

Number aged 0 years and under 1 at 30th June, 1966

minus

$\begin{Bmatrix} \text{15 per cent of deaths during 1966--67 aged 0 years and under 1} \\ \text{One-half of deaths during 1966--67 aged 1 year and under 2} \end{Bmatrix}$

To take net migration into account, we use the same reasoning and we shall have

Number aged 15 years and under 16 at 30th June, 1967

equals

Number aged 14 years and under 15 at 30th June, 1966

minus

$\begin{Bmatrix} \text{One-half of deaths during 1966--67 aged 14 years and under 15} \\ \text{One-half of deaths during 1966--67 aged 15 years and under 16} \end{Bmatrix}$

plus

$\begin{Bmatrix} \text{One-half of overseas arrivals during 1966--67 aged 14 years and under 15} \\ \text{One-half of overseas arrivals during 1966--67 aged 15 years and under 16} \end{Bmatrix}$

minus

$\begin{Bmatrix} \text{One-half of overseas departures during 1966--67 aged 14 years and under 15} \\ \text{One-half of overseas departures during 1966--67 aged 15 years and under 16} \end{Bmatrix}$

It will be observed that when the numbers in each age group are added through to give total population at 30th June, 1967, this will agree with the total obtained by adding natural increase and net migration for 1966–67 to the population at 30th June, 1966, as illustrated in Table 15.3 above. It should be also noted that while it is necessary to work here in single age groups, age distributions are frequently quoted in quinquennial age groups.

As an example of the above method, we estimate the age distribution of a population at 31st December, 1967, given the following data—

Age Last Birthday	Population 31/12/66	Deaths in 1967	Overseas Arrivals in 1967	Overseas Departures in 1967	Net Migration in 1967 (a)
Under 1 year	90,000	2,600	1,000	300	700
1 year	89,000	250	1,100	300	800
2 years	93,000	160	1,300	450	850
3 years	77,000	120	1,500	300	1,200
4 years	79,000	80	1,600	200	1,400
5 years	75,000	70	1,800	350	1,450
etc.	etc.	etc.	etc.	etc.	etc.

Births in 1967 = 95,000

(a) Overseas arrivals *less* overseas departures.

We shall have

Table 15.7

INTERCENSAL ESTIMATE OF AGE DISTRIBUTION OF A POPULATION

Population 31/12/66		Half Deaths in Given Age Group (2)	Half Deaths in Next Age Group (3)	Half Net Migrants in Given Age Group (4)	Half Net Migrants in Next Age Group (5)	Population 31/12/67	
Age last Birthday	Numbers (1)					Age last Birthday	Numbers (a) (6)
Births 1967 Under 1 year	95,000		2,210 (b)		350	Under 1 year	93,140
Under 1 year .	90,000	390 (c)	125	350	400	1 year	90,235
1 year .	89,000	125	80	400	425	2 years	89,620
2 years .	93,000	80	60	425	600	3 years	93,885
3 years .	77,000	60	40	600	700	4 years	78,200
4 years .	79,000	40	35	700	725	5 years	80,350
5 years .	75,000	etc.	etc.	etc.	etc.	etc.	etc.
etc.	etc.						

(a) Col. (6) = col. (1) − col. (2) − col. (3) + col (4) + col. (5).
(b) 85 per cent of infant deaths.
(c) 15 per cent of infant deaths.

Factors Determining the Age Distribution

The births, deaths and migrations of the past determine the current age distribution. It is important to realize that current age distributions are related in a quasi-mathematical fashion to past distributions and that future age distributions are related to current ones. We can

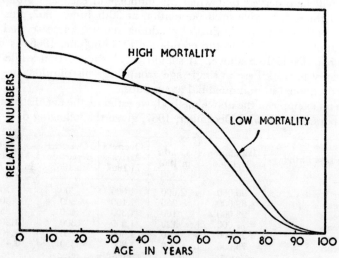

FIG. 15.5. AGE DISTRIBUTION FOR A CONSTANT STREAM OF BIRTHS

High mortality based on mortality of Australian males, 1881–90. Low mortality based on mortality of Australian males, 1960–62.

illustrate the way in which age distributions are built up by considering some hypothetical examples. These are illustrated in Figs. 15.5 to 15.7 which indicate the general shapes of the relative age distributions.

If the number of births per annum has been constant for a long time, in the absence of migration, the age distribution will depend on the incidence of mortality. The numbers in age groups will become progressively smaller the older the age group, as mortality takes its toll. The higher the rate of mortality the more rapidly will the numbers in the upper age groups tail off, and hence the younger the population on the average. This is pictured in Fig. 15.5.

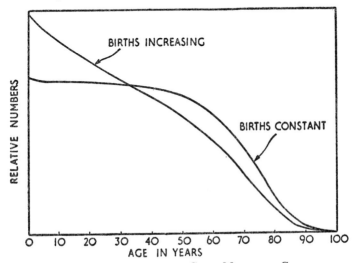

Fig. 15.6. Age Distribution for Given Mortality Conditions
Mortality based on mortality of Australian males, 1960–62. Increasing births are increasing at 10 per 1,000 per annum.

If the annual number of births has been steadily increasing over the past, the younger age groups will contain a relatively greater proportion of the population, since they will be the survivors of relatively greater numbers of births. The more rapid the rate of growth, the younger will be the population on the average. This is shown in Fig. 15.6.

If the annual number of births has been steadily decreasing over the past, the older age groups will be the survivors of relatively greater numbers of births. This will result in a tendency for the numbers in age groups to increase with age, but mortality will sooner or later reduce the older age groups, so that the age distribution will become humped, as in Fig. 15.7.

Wars and migrations introduce elements of irregularity into age distributions. The immediate effect of a war is to "bite" into the

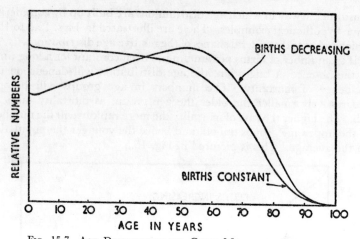

Fig. 15.7. Age Distribution for Given Mortality Conditions

Mortality based on mortality of Australian males, 1960-62. Decreasing births are decreasing at 10 per 1,000 per annum.

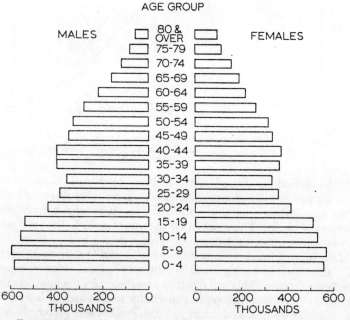

Fig. 15.8. Age Distribution of Males and Females, Australia, 30th June, 1966

Source: Commonwealth Bureau of Census and Statistics, Australia: *1966 Census Bulletins*, No. 9.3, pp. 6, 8 and No. 9.7, pp. 5, 7

numbers of males in the ages of from about 20 to 30 years. This bite will then move forward in the age distribution as time passes and the original 20 to 30 year age group grows older. The same holds for emigration, which usually is concentrated in the age groups of young active people. Immigration has the converse effect of adding a "lump" to the age distribution.

The customary diagrammatic method of illustrating sex and age distribution is shown in Fig. 15.8. The diagram is called a *population silhouette*. It is a histogram with the axes reversed.

The relatively small numbers in the age groups 25–29 years and 30–34 years reflect the substantial drop in the number of births which took place in the 1930s.

Factors Determining the Sex Distribution

The sex distribution of a population is measured by the *masculinity* of the population. This is defined as

$$\text{Masculinity} = \text{number of males per 100 females}$$

Masculinity can be measured for the whole population or for age groups. In the absence of wars and migration, the masculinity of a particular age group is determined by the masculinity of births and deaths. Masculinity of births is measured by the number of male births per 100 female births, and is usually about 105. But male mortality is heavier than female mortality, so that as age rises the males gradually lose their preponderance. In Australia, for example, other things being equal, the sexes would break even at about age 50 to 55, and thereafter the females would be preponderant. For the whole population, therefore, masculinity depends on the age distribution and hence on the rate of growth of the population. Normally the sexes are fairly evenly balanced. The immediate effect of wars is to reduce masculinity mainly in the 20 to 30 age group. As the years go by, this deficiency works its way up the age distribution. The immediate effect of immigration is to increase masculinity in the 20–50 age group, since migrants tend to be predominantly male. These considerations are illustrated in Table 15.8 on the next page.

Significance of Sex and Age Distribution

The sex and age distribution of a population is of considerable importance from both social and economic points of view, and it is subject to quite substantial changes with the passage of time. The significance of sex and age distribution is illustrated in Tables 15.9 to 15.11 below which refer to Australian experience. The first shows the highly abnormal sex distribution of Australia in the days of its early development and the continuous drop in masculinity which has been

Table 15.8

MASCULINITY OF THE AUSTRALIAN POPULATION
CENSUS 30TH JUNE, 1966

Age	Males per 100 Females	Comment
0–4	105	Masculinity at birth = 105
5–9	105	
10–14	105	
15–19	105	
20–24	105	
25–29	106	1948–1966 Migration
30–34	107	
35–39	108	
40–44	105	
45–49	103	1939–1945 War
50–54	102	
55–59	103	
60–64	98	
65–69	83	Heavier male mortality
70–74	72	
75 and over	62	
All Ages	101	

Source: *See* Table 15.6

taking place from pioneering days onwards. This has had a profound influence on society. The second shows how the ratio of active males to all males has changed over the past fifty years, due to improvements in longevity and to the fall and partial recovery in the birth rate. This ratio has an important effect on the standard of living of the country, since it influences the volume of production relative to the

Table 15.9

MASCULINITY OF THE AUSTRALIAN POPULATION
1820–1967

Year	Males per 100 Females
1820	244
1840	202
1860	140
1880	117
1900	111
1920	103
1940	102
1960	102
1967	102

Source: Commonwealth Bureau of Census and Statistics, Australia: *Demography Bulletin*, No. 67, 1949, pp. 150–153, and No. 78, 1960, pp. 170–17 and *Australian Demographic Review*, No. 248, p. 4

number of consumers in the economy. The third table shows how the ratio of reproductive females to all females has changed. The substantial change in the proportion of immature females has important implications for the reproductive capacity of the population in the future.

Table 15.10

AGE DISTRIBUTION OF MALES AT AUSTRALIAN CENSUSES

Age Group (years)	1891	1933	1966
	per cent	per cent	per cent
0–14: Young dependants	34·8	27·5	29·9
15–64: Active producers	62·0	66·1	63·0
65 and over: Old dependants	3·2	6·4	7·1
All Ages	100·0	100·0	100·0

Source: Commonwealth Bureau of Census and Statistics, Australia: *Census of the Commonwealth of Australia*, 30th June, 1933, *Statistician's Report*, p. 69 and Vol. I, p. 717; *1966 Census Bulletin*, No. 9.3, 1968, p. 3

Table 15.11

AGE DISTRIBUTION OF FEMALES AT AUSTRALIAN CENSUSES

Age Group (years)	1891	1933	1966
	per cent	per cent	per cent
0–14: Immature	39·4	27·4	28·9
15–44: Reproductive	47·1	47·5	41·2
45 and over: Sterile	13·5	25·1	29·9
All Ages	100·0	100·0	100·0

Source: Commonwealth Bureau of Census and Statistics, Australia: *Census of the Commonwealth of Australia*, 30th June, 1933, *Statistician's Report*, p. 70, and Vol. I, p. 717; *1966 Census Bulletin*, No. 9.3, 1968, p. 3

15.4. BIRTH AND DEATH RATES AND RATES OF INCREASE

Crude Birth Rate

The annual *crude birth rate* is defined—

$$\text{Crude birth rate} = \frac{\text{annual births}}{\text{annual mean population}} \times 1,000$$

In this measure the births are related to the *mean* population and not to the population at a particular date. The crude birth rate of a given year tells us at what rate births have augmented the population over the course of the year. For 1967 for Australia we have—

$$\text{Crude birth rate} = \frac{229{,}296}{11{,}817{,}413} \times 1{,}000 = 19 \cdot 40 \text{ per } 1{,}000 \text{ per annum}$$

The crude birth rate usually lies between 10 and 55 per 1,000. In Western European type countries it has fallen very considerably over the past eighty years, although it has risen quite appreciably during the past decade. For crude birth rates of various countries see Table 15.12, p. 422.

The level of the crude birth rate is determined by
 (i) the sex and age distribution of the population; and
 (ii) the fertility of the population, i.e. the average rate of child-bearing of females.

A relatively high crude birth rate can be recorded if the sex and age distribution is favourable even though fertility is low, e.g. countries with a relatively large proportion of population in the 15–50 years age groups will have a relatively high crude birth rate, other things being equal. We consider how to measure fertility in section 15.7 below.

Crude Death Rate

The annual *crude death rate* is defined—

$$\text{Crude death rate} = \frac{\text{annual deaths}}{\text{annual mean population}} \times 1{,}000$$

The crude death rate for a given year tells us at what rate deaths have depleted the population over the course of the year. We can calculate the crude death rate for males and females separately. For 1967 for Australia we have—

$$\text{Crude death rate (males)} = \frac{57{,}508}{5{,}950{,}220} \times 1{,}000$$

$$= 9 \cdot 66 \text{ per } 1{,}000 \text{ per annum}$$

$$\text{Crude death rate (females)} = \frac{45{,}195}{5{,}867{,}193} \times 1{,}000$$

$$= 7 \cdot 70 \text{ per } 1{,}000 \text{ per annum}$$

$$\text{Crude death rate (persons)} = \frac{102{,}703}{11{,}817{,}413} \times 1{,}000$$

$$= 8 \cdot 69 \text{ per } 1{,}000 \text{ per annum}$$

The crude death rate usually lies between 8 and 30 per 1,000. The female rate is generally lower than the male rate. In most countries crude death rates have fallen substantially over the past half century or so. For crude death rates of various countries, see Table 15.12, p. 422.

The level of the crude death rate is determined by
  (i) the sex and age distribution of the population; and
  (ii) the mortality of the population, i.e. the average longevity of the population.

An old population can exhibit a relatively high crude death rate even if longevity is high (i.e. mortality is low). We consider how to measure mortality in section 15.5 below.

Crude Rate of Natural Increase

The annual *crude rate of natural increase* is defined—

$$\begin{aligned}\text{Crude rate of natural increase} &= \frac{\text{annual natural increase}}{\text{annual mean population}} \times 1{,}000 \\ &= \frac{\text{annual births} - \text{annual deaths}}{\text{annual mean population}} \times 1{,}000 \\ &= \text{crude birth rate} - \text{crude death rate}\end{aligned}$$

The crude rate of natural increase for a given year tells us at what rate natural increase has augmented the population over the course of the year. For 1967 for Australia we have—

$$\begin{aligned}\text{Crude rate of natural increase} &= 19\cdot 40 - 8\cdot 69 \\ &= 10\cdot 71 \text{ per } 1{,}000 \text{ per annum}\end{aligned}$$

The crude rate of natural increase varies considerably from country to country. It can be negative (natural decrease, i.e. excess of deaths over births), but it is unlikely to be much higher than about 25 per 1,000. For crude rates of natural increase of various countries, see Table 15.12, p. 422. The level of the crude rate of natural increase is dependent on those factors determining the crude birth and death rates. A population with a relatively large proportion in the age group 15–50 years is likely to have a comparatively high crude rate of natural increase, other things being equal, for its crude birth rate will tend to be high and its crude death rate low.

Rate of Net Migration

The annual *rate of net migration* is defined—

$$\begin{aligned}\text{Rate of net migration} &= \frac{\text{overseas arrivals} - \text{overseas departures}}{\text{annual mean population}} \times 1{,}000 \\ &= \frac{\text{annual net migration}}{\text{annual mean population}} \times 1{,}000\end{aligned}$$

The rate of net migration for a given year tells us at what rate net migration has augmented the population over the course of the year. For 1967 for Australia we have—

$$\text{Rate of net migration} = \frac{637{,}170 - 545{,}261}{11{,}817{,}413} \times 1{,}000$$

$$= 7\cdot78 \text{ per } 1{,}000 \text{ per annum}$$

This rate varies a great deal according to world economic and political conditions and government policy. It can, of course, be negative.

Rate of Total Increase

The annual *rate of total increase* is defined—

$$\text{Rate of total increase}$$
$$= \frac{\text{annual total increase}}{\text{annual mean population}} \times 1{,}000$$
$$= \frac{\text{annual natural increase} + \text{annual net migration}}{\text{annual mean population}} \times 1{,}000$$
$$= \text{crude rate of natural increase} + \text{rate of net migration}$$

The rate of total increase for a given year tells us at what rate the population has increased over the year. For 1967 for Australia we have

$$\text{Rate of total increase} = 10\cdot71 + 7\cdot78$$
$$= 18\cdot49 \text{ per } 1{,}000 \text{ per annum}$$

Rate of Population Growth

The annual *rate of population growth* is defined—

$$\text{Rate of population growth}$$
$$= \frac{\text{annual total increase}}{\text{population at beginning of year}} \times 1{,}000$$
$$= \frac{\text{population at end of year} - \text{population at beginning of year}}{\text{population at beginning of year}} \times 1{,}000$$

The rate of population growth for a given year tells us by what proportion the population has increased over the year. Whereas the rate of total increase measures the *average rate* at which the population has been increasing over the year, the annual rate of population growth measures the *proportion* by which the population has grown over the year. Numerically the difference between the two rates is always small and the rate of population growth is always a little greater than the rate of total increase. Mathematically speaking, the rate of population growth is the rate of total increase compounded continuously over a year. For 1967 for Australia we have—

$$\text{Rate of population growth} = \frac{11{,}928{,}889 - 11{,}710{,}387}{11{,}710{,}387} \times 1{,}000$$

$$= 18\cdot 66 \text{ per 1,000 per annum,}$$

$$\text{or } 1\cdot 866 \text{ per cent}$$

Sometimes we want to ascertain the average annual rate of population growth which has obtained over a period of years. Suppose we know that the population at the end of the year 0 is P_0 and the population at the end of year t (i.e. t years later) is P_t, and ask what the average annual rate of growth has been. The question is: At what annual rate of growth will P_0 accumulate to P_t after t years? This is simply a problem in compound interest. If i is the rate of growth in decimals then we have—

$$P_t = P_0(1 + i)^t$$

Taking logarithms,

$$\log P_t = \log P_0 + t \log (1 + i)$$

i.e.

$$\log (1 + i) = \frac{\log P_t - \log P_0}{t}$$

and i can be obtained from the anti-logarithm.

For example—
The population of Australia at 30th June, 1947, was 7,579,358. By 30th June, 1965, it had grown to 11,333,119. What was the average annual rate of population growth over this period?

We have $\quad 11{,}333{,}119 = 7{,}579{,}358 (1 + i)^{18}$

hence $\quad \log 11{,}333{,}119 = \log 7{,}579{,}358 + 18 \log (1 + i)$

i.e. $\quad \log (1 + i) = \dfrac{7\cdot 054345 - 6\cdot 879632}{18}$

$$= 0\cdot 009706$$

and $\quad 1 + i = 1\cdot 0226$

The required average rate is 22·6 per 1,000 per annum.

Infant Mortality Rate

The *infant mortality rate* is defined—

$$\text{Infant mortality rate} = \frac{\text{annual infant deaths}}{\text{annual births}} \times 1{,}000$$

The rate approximately measures for a given year the chances of a birth failing to survive one year of life. Still-births are not included in the infant deaths. The rate can be calculated for males and females separately. For 1967 for Australia we have—

Infant mortality rate (males) $= \dfrac{2{,}421}{117{,}680} \times 1{,}000$

$= 20 \cdot 57$ per 1,000 per annum

Infant mortality rate (females) $= \dfrac{1{,}766}{111{,}616} \times 1{,}000$

$= 15 \cdot 82$ per 1,000 per annum

Infant mortality rate (all births) $= \dfrac{4{,}187}{229{,}296} \times 1{,}000$

$= 18 \cdot 26$ per 1,000 per annum

The infant mortality rate varies considerably according to time and place. In countries with high standards of maternal and infant welfare it is as low as 15 to 20 per 1,000, but in some underdeveloped countries it is still well over 100 per 1,000. In many countries it has fallen spectacularly over the past sixty years or so. The male rate is appreciably higher than the female rate. For infant mortality rates of various countries, see Table 15.12, p. 422.

Ideally we wish to measure by the infant mortality rate the probability[1] at birth of a child dying before attaining the age of one year, according to the mortality conditions of a given year. The deaths of infants in any one year can be of children born in the given year or in the preceding year and are influenced by the trend of births between the two years. For example, if last year's births were much lower than this year's, this year's infant deaths will be lower than otherwise, and if we relate this year's infant deaths to this year's births, we shall understate the true infant mortality position for this year. One way of handling this problem[2] is to regard the probability at birth of a child surviving to the age of one year as being compounded of the probability of a child surviving to the end of the calendar year in which it was born and the probability of a child who has survived to the end of the calendar year in which it was born surviving to the age of one. When both these component probabilities are based on the mortality conditions of a given year we have—

$$p_0 = \dfrac{B^t - D^{t(t)}}{B^t} \times \dfrac{B^{t-1} - D^{t-1(t-1)} - D^{t(t-1)}}{B^{t-1} - D^{t-1(t-1)}}$$

[1] For the definition of "probability," see section 6.2, p. 82.
[2] See Valaoras, V. G.: "Refined Rates for Infant and Childhood Mortality," *Population Studies*, Vol. IV, No. 3, 1950, p. 253.

where p_0 is the probability at birth of surviving to the age of one year according to the mortality conditions of year t,
B^t is the number of births in year t,
$D^{t(t)}$ is the number of infant deaths in year t of children born in year (t).

The numerator of the first term in the above formula is the number of births in year t surviving to the end of year t, and that of the second term is the number of births surviving from the previous calendar year who survive to the age of one in year t.

Since p_0 is the probability at birth of surviving to the age of one, we can write

$$q_0 = 1 - p_0$$

where q_0 is the probability at birth of a child dying before attaining the age of one year.

In order to compute q_0 as defined above, it is necessary to have the infant deaths of a particular year classified according to the calendar year of birth of the infants. Frequently this information is not available. In such a situation, it is often possible to estimate a *separation factor* by which the infant deaths in any one year can be separated into those of children born in the current and those of children born in the preceding year. Evidently the proportion of infant deaths of a particular year which is due to births of the current year will be higher the lower the mean age at death of infants. This mean age is generally lower the lower the rate of infant mortality, since low rates of infant mortality are due more to low rates of mortality after the first month of life than within the first month. For Australia the separation factor is at present about 0·85. In 1966 and 1967 male births were 114,530 and 117,680 respectively, and male infant deaths were 2,328 and 2,421. Accordingly, the refined infant mortality rate for 1967 is—

$$q_0 = 1 - \left[\frac{117{,}680 - (0\cdot85)2{,}421}{117{,}680} \times \frac{114{,}530 - (0\cdot85)2{,}328 - (0\cdot15)2{,}421}{114{,}530 - (0\cdot85)2{,}328} \right]$$

$= 0\cdot02066$, or 20·66 per 1,000 per annum.

It will be noted that the crude rate is rather lower than the refined rate, due to the rise in the number of births occurring between 1966 and 1967.

The use of a separation factor is not entirely satisfactory, since the factor itself ought to be adjusted for changes in the rate of births. In principle, refined rates of infant mortality are to be preferred to crude rates, but, in practice, such procedures are most important for countries in which the mean age of infants at death is appreciably higher than in Australia. Furthermore, one would not place a great deal of reliance on the infant mortality rate for one year, but would prefer to calculate the rate for, say, three years combined, by relating three years' deaths to three years' births. In this case, in countries like Australia, about 95 per cent of the infant deaths taking place in the three-year period will be deaths of children born in the same period, and the crude measure can be regarded as fairly satisfactory. For example, for males for Australia 1960–62 we have—

$$\text{Infant mortality rate} = \frac{\text{infant deaths during 1960, 1, 2}}{\text{births during 1960, 1, 2}} \times 1{,}000$$

$$= \frac{2{,}651 + 2{,}690 + 2{,}790}{118{,}415 + 123{,}112 + 122{,}110} \times 1{,}000$$

$$= 22 \cdot 36 \text{ per 1,000 per annum}$$

This will closely approximate[1] the probability at birth of a child dying within the first year of life according to the mortality conditions of 1960–62, i.e. on the average out of every 1,000 male births subjected to the mortality conditions of Australia, 1960–62, 22 will not survive to the end of their first year of life.

Neo-natal mortality Rate

The *neo-natal mortality rate* is defined—

Neo-natal mortality rate
$$= \frac{\text{annual deaths of infants under the age of 1 month}}{\text{annual births}} \times 1{,}000$$

The rate measures for a given year the chance of a birth failing to survive one month of life. For 1967 for Australia we have:

$$\text{Neo-natal mortality rate (males)} = \frac{1{,}795}{117{,}680} \times 1{,}000$$

$$= 15 \cdot 25 \text{ per 1,000 per annum}$$

$$\text{Neo-natal mortality rate (females)} = \frac{1{,}258}{111{,}616} \times 1{,}000$$

$$= 11 \cdot 27 \text{ per 1,000 per annum}$$

$$\text{Neo-natal mortality rate (all births)} = \frac{3{,}053}{229{,}296} \times 1{,}000$$

$$= 13 \cdot 31 \text{ per 1,000 per annum}$$

The neo-natal mortality rate is a component of the infant mortality

---

[1] Using a more complicated procedure, the Commonwealth of Australia Actuary has estimated this probability at 22·39 per 1,000. See Commonwealth Bureau of Census and Statistics, Australia: *Census of the Commonwealth of Australia, Australian Life Tables*, 1960–62, pp. 10, 25.

rate. For Australia, for example, it covers about 70 per cent of it—in other words, most infant deaths occur within the first month of life. In fact, the neo-natal mortality rate represents to a very large extent the hard core of infant mortality. Thus, over the past sixty years in Australia, the infant mortality rate has fallen from above 80 to about 18 per 1,000, whereas the neo-natal mortality rate has fallen from above 30 to about 13 per 1,000—the greatest part of this latter fall occurring in the past twenty years. Of the neo-natal deaths most occur within the first week of life, and the mean age of neo-natal deaths is only 4 or 5 days, hence almost all neo-natal deaths are derived from births occurring in the same year, and the neo-natal mortality rate is not subject to the same limitations as the uncorrected infant mortality rate.

Maternal Mortality Rate

The *maternal mortality rate* is defined—

$$\text{Maternal mortality rate} = \frac{\text{annual maternal deaths}}{\text{confinements}} \times 1,000$$

The rate measures the probability of a woman dying from childbirth. For 1967 for Australia we have—

$$\text{Maternal mortality rate} = \frac{53}{226,983} \times 1,000$$

$$= 0 \cdot 23 \text{ per } 1,000 \text{ per annum}$$

Like the infant mortality rate this rate depends largely on the standard of maternal welfare in the country under consideration.

Crude Marriage Rate

The *crude marriage rate* is defined—

$$\text{Crude marriage rate} = \frac{\text{annual marriages}}{\text{annual mean population}} \times 1,000$$

The crude marriage rate for a given year tells us at what rate marriages have been taking place over the year. For 1967 for Australia we have—

$$\text{Crude marriage rate} = \frac{100,000}{11,817,413} \times 1,000$$

$$= 8 \cdot 46 \text{ per } 1,000 \text{ per annum}$$

The level of the crude marriage rate is determined by
(i) the sex and age distribution of the population; and
(ii) the nuptiality of the population, i.e. the propensities of people to marry.

Hence a population which has an evenly-balanced sex distribution and which has a relatively high proportion of unmarried persons in the younger age groups will exhibit a high crude marriage rate, other things being equal. Techniques for measuring nuptiality are beyond the scope

Table 15.12

BIRTH AND DEATH RATES FOR SELECTED COUNTRIES
1920–24, 1935–39, 1958 and 1966

Country and Period	Crude Birth Rate per 1,000 p.a.	Crude Death Rate per 1,000 p.a.	Crude Rate of Natural Increase per 1,000 p.a.	Infant Mortality Rate per 1,000 p.a.
Australia—				
1920–24	24·4	9·8	14·6	61·0
1935–39	17·2	9·6	7·6	39·1
1958	22·6	8·5	14·1	20·5
1966	19·3	9·0	10·3	18·2
India—				
1920–24	33·0	26·8	6·2	184·2
1935–39	33·8	22·6	11·2	161·6
1951–61	41·7	22·8	18·9	139·0
Italy—				
1920–24	30·1	17·5	12·6	128·8
1935–39	23·2	13·9	9·3	102·7
1958	17·9	9·4	8·5	48·2
1966	18·9	9·5	9·4	34·3
Japan—				
1920–24	35·0	23·0	12·0	164·7
1935–39	29·2	17·4	11·8	110·4
1958	18·0	7·5	10·5	34·6
1966	13·7	6·8	6·9	19·3
Netherlands—				
1920–24	26·7	11·0	15·7	74·4
1935–39	20·3	8·7	11·6	37·4
1958	21·1	7·5	13·6	17·2
1966	19·2	8·1	11·1	14·7
Sweden—				
1920–24	20·3	12·4	7·9	61·4
1935–39	14·5	11·7	2·8	43·2
1958	14·2	9·6	4·6	15·8
1966	15·8	10·0	5·8	12·6
United Kingdom—				
1920–24	21·7	12·5	9·2	79·2
1935–39	15·3	12·2	3·1	58·5
1958	16·8	11·7	5·1	23·5
1966	17·9	11·8	6·1	19·6
United States—				
1920–24	22·8	12·0	10·8	76·7
1935–39	17·1	11·0	6·1	53·2
1958	24·3	9·5	14·8	26·9
1966	18·5	9·5	9·0	23·4

Source: United Nations:
Demographic Year Book, 1954, pp. 252, 516, 588; 1959, pp. 209–17, 547–53, 599–605; 1966, pp. 116–19; and *Population and Vital Statistics Report*, 1968, pp. 17, 22–24

Table 15.13

POPULATION, AND BIRTH, DEATH AND MARRIAGE RATES AUSTRALIA, 1861–1967

Period	Population at Middle of Period '000	Crude Birth Rate per 1,000 p.a.	Crude Death Rate per 1,000 p.a.	Crude Rate of Natural Increase per 1,000 p.a.	Rate of Net Migration per 1,000 p.a.	Rate of Population Growth per 1,000 p.a. (b)	Infant Mortality Rate per 1,000 p.a.	Crude Marriage Rate per 1,000 p.a.
1861–70	1,390	41.0	16.6	24.3	12.1	37.0	(c)	8.0
1871–80	1,898	36.2	15.7	20.5	10.0	30.8	120.8	7.2
1881–90	2,695	35.2	15.3	19.9	14.2	35.1	122.2	7.9
1891–1900	3,492	30.0	13.0	17.0	0.7	18.0	110.4	6.7
1901–10	4,033	26.5	11.3	15.3	1.0	16.3	87.3	7.5
1911–20	4,969	26.6	10.8	15.8	4.2	20.4	67.5	8.3
1921–30	6,003	22.4	9.4	13.0	5.2	18.5	54.9	7.8
1931–40	6,756	17.2	9.3 (a)	7.9	0.5	8.6	40.0	8.3
1941–50	7,430	21.8	9.9 (a)	12.0	4.8	16.5	31.1	9.9
1951–60	9,313	22.7	9.0	13.7	8.8	22.7	22.2	7.9
1967	11,810	19.4	8.7	10.7	7.8	18.7	18.3	8.5

(a) Exclusive of deaths of defence personnel, September, 1939 to 30th June, 1947.
(b) This column is not the sum of the two preceding columns. It expresses the rate of growth per annum over the period in relation to the population at the beginning of the period (see p. 417 above). Moreover, it is based on census figures, while the two preceding columns are based on annual registrations.
(c) Not available.

Source: Commonwealth Bureau of Census and Statistics, Australia: *Demography Bulletin*, No. 67, 1949, pp. 154–5; No. 78, 1960, pp. 12, 32, 51, 69, 152, 172; and *Australian Demographic Review*, No. 248, pp. 3–10

of this chapter, but they are analogous to those for measuring mortality.[1] It should be noted, however, that nuptiality varies according to economic conditions, and hence the crude marriage rate tends to be high in booms and low in slumps. In addition the crude marriage rate tends to increase in times of war.

Table 15.12 sets out some of the rates referred to above for selected countries. Table 15.13 shows more detail for Australia.

15.5. Measurement of Mortality

Our object is to find a method of expressing the level of mortality in a particular area during a given year or period of years. Suppose, for example, we wish to examine the mortality conditions to which Australian males were subject in the year 1967. The simplest way to summarize these mortality conditions is to calculate the age to which Australian males can, on the average, expect to live if they are subjected to 1967 mortality conditions. We can start with a hypothetical group of 1,000 male births (known as a *cohort* of births) and estimate the numbers which will survive to every age, if they are subjected to the mortality conditions under consideration. Then we can say, for example, that out of our initial 1,000, 974 will reach the age of 10, 965 the age of 20 and so on, and that the mean age at which they will die is 67·8 years. This is a simple and concise way of expressing the conditions of mortality (or, in other words, of longevity) under which Australian males lived in 1967. The technique suggested above gives rise to what is known as a *life table*.

Construction of a Life Table

The data required for the construction of a life table are the age distribution of the population during the period under consideration and the number of deaths occurring during that period, distributed according to age. We start with a hypothetical cohort of births. These births are, of course, exactly aged 0. We wish to trace through the number of these births which will survive to the various ages. We write l_x for the number of the initial cohort of births which will survive to the exact age of x. For short we call l_x survivors at age x. If we start with 1,000 births, then clearly $l_0 = 1,000$. As x gets larger, l_x gets smaller, until finally, when $x = 100$ years or so, $l_x = 0$. We also write p_x for the *probability at age x of surviving one year to reach age $(x + 1)$*. By this we mean that if we have, say, 1,000 males aged x exactly, we should expect, on the average, $1,000 \times p_x$ of them to reach their next birthday, if they were subject to the mortality conditions under consideration. All the p_x's are fractions less than unity. If we start with l_0 new-born

[1] See, for example, Glass, D. V.: *Population Policies and Movements in Europe* (Oxford University Press 1940), Appendix, pp. 399–405.

babies, the number which will survive to the exact age of 1 year will be $l_1 = l_0 \times p_0$, the number which will survive to the exact age of 2 years will be $l_2 = l_1 \times p_1$, and so on. In general then, we have—
$$l_{x+1} = l_x \times p_x$$
It follows that in order to calculate the l_x's we must calculate the p_x's.

We have already touched on p_0 in discussing the infant mortality rate. It was pointed out that this rate was an approximation to the probability at birth of dying within one year. Since one must either die or survive we must have—
$$p_0 = 1 - \text{infant mortality rate}$$
where the infant mortality rate is expressed in decimals.

The subsequent p_x's can be computed as follows. For any age group x years and under $(x + 1)$, we know for the actual population the mean number of males for the year under consideration. Let this number be designated P_x. Similarly, we know the actual number of deaths in the various age groups which took place during the year under consideration. For the age group x years and under $(x + 1)$, let this be designated D_x. With these data we define *specific mortality rates*. The specific mortality rate for the age group x years and under $(x + 1)$ is the ratio of the annual number of deaths in that age group to the mean population in that age group, that is
$$m_x = \frac{D_x}{P_x}$$
where m_x is the specific mortality rate for the age group x years and under $(x + 1)$. This rate represents that rate at which persons of a particular age group are dying, throughout the year under consideration. The specific mortality rates represent the basic conditions of mortality and are readily calculable.

If from our initial hypothetical group of births l_0, there are l_x survivors at age x and l_{x+1} at age $(x + 1)$, the mean number of survivors in the age group x years and under $(x + 1)$ will be $\frac{1}{2}(l_x + l_{x+1})$, provided that the deaths take place evenly over the year. The number of deaths in this age group will be given by $(l_x - l_{x+1})$. If the l_x values are to reflect the mortality conditions of the population, this number of deaths must be equal to the number which would take place by applying to the mean number in the age group the relevant specific mortality rate. Hence we must have
$$l_x - l_{x+1} = \tfrac{1}{2}(l_x + l_{x+1}) \times m_x$$
but
$$l_{x+1} = l_x \times p_x$$
hence
$$1 - p_x = \tfrac{1}{2}(1 + p_x) \times m_x$$
i.e.
$$p_x = \frac{1 - \tfrac{1}{2}m_x}{1 + \tfrac{1}{2}m_x}$$

Consequently, given the m_x's we can readily calculate the p_x's and hence the l_x's. Actually we can calculate the p_x's direct from the raw data, for

$$p_x = \frac{1 - \frac{1}{2}\frac{D_x}{P_x}}{1 + \frac{1}{2}\frac{D_x}{P_x}} = \frac{P_x - \frac{1}{2}D_x}{P_x + \frac{1}{2}D_x}$$

It is possible to interpret the above formula for p_x in a fairly simple way by assuming as an approximation that all those in the age group x and under $(x + 1)$ years in the actual population are exactly aged $(x + \frac{1}{2})$ years at the middle of the year under consideration. Then P_x would represent the number aged $(x + \frac{1}{2})$ years at the middle of the year. Since D_x is the number of those aged x and under $(x + 1)$ who die over the course of the year, then, provided that the deaths take place evenly over the year, half the deaths will occur between the beginning and the middle of the year and half will occur between the middle and the end of the year. Consequently, $P_x + \frac{1}{2}D_x$ will be the number of males aged exactly x at the beginning of the year, and $P_x - \frac{1}{2}D_x$ will be the number of them who have survived to the end of the year to attain the exact age of $(x + 1)$. It follows that the probability at age x of surviving one year will be given, as above, by

$$p_x = \frac{P_x - \frac{1}{2}D_x}{P_x + \frac{1}{2}D_x}$$

In practice the calculation of the l_x's will be rather tedious, for the p_x's will have to be calculated for about 100 ages and then successively multiplied. A great simplification can be achieved (without much loss in accuracy) by working in quinquennial age groups, i.e. by using population and deaths classified into five-year groups and calculating every fifth l_x only. We write $_5p_x$ for the probability at age x of surviving five years to age $(x + 5)$, so that

$$l_{x+5} = l_x \times {_5p_x}$$

The specific mortality rates are calculated for quinquennial age groups, so that m_x now stands for the rate at which persons aged x and under $(x + 5)$ die in any given year. Following the preceding argument we must have

$$l_x - l_{x+5} = \tfrac{1}{2}(l_x + l_{x+5}) \times 5m_x$$

The multiplier is $5m_x$, because m_x is an annual rate, and in surviving from age x to $(x + 5)$, five years must be lived through. Accordingly,

$$_5p_x = \frac{1 - 2\tfrac{1}{2}m_x}{1 + 2\tfrac{1}{2}m_x} = \frac{P_x - 2\tfrac{1}{2}D_x}{P_x + 2\tfrac{1}{2}D_x}$$

where P_x and D_x here refer to the population and deaths respectively in the age group x years and under $(x + 5)$. The abridged quinquennial

DEMOGRAPHY

method is satisfactory so long as deaths are fairly evenly distributed over the quinquennial age groups. This condition holds sufficiently well except for the first few years of life.[1] Accordingly, we calculate l_1, as before, from

$$l_1 = l_0 \times p_0$$

Table 15.14

CONSTRUCTION OF ABRIDGED LIFE TABLE—AUSTRALIAN MALES, 1967

CALCULATION OF PROBABILITIES OF SURVIVING

Age Group (years) x	Population at 30/6/67 P_x	Deaths During 1967 D_x	$P_x - 2\tfrac{1}{2}D_x$	$P_x + 2\tfrac{1}{2}D_x$	$_5p_x = \dfrac{P_x - 2\tfrac{1}{2}D_x}{P_x + 2\tfrac{1}{2}D_x}$
Births	117,700	2,421 (a)			0·97943 (b)
1–4	474,900	462	473,976 (c)	475,824 (d)	0·99612
5–9	613,300	268	612,630	613,970	0·99782
10–14	568,200	236	567,610	568,790	0·99793
15–19	538,300	698	536,555	540,045	0·99354
20–24	477,000	806	474,985	479,015	0·99159
25–29	399,400	609	397,878	400,922	0·99241
30–34	364,300	597	362,808	365,792	0·99184
35–39	392,200	907	389,932	394,468	0·98850
40–44	399,700	1,560	395,800	403,600	0·98067
45–49	355,300	2,232	349,720	360,880	0·96908
50–54	325,300	3,279	317,102	333,498	0·95084
55–59	283,800	4,860	271,650	295,950	0·91789
60–64	222,000	6,055	206,862	237,138	0·87233
65–69	165,400	7,227	147,332	183,468	0·80304
70–74	115,100	7,576	96,160	134,040	0·71740
75–79	80,400	7,817	60,858	99,942	0·60893
80–84	39,000	5,683	24,792	53,208	0·46594
85–89	14,200	2,988	6,730	21,670	0·31057
90–94	3,300	1,005	788	5,812	0·13558
95–99	500	222	−55 (e)	1,055	0·00000

(a) Infant deaths

(b) $1 - \dfrac{\text{infant deaths}}{\text{births}}$

(c) $P_1 - 2D_1$

(d) $P_1 + 2D_1$

(e) The negative value occurs because deaths in this age group are not spread evenly over the five year range but are concentrated near the lower end. This value is treated as if it were zero, giving a probability of surviving from 95 to 100 years of zero.

Source: Commonwealth Bureau of Census and Statistics, Australia: *Australian Demographic Review*, No. 245, p. 4; 246, p. 8 and 247, p. 4

Note. So that as up-to-date an example as possible may be given, preliminary estimates of the age distribution of the population have been used. These estimates are rounded to the nearest thousand, and consequently the ratios in the above table are not accurate to the number of decimal places quoted. The figures for the distribution of population over 75 years and for that of deaths over 85 years are the authors' estimates.

[1] Also except for the last few years of life, but the tail end of the life table is relatively unimportant.

and l_5 from
$$l_5 = l_1 \times {}_4p_1$$

where
$${}_4p_1 = \frac{1 - 2m_1}{1 + 2m_1} = \frac{P_1 - 2D_1}{P_1 + 2D_1}$$

and P_1 and D_1 here refer to population and deaths respectively in the age group 1 year and under 5.

The survivor values, l_x, constitute the principal element of the life table. As an example, an abridged life table for Australian males for 1967 is constructed in Tables 15.14 and 15.15.

Table 15.15

CONSTRUCTION OF ABRIDGED LIFE TABLE—AUSTRALIAN MALES, 1967
CALCULATION OF SURVIVORS

Exact Age (years) x	Survivors (a) l_x
0	(b) 1,000
1	1,000 × 0·97943 = 979
5	979·43 × 0·99612 = 976
10	975·63 × 0·99782 = 974
15	973·50 × 0·99793 = 971
20	971·49 × 0·99354 = 965
25	965·21 × 0·99159 = 957
30	957·09 × 0·99241 = 950
35	949·83 × 0·99184 = 942
40	942·08 × 0·98850 = 931
45	931·25 × 0·98067 = 913
50	913·24 × 0·96908 = 885
55	885·00 × 0·95084 = 842
60	841·50 × 0·91789 = 772
65	772·40 × 0·87233 = 674
70	673·79 × 0·80304 = 541
75	541·08 × 0·71740 = 388
80	388·17 × 0·60893 = 236
85	236·37 × 0·46594 = 110
90	110·13 × 0·31057 = 34
95	34·20 × 0·13558 = 5
100	4·64 × 0·00000 = 0

(a) $l_{x+5} = l_x \times {}_5p_x$
(b) Figures hereunder are taken from last column in Table 15.14

The interpretation of the l_x column is as follows. Starting with 1,000 male births subject to the conditions of mortality affecting males in Australia, 1967, as expressed in the specific mortality rates, we trace through their survivorship. Twenty-one die before reaching age 1, and 979 survive; of these 3 die before reaching age 5, and 976 survive, and

DEMOGRAPHY

so on. From the l_x column we can readily derive the probability at birth of surviving to an exact age, by dividing l_x by l_0, e.g. the probability at birth of surviving to age 25 is $957 \div 1,000 = 0.957$. On the other hand, the $_5p_x$ column tells us the proportion of males aged x which we can expect to reach age $(x + 5)$, if they are subject to the given mortality conditions. The l_x column gives a picture of the mortality conditions under consideration, but it would be useful to have a summary measure. Our original aim was to calculate the age to which Australian males, on the average, can expect to live if they are subject to the mortality conditions of 1967. This calculation can now be made.

Mean Expectation of Life at Birth

The mean age to which, on the average, the 1,000 male births can expect to live is called the *mean expectation of life at birth* and is, of course, the same thing as the mean age at death of these 1,000 births. This can be calculated by computing the mean age at which the deaths in the life table occur. If we write d_x for the deaths occurring in the life table between the age of x and $x + 5$ we shall have

$$d_x = l_x - l_{x+5}$$

Thus, 21 will die between age 0 and 1; 3 between ages 1 and 5; 2 between ages 5 and 10, and so on. As has already been pointed out (p. 406), deaths in the first year of life have a mean age at death of about 0·15 years, but for subsequent years deaths may be assumed to be evenly distributed. Hence we shall have 21 deaths centred at 0·15 years, 3 at 3 years, 2 at $7\frac{1}{2}$ years, etc. The mean age at death of the whole original 1,000 births can thus be readily calculated from this frequency distribution of age at death. In our example it comes to 67·8 years.

However, it is preferable to proceed in another way, which leads ultimately to more information. We may ask: How many years on the average will the original 1,000 births live? As has been pointed out above, this is precisely the same as the mean age at which they will die. How many years will the original 1,000 births live between them? First, how many years will they live in their first year of life? The mean age of those who die in that year is approximately 0·15, so that the whole l_0 will live on the average for 0·15 years and l_1 of them will live a further 0·85 years. Hence

$$L_0 = 0{\cdot}15\, l_0 + 0{\cdot}85\, l_1$$

where L_0 is the number of years lived by the original l_0 births between the ages of 0 and 1. The mean age at death between the ages of 1 and 5 can be taken as 3 years; hence all the l_1 will live on the average for 2 years between the ages of 1 and 5, and l_5 of them will live another 2 years. Hence

$$L_1 = 2l_1 + 2l_5$$

where L_1 is the number of years lived by the original l_0 births between the ages of 1 and 5. Similarly

$$L_5 = 2\tfrac{1}{2} l_5 + 2\tfrac{1}{2} l_{10}$$

and, in general

$$L_x = 2\tfrac{1}{2}(l_x + l_{x+5})$$

where L_x is the number of years lived by the original l_0 births between the ages of x and $(x + 5)$. The total number of years lived by the original births will then be given by $L_0 + L_1 + L_5 + \ldots = \Sigma L_x$, and the number per birth by $\dfrac{\Sigma L_x}{l_0}$. This is the mean expectation of life at birth and is written

$$e_0^0 = \frac{\Sigma L_x}{l_0}$$

The mean expectation of life at birth is the best overall measure of the mortality of a population at a given time. For Australian males in 1967 the mean expectation of life at birth was 67·8 years. This means that on the average, new-born male babies can expect to live 67·8 years, if throughout their lifetime they are subject to the same mortality conditions as operated in Australia in 1967.

We can calculate the mean expectation of life at ages other than at birth. Thus the mean expectation of life at age 25 is the number of years which the original births will live after attaining the age of 25 divided by the number who survive to 25. In general, we can write the mean expectation of life at age x

$$e_x^0 = \frac{L_x + L_{x+5} + \ldots}{l_x} = \frac{\sum\limits_{i=x} L_i}{l_x}$$

This mean expectation tells us the number of years a person can expect to live after attaining the age of x. For example, for Australian males in 1967 the mean expectation of life at age 25 was 45·5 years, so that having attained age 25 a man can expect to live to 70·5 years if he is subjected for the remainder of his life to the 1967 conditions. Thus if persons attain the age x, they can expect, on the average, to live to $(x + e_x^0)$ years. Clearly $(x + e_x^0)$ will be greater than e_0^0, for once people survive x years they will have overcome the hazards of life in those x years and can expect to live to a riper age than those just starting out on life. Table 15.16 sets out the calculation of e_0^0 and e_x^0.

Table 15.16

CONSTRUCTION OF ABRIDGED LIFE TABLE—AUSTRALIAN MALES, 1967
CALCULATION OF MEAN EXPECTATION OF LIFE

Exact Age (years) x	Survivors l_x (a)	L_x (b)	$\sum_{i=x} L_i$ (c)	$e_x^0 = \dfrac{\sum_{i=x} L_i}{l_x}$ (d)
0	1,000		67,782	67·8
		982		
1	979		66,800	68·2
		3,910		
5	976		62,890	64·4
		4,875		
10	974		58,015	59·6
		4,862·5		
15	971		53,152·5	54·7
		4,840		
20	965		48,312·5	50·1
		4,805		
25	957		43,507·5	45·5
		4,767·5		
30	950		38,740	40·8
		4,730		
35	942		34,010	36·1
		4,682·5		
40	931		29,327·5	31·5
		4,610		
45	913		24,717·5	27·1
		4,495		
50	885		20,222·5	22·9
		4,317·5		
55	842		15,905	18·9
		4,035		
60	772		11,870	15·4
		3,615		
65	674		8,255	12·2
		3,037·5		
70	541		5,217·5	9·6
		2,322·5		
75	388		2,895	7·5
		1,560		
80	236		1,335	5·7
		865		
85	110		470	4·3
		360		
90	34		110	3·2
		97·5		
95	5		12·5	2·5
		12·5		
100	0		0	0·0

(a) Taken from last column in Table 15.15.
(b) The number of years lived by the original 1,000 births between the stated age and the next higher age.
(c) The number of years lived by the original 1,000 births after attaining the stated age. It is the L_x column accumulated upwards.
(d) The quotient of the $\sum L_i$ and l_x columns.

We may now set out our abridged life table in detail:

Table 15.17
ABRIDGED LIFE TABLE—AUSTRALIAN MALES, 1967

Exact Age (years) x	l_x (a)	d_x (b)	p_x (c)	q_x (d)	e_x^0 (e)
0	1,000	21	0·979	0·021	67·8
1	979	3	0·996	0·004	68·2
5	976	2	0·998	0·002	64·4
10	974	3	0·998	0·002	59·6
15	971	6	0·994	0·006	54·7
20	965	8	0·992	0·008	50·1
25	957	7	0·992	0·008	45·5
30	950	8	0·992	0·008	40·8
35	942	11	0·989	0·011	36·1
40	931	18	0·981	0·019	31·5
45	913	28	0·969	0·031	27·1
50	885	43	0·951	0·049	22·9
55	842	70	0·918	0·082	18·9
60	772	98	0·872	0·128	15·4
65	674	133	0·803	0·197	12·2
70	541	153	0·717	0·283	9·6
75	388	152	0·609	0·391	7·5
80	236	126	0·466	0·534	5·7
85	110	76	0·311	0·689	4·3
90	34	29	0·136	0·864	3·2
95	5	5	0·000	1·000	2·5
100	0				0

(a) Taken from last column in Table 15.15. The l_x column gives the number of survivors of the original l_0 births. The ratios l_x/l_0 (i.e. $l_x \div 1,000$) give the probabilities at birth of surviving to the exact age of x.

(b) By definition $d_x = l_x - l_{x+5}$. The d_x column gives the number of deaths of the original l_0 births between the age of x and the next highest age shown. The ratios d_x/l_0 (i.e. $d_x \div 1,000$) give the probabilities at birth of dying between the age of x and the next highest age shown.

(c) Taken from last column in Table 15.14. The p_x column gives the probabilities at age x of surviving to the next highest age shown. By definition $l_{x+5} = l_x \times {}_5p_x$.

(d) By definition $q_x = 1 - p_x = d_x/l_x$. The q_x column gives the probabilities at age x of dying between the age of x and the next highest age shown.

(e) Taken from last column in Table 15.16. The e_x^0 column gives the mean expectation of life at age x when that age has been exactly attained.

For most demographic purposes an abridged life table is sufficiently accurate. It loses some accuracy in the later years of survivorship, since the assumption that deaths are spread out evenly over the quinquennial age groups becomes less valid. Complete life tables which are necessary for actuarial purposes are calculated in single age groups and involve complicated techniques for smoothing out errors in the basic data.

In most countries complete life tables are calculated whenever there is a census. Abridged life tables can be quite rapidly calculated and are particularly useful for intercensal years for which complete tables do not exist. Generally, however, it is unwise to base a life table on one year's data. The Commonwealth of Australia Actuary, for example, has been in the habit of using three years' data, e.g. the latest Official Life Table is for 1960–62. Mortality experience of one year alone may be rather unreliable. The year 1966 for Australia is a good example. In that year, mortality in Australia was, in fact, somewhat higher than in either 1965 or 1967. This can be seen by examining the crude death rate which for males was 8·79, 8·99, 8·69 per 1,000 for the years 1965, 1966 and 1967 respectively.

Life tables are drawn up for given areas at certain times. They are based on the mortality conditions existing in the given area at the time under consideration. These mortality conditions are expressed in detail in the specific mortality rates. Given these rates, which can readily be computed for the population with which we are concerned, we can calculate the life table.

15.6. Applications of the Life Table

Computation of Probabilities of Surviving and Dying

It is possible to derive from the life table the probabilities of more or less complex events happening. Thus the probability at birth of dying between the ages of 20 and 30 years will be given by the number of the original births dying between the ages of 20 and 30, divided by the number of original births, i.e. in our example by $15 \div 1,000 = 0.015$. This probability tells us that on the average out of every 1,000 male births subject to the Australian mortality of 1967, 15 will die between the ages of 20 and 30 years. Similarly, if we wish to estimate the probability at birth of dying between the ages of 25 and 33 years, we shall have to estimate the number of the original births dying between these ages. We know that 7 die between 25 and 30 and to ascertain those dying between 30 and 33 we must interpolate within the age group 30 to 35, i.e. the required number will be estimated as $3/5 \times 8 = 5$. The required probability will then be $(7 + 5) \div 1,000 = 0.012$.

Again, if we want the probability at age 20 of dying before reaching 30, we shall calculate the ratio of deaths between 20 and 30 to survivors at age 20, i.e. $15 \div 965 = 0\cdot016$. This means that out of every 1,000 males who reach 20, on the average 16 will die before reaching 30 years. Finally, we may consider a case involving two lives. What is the probability that a man aged 30 and a man aged 50 will both survive 10 years? The answer for our example is $\frac{931}{950} \times \frac{772}{885} = 0\cdot855$. This holds, of course, only if the probabilities of surviving of the two men are independent.

Life Assurance

The life table was developed primarily to meet the needs of life assurance offices. It forms the basis for calculations of the premiums necessary to purchase various amounts of life assurance. Actually these calculations are very complex, but the underlying principles are simple. For example, according to 1967 mortality, what annual premium would an Australian have to pay on a full-life policy worth $100 if his life was assured at birth, assuming that the assurance office earns no income on its funds? Let the premium be x per annum. Since a male on the average can be expected to live 67·8 years, over his lifetime a man will have paid $x \times 67\cdot8$ in premiums. This will have to equal the value of the policy, $100, so that the annual premium must be $100 \div 67\cdot8 = \$1\cdot47$. If the policy was taken out at age 20, then total premiums paid will be $x \times 50\cdot1$, for 50·1 years is the expectation of life at 20, and the annual premium must be $2. If the policy were an endowment policy, taken out at, say, 20 and payable at 30 or prior death, we should proceed in a somewhat different fashion. From Table 15·16, we know that the 965 survivors at age 20 live $4{,}805 + 4{,}767\cdot5$ years between them between the ages of 20 and 30. Consequently, on the average a total of $x \times (9{,}572\cdot5 \div 965)$ premiums will be collected, and hence the annual premium must be $100 \div 9\cdot920 = \$10\cdot08$.

Mortality Due to Specific Causes[1]

It is of interest to know the relative importance of various causes of death. Data are usually available showing deaths distributed according to specific causes, and a percentage distribution can be readily calculated. Such a distribution is not, however, very satisfactory, since causes differ in their age-incidence. For example, the younger groups are susceptible to transport accidents, whereas heart disease attacks the old age groups. Consequently a young population will have a relatively high proportion of accidental deaths and a

[1] See Kuczynski, R. R.: *The Measurement of Population Growth* (Sidgwick and Jackson, 1935), pp. 194–5.

relatively low one of deaths from heart disease. What we really want to ascertain are the probabilities at given ages of dying from a specific cause. With the help of a life table these can be calculated.

As an illustration there are worked out below the probabilities at birth of dying from transport accidents and heart disease for Australian males in 1967. First, it is necessary to calculate the ratios which deaths from these causes bear to total deaths in the various age groups. Decennial groups will suffice for this purpose.

Table 15.18

DEATHS FROM MOTOR VEHICLE ACCIDENTS AND HEART DISEASE
AUSTRALIAN MALES, 1967

Age Group (years)	All Causes (a)	Motor Vehicle Accidents		Heart Disease	
		Numbers (a)	Proportion of Deaths from All Causes	Numbers (a)	Proportion of Deaths from All Causes
0–5	2,884	74	0·02566	9	0·00312
5–14	504	126	0·25000	6	0·01190
15–24	1,504	807	0·53657	30	0·01995
25–34	1,206	343	0·28441	91	0·07546
35–44	2,468	293	0·11872	634	0·25689
45–54	5,512	263	0·04771	2,329	0·42253
55–64	10,917	244	0·02235	5,079	0·46524
65–74	14,806	186	0·01256	6,472	0·43712
75 and over	17,707	152	0·00858	7,225	0·40803
All Ages	57,508	2,488	0·04326	21,875	0·38038

(*a*) Twelve deaths from all causes, two from heart disease and one from motor vehicle accidents, age not stated, have been allocated proportionately.

Source: Commonwealth Bureau of Census and Statistics, Australia: *Australian Demographic Review*, No. 247, pp. 22, 24

From the life table we know that out of the 1,000 original births 24 will die between ages 0 and 5. From the above we can estimate that of these deaths $24 \times 0·02566 = 0·62$ will be due to transport accidents and $24 \times 0·00312 = 0·07$ to heart disease. If we do this for all age groups, we can readily calculate the numbers of the original 1,000 births which will die due to these specific causes. This is done in Table 15.19.

From the table we see that, according to the Australian mortality conditions of 1967, out of the 1,000 original births 31 will ultimately die from transport accidents. Accordingly, the probability at birth of a male dying from transport accidents is 0·031. Likewise, the probability at birth of dying from heart disease is 0·400. These compare with the actual proportions of deaths from these causes in 1967, shown in Table 15.18 as 0·043 and 0·380 respectively. If we look at these

Table 15.19
Calculation of Probabilities at Birth of Dying from Motor Vehicle Accidents and Heart Disease, Australian Males, 1967

Age Group (years)	Deaths of Original 1,000 Births in Life Table (1)	Proportion due to Motor Vehicle Accidents (2)	Numbers of Deaths in Life Table due to Motor Vehicle Accidents (3)	Proportion due to Heart Disease (4)	Number of Deaths in Life Table due to Heart Disease (5)
0–5	24	0·02566	0·62	0·00312	0·07
5–14	5	0·25000	1·25	0·01190	0·06
15–24	14	0·53657	7·51	0·01995	0·28
25–34	15	0·28441	4·27	0·07546	1·13
35–44	29	0·11872	3·44	0·25689	7·45
45–54	71	0·04771	3·39	0·42253	30·00
55–64	168	0·02235	3·75	0·46524	78·16
65–74	286	0·01256	3·59	0·43712	125·02
75 and over	388	0·00858	3·33	0·40803	158·32
All ages	1,000		31·15		400·49

Col. (1): From third column of Table 15.17 above.
Cols. (2) and (4): From Table 15.18 above.
Col. (3) = col. (1) × col. (2).
Col. (5) = col. (1) × col. (4).

latter figures only, we might be inclined to say that the incidence of heart disease was 9 times that of transport accidents, whereas in fact a man is about 13 times more likely to die from the former than from the latter cause over his lifetime. This apparent discrepancy is due to the relatively high percentage of the Australian population in the younger age groups. This conflict occurs because the two sets of figures measure different things, the one measures the actual incidence of the causes in 1967, the other the probability at birth of a man dying from the causes throughout his lifetime.

It is important to realize that a rise, say, in the probability at birth of a male dying from heart disease over his lifetime does not necessarily indicate that males are becoming more susceptible to death from that cause. Since people must die from one cause or another sooner or later, a fall in the rates at which males of given ages die from certain diseases must, through increasing the number surviving those diseases, result in an increase in the proportion dying from other diseases, even though the rates, at which males of given ages die from those other diseases, remain constant. It is the complex of such age-specific rates (i.e. number of deaths of males aged x from a certain cause per 1,000 of the population of males aged x) to which we must look if we wish to examine the susceptibility of people to death from certain causes.

Stationary Population

So far we have viewed the life table as the life history of an initial group of births subject to given mortality conditions. We have taken 1,000 births and traced out how many of these will survive to various ages. Now, suppose that 1,000 births take place every year, spread out evenly over the year, and that these births are subject to the mortality conditions under consideration. These births will result in the building up of a population. We could take a count of that population at the end of any year, say. If we do so, how many people shall we find in the age group x years and under $(x + 5)$? Take a particular age group, say 50 years and under 55 in the year 2000. Those in that age group at the 31st December, 2000 must have been born during 1950, 1949, 1948, 1947 or 1946. Of the 5,000 births during these years, $5l_{50}$ will survive to attain the exact age of 50 during the years 1996–2000, and $5(l_{50} - l_{55})$ will die between the exact ages of 50 and 55 during the years 1996–2005. By the end of the year 2000, half of these latter deaths will have taken place, so that the number in the age group as at 31st December, 2000 will be $5l_{50} - 2\frac{1}{2}(l_{50} - l_{55}) = 2\frac{1}{2}(l_{50} + l_{55})$. This number will be the same whether or not the count is made at 31st December, 2000, or any other year, provided only that there have been 1,000 births per annum in the past and they have been subject to the given mortality conditions.

We can generalize the above argument by saying that a population built up from a constant stream of births per annum subject to given mortality conditions will have $2\frac{1}{2}(l_x + l_{x+5})$ persons in the age group x years and under $(x + 5)$, whenever a count of the population is made. Since the population will have the same number in each age group whenever measured, it must contain the same total number of persons, i.e. it must be *stationary* in numbers. Consequently, if there are 1,000 births per annum there will also be 1,000 deaths per annum.

The *stationary population* of a particular life table is the population which would be built up if there were 1,000 births per annum continuously subject to the mortality conditions of which the life table is an expression. The age distribution of the stationary population for the age group x years and under $(x + 5)$ is given by $2\frac{1}{2}(l_x + l_{x+5})$, which is in fact the L_x which we have already calculated and which represents the number of years lived by an original 1,000 births between the ages of x years and $(x + 5)$. Consequently, the age distribution of the stationary population in quinquennial age groups is defined by—

$$L_x = 2\frac{1}{2}(l_x + l_{x+5})$$

Thus we can use a life table to indicate either the life history of a group of births or the age distribution of the population which would be built up from a constant stream of births, provided the births (in both cases) are subject to the given mortality conditions.

The concept of a stationary population can perhaps best be appreciated by considering a hypothetical single age life table—

Age Group (years)	Population as at End of			
	Year 0	Year 1	Year 2	Year 3
0 and under 1	973	973	973	973
1 ,, ,, 2	966	966	966	966
2 ,, ,, 3	964	964	964	964
3 ,, ,, 4	962	962	962	962
4 ,, ,, 5	961	961	961	961
.	.	.	.	.
All Ages	66,070	66,070	66,070	66,070

During year 3, for example, there are 1,000 births; of these 973 are surviving in the age group 0 and under 1 at the end of the year. In the age group 1 and under 2 there are 966 persons being survivors of the 973 in the preceding age group in the previous year, and so on. Furthermore, in year 3 total deaths are given by infant deaths (1,000 − 973) plus the depletions of the age groups from the preceding year (973 − 966 and 966 − 964, etc.). Clearly, total deaths must add to 1,000.

Referring back to Table 15.16 containing the L_x values, we see that 1,000 male births per annum would ultimately result in a total stationary population of 67,782, if they were subject to 1967 mortality. Moreover, the numbers in each group would be stationary; for example, there would always be 4,317 males in the age group 50 and under 55, whenever a count was made.

Comparisons of Mortality Conditions

The mean expectation of life at birth is the best general overall index of mortality. It varies considerably according to place and time, ranging between about 20 and 70 years. In most countries it has risen steadily over the past half-century or so, largely due to the decline in infantile mortality. The female expectation of life is usually higher than the male, except where maternal mortality is high. The tables below set out the mean expectation of life at birth for various countries.

Table 15.20

MEAN EXPECTATION OF LIFE AT BIRTH—AUSTRALIA, 1881–1962

Period	Males	Females
	(years)	(years)
1881–90	47·20	50·84
1891–1900	51·06	54·76
1901–10	55·20	58·84
1920–22	59·15	63·31
1932–34	63·48	67·14
1946–48	66·07	70·63
1953–55	67·14	72·75
1960–62	67·92	74·18

Source: Commonwealth Bureau of Census and Statistics, Australia: *Year Book*, No. 46, 1960, p. 352; No. 53, 1967, p. 234

Table 15.21

MEAN EXPECTATION OF LIFE AT BIRTH FOR SELECTED COUNTRIES

Country and Period	Males	Females
	(years)	(years)
India—		
1921–31	26·91	26·56
1941–50	32·45	31·66
1951–60	41·89	40·55
Italy—		
1921–22	49·27	50·75
1930–32	53·76	56·00
1954–57	65·75	70·02
1960–62	67·24	72·27
Japan—		
1921–25	42·06	43·20
1935–36	46·92	49·63
1958	64·98	69·58
1965	67·73	72·95
Netherlands—		
1921–30	61·9	63·5
1931–40	65·7	67·2
1953–55	71·0	73·9
1961–65	71·1	75·9
Sweden—		
1921–30	60·97	63·16
1931–40	63·76	66·13
1956	70·92	74·35
1961–65	71·60	75·70
United Kingdom (*a*)—		
1920–22	55·62	59·58
1930–32	58·74	62·88
1958	67·95	73·69
1963–65	68·10	74·20
United States—		
1919–21	55·50	57·40
1939–41	61·60	65·89
1958	66·40	72·70
1965	66·80	73·70

(*a*) England and Wales

Source: United Nations: *Demographic Year Book*, 1959, pp. 640–4 and 1966, pp. 116–19

When a more detailed comparison of mortality is required, comparison of the probabilities of dying within, say, 5 or 10 years at different ages, or of the mean expectation of life at different ages, or of the number of survivors at different ages is useful. Direct comparison can also be made of specific mortality rates for different age groups. Tables 15.22 to 15.25 give examples.

Table 15.22

PROBABILITY OF DYING WITHIN 10 YEARS AT STATED AGES ($_{10}q_x$)—AUSTRALIAN MALES

Exact Age (years)	1881–90	1901–10	1920–22	1932–34	1946–48	1960–62	1960–62 as Ratio to 1881–90
0	0.203	0.134	0.106	0.068	0.044	0.029	0.14
10	0.039	0.025	0.019	0.015	0.011	0.009	0.23
20	0.080	0.043	0.034	0.024	0.017	0.015	0.19
30	0.091	0.061	0.046	0.034	0.023	0.020	0.22
40	0.133	0.101	0.080	0.063	0.054	0.047	0.35
50	0.218	0.168	0.147	0.137	0.136	0.125	0.57
60	0.372	0.326	0.301	0.284	0.297	0.291	0.78
70	0.618	0.626	0.580	0.556	0.564	0.551	0.89
80	0.873	0.885	0.885	0.868	0.862	0.846	0.97

Source: Derived from Commonwealth Bureau of Census and Statistics, Australia: *Demography Bulletin*, No. 64, 1946, p. 180, No. 74, 1956, p. 204; No. 77, 1959, p. 193 and No. 83, 1965, p. 146

Table 15.23

EXPECTATION OF LIFE AT STATED AGES (e_x^0) AUSTRALIAN MALES

Exact Age (years)	1881–90	1901–10	1920–22	1932–34	1946–48	1960–62	1960–62 as Ratio to 1881–90
0	47.2	55.2	59.1	63.5	66.1	67.9	1.44
10	48.9	53.5	56.0	58.0	59.0	59.9	1.22
20	40.6	44.7	47.0	48.8	49.6	50.4	1.24
30	33.6	36.5	38.4	39.9	40.4	41.1	1.22
40	26.5	28.6	30.1	31.1	31.2	31.8	1.20
50	19.7	21.2	22.2	22.8	22.7	23.1	1.17
60	13.8	14.3	15.1	15.6	15.4	15.6	1.13
70	8.8	8.7	9.3	9.6	9.6	9.8	1.11
80	5.1	5.0	5.0	5.2	5.4	5.6	1.10

Source: Commonwealth Bureau of Census and Statistics, Australia: *Demography Bulletin*, No. 64, 1946, p. 180; No. 77, 1959, p. 196 and No. 83, 1965, p. 147

The crude death rate is sometimes used as an index of mortality. As such it can be extremely misleading. As has already been pointed out, it depends not only on mortality conditions but also on the age

Table 15.24
Number of 1,000 Births Surviving at Stated Ages (l_x)
Australian Males

Exact Age (years)	1881–90	1901–10	1920–22	1932–34	1946–48	1960–62	1960–62 as Ratio to 1881–90
0	1,000	1,000	1,000	1,000	1,000	1,000	1·00
10	797	866	894	932	956	971	1·22
20	766	845	877	918	946	962	1·26
30	705	808	847	896	930	947	1·34
40	641	759	808	865	908	929	1·45
50	556	682	743	811	859	885	1·59
60	435	568	634	700	743	775	1·78
70	273	383	443	501	522	549	2·01
80	104	143	186	222	228	247	2·38
90	13	17	21	29	31	38	2·92

Source: Commonwealth Bureau of Census and Statistics, Australia: *Demography Bulletin*, No. 64, 1946, p. 179; No. 74, 1956, p. 203; No. 77, 1959, p. 193; and No. 83, pp. 145–61

Table 15·25
Male Specific Mortality Rates
New South Wales and South Australia, 1960–62

Age Group (years)	Specific Mortality Rates per 1,000 per annum		Ratio of S.A. to N.S.W.
	New South Wales	South Australia	
Under 1	23·6	22·9	0·97
1–4	1·2	1·2	1·00
5–9	0·5	0·5	1·00
10–14	0·5	0·5	1·00
15–19	1·2	1·1	0·92
20–24	1·6	1·4	0·88
25–29	1·5	1·6	1·07
30–34	1·7	1·6	0·94
35–39	2·4	2·1	0·88
40–44	3·9	3·4	0·87
45–49	6·4	5·4	0·84
50–54	10·8	9·2	0·85
55–59	17·8	15·7	0·88
60–64	28·0	24·4	0·87
65–69	43·4	37·8	0·87
70–74	65·9	59·5	0·90
75–79	97·9	88·7	0·91
80–84	148·5	135·8	0·92
85 and over	249·8	227·4	0·91

Source: Commonwealth Bureau of Census and Statistics, Australia: *Year Book*, No. 52, 1966, p. 253

distribution of the population. In particular, a population with a preponderance in the ages 15 to 50 will tend to have a lower crude death rate than otherwise, and a growing population will tend to have a lower one than a declining population. The age distribution of a population depends on its past history of births, deaths and migrations. Hence, the particular value which the crude death rate takes is largely fortuitous. However, the mean expectation of life at birth is clear of such a charge. For example, the crude death rate in Australia, 1901–10, was 11·3 per 1,000 per annum, and in France for 1928–33 it was 16·3 per 1,000, but in both countries at those periods the mean expectation of life was approximately 57 years. The higher crude death rate for France was only a reflection of an older population, and no indication that one's chances of longevity were less in France.[1]

There is also a measure closely connected with e_0^0, which is sometimes used. If given mortality conditions are applied to a constant stream of births, they give rise to a stationary population uniquely determined by these mortality conditions themselves (see p. 437 above). The death rate of such a population is known as the *true death rate*, and it can be used as an index of mortality. The total stationary population is given by ΣL_x. If it has been built up from 1,000 births per annum, there will be 1,000 deaths per annum. Accordingly,

$$\text{True death rate} = \frac{1,000}{\Sigma L_x} \times 1,000$$

Since $e_0^0 = \frac{\Sigma L_x}{1,000}$, the true death rate is simply the reciprocal of the mean expectation of life. It is the crude death rate of the stationary population and is unaffected by the age distribution of the actual population. For 1967 for Australian males we have—

$$\text{True death rate} = \frac{1,000}{67,782} \times 1,000 = 14\cdot75 \text{ per annum}$$

The true death rate is sometimes used instead of the mean expectation of life when comparing mortality conditions. But the concept behind e_0^0 would seem to be simpler, and for that reason e_0^0 should be preferred. However, the true death rate can be very useful in interpreting the crude death rate. For example, in 1967 Australian males had a crude death rate of 9·66 per 1,000 per annum, whereas the true death rate was 14·75. Past growth of and past higher mortality in the Australian population had produced by 1967 an age distribution younger than that of the stationary population implied in the mortality conditions of which the true death rate of 14·75 is an expression. This accounts for the difference in the two death rates. How can we expect

---

[1] A similar example is given in section 1.3, pp. 4–5 above.

the crude death rate to move in the future? If mortality remains at about the 1967 level, and if the population ultimately becomes stationary, the actual age distribution will become older. Consequently the crude death rate must rise ultimately to 14·75 per 1,000. Such a rise would not indicate any increase in mortality. The crude death rate will, however, remain lower than the true death rate if the population is increasing, for then the actual population will be younger than the stationary one implied in the 1967 mortality conditions, and the more rapidly the population is increasing the lower will be the crude death rate relatively to the true rate. On the other hand, it is unlikely that the Australian population will increase rapidly for ever. This implies that the crude death rate will rise in the future unless mortality falls substantially. Such a rise should not cause concern, for it does not in itself mean any deterioration in mortality conditions. In fact, if the Australian population did become stationary, mortality would have to fall beyond all reasonable hopes to maintain the crude death rate at 9·66 per 1,000. For such a death rate in a stationary population implies a mean expectation of life at birth of 1,000 ÷ 9·66 = 103·5 years.

Population Projections

Since the persons in the age group x and under $(x + 5)$ at a particular date are the survivors of those aged $(x - 5)$ and under x five years earlier, it is possible to project forward age groups by applying to existing age groups appropriate *survivorship ratios*. Thus, suppose in 1967 it is required to estimate the number of males aged 60 and under

Table 15.26

PROJECTING AGE GROUPS

At 30th June, 1967		Survivorship Ratio (b)	At 30th June, 1977	
Age Group (years)	Number (a)		Age Group (years)	Number (c)
50 and under 55	325,300	$\frac{3,615}{4,317\cdot 5} = 0\cdot 8373$	60 and under 65	272,400
55 ,, ,, 60	283,800	$\frac{3,037\cdot 5}{4,035} = 0\cdot 7528$	65 ,, ,, 70	213,600
60 ,, ,, 65	222,000	$\frac{2,322\cdot 5}{3,615} = 0\cdot 6425$	70 ,, ,, 75	142,600
Total . .	831,100		Total . .	628,600

(a) Source: Commonwealth Bureau of Census and Statistics, Australia: *Australian Demographic Review*, No. 246, p. 8.
(b) From L_x column of Table 15.16 above.
(c) The number at 30th June, 1967, multiplied by the appropriate survivorship ratio.

75 at 30th June, 1977, in Australia. These males will be the survivors of those aged 50 and under 65 at 30th June, 1967. From the 1967 life table (Table 15.16) we know that the number of males aged 50 and under 55 in the stationary population which would result from a constant stream of 1,000 births per annum is 4,317·5. In ten years' time these will be aged 60 and under 65, and will have been reduced in number to 3,615. Hence 3,615 ÷ 4,317·5 of the 50–54 age group will survive ten years to become the 60–64 group, provided they are subject to 1967 mortality conditions. Accordingly we proceed as given in Table 15.26.

Such a population projection will be accurate provided that there are no radical changes in mortality conditions and that migration is negligible. Of course, we might guess at probable migration and add it in. The example illustrates the projection into the future only of already existing age groups. Projection of the whole age distribution is much more difficult, and much less reliable, since some of the future age groups will not yet be in existence and future births must be estimated. It is possible to make such estimates and also to allow for changes in mortality, but these questions will not be dealt with here.[1]

Limitations of the Life Table

The life table suffers from one important limitation. Whilst it reflects the actual mortality experience of a given year or period of years, it does not reflect the actual mortality experience of any group of births. Thus the births of 1900 are subject in their first year to the mortality conditions of 1900–01, in their second year to those of 1901–02, in their third year to those of 1902–03, and so on. Since mortality is selective in the biological sense, relatively low rates for the higher age groups in, say, 1967 may be due to the selective force of higher than current mortality rates in the earlier ages of those higher age groups. Consequently it is by no means certain that the current mortality rates of 1967 could apply throughout to a generation born in 1967. Although we say that the mean expectation of life at birth for Australian males is now about 67·8 years, this does not mean that persons born now will, on the average, actually live to that age; it only means that if persons born now were subject to existing mortality conditions *throughout their whole lives*, that is the age to which they could expect to live on the average. Consequently, the life table and all the measures associated with it are strictly *hypothetical*. This does not prevent the life table being very useful, nor does it prevent the mean expectation of life at birth being the best available index of overall mortality, but it is a limitation which should always be borne in mind.

[1] See Cox, P. R.: *Demography* (Cambridge University Press, 1959), Chap. 9, p. 204; United Nations: *Methods for Population Projections by Sex and Age* (New York, 1956).

DEMOGRAPHY

15.7. MEASUREMENT OF FERTILITY

In demography the term *fertility* refers to the actual production of children. Fertility must be distinguished from *fecundity* which refers to the capacity to bear children. Fecundity sets an upper limit to fertility. The simplest way to summarize the fertility conditions of a particular area during a given period is to calculate the mean number of children which females living right through their child-bearing period will (on the average) bear, if they are subject to the fertility conditions holding in the particular area during the given period. Such a measure is known as the *total fertility rate*. The full child-bearing period of a female is usually taken to be the span from 15 years of age to 50.

Total Fertility Rate

The data required for the calculation of the total fertility rate are the female age distribution and the number of births according to the age of mother for the period under consideration. Starting with, say, 1,000 females aged 15 years exactly, our object is to find out the number of children they will have if they all live right through to 50 and are subject to the particular fertility conditions. Consider the fertility conditions for Australia in 1967, for example. At 30th June, 1967, there were 104,800 females aged 15 and under 16 years, 103,800 aged 16 and under 17 years, 103,000 aged 17 and under 18 years, and so on. During 1967 there were 450, 1,861 and 4,419 births respectively to females in the above three age groups. If we start with 1,000 females exactly aged 15 and subject them to 1967 Australian fertility conditions, we shall expect them to have $450/104,800 \times 1,000 = 4$ births during the year in which they pass from 15 to 16 years old, $1,861/103,800 \times 1,000 = 18$ births during the year in which they pass from 16 to 17 years old, and $4,419/103,000 \times 1,000 = 43$ births during the year in which they pass from 17 to 18 years old, and so on. At the exact age of 15 they will have borne 0 children, by the age of 16 they will have borne 4 children, by the age of 17 they will have borne $4 + 18 = 22$ children, by the age of 18 they will have borne $4 + 18 + 43 = 65$ children, and so on. If we continue this process, we can readily ascertain the total number of children which the 1,000 females will bear over their whole child-bearing period.

We now define the *specific fertility rate* for the age group x years and under $(x + 1)$—

Specific fertility rate

$$= \frac{\text{annual births to females aged } x \text{ and under } (x + 1)}{\text{mean number of females aged } x \text{ and under } (x + 1)} \times 1,000$$

The specific fertility rate for a particular age group is the rate per

1,000 per annum at which the females in that age group produce offspring. It follows that the sum of the specific fertility rates for the age groups between 15 and 50 will give the total number of children born to 1,000 females living right through their child-bearing period and subject to the given fertility conditions as expressed in the specific fertility rates. The mean number of children born per female will be given by dividing this total by 1,000, and this is the total fertility rate. Accordingly the total fertility rate is the mean number of children which a female aged 15 can expect to bear if she lives until at least the age of 50, provided that she is subject to the given fertility conditions over the whole of her child-bearing period. The total fertility rate for a particular area during a given period is a summary measure of the fertility conditions operating in that area during that period.

In order to calculate the total fertility rate we shall have to calculate 35 specific fertility rates and then add them. In practice, we can shorten this procedure by working in quinquennial age groups. We define the specific fertility rate for group x years and under $(x + 5)$—

Specific fertility rate
$$= \frac{\text{annual births to females aged } x \text{ and under } (x + 5)}{\text{mean number of females aged } x \text{ and under } (x + 5)} \times 1{,}000$$

Such a specific fertility rate is the rate per 1,000 per annum at which the females in the particular age group produce offspring. Thus, for

Table 15.27
CALCULATION OF TOTAL FERTILITY RATE—AUSTRALIA, 1967

Age Group (years)	Female Population at 30th June, 1967 (1)	Births by Age of Mother (2)	Specific Fertility Rates per 1,000 per annum (d) (3)
15–19	513,200	24,828 (a)	48·38
20–24	453,500	77,448	170·78
25–29	373,600	69,111	184·99
30–34	341,500	35,099	102·78
35–39	362,000	17,322	47·85
40–44	378,600	5,110	13·50
45–49	346,000	378 (b)	1·09
Total		229,296 (c)	569·37

(a) Includes births to females under 15 years.
(b) Includes births to females over 50 years.
(c) Twenty-three births, age of mother not stated, have been allocated proportionately.
(d) Col. (3) = (col. (2) ÷ col. (1)) × 1,000.

Source: Commonwealth Bureau of Census and Statistics, Australia: *Australian Demographic Review*, No. 245, p. 8

Total fertility rate = (569·37 × 5) ÷ 1,000 = 2·847

Australia in 1967, there were 513,200 females aged 15–19 at 30th June, 1967. This figure must be used as an estimate of the mean number in the age group. These females gave birth to 24,828 children. Hence the specific fertility rate for the age group 15–19 was $24{,}828/513{,}200 \times 1{,}000 = 48\cdot38$ per 1,000 per annum. Accordingly, 1,000 females exactly aged 15 would by the time they reached 20 have borne $48\cdot38 \times 5 = 241\cdot90$ children. It is necessary to multiply by 5 since the specific fertility rate is a rate per annum and by the time the females reach the age of 20 they will have spent 5 years in the age group 15–19. It follows that, if we add the quinquennial specific fertility rates and multiply by 5, we shall have the total number of children which 1,000 females aged 15 will bear over their lifetimes. A calculation based on quinquennial age groups involves only one-fifth of the arithmetic of one based on single age groups and is very nearly as accurate. The calculation of the total fertility rate for Australia, 1967, is shown in Table 15.27.

In the table below is shown the number of births which 1,000 females will have borne by the time they reach certain ages. The table is in a sense analogous to a life table.

Table 15.28

Exact Age (years)	Total Births Produced per 1,000 Females Aged 15 by Stated Ages (a)
15	0
20	242
25	1,096
30	2,021
35	2,535
40	2,774
45	2,842
50	2,847

(a) Last column of Table 15.27 multiplied by 5 and accumulated downwards

A total fertility rate of 2·847 for Australia in 1967 means that on the average a female aged 15 could expect to produce 2·847 births over the course of her lifetime if she were subject to 1967 Australian fertility conditions, but *not* subject to mortality over her child-bearing period.

Gross Reproduction Rate

The total fertility rate refers to the number of children which a female can expect to produce. A more significant figure is the number of *female* children. For this will give an indication of the number of females which a female will produce over her lifetime to replace herself. The total fertility rate can be calculated in terms of female births only, by restricting the births in the specific fertility rates to female births. Such a calculation leads to a measure called the *gross reproduction rate*. The gross reproduction rate measures the mean number of female children which a female aged 15 can expect to bear if she lives right through the child-bearing period and is subject to the given fertility conditions. It follows that the gross reproduction rate measures the mean number of female children which will be born to a newly-born female who is subject to the given fertility conditions throughout her lifetime, but is *not* subject to mortality.

The gross reproduction rate can be calculated directly by the method referred to above, or very nearly as accurately by multiplying the total fertility rate by the proportion of all births which are female births. Thus

Gross reproduction rate

$$= \text{total fertility rate} \times \frac{\text{number of female births}}{\text{number of births}}$$

For Australia, for 1967, we have—

$$\text{Gross reproduction rate} = 2.847 \times \frac{111{,}616}{229{,}296}$$

$$= 1.386$$

The gross reproduction rate is used as a measure of the fertility of a population. As such it has certain limitations which are discussed later (see pp. 462–4 below). However, it is superior to the crude birth rate as such a measure, because, as has already been pointed out, the latter rate depends to some extent on the sex and age distribution of the population and this distribution may be favourable to a high or low number of births irrespective of current fertility conditions. On the other hand, the gross reproduction rate depends only on the fertility conditions current to the period under consideration. Furthermore, the gross reproduction rate corresponds more closely to what we mean by "fertility" than the crude birth rate, which is simply the rate at which the population is augmenting its numbers through births.

The gross reproduction rate is useful for comparing fertility in different areas or in the same area at different times. In the latter case, if the periods to be compared are close to one another (e.g. adjacent years), movements in the gross reproduction rates and the crude birth rate usually closely correspond, for the sex and age distribution cannot undergo very substantial changes over very short periods. However, for comparisons over longer periods the two measures can give different impressions, and the crude birth rate can be misleading. For example, in Australia in 1881 the crude birth rate was 35·26 per 1,000. In 1891 it was 34·47 per 1,000, a fall of only 2 per cent. However, over this period the proportion of females aged between 20 and 45 years in the total population rose appreciably from 14½ per cent to 16 per cent. Consequently for this reason alone the crude birth rate in 1891 would overstate the level of fertility relatively to that in 1881. In fact, the gross reproduction rates for the two years were 2·65 and 2·30 respectively, a fall of 13 per cent. The fall in fertility was very much greater than that revealed by the fall in the crude birth rate.

The gross reproduction rate could, in theory, range from 0 to about 5. Values of gross reproduction rates actually recorded range from 0·80 in Austria for 1933 to 3·65 in the Ukraine, 1896–97.[1] The corresponding total fertility rates were approximately 1·64 and 7·48. The gross reproduction rate in England and Wales, for example, was 1·31 in 1921, 0·92 in 1931 and 1·39 in 1963. The course of the gross reproduction rate in Australia is given in Table 15.29. The table reflects the decline in fertility which took place from the latter part of the nineteenth century until the Second World War and the subsequent rise. This movement is characteristic of most Western European type countries. Recent gross reproduction rates for a number of countries are given in Table 15.33 on p. 453 below.

Table 15.29

GROSS REPRODUCTION RATES
AUSTRALIA, 1881–1967

1881	.	.	.	2·65 (*a*)
1891	.	.	.	2·30 (*a*)
1901	.	.	.	1·74 (*a*)
1911	.	.	.	1·71
1921	.	.	.	1·51
1931	.	.	.	1·14
1941	.	.	.	1·15
1951	.	.	.	1·49
1961	.	.	.	1·73
1967	.	.	.	1·39

(*a*) Approximate only
Source: Commonwealth Bureau of Census and Statistics, Australia:
Year Book, No. 40, 1954, p. 390 and *Australian Demographic Review*, No. 245, p. 8

15.8. MEASUREMENT OF POPULATION REPLACEMENT

We have discussed the measurement of mortality and fertility separately. We now combine the two and discuss the measurement of population replacement and population growth. Apart from migration, population growth depends on the balance between mortality and fertility. The central questions in demography are: Would a given population replace itself if the fertility and mortality conditions currently holding continued to hold indefinitely? Would it grow or decline? At what rate? In answering such questions we leave aside migration,

[1] See Kuczynski: *op. cit.*, p. 126.

since we are interested in whether or not the population will grow from its own resources.

The gross reproduction rate measures the mean number of female children which will be born to a newly-born female who is subject to the given fertility conditions throughout her lifetime but is *not* subject to mortality. If the rate is exactly unity, this means that, on the average, each female will produce sufficient female offspring to replace herself *only* if she is not subject to mortality. Since, in fact, some females die before they reach the end of their child-bearing period and therefore before they will have produced the mean number of *one* female replacement, a gross reproduction rate of unity will not ensure population replacement. It follows from this that in order to measure population replacement we must calculate the mean number of female children which will be born to a newly-born female who is subject to the given fertility *and* mortality conditions throughout her lifetime. Such a measure is known as the *net reproduction rate*. The fertility conditions are expressed by the specific fertility rates and the mortality conditions by the specific mortality rates as combined into a life table.

Net Reproduction Rate

Consider the calculation of the net reproduction rate for Australia, 1967. We start with 1,000 female births. The relevant portion of the life table is given below.

Table 15.30
LIFE TABLE—AUSTRALIAN FEMALES, 1967

Exact Age (years) x	l_x	L_x
0	1,000	
15	978	4,885·0
20	976	4,872·5
25	973	4,855·0
30	969	4,835·0
35	965	4,807·5
40	958	4,762·5
45	947	4,690·0
50	929	

Source: Computed by the authors using same method as in Tables 15.14–15.16 above

Of the initial 1,000 female births, 978 will reach the age of 15. These 978 will live 4,885 years between them between the ages of 15 and 20.

DEMOGRAPHY 451

But we know from the specific fertility rates (see Table 15.27, p. 446) that females between the ages of 15 and 20 have children at the rate of 48·38 per 1,000 per annum. Consequently, in the five years between 15 and 20 we can expect our initial 1,000 births to produce 4,885 × 48·38/1,000 = 236·34 offspring. Again, between the ages of 20 and 25 our initial 1,000 births will live 4,872·5 years between them and can be expected to produce a further 4,872·5 × 170·78/1,000 = 832·13 offspring. Proceeding in this way, we can readily calculate the total number of offspring which will be produced. If we multiply this figure by the ratio of female births to all births, we get the total number of female offspring born to the initial 1,000 female births over their lifetimes. By dividing by 1,000, we get the mean number per female birth, and this is the net reproduction rate. The calculations are shown below.

Table 15.31
CALCULATION OF NET REPRODUCTION RATE—AUSTRALIA, 1967

Age Group (years)	Specific Fertility Rate per 1,000 per annum (1)	Years Lived by 1,000 Female Births (2)	Births per 1,000 Female Births (3)
15–19	48·38	4,885	236·34
20–24	170·78	4,872·5	832·13
25–29	184·99	4,855	898·13
30–34	102·78	4,835	496·94
35–39	47·85	4,807·5	230·04
40–44	13·50	4,762·5	64·29
45–49	1·09	4,690	5·11
Total			2,762·98

Col. (1): From Table 15.27 above.
Col. (2): From Table 15.30 above.
Col. (3) = (Col. (1) × Col. (2)) ÷ 1,000.

$$\text{Net reproduction rate} = \left(2{,}762 \cdot 98 \times \frac{111{,}616}{229{,}296}\right) \div 1{,}000 = 1 \cdot 345$$

In Table 15.32 on p. 452 is shown the number of births which 1,000 females will have borne by the time they reach certain ages.

The net reproduction rate measures the mean number of female children which will be born to a newly-born female who is subject to the given fertility and mortality conditions throughout her lifetime. If the net reproduction rate equals unity, then throughout her lifetime each female will, on the average, produce one female to replace herself. Under such conditions the population will *ultimately* become stationary.

Table 15.32

Exact Age (years)	Total Births Produced per 1,000 Female Births at Stated Ages (a)
15	0
20	236
25	1,068
30	1,967
35	2,464
40	2,694
45	2,758
50	2,763

(a) Last column of Table 15.31 accumulated downwards

We emphasize *ultimately* because the age distribution of the population may be such as to ensure a current excess of births over deaths (see p. 454 below), but ultimately, if the given fertility and mortality conditions continue, each female will just replace herself and the population must become stationary, for each female on her death will have left only one female to take her place. If the net reproduction rate is greater than unity, the population will ultimately increase if the given conditions continue, because each female on her death will have left more than one female to take her place. If the net reproduction rate is less than unity, the population will ultimately decrease.

The net reproduction rate is a measure of the *replacement potential* of the population. It tells us what will happen to the population ultimately if the given conditions upon which it has been based continue to hold. It is thus a measure which is both *potential* and *hypothetical*—potential, because it does not tell us whether the population is currently increasing or decreasing but whether the population will ultimately increase or decrease; and hypothetical, because it does not tell us what will actually happen but only what will happen if certain conditions continue to hold. It is in no sense a forecast of what can be anticipated, only a projection of what will happen if certain conditions continue to hold. Only in so far as it is likely that these conditions will continue to hold has it any value for prognostication.

The net reproduction rate could, in theory, range from 0 to about 5. Values of net reproduction rates as low as 0·67 (for Austria, 1933) and as high as 1·96 (for Ukraine, 1896–97) have been recorded.[1] The net reproduction rates of most Western European type countries had fallen from about 1·5 in the latter part of the nineteenth century to rather less than unity in the 1930s. Since the Second World War the rates have

---

[1] Kuczynski: *op. cit.*, p. 214.

risen. Thus, the rate for England and Wales was 1·12 in 1921, 0·81 in 1931 (it reached a minimum of 0·77 in 1933) and 1·34 in 1963. To-day there is a fairly wide dispersion of rates as Table 15.33 shows, and the rates vary a good deal from year to year (see p. 463 below). Table 15.33 gives recent gross and net reproduction rates for a number of countries. Table 15.34 sets out the course of the rates for Australia over a long period.

Table 15.33

GROSS AND NET REPRODUCTION RATES OF SELECTED COUNTRIES

Country	Period	Reproduction Rates	
		Gross	Net
Australia	1965 (a)	1·44	1·40
Belgium	1963	1·30	1·25
England and Wales	1963	1·39	1·34
France	1964	1·42	1·37
Ireland, Republic of	1963	1·90	1·74
Japan	1963	0·96	0·92
Netherlands	1964	1·55	1·50
New Zealand	1964	1·83	1·77
Norway	1963	1·42	1·38
Portugal	1960	1·52	1·33
Scotland	1964	1·49	1·44
Sweden	1964	1·21	1·18
Switzerland	1962	1·20	1·16
United States	1963	1·62	1·56

(a) Based on 1960–62 mortality experience. Excludes full-blood Aborigines.

Source: Commonwealth Bureau of Census and Statistics, Australia: *Year Book*, No. 53, 1967, p. 223

Table 15.34

NET REPRODUCTION RATES
AUSTRALIA, 1881–1967

1881	1·88 (a)
1891	1·73 (a)
1901	1·39 (a)
1911	1·42
1921	1·31
1931	1·04
1941	1·05
1951	1·41
1961	1·67
1967	1·34

(a) Approximate only

Source: Commonwealth Bureau of Census and Statistics, Australia:
Year Book, No. 40, 1954, p. 390 and *Australian Demographic Review*, No. 245, p. 8
Note that the above figures are based on the fertility *and* mortality of the year under consideration

If the net reproduction rate of a country is below unity, the population will ultimately decrease and die out unless fertility is raised and/or mortality lowered. The fact that the population is currently increasing is irrelevant. For example, in Australia in 1934 the net reproduction rate was 0·939, although there was an excess of births over deaths in that year at the rate of 7·07 per 1,000 of the population. But in 1934 there was a relatively high proportion of females in the child-bearing ages and a relatively low proportion of the population in the ages of high mortality. This state of affairs can be accounted for by the relatively high fertility and mortality of former years. With 1934 mortality and fertility conditions obtaining, this situation could not continue and the age distribution of the population would change, thus converting the current excess of births into an excess of deaths. This would happen inevitably, provided the 1934 conditions of mortality and fertility persisted into the future (see p. 458).

The margin between the gross and net reproduction rate indicates the extent to which mortality offsets fertility in population replacement. The higher the mortality in the ages below 50 the greater the margin. This can be seen in Table 15.33 above. The gross reproduction rate shows what the net reproduction rate would be if there were no female mortality up to the age of 50. If the gross reproduction rate is below unity, then no improvement in mortality can raise the net reproduction rate to replacement level. If the population is not to die out ultimately, fertility must be raised. The margin between the gross and net reproduction rate has narrowed greatly over the past seventy or eighty years with the great reductions in mortality. This is shown for Australia in Table 15.35.

Table 15.35

TOTAL FERTILITY, GROSS REPRODUCTION AND NET REPRODUCTION RATES—AUSTRALIA, 1881 AND 1967

	Total Fertility Rate	Gross Reproduction Rate	Net Reproduction Rate	Ratio of Net to Gross Reproduction Rate
1881	5·43	2·65	1·88	0·71
1967	2·85	1·39	1·34	0·96

If 1881 conditions of mortality had held in 1967, the net reproduction rate would have been only approximately 1·00.

True Rate of Natural Increase

The net reproduction rate can be interpreted as the ratio between two successive generations. For example, in Australia in 1967 the net reproduction rate was 1·345. If 1,000 female births will produce 1,345 female births over their lifetimes, then 1,345:1,000 represents the ratio of the second generation to the first. Consequently, the net reproduction rate tells us the rate at which the population will ultimately turn over per generation. Hence, ultimately each generation will see the Australian population multiplied by 1·345, or, in other words, ultimately the population will increase by 345 per 1,000 per generation.

Since the net reproduction rate gives a per generation rate of growth, if the length of a generation is known, a per annum rate can be obtained. If we write R for the net reproduction rate, T for the length of a generation in years, and r for the per annum rate of growth, we shall have—

$$(1 + r)^T = R$$

If we know T and R, we can readily find r (see section 15.4, p. 417).

The precise definition of the length of a generation is rather complex, but it can be interpreted approximately as the mean interval between the birth of a female and the birth of her children. Given an initial 1,000 female births this latter is the mean age at which these births are themselves confined. This mean age can be readily calculated. For Australia 1967 we have—

Table 15.36

CALCULATION OF APPROXIMATE LENGTH OF A GENERATION

Age Group (years)	Mid-point of Class Interval Deviated from Arbitrary Origin of 27·5, in Class Interval Units x'	Births per 1,000 Female Births (a) f	fx'
15–19	− 2	237	−474
20–24	− 1	832	−832
25–29	0	898	0
30–34	1	497	497
35–39	2	230	460
40–44	3	64	192
45–49	4	5	20
Total		2,763	−137

(a) From Table 15.31

Mean age at confinement $= 27\cdot5 + \left(\dfrac{-137}{2,763} \times 5\right)$

$= 27\cdot25$ years

It follows that we can now compute r approximately from the above formula. For 1967 for Australia, we have—

$$(1 + r)^{27 \cdot 25} \approx 1 \cdot 345$$

i.e.
$$\log (1 + r) \approx \frac{\log 1 \cdot 345}{27 \cdot 25}$$

$$\approx \frac{0 \cdot 12872}{27 \cdot 25}$$

$$\approx 0 \cdot 00472$$

i.e. $\qquad 1 + r \approx 1 \cdot 01093$

i.e. $\qquad r \approx 0 \cdot 01093$ or $10 \cdot 9$ per 1,000 per annum

It can be demonstrated mathematically that any population irrespective of its age distribution will, if subjected constantly to given fertility and mortality conditions, as expressed by the specific fertility and mortality rates, ultimately (in practice after about sixty to seventy years) produce a flow of births increasing (or decreasing) at a constant rate per annum.[1] It follows that, with given probabilities of surviving, the number in any age group will increase at the same constant rate per annum, and hence the total population will similarly increase. This rate of increase is known as the *true rate of natural increase* and

---

[1] Consider a population of any given age distribution at a particular time, provided only that it contains some individuals in the reproductive age groups. We limit our attention to females only. Assume that there is no migration and that the given fertility and mortality conditions obtain indefinitely into the future. We treat these conditions as continuous functions of age, representing them by $b(x)$ and $l(x)$, where $b(x)$ is the rate at which females produce female offspring at age x (specific fertility rates) and $l(x)$ is the probability at birth of a female surviving to the exact age of x (survivor column of the life table).

After the passage of about fifty years all females of reproductive age in the population will be the offspring of the original population. The number of such females of age x at time t will be given by $B(t - x)l(x)$, where $B(t - x)$ is the number of female births which took place at time $(t - x)$, i.e. x years ago. The number of female births to females of age x at time t will consequently be $B(t - x)l(x)b(x)$. It follows that the total number of female births taking place at time t will be given by—

$$B(t) = \int_0^\infty B(t - x)l(x)b(x)dx$$

This integral equation will be satisfied by a function of the form—

$$B(t) = \sum_{i=0}^{\infty} Q_i e^{r_i t}$$

where $e = 2 \cdot 71828$, the Q_i's are constants determined by the initial age distribution and the r_i's are the roots of the characteristic equation—

$$1 = \int_0^\infty e^{-rx} l(x)b(x)dx \qquad (*)$$

corresponds to the r we have defined above. The true rate of natural increase is the rate at which a population subjected to given fertility and mortality conditions will ultimately grow. Furthermore, since the numbers in all age groups will ultimately increase at the same rate, the relative age distribution of the population will ultimately become stable, i.e. there will be the same proportion of persons in each age group year in and year out. Such a population is called a *stable population*. A stationary population is a particular type of stable population, namely one in which the rate of increase is zero.

The interpretations of the true rate of natural increase and the net reproduction rate are the same, because the true rate of natural increase is simply the per annum rate corresponding to the net reproduction rate, which is a per generation rate. The true rate of natural increase is, like the net reproduction rate, a potential and hypothetical measure. It tells us at what rate per annum the population will *ultimately* grow, *if* the given fertility and mortality conditions continue to hold. Thus the

---

But since $l(x)b(x) \geqslant 0$ for all x, the right-hand side of (*) is monotonic decreasing in r, so that (*) has only one real root, say r_0. If $r = u + iv$ be a complex root, then by taking real parts of both sides of (*) we have—

$$1 = \int_0^\infty e^{-ux} \cos vx \, l(x)b(x)dx$$

and since $\cos vx < 1$, then $u < r_0$. Hence r_0 is greater than the real parts of any of the complex roots. It follows that

$$B(t) = \sum_{i=0}^\infty Q_i e^{r_i t}$$

is made up of one aperiodic term (corresponding to the one real root) and a number of periodic terms (corresponding to the complex roots), but the aperiodic term is dominant, so that, for large t, $B(t)$ will tend asymptotically towards

$$B(t) = Q_0 e^{r_0 t}$$

where r_0 is the one real root of (*), and corresponds to 'r' in the text, and Q_0 is a constant.

This demonstrates that any given population, if subjected continuously to given fertility and mortality conditions, will ultimately give rise to a stream of births increasing or decreasing at a constant rate.

The equation (*) cannot be solved exactly for its real root, and r can be obtained by approximation only. The formula given in the text is such an approximation, based on an approximation to the length of a generation. A more accurate approximation can be obtained by solving

$$\log_e R - rk_1 + \frac{r^2}{2}k_2 = 0$$

where k_1 is the mean age of mothers at confinement as defined in the text and k_2 is the corresponding variance of age of mothers at confinement. (See Karmel, P. H.: "The Relations Between Male and Female Nuptiality in a Stable Population," *Population Studies*, Vol. I, 1947–48, p. 359, footnote.)

The above theory is due to A. J. Lotka. For an early formulation, see Lotka, A. J., and Sharpe, F. R.: "A Problem in Age Distribution," *Philosophical Magazine*, Vol. 21, 1911, p. 435. The complete theory of the stable population is set out in Lotka, A. J., *Analyse démographique avec application particulière à l'espèce humaine*, Actualités scientifiques et industrielles 780 (Hermann, Paris, 1939).

fertility and mortality conditions of Australia in 1967 were such that if they continued to operate, ultimately (perhaps by about the year 2030) the Australian population would grow, migration apart, at 10·9 per 1,000 per annum.

The true rate of natural increase can differ quite considerably from the crude rate of natural increase defined in section 15.4, p. 415 above. This is because the crude rate depends not only on fertility and mortality conditions, but also on the sex and age distribution inherited from the past. A good example of contrasts in crude and true rates is provided by Australia in 1934. The net reproduction rate was 0·939, and the corresponding true rate of natural increase was $-2\cdot2$ per 1,000 per annum, i.e. a rate of decrease. On the other hand, the crude rate was 7·07 per 1,000. The Australian population was actually increasing, but the underlying fertility and mortality conditions were such that ultimately the population would decrease unless these conditions changed (see p. 454 above). In fact, to provide an ultimate rate of increase of 7·07 per 1,000, fertility would have had to be raised by some 25–30 per cent. This underlines the fact that statistics cannot always be taken at their face value and that it is frequently necessary to go further than the more obvious data and measures. In considering population replacement the net reproduction rate or the true rate of natural increase must be examined. For some purposes the one is more convenient, for other purposes the other. They both lead to identical conclusions.

The table below sets out the crude rate of natural increase and an approximation to the value of the true rate for Australia, 1881–1967.

Table 15.37

CRUDE AND TRUE RATES OF NATURAL INCREASE
AUSTRALIA, 1881–1967

	Crude Rate per 1,000 per annum	True Rate per 1,000 per annum
1881	20·6	23·4
1891	19·6	20·0
1901	14·9	11·6
1911	16·5	12·4
1921	15·0	9·4
1931	9·5	1·4
1941	8·9	2·4
1951	13·2	12·3
1961	14·4	18·6
1967	10·7	10·9

Source: Crude rates: Commonwealth Bureau of Census and Statistics, Australia: *Demography Bulletin*, No. 78, 1960, p. 179 and *Australian Demographic Review*, Nos. 245, p. 3 and 247, p. 3
True rates: computed by the authors

Stable Population

In order to illustrate the arguments of the preceding paragraphs a hypothetical (and highly imaginative) example is worked below. We suppose that we have a population of one sex, with an upper limit to life of 5 years. All births take place on the 31st December, so that at the end of any year the population is aged exactly 1, 2, 3, 4 or 5 years. At the end of year 0, the numbers at these five ages are assumed to be 1,000, 1,500, 1,500, 1,500, 1,000 respectively. The mortality conditions are defined by supposing that no deaths occur before the age of 5, and when persons turn 5 they die immediately. Consequently, the mean expectation of life is 5 years. The fertility conditions are defined as follows: no births to persons aged 1 and 5, and births at the rate of 300 per 1,000 per annum to those aged 2, 3 and 4. It follows that the gross reproduction rate is $(300 + 300 + 300) \div 1,000 = 0.9$ and, since no deaths occur before the end of the child-bearing period, the net reproduction rate is also 0·9. The approximate length of a generation will be 3 years (3 years being the mean age at confinement) and hence the true rate of natural increase will be given approximately by $(1 + r)^3 \approx 0.9$, i.e. $r \approx -34.5$ per 1,000 per annum. The number of births occurring at 31st December, year 0, will be $(1,500 + 1,500 + 1,500) \times 0.3 = 1,350$, and the number of deaths will be the 1,000 exactly aged 5. At the end of year 1, there will be 1,350 aged 1, and the number aged 2 will be the 1,000 aged 1 in the previous year, and so on. In year 0, the crude birth rate will be $1,350/6,500 \times 1,000 = 207$ per 1,000, the crude death rate will be $1,000/6,500 \times 1,000 = 154$ per 1,000 and the crude rate of natural increase will be $350/6,500 \times 1,000 = 54$ per 1,000. In Table 15.38 the age distribution and the numbers of births and deaths occurring as at the ends of year 0 to year 23 inclusive are set out.

In Table 15.39 the crude birth rate, crude death rate and crude rate of natural increase are set out, all measured as at the end of years.

There are a number of interesting features about Table 15.39. Although the fertility and mortality conditions remain constant over the whole period, the crude birth and crude death rates fluctuate a great deal until about year 10. In general, the crude birth rate falls, and the crude death rate rises over the whole period. The high crude birth rate and low crude death rate in year 0 is due to a particularly favourable age distribution with a preponderance of persons in the child-bearing ages. Furthermore, although the crude rate of natural increase is $+ 53.9$ per 1,000 in year 0, the population will ultimately decline. By about the year 20 a steady rate of decline of 34·4 per 1,000 sets in. This is close to the approximate value for the true rate of natural increase of -34.5 per 1,000 computed above. The misleading character

Table 15.38
AGE DISTRIBUTION, BIRTHS AND DEATHS OF HYPOTHETICAL POPULATION

Age (years)	Year							
	0	1	2	3	4	5	6	7
1	1,000	1,350	1,200	1,155	1,065	1,112	1,026	1,000
2	1,500	1,000	1,350	1,200	1,155	1,065	1,112	1,026
3	1,500	1,500	1,000	1,350	1,200	1,155	1,065	1,112
4	1,500	1,500	1,500	1,000	1,350	1,200	1,155	1,065
5	1,000	1,500	1,500	1,500	1,000	1,350	1,200	1,155
Total Population	6,500	6,850	6,550	6,205	5,770	5,882	5,558	5,358
Births	1,350	1,200	1,155	1,065	1,112	1,026	1,000	961
Deaths	1,000	1,500	1,500	1,500	1,000	1,350	1,200	1,155

Age (years)	Year							
	8	9	10	11	12	13	14	15
1	961	941	896	871	839	812	782	757
2	1,000	961	941	896	871	839	812	782
3	1,026	1,000	961	941	896	871	839	812
4	1,112	1,026	1,000	961	941	896	871	839
5	1,065	1,112	1,026	1,000	961	941	896	871
Total Population	5,164	5,040	4,824	4,669	4,508	4,359	4,200	4,061
Births	941	896	871	839	812	782	757	730
Deaths	1,065	1,112	1,026	1,000	961	941	896	871

Age (years)	Year							
	16	17	18	19	20	21	22	23
1	730	705	681	658	635	613	592	572
2	757	730	705	681	658	635	613	592
3	782	757	730	705	681	658	635	613
4	812	782	757	730	705	681	658	635
5	839	812	782	757	730	705	681	658
Total Population	3,920	3,786	3,655	3,531	3,409	3,292	3,179	3,070
Births	705	681	658	635	613	592	572	552
Deaths	839	812	782	757	730	705	681	658

Table 15.39
Birth and Death Rates for Hypothetical Population

Year	Crude Birth Rate per 1,000 p.a.	Crude Death Rate per 1,000 p.a.	Crude Rate of Natural Increase per 1,000 p.a. (a)
0	208	154	53·9
1	175	219	− 43·8
2	176	229	− 52·7
3	172	242	− 70·1
4	193	173	19·4
5	174	230	− 55·1
6	180	216	− 36·0
7	179	216	− 36·2
8	182	206	− 24·0
9	178	221	− 42·8
10	181	213	− 32·1
11	180	214	− 34·5
12	180	213	− 33·1
13	179	216	− 36·5
14	180	213	− 33·1
15	180	214	− 34·7
16	180	214	− 34·2
17	180	214	− 34·6
18	180	214	− 34·0
19	180	214	− 34·6
20	180	214	− 34·3
21	180	214	− 34·4
22	180	214	− 34·3
23	180	214	− 34·5

(a) Crude rate of natural increase = crude birth rate − crude death rate, but the crude rates of natural increase are taken to an additional decimal place.

of the crude rates is clearly illustrated. By about the year 20, not only have the various crude rates stabilized, but the population has achieved a stable age distribution, as can be seen in the table below.

Table 15.40
Percentage Age Distribution of Hypothetical Population

Age (years)	Year 0	Year 10	Year 20	Year 21	Year 22	Year 23
1	15·4	18·6	18·6	18·6	18·6	18·6
2	23·1	19·5	19·3	19·3	19·3	19·3
3	23·1	19·9	20·0	20·0	20·0	20·0
4	23·1	20·7	20·7	20·7	20·7	20·7
5	15·4	21·3	21·4	21·4	21·4	21·4
Total	100·0	100·0	100·0	100·0	100·0	100·0

This example illustrates the proposition that a population, irrespective of its initial age distribution, subjected constantly to given fertility and mortality conditions, will ultimately attain a stable age distribution and will increase or decrease at a constant rate per annum. A fair degree of stability is reached after about three generations. It can now be seen why the population in year 0 shows a substantial increase, whereas in fact it will ultimately decrease if the given fertility and mortality conditions continue to hold. In year 0 some 69 per cent of the population is in the child-bearing ages, whereas ultimately only 60 per cent will be in those ages. This favours a high number of births. On the other hand only 15 per cent are of the age subject to mortality, whereas ultimately some 21 per cent will be of that age. This favours a low number of deaths.

Limitations of the Gross and Net Reproduction Rates

Finally attention must be drawn to certain limitations of the gross and net reproduction rates and the true rate of natural increase. These three measures are based on fertility conditions as expressed by the specific fertility rates. In so far as these rates do not properly represent the underlying fertility conditions of the population at the time under consideration, the gross and net reproduction rates and the true rate of natural increase will be defective. The underlying fertility conditions depend basically on two factors. The first may be called the *propensity to marry* and the second the *propensity to have families of various sizes*.

It may be the case, for example, that 90 per cent of all males will marry at some time or another during their lifetimes and that on the average married couples plan to have families of three children. If for any reason marriages are abnormally high in any particular year (e.g. due to high prosperity causing people to marry earlier than usual), then there will be an abnormal proportion of newly-weds among the females. This will lead to relatively high specific fertility rates, and these rates cannot be said to reflect fertility conditions truly. For if an abnormally high number of marriages occurs for a few years, it is inevitable that an abnormally low number must occur in the near future, since the pool of marriageable people will have been reduced. Consequently an abnormally high marriage rate will inflate specific fertility rates now and will inevitably exercise a depressing effect on them in the future, even though the propensity to marry and the propensity to have children in the broader sense have remained unchanged.

Similarly, if the sex distribution is unbalanced, this will lead to abnormal behaviour in the marriage rate. Thus, a substantial excess of males (due, say, to migration) will lead to a greater proportion of married females than normal and will inflate specific fertility rates. For these rates are the rates at which *all* females produce children, and the greater the proportion of females married the higher they will be.

But with the passage of time, other things being equal, the excess of males will disappear and the specific fertility rates will inevitably fall.[1]

Furthermore, married couples will be influenced by a variety of factors in determining the spacing of their families. Thus in times of economic depression couples may postpone having their children and in times of prosperity have them earlier than they would otherwise do. Consequently, an economic boom may lead to a rise in specific fertility rates even though the propensity to have families of various sizes remains unchanged, the rise in specific fertility rates later being followed by a fall.

From this discussion it can be seen that short-term fluctuations in specific fertility rates, and hence in the gross and net reproduction rates and the true rate of natural increase which are based on them, may occur even though the underlying propensities to marry and to have children remain unchanged. Consequently, great care must be taken in interpreting fluctuations in these rates.

Two examples may be taken to illustrate this. From about 1880 to 1930 there was a steady downward trend in the net reproduction rate in Australia, a rate of 1·88 in 1881 falling to 1·04 in 1931. By 1934 the net reproduction rate was as low as 0·94. From that year onwards it rose, until in 1947 it reached 1·42. Was this rise a reversal of the previous trend? There are reasons for doubting that it was wholly this. During the depression of the early '30s, marriages were postponed and additions to families delayed. These abnormally depressed the net reproduction rate. With economic recovery between 1935 and 1939 these marriages and births took place, thus raising the net reproduction rate. Then came the Second World War and a boom in marriages. The war and post-war expansion brought full employment and economic prosperity. These factors led to the bringing forward of marriages and births and hence raised the net reproduction rate further. Consequently, the upward movement in the rate between 1934 and 1947 could possibly be explained in these terms and could have occurred without any fundamental change in the underlying fertility conditions. In fact, research, which makes due allowance for the factors referred to above, suggests that between 1934 and 1942, fertility, if anything, fell a little but that between 1943 and 1947 it rose somewhat, although by no means by as much as the net reproduction rate suggests.[2] Since

---

[1] See Hajnal, J.: "Aspects of Recent Trends in Marriage in England and Wales," *Population Studies*, Vol. I, 1947–48, p. 72; Karmel, P. H.: "The Relations between Male and Female Net Reproduction Rates," *Population Studies*, Vol. I, 1947–48, p. 249, "The Relations between Male and Female Nuptiality in a Stable Population," *Population Studies*, Vol. I, 1947–48, p. 353, and "An Analysis of the Sources and Magnitudes of Inconsistencies between Male and Female Net Reproduction Rates in Actual Populations," *Population Studies*, Vol. II, 1948–49, p. 240.

[2] See Karmel, P. H.: "Fertility and Marriages—Australia, 1933–42," *Economic Record*, Vol. XX, 1944, p. 74, and "Population Replacement—Australia, 1947," *Economic Record*, Vol. XXV, 1949, p. 83.

1947 fertility had been increasing to reach a peak in 1961, after which it declined again towards the 1947 level.

A second example compares the net reproduction rate in Australia in 1881 with that in 1921. In 1881 the rate was 1·88, and in 1921, 1·31, but in 1881 the masculinity of the Australian population was 118 males per 100 females, whereas in 1921 it was 103. It follows that the proportion of females married in 1881 was abnormally high and that the net reproduction rate for that year was rather too high to be a correct reflection of the underlying conditions. Consequently the decline in the rate between 1881 and 1921 exaggerates the fall in fertility.

One way of partially overcoming the difficulties illustrated above is to pay attention not so much to the current specific fertility rates obtaining in a particular year, but rather to the reproductive behaviour of a particular generation of females over their lifetime. Thus we may trace out the actual marriage and child-bearing experience of females born in a particular year.[1] The main limitation in such a study of *generation fertility* is that it is never possible to be up to date, since, at any particular time, the more recent generations of females will not have completed their reproductive span. Nevertheless, this sort of analysis enables short-term fluctuations in current fertility rates to be placed in their proper perspective.

[1] See, for example, Whelpton, P. K.: *Cohort Fertility—Native White Women in the United States* (Princeton University Press, 1954).

APPENDIXES

APPENDIX A

STATISTICAL TABLES

Table I[1]
AREAS UNDER THE NORMAL PROBABILITY CURVE

$T = \dfrac{X - \mu}{\sigma}$	·00	·01	·02	·03	·04	·05	·06	·07	·08	·09
0·0	·0000	·0040	·0080	·0120	·0159	·0199	·0239	·0279	·0319	·0359
0·1	·0398	·0438	·0478	·0517	·0557	·0596	·0636	·0675	·0714	·0753
0·2	·0793	·0832	·0871	·0910	·0948	·0987	·1026	·1064	·1103	·1141
0·3	·1179	·1217	·1255	·1293	·1331	·1368	·1406	·1443	·1480	·1517
0·4	·1554	·1591	·1628	·1664	·1700	·1736	·1772	·1808	·1844	·1879
0·5	·1915	·1950	·1985	·2019	·2054	·2088	·2123	·2157	·2190	·2224
0·6	·2257	·2291	·2324	·2357	·2389	·2422	·2454	·2486	·2518	·2549
0·7	·2580	·2612	·2642	·2673	·2704	·2734	·2764	·2794	·2823	·2852
0·8	·2881	·2910	·2939	·2967	·2995	·3023	·3051	·3078	·3106	·3133
0·9	·3159	·3186	·3212	·3238	·3264	·3289	·3315	·3340	·3365	·3389
1·0	·3413	·3438	·3461	·3485	·3508	·3531	·3554	·3577	·3599	·3621
1·1	·3643	·3665	·3686	·3708	·3729	·3749	·3770	·3790	·3810	·3830
1·2	·3849	·3869	·3888	·3907	·3925	·3944	·3962	·3980	·3997	·4015
1·3	·4032	·4049	·4066	·4083	·4099	·4115	·4131	·4147	·4162	·4177
1·4	·4192	·4207	·4222	·4236	·4251	·4265	·4279	·4292	·4306	·4319
1·5	·4332	·4345	·4357	·4370	·4382	·4394	·4406	·4418	·4430	·4441
1·6	·4452	·4463	·4474	·4485	·4495	·4505	·4515	·4525	·4535	·4545
1·7	·4554	·4564	·4573	·4582	·4591	·4599	·4608	·4616	·4625	·4633
1·8	·4641	·4649	·4656	·4664	·4671	·4678	·4686	·4693	·4699	·4706
1·9	·4713	·4719	·4726	·4732	·4738	·4744	·4750	·4758	·4762	·4767
2·0	·4773	·4778	·4783	·4788	·4793	·4798	·4803	·4808	·4812	·4817
2·1	·4821	·4826	·4830	·4834	·4838	·4842	·4846	·4850	·4854	·4857
2·2	·4861	·4865	·4868	·4871	·4875	·4878	·4881	·4884	·4887	·4890
2·3	·4893	·4896	·4898	·4901	·4904	·4906	·4909	·4911	·4913	·4916
2·4	·4918	·4920	·4922	·4925	·4927	·4929	·4931	·4932	·4934	·4936
2·5	·4938	·4940	·4941	·4943	·4945	·4946	·4948	·4949	·4951	·4952
2·6	·4953	·4955	·4956	·4957	·4959	·4960	·4961	·4962	·4963	·4964
2·7	·4965	·4966	·4967	·4968	·4969	·4970	·4971	·4972	·4973	·4974
2·8	·4974	·4975	·4976	·4977	·4977	·4978	·4979	·4980	·4980	·4981
2·9	·4981	·4982	·4983	·4984	·4984	·4984	·4985	·4985	·4986	·4986
3·0	·4987	·4987	·4987	·4988	·4988	·4988	·4989	·4989	·4989	·4990
3·1	·4990	·4991	·4991	·4991	·4992	·4992	·4992	·4992	·4993	·4993

[1] This table is reproduced from Rugg, H. O.: *Statistical Methods Applied to Education*, published by Houghton Mifflin Company, Boston, U.S.A., by permission of the publishers.

APPENDIX A

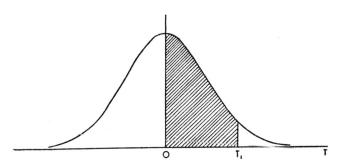

Note. The figures in the body of the table opposite give proportions of total area under the curve lying between the central ordinate and the value of $T = \dfrac{X - \mu}{\sigma}$ under consideration. These correspond in the diagram above to the ratio of the hatched area to the total area. They give the probability that T will lie between 0 and T_1, i.e. the probability that X will lie between μ and X_1.

Table II[1]
VALUES OF t

ν	Level of Significance (P)											
	·9	·8	·7	·6	·5	·4	·3	·2	·1	·05	·02	·01
1	·158	·325	·510	·727	1·000	1·376	1·963	3·078	6·314	12·706	31·821	63·657
2	·142	·289	·445	·617	·816	1·061	1·386	1·886	2·920	4·303	6·965	9·925
3	·137	·277	·424	·584	·765	·978	1·250	1·638	2·353	3·182	4·541	5·841
4	·134	·271	·414	·569	·741	·941	1·190	1·533	2·132	2·776	3·747	4·604
5	·132	·267	·408	·559	·727	·920	1·156	1·476	2·015	2·571	3·365	4·032
6	·131	·265	·404	·553	·718	·906	1·134	1·440	1·943	2·447	3·143	3·707
7	·130	·263	·402	·549	·711	·896	1·119	1·415	1·895	2·365	2·998	3·499
8	·130	·262	·399	·546	·706	·889	1·108	1·397	1·860	2·306	2·896	3·355
9	·129	·261	·398	·543	·703	·883	1·100	1·383	1·833	2·262	2·821	3·250
10	·129	·260	·397	·542	·700	·879	1·093	1·372	1·812	2·228	2·764	3·169
11	·129	·260	·396	·540	·697	·876	1·088	1·363	1·796	2·201	2·718	3·106
12	·128	·259	·395	·539	·695	·873	1·083	1·356	1·782	2·179	2·681	3·055
13	·128	·259	·394	·538	·694	·870	1·079	1·350	1·771	2·160	2·650	3·012
14	·128	·258	·393	·537	·692	·868	1·076	1·345	1·761	2·145	2·624	2·977
15	·128	·258	·393	·536	·691	·866	1·074	1·341	1·753	2·131	2·602	2·947
16	·128	·258	·392	·535	·690	·865	1·071	1·337	1·746	2·120	2·583	2·921
17	·128	·257	·392	·534	·689	·863	1·069	1·333	1·740	2·110	2·567	2·898
18	·127	·257	·392	·534	·688	·862	1·067	1·330	1·734	2·101	2·552	2·878
19	·127	·257	·391	·533	·688	·861	1·066	1·328	1·729	2·093	2·539	2·861
20	·127	·257	·391	·533	·687	·860	1·064	1·325	1·725	2·086	2·528	2·845
21	·127	·257	·391	·532	·686	·859	1·063	1·323	1·721	2·080	2·518	2·831
22	·127	·256	·390	·532	·686	·858	1·061	1·321	1·717	2·074	2·508	2·819
23	·127	·256	·390	·532	·685	·858	1·060	1·319	1·714	2·069	2·500	2·807
24	·127	·256	·390	·531	·685	·857	1·059	1·318	1·711	2·064	2·492	2·797
25	·127	·256	·390	·531	·684	·856	1·058	1·316	1·708	2·060	2·485	2·787
26	·127	·256	·390	·531	·684	·856	1·058	1·315	1·706	2·056	2·479	2·779
27	·127	·256	·389	·531	·684	·855	1·057	1·314	1·703	2·052	2·473	2·771
28	·127	·256	·389	·530	·683	·855	1·056	1·313	1·701	2·048	2·467	2·763
29	·127	·256	·389	·530	·683	·854	1·055	1·311	1·699	2·045	2·462	2·756
30	·127	·256	·389	·530	·683	·854	1·055	1·310	1·697	2·042	2·457	2·750
∞	·126	·253	·385	·524	·674	·842	1·036	1·282	1·645	1·960	2·326	2·576

[1] This table is reproduced from Fisher, R. A. and Yates. F.: *Statistical Tables for Biological, Agricultural and Medical Research*, published by Hafner Publishing Co., New York, by permission of the authors and publishers.

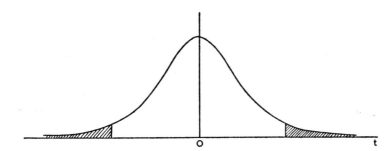

Note. The body of the table gives, for particular numbers of degrees of freedom (ν), the value of t which cuts off tails together covering a proportion of area equal to the required level of significance (P). The last row of the table for $\nu = \infty$, shows values of t corresponding to the normal distribution, which the t-distribution approximates for large ν.

Table III[1]

VALUES OF χ^2

ν	Level of Significance (P)												
	·99	·98	·95	·90	·80	·70	·50	·30	·20	·10	·05	·02	·01
1	·0002	·0006	·0039	·0158	·0642	·148	·455	1·074	1·642	2·706	3·841	5·412	6·635
2	·0201	·0404	·103	·211	·446	·713	1·386	2·408	3·219	4·605	5·991	7·824	9·210
3	·115	·185	·352	·584	1·005	1·424	2·366	3·665	4·642	6·251	7·815	9·837	11·341
4	·297	·429	·711	1·064	1·649	2·195	3·357	4·878	5·989	7·779	9·488	11·668	13·277
5	·554	·752	1·145	1·610	2·343	3·000	4·351	6·064	7·289	9·236	11·070	13·388	15·086
6	·872	1·134	1·635	2·204	3·070	3·828	5·348	7·231	8·558	10·645	12·592	15·033	16·812
7	1·239	1·564	2·167	2·833	3·822	4·671	6·346	8·383	9·803	12·017	14·067	16·622	18·475
8	1·646	2·032	2·733	3·490	4·594	5·527	7·344	9·524	11·030	13·362	15·507	18·168	20·090
9	2·088	2·532	3·325	4·168	5·380	6·393	8·343	10·656	12·242	14·684	16·919	19·679	21·666
10	2·558	3·059	3·940	4·865	6·179	7·267	9·342	11·781	13·442	15·987	18·307	21·161	23·209
11	3·053	3·609	4·575	5·578	6·989	8·148	10·341	12·899	14·631	17·275	19·675	22·618	24·725
12	3·571	4·178	5·226	6·304	7·807	9·034	11·340	14·011	15·812	18·549	21·026	24·054	26·217
13	4·107	4·765	5·892	7·042	8·634	9·926	12·340	15·119	16·985	19·812	22·362	25·472	27·688
14	4·660	5·368	6·571	7·790	9·467	10·821	13·339	16·222	18·151	21·064	23·685	26·873	29·141
15	5·229	5·985	7·261	8·547	10·307	11·721	14·339	17·322	19·311	22·307	24·996	28·259	30·578
16	5·812	6·614	7·962	9·312	11·152	12·624	15·338	18·418	20·465	23·542	26·296	29·633	32·000
17	6·408	7·255	8·672	10·085	12·002	13·531	16·338	19·511	21·615	24·769	27·587	30·995	33·409
18	7·015	7·906	9·390	10·865	12·857	14·440	17·338	20·601	22·760	25·989	28·869	32·346	34·805
19	7·633	8·567	10·117	11·651	13·716	15·352	18·338	21·689	23·900	27·204	30·144	33·687	36·191
20	8·260	9·237	10·851	12·443	14·578	16·266	19·337	22·775	25·038	28·412	31·410	35·020	37·566
21	8·897	9·915	11·591	13·240	15·445	17·182	20·337	23·858	26·171	29·615	32·671	36·343	38·932
22	9·542	10·600	12·338	14·041	16·314	18·101	21·337	24·939	27·301	30·813	33·924	37·659	40·289
23	10·196	11·293	13·091	14·848	17·187	19·021	22·337	26·018	28·429	32·007	35·172	37·968	41·638
24	10·856	11·992	13·848	15·659	18·062	19·943	23·337	27·096	29·553	33·196	36·415	40·270	42·980
25	11·524	12·697	14·611	16·473	18·940	20·867	24·337	28·172	30·675	34·382	37·652	41·566	44·314
26	12·198	13·409	15·379	17·292	19·820	21·792	25·336	29·246	31·795	35·563	38·885	42·856	45·642
27	12·879	14·125	16·151	18·114	20·703	22·719	26·336	30·319	32·912	36·741	40·113	44·140	46·963
28	13·565	14·847	16·928	18·939	21·588	23·647	27·336	31·391	34·027	37·916	41·337	45·419	48·278
29	14·256	15·574	17·708	19·768	22·475	24·577	28·336	32·461	35·139	39·087	42·557	46·693	49·588
30	14·953	16·306	18·493	20·599	23·364	25·508	29·336	33·530	36·250	40·256	43·773	47·962	50·892

[1] This table is reproduced from Fisher, R. A. and Yates, F.: *Statistical Tables for Biological, Agricultural and Medical Research*, published by Hafner Publishing Co., New York, by permission of the authors and publishers.

APPENDIX A

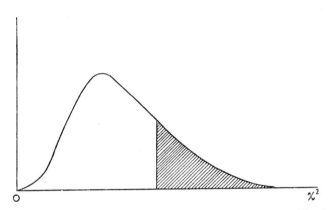

Note. The body of the table gives, for particular numbers of degrees of freedom (ν), the value of χ^2 which cuts off a tail covering a proportion of area equal to the required level of significance (P). For large values of ν, the statistic $\sqrt{2\chi^2}$ is approximately normally distributed about a mean $\sqrt{2\nu-1}$ with a standard deviation of unity.

APPENDIX B

SOURCES OF AUSTRALIAN STATISTICS[1]

1. GENERAL

IN Australia most governmentally collected statistics are published by the Central Office of the Commonwealth Bureau of Census and Statistics on a Commonwealth and State basis. In addition, each State office of the Bureau publishes in respect of its own State a series of publications similar to that of the Central Office, e.g. the *Quarterly Abstract of South Australian Statistics*, but which contain greater detail for the State, and in some instances information for sub-divisions of the State. Prior to 1957, all States, except Tasmania, maintained separate statistical offices, which, although independent of the Commonwealth Bureau, collaborated closely with it However, during 1957 and 1958 the statistical offices and services of the Commonwealth and States were amalgamated to form an integrated statistical service. In addition to statistical publications proper, other governmentally collected statistics are to be found in such publications as the *Report of the Commissioner of Taxation*, the *Budget Papers*, other financial statements of the Commonwealth and State governments, the reports of government business undertakings, the publications of the Commonwealth Bureau of Agricultural Economics and of the Industries Division of the Department of Trade and Industry, etc. There are few private statistical publications, but mention should be made of the *Reserve Bank of Australia Statistical Bulletin*, of the *Australian Insurance and Banking Record*, and of the company statistics included in the *Official Record* of the Stock Exchange of Melbourne, in the Sydney Stock Exchange *Investment Service* and in Jobson's *Investment Digest*.

The following are the major regular printed publications of the Central Office of the Commonwealth Bureau of Census and Statistics. Preliminary summary statements of some of these publications are also issued.

General Publications

Official Year Book of the Commonwealth of Australia (annual)
Pocket Compendium of Australian Statistics (annual)
Quarterly Summary of Australian Statistics (quarterly)

[1] This appendix was written in October, 1968, and relates to statistical collections, publications and procedures current at that time. The scope of Australian official statistics is being continuously expanded; and revisions in existing procedures are made from time to time. Reference should, therefore, be made to the official publications to determine whether, in any particular instance, the material contained in this appendix requires modification. The list of publications in section 1 was up to date as at April, 1968.

APPENDIX B 473

Monthly Review of Business Statistics (monthly)
Digest of Current Economic Statistics (monthly)

Other Publications

BUILDING—
  Building and Construction (annual)
  Building Approvals (monthly)
  Quarterly Building Statistics (quarterly)

DEMOGRAPHY—
  Overseas Arrivals and Departures, Part 1, Permanent and Long-term Movement (quarterly)
  Overseas Arrivals and Departures, Part 2, Total and Short-term Movement (quarterly)
  Overseas Arrivals and Departures (monthly)
  Births (annual)
  Causes of Death (annual)
  Deaths (annual)
  Australian Demographic Review (quarterly)
  Demography (annual)
  Marriages (annual)
  Projections of the Population (1966 to 1986) (irregular)

FINANCE—
  Banking and Currency (annual)
  Banking Statistics (monthly)
  Registered Building Societies (annual)
  Capital and Maintenance Expenditure by Private Businesses in Australia (half-yearly)
  Capital Expenditure by Private Businesses in Australia (half-yearly)
  New Capital Raisings by Companies in Australia (quarterly)
  State, Territory and Local Government Authorities' Finance and Government Securities (annual)
  Commonwealth Finance (annual)
  Finance Companies (annual)
  Finance Companies, Australia (monthly)
  Insurance and Other Private Finance (other than Banking and Currency) (annual)
  Insurance, Fire, Marine and General (annual)
  Life Insurance Statistics (annual)
  Life Insurance Statistics (monthly)
  Instalment Credit for Retail Sales (quarterly)
  Overseas Investment (annual)

FINANCE—*(Cont'd)*

　Government Pension and Superannuation Schemes (annual)
　Survey of Selected Private Pension Funds (annual)
　Savings Bank Statistics (monthly)
　Commonwealth Taxation Assessments (annual)
　Commonwealth, State and Territory Taxation Collections (annual)
　Unit Trusts, Land Trusts and Mutual Funds (quarterly)

LABOUR, WAGES AND EMPLOYMENT—

　Survey of Weekly Earnings and Hours (annual)
　Employment and Unemployment (monthly)
　Industrial Disputes (quarterly)
　Labour Report (annual)
　Labour Turnover (half-yearly)
　Survey of Leavers from Schools, Universities or Other Educational Institutions (annual)
　Survey of Multiple Jobholding, Australia (irregular)
　Minimum Rates of Wage, and Prescribed Hours of Work—Adult Male and Female Occupations, Sydney and Melbourne (half-yearly)

NATIONAL ACCOUNTS—

　Australian National Accounts—National Income and Expenditure (annual)
　National Income and Expenditure, Quarterly Estimates (quarterly)

OVERSEAS TRADE AND BALANCE OF PAYMENTS—

　Balance of Payments (half-yearly)
　Balance of Payments, Quarterly Summary (quarterly)
　Australian Export Commodity Classification (annual)
　Australian Exports (annual)
　Australian Produce, Exports of Wool (monthly)
　Australian Import Commodity Classification (annual)
　Australian Imports (annual)
　Imports Cleared for Home Consumption (annual)
　Imports of Assembled New Passenger Motor Cars (other than Buses or Special Vehicles) (monthly)
　Overseas Trade, Part 1—Exports (monthly)
　Overseas Trade, Part 2—Imports (monthly)
　Overseas Trade (annual)
　Overseas Trade, Exports by Commodity Divisions, Australia (monthly)
　Overseas Trade, Imports by Commodity Divisions, Australia (monthly)

APPENDIX B   475

OVERSEAS TRADE AND BALANCE OF PAYMENTS—(*Cont'd*.)
Overseas Trade, Major Groups of Countries (quarterly)
Trade of Australia with Eastern Countries (annual)

PRICES AND PRICE INDEXES—
Consumer Price Index (quarterly)
Export Price Index (monthly)
Average Retail Prices of Selected Food and Grocery Items (quarterly)
Wholesale Price (Basic Materials and Foodstuffs) Index (monthly)
Wholesale Prices—Price Index of Electrical Installation Materials (quarterly)

PRIMARY INDUSTRIES—
Aerial Agriculture Operations (quarterly)
Bee Farming (annual)
Dairying Industry (half-yearly)
Dairying Industry (monthly)
Farm Machinery on Rural Holdings (annual)
Fisheries (annual)
Food Production and the Apparent Consumption of Foodstuffs and Nutrients (annual)
Fruit Growing Industry (annual)
Gold Mining Industry (annual)
Livestock Numbers (annual)
Meat Industry (monthly)
Mineral Exploration (annual)
The Australian Mineral Industry (quarterly)
Minerals and Mineral Products (annual)
Minerals and Mineral Products (monthly)
Mining and Quarrying (annual)
Overseas Participation in Australian Mining Industry (annual)
Non-rural Primary Industries (annual)
Value of Production (annual)
Value of Production, and Indexes of Price and Quantum of Farm Production (annual)
Classification of Rural Holdings, by Size and Type of Activity (irregular)
Rural Industries (annual)
Rural Land Use and Crop Production (annual)
Apparent Consumption of Tea and Coffee (annual)
Tractors on Rural Holdings (triennial)
Wheat Industry (annual)
Wholesale Sales and Stocks of Wine and Brandy (annual)
Wool Production and Utilization (annual)

RETAIL TRADE—

Census of Retail Establishments and Other Services (irregular)
Retail Sales of Goods (quarterly)

SECONDARY INDUSTRIES—

New Agricultural Machinery Statistics (quarterly)
Sales of New Construction Machinery and Attachments (quarterly)
Principal Statistics of Factories by Class of Industry (annual)
Summary of Principal Statistics of Factories (annual)
Indexes of Factory Production (annual)
Principal Factory Products (annual)
Manufacturing Industries (bulletins for 35 individual industries) (annual)
Manufacturing Industry (annual)
Packaging and Associated Items (annual)
Production Statistics (monthly)
Production Summaries (bulletins for 59 individual industries) (monthly)
Secondary Industries, Materials Used and Articles Produced in Factories (annual)
Textile Industries, Machines Installed (annual)
New Tractors, Receipts, Sales and Stocks (quarterly)

SOCIAL STATISTICS—

Divorce (annual)
Divorce (quarterly)
School Enrolments (annual)
Schools (annual)
University Statistics, Part 1. Students and Degrees Conferred (annual)
University Statistics, Part 2. Staff and Libraries (annual)
University Statistics, Part 3. Finance (annual)

TRANSPORT AND COMMUNICATION—

Motor Vehicle Registrations (annual)
Motor Vehicle Registrations (monthly)
Census of Motor Vehicles (irregular)
Road Traffic Accidents Involving Casualties (quarterly)
Overseas Shipping Cargo (monthly)
Transport and Communication (annual)

Population census results are published in mimeographed form as they become available, and the full results for the 1954 and 1961 censuses have been published in separate parts and in complete volumes for each State, the Commonwealth and the Territories. The *Statistician's Report* on the Census, which contains summary tables and textual comment, is published separately and included in the volume for Australia. Since Federation, censuses have been conducted in 1911, 1921, 1933, 1947, 1954, 1961 and 1966. Life tables, calculated on the basis of census results, are also published separately, the latest being for the period 1960–62. The 1966 Census results are at present being issued in mimeographed form.

As well as collecting regular statistics and conducting periodic population censuses, the Commonwealth Bureau conducts special surveys from time to time. These include the Censuses of Retail Establishments 1947–48, 1948–49, 1952–53, 1956–57 and 1961–62, and the censuses of Motor Vehicles 1947–48, 1955 and 1962, the results of which are published in printed and/or mimeographed form.

The above lists cover all the major Australian statistical publications with the exception of certain migration statistics which are published in the quarterly statistical bulletin of the Department of Immigration, *Australian Immigration: Consolidated Statistics* and certain statistics of employment and unemployment which are published monthly by the Department of Labour and National Service in a press release entitled *News Release* and in a bulletin entitled *Commonwealth Employment Service Statistics*. In addition, the Bureau of Agricultural Economics issues reports of a semi-statistical character on the situations in various primary industries (*The Wool Situation, The Wheat Situation*, etc.), and the Department of Trade and Industry conducts a half-yearly survey on manufacturing activity known as *Survey of Manufacturing Activity in Australia*. Since 1956 the Treasury has been issuing quarterly the *Treasury Information Bulletin* which contains up-to-date statistics on a number of variables of current economic interest, and a series of supplements to this (e.g. *National Accounting Estimates of Public Authority Expenditure, Private Overseas Investment in Australia*, etc.), as well as an annual survey of the economy called *The Australian Economy*.

The most generally useful of the monthly publications are the *Monthly Review of Business Statistics* and the *Digest of Current Economic Statistics*. They both give information over a wide field. The latter is produced in the month following that to which the statistics relate and presents data in a convenient tabular and graphical form. The former is less up to date but has a more extensive coverage. Greater detail is provided in the specialized annual publications, but they can never be quite up to date, on account of the work involved in preparing them and in the subsequent printing processes. A summarized survey of the whole field with textual comment is provided by the *Official Year Book of the Commonwealth*, which is very useful although inevitably published with some time-lag. All persons working in the field of economics or commerce and industry in Australia should be familiar with, at least, the annual principal publications listed above, the *Monthly Review*, the *Digest*, the *Year Book*, the *Statistician's Report* on the latest census, the *Australian National Accounts* and the annual budget paper on and the quarterly estimates of *National Income and Expenditure*.

An important recent addition to the publications of the Commonwealth Bureau is *Seasonally Adjusted Indicators*. This publication, in its 1968 issue, contains sixty-six selected series of statistical indicators of Australian economic activity in tabular and graphical form, together with explanatory matter on "Seasonal adjustment." The seasonally-adjusted series are up-dated in the *Monthly Review* and *Digest*.

With the exception of the *Statistician's Report on the Census*, the *Year Book*, the *Labour Report*, the *Australian National Accounts*, the *Australian Balance of Payments* and the budget paper on *National Income and Expenditure*, the printed publications provide little textual comment on the statistics. It is most important, however, that anyone using the publications should be aware of the precise meaning of the various series contained in them. Definitions and other explanatory matter are contained in most of the publications, and these should always be referred to, as should any footnotes to the tables.

REFERENCES

Official Year Book of the Commonwealth, No. 54, 1968 (Commonwealth Bureau of Census and Statistics, Canberra), Chap. 30, pp. 1233–1247.

Publications of the Commonwealth Bureau of Census and Statistics, April, 1968 (Canberra, Australia), with catalogue of major publications and subject index designed to show the Bureau publications in which information on various subjects, dealt with by the Bureau, is to be found. This publication lists also the publications issued by the State offices of the Bureau.

2. SAMPLE SURVEYS

The Commonwealth Bureau of Census and Statistics undertakes some of its statistical collections on a sample basis. Apart from *ad hoc* surveys organized to obtain special information, the Bureau conducts regular sample surveys to provide continuing information in a number of fields.

Regular sample surveys are as follows—

Surveys of Retail Establishments—
    Quarterly Sales Sample
    Monthly Sales Sub-Sample
    Hire Purchase and Instalment Credit Survey (quarterly)
Business Surveys—
    Survey of Stocks (annual)
    Survey of Capital Expenditure (quarterly)

APPENDIX B 479

Survey of Labour Turnover (six-monthly)
Income Tax Survey (annual)
Surveys of Employment, Labour and Earnings—
  Local Government Employment Survey (monthly)
  Survey of Earnings and Hours (annual)
Production Surveys—
  New South Wales Survey of Wheat Acreage (annual)
  New South Wales Survey of Wheat Production (annual)
  Wool Clip Survey (monthly)
  Tasmanian Timber Survey (monthly)
  Survey of Aerated and Carbonated Waters, Cordials and Syrups (monthly)
  Survey of Men's, Youths' and Boys' Outerwear (monthly)
  Survey of Women's Outerwear (monthly)
Overseas Trade Surveys—
  Survey of Overseas Imports (monthly)
  Survey of Import Values (quarterly)
Population Survey (quarterly)
Survey of Agricultural and Pastoral Expenditure (annual)

These surveys involve sampling designs of varying degrees of complexity. In order to illustrate some of the aspects of sample design, a brief discussion of the Survey of Retail Establishments is set out below.

The purpose of the regular sample survey is to provide quarterly estimates of retail sales in Australia and for each State by type of commodity sold. Censuses of Retail Establishments are taken at intervals of several years, the Census of Retail Establishments conducted for the year ended 30th June, 1962 being the basis of the current survey. This census gives for States and areas within States a classification of aggregate retail sales by description of store, by commodity group and by size of turnover. This enables a detailed stratification of the population to be made. The population of stores is stratified by States (six strata), by area (metropolitan and country—two strata), by description of store (fourteen strata) and by size of annual turnover (six strata), making a total of 1,008 distinct strata or cells. One such cell would be, for example, New South Wales, metropolitan, non-chain grocers, of size $20,000–$89,999.

All told there are approximately 140,000 retail stores. The sample now in use consists of about 11,000 stores. The sample is optimally allocated between cells with the object of producing estimates of comparable precision for each commodity within each State. The population variances used in the optimum allocations are those appropriate to ratio estimates. In accordance with the optimum design all chain stores and the cells containing the largest firms, i.e. those with a turnover in 1961–62 of $200,000 or more in Tasmania and South Australia and of $500,000 or more in the other States, are fully enumerated on account of their importance and small numbers. A simple random sample is selected from each of the remaining cells.

The survey is made quarterly, and stores once selected remain in the sample until the sample is re-designed following the next census. This simplifies administration and increases the co-operation of the stores. In any particular quarter non-response to questionnaires is somewhat less than 2 per cent in all States, and this is due mainly to unavoidable circumstances. The ratio method of estimation is used to estimate the value of retail sales. This method is used in several ways in this survey, but the general principle involved is illustrated as follows. For an individual cell, we put Z_i for the total value of sales recorded in the 1961–62 census for cell i, z_i for the value of sales recorded in the census for stores now in the sample, x_i for the value of sales reported currently by stores in the sample, and X_i for the estimated current total value of sales for cell i. Then we have

$$X_i = Z_i \frac{x_i}{z_i}$$

This method of estimation is equivalent to assuming that for an individual cell the value of retail sales of non-sample stores moves in the same ratio as those of sample stores. Since the sampling in each cell is random, this assumption is reasonable, subject, of course, to the usual errors of sampling. The above formula gives an estimate for an individual cell. Estimates of broader classifications are made by aggregating individual cell estimates or by aggregating cells before estimation.

The accuracy of the estimates derived from the sample can be judged by obtaining the 95 per cent confidence limits for the population values being estimated. Thus, for total turnover for Australia, the range of these limits is of the order of $\pm 0\cdot 8$ per cent of total turnover, and for the turnover in each commodity group for Australia $\pm 2\cdot 5$ per cent of turnover in the group. For State estimates the figures are ± 2 per cent and ± 6 per cent respectively. (The more detailed estimates naturally have wider confidence limits because they are based on a smaller number of stores.) For example, sales of groceries in Australia for the March Quarter, 1968, were estimated at $306·9 million. This is a sample estimate, and we can be 95 per cent confident that the population value lies in the range $306·9 million $\pm 2\cdot 5$ per cent, i.e. $299·2 million to $314·6 million.

There are certain difficult problems in a survey of this kind, connected with the passage of time. As time moves on from the benchmark of the census the relationship between current sales and census sales becomes less and less precise. One consequence of this is that the size stratification becomes less exact. This does not of itself introduce bias, but it does increase the sampling error. This can be allowed for by increasing the size of sample over time. However, to limit the increase in the size of the sample thus arising, a large sample was collected in June 1967. This sample collected data on Retail Sales for the year 1966–67 from 27,000 stores and was used to provide an updated benchmark, itself on a sample basis, on which to base future quarterly samples. The estimate from this large sample is a ratio estimate using 1962 Census figures as benchmarks, and the quarterly sample is thus a ratio estimate based upon a ratio estimate. The first quarterly sample based on this large sample was held in June 1968. Another serious problem is that some stores go out of business and new stores are created. If the number of stores as a whole is increasing over time, estimates of retail trade based on the above methods will tend to be downwardly biased, unless some account is taken of the increase in the population relative to the sample. Furthermore, if new stores are not introduced into the sample, the sample will become smaller as sample stores go out of business from time to time. Allowance for the births and deaths of stores is made by taking a census of new stores each September and grafting these onto the frame, and by removing defunct stores from the frame. Apart from these difficulties which are associated with the sampling procedure, there are, of course, all the census problems of definition and classification of retail stores, e.g. whether a bootmaker is a retail store, how a departmental store is to be classified, etc.

3. National Income Statistics

For Australia, unofficial estimates of national income date back to 1887. The most recent and most comprehensive of these unofficial estimates were made in 1938 by Colin Clark and J. G. Crawford in *The National Income of Australia* (Angus and Robertson, Sydney, 1938). During the Second World War, however, more detailed estimates of the items entering into the social accounts were found necessary for the formulation of wartime economic policy, and in 1945 detailed official estimates of national income and allied aggregates were published in a paper accompanying the Commonwealth Budget. These estimates referred to the financial years 1938–39 to 1944–45 inclusive. Since 1945 the paper on *National Income and Expenditure* has been an annual accompaniment to the Budget. In 1963, *Australian National Accounts, National Income and Expenditure 1948–49* to *1961–62* introduced a number of changes in the structure and presentation of the national accounts, and in the conceptual basis and definitions of the principal aggregates. Successive issues of this publication provide the main national accounting reference, giving full details for years since 1948–49.

Further details relating to the balance of payments are given in the half-yearly bulletin, *The Australian Balance of Payments*.

Clark and Crawford's estimates of national income were made by the production method. Reasonably complete data for primary and secondary production were available in the Primary and Secondary Industries Bulletins, but the contributions of tertiary industries to gross national product had to be based on flimsy data. In all, perhaps about two-thirds of the content of their estimates was based on reliable data. They also made some estimates by the incomes received method. Their two sets of estimates were in fairly close agreement, being about 6 per cent apart. However, the data available for the incomes received estimates were very scanty, and only about one-third of the content of their estimates was well based. Since the early 1940s, due to the introduction of pay-roll tax resulting in a substantially comprehensive collection of statistics on wages and salaries paid, to the lowering of the exemption limit for payment of income tax resulting in a greater cover of incomes, and to improved tabulation methods, the data available for estimating national income by the incomes received method have vastly improved. Official estimates of national income are in the main derived by the incomes received method. These estimates are based mainly on pay-roll tax tabulations, personal and company income tax statistics and annual collections of statistics of the value of farm production.

Estimates of gross national product by the expenditure method are also made. These are adjusted to equality with those made by the incomes received method by means of an item entitled "statistical discrepancy," which is, in practice, shown on the expenditure side of the account. These estimates are derived mainly from censuses and sample surveys of retail trade establishments (for personal consumption); the accounts of public authorities (for government expenditure); sample surveys of businesses, statistics of dwelling construction, statistics of motor vehicle registrations, and factory statistics (for gross private investment); and oversea trade statistics (for exports and imports).

Estimates of the many detailed items published in the social accounts are based on a wide variety of sources as well as on those referred to in the preceding two paragraphs.

The main tables of the Australian social accounts for 1967–68 are reproduced below. These tables are not set out in the highly systematized form of Chapter XII, above. This is in part due to the limitations of the statistics available to estimate the various items. But the form of accounts of Chapter XII can be conceived as lying behind the published figures. Table 1 is the consolidated national production account, although the left-hand side is in terms of factor payments

APPENDIX B 483

THE AUSTRALIAN NATIONAL ACCOUNTS 1967–68[1]

(Items in brackets are tentative estimates and all items are subject to revision to some degree)

Table 1
NATIONAL PRODUCTION ACCOUNT

	$m		$m
Wages, salaries and supplements	12,464	Net current expenditure on goods and services—	
Gross operating surplus of trading enterprises—		Personal consumption	14,583
Companies	(3,615)	Financial enterprises	346
Unincorporated enterprises	(3,626)	Public authorities	3,028
Dwellings owned by persons	1,131	Gross fixed capital expenditure—	
Public enterprises	758	Private	(3,968)
		Public enterprises	1,307
Gross National Product at Factor Cost	21,594	Public authorities	986
Indirect taxes less subsidies	2,620	Increase in value of stocks	(311)
		Statistical discrepancy	303
Gross National Product	24,214	Gross National Expenditure	24,832
Imports of goods and services	4,159	Exports of goods and services	3,541
National Turnover of Goods and Services	28,373	National Turnover of Goods and Services	28,373

NOTE

National turnover of goods and services represents the total flow of final goods and services through the Australian market. It is made up of domestic production (gross national product) and goods and services supplied from overseas (imports). These goods and services are absorbed by domestic expenditure (gross domestic expenditure) and expenditure from overseas (exports).

instead of values added, and so includes some items which are shown in Chapter XII in the distribution accounts. Gross national product is estimated by the incomes received method in the left-hand side of the account and by the expenditure method in the right-hand side. Balance is achieved by including the balancing item "statistical discrepancy." Tables 2, 3, 4, 5 and 6 are the current accounts of the five sectors. Table 6 is in fact a statement of the balance of international payments on current account. Table 7 is the consolidated national capital account.

In 1960, quarterly publication of the social accounts commenced under the title of *Quarterly Estimates of National Income and Expenditure*. These estimates cover most of the items in the annual publication and include more details for some items. Naturally, they are not as reliable as the annual figures, as they must in many cases be based on less complete information. However, they are of crucial importance for the analysis of short-term fluctuations in economic activity. They are being produced with a relatively short lag (about eight weeks after the end of the quarter).

[1] Source: Commonwealth of Australia, *National Income and Expenditure*, 1967–68 (Government Printer, Canberra), pp. 11–15.

Table 2

TRADING ENTERPRISES INCOME APPROPRIATION ACCOUNT
(Including public enterprises, fire and general insurance, farms, professional businesses and ownership of dwellings)

	$m		$m
Depreciation allowances	2,019	Gross operating surplus	(9,130)
Interest, etc., paid	1,205	Interest, etc., received	
Company income—		Dividends received	187
Income tax payable		Undistributed income accruing from overseas	
Dividends payable	(2,335)		
Undistributed income			
Unincorporated enterprises income	(2,622)		
Personal income from dwelling rent	592		
Public enterprises income	544		
Total Outlay	9,317	Total Receipts	9,317

Table 3

FINANCIAL ENTERPRISES INCOME APPROPRIATION ACCOUNT
(Including banks, life assurance offices, superannuation funds and mortgage lending agencies)

	$m		$m
Depreciation allowances	30	Interest, etc., received	
Net current expenditure on goods and services	346	Dividends received	1,579
Interest paid	497	Undistributed income accruing from overseas	
Company income—			
Income tax payable			
Dividends payable	(152)		
Undistributed income			
Public enterprises income	168		
Retained investment income of life insurance funds, etc.	386		
Total Outlay	1,579	Total Receipts	1,579

APPENDIX B

Table 4
PERSONAL CURRENT ACCOUNT

	$m		$m
Personal consumption	14,583	Wages, salaries and supplements	12,464
Interest paid	211	Interest, etc., received	663
Income tax payable	2,081	Dividends	(431)
Estate and gift duties paid	182	Unincorporated enterprises income—	
Remittances overseas	84	Farm	867
Saving	(1,111)	Other	(1,755)
		Income from dwelling rent	592
		Remittances from overseas	158
		Cash benefits from public authorities	1,322
Total Outlay	18,252	Total Receipts	18,252

Table 5
PUBLIC AUTHORITIES CURRENT ACCOUNT

	$m		$m
Net current expenditure on goods and services	3,028	Indirect taxes	2,787
Subsidies	167	Income tax, estate and gift duties	3,213
Interest, etc., paid	632	Interest, etc., received	101
Overseas grants and contributions	159	Public enterprises income	712
Cash benefits to persons	1,322		
Grants towards private capital expenditure	49		
Devaluation compensation	21		
Surplus on current account	1,435		
Total Outlay	6,813	Total Receipts	6,813

Table 6
OVERSEAS CURRENT ACCOUNT

	$m		$m
Exports of goods and services	3,541	Imports of goods and services	4,159
Interest, etc., received from overseas	114	Interest, etc., paid overseas	394
Dividends receivable from overseas		Dividends payable and profits remitted overseas	
Undistributed income accruing from overseas	17	Undistributed income accruing to overseas residents	115
Personal remittances from overseas	158	Personal remittances overseas	84
Overseas balance on current account	1,081	Public authority grants and contributions	159
Total Debits to Non-residents	4,911	Total Credits to Non-residents	4,911

Table 7
NATIONAL CAPITAL ACCOUNT

	$m		$m
Gross fixed capital expenditure—		Depreciation allowances	2,049
Private	(3,968)	Increase in income tax and dividend provisions	(33)
Public enterprises	1,307	Undistributed company income accruing to residents	(710)
Public authorities	986	Retained investment income of life insurance funds, etc.	386
Increase in value of stocks	(311)	Personal saving	(1,111)
Total Capital Expenditure	6,572	Public authority grants towards private capital expenditure	49
Statistical discrepancy	303	Devaluation compensation	21
		Public authorities surplus on current account	1,435
		Deficit on current account from overseas—	
		Withdrawal from overseas monetary reserves	−78
		Net apparent capital inflow	1,159
Total Finance for Capital Expenditure	6,875	Total Finance for Capital Expenditure	6,875

The methods and sources used for estimating the published figures of national income and of the various items in the social accounts have not been published in any detail. However, a general idea of the methods and sources of data can be gained by reference to—

Papers presented at the Conference of British Commonwealth Statisticians, 1951 (Government Printer, Canberra), pp. 165–9.

Youngman, D. V.: *The Estimation of Farm Income* (Paper read to Section G of the Australian and New Zealand Association for the Advancement of Science, Sydney, 1952) (Mimeographed).

Haig, B. D.: *Quarterly Estimates of National Income and Expenditure* (Paper read to Section G of the Australian and New Zealand Association for the Advancement of Science, Brisbane, 1960) (Mimeographed).

Haig, B. D. and McBurney, S. S.: *The Interpretation of National Income Estimates* (A.N.U. Press, 1968).

Australian National Accounts, National Income and Expenditure, latest issue (Commonwealth Bureau of Census and Statistics, Canberra).

National Income and Expenditure, latest issue (Government Printer, Canberra).

The Australian Balance of Payments, latest issue (Commonwealth Bureau of Census and Statistics, Canberra).

Quarterly Estimates of National Income and Expenditure, latest issue (Commonwealth Bureau of Census and Statistics, Canberra).

In 1961 the Reserve Bank of Australia published flow-of-funds statements for Australia 1953–54 to 1957–58. The most recent issue

APPENDIX B   487

covers the years 1961–62 to 1965–66. These statements concentrate on transactions in financial claims. They show the sources and uses of funds classified into some twenty types of financial transactions for twelve sectors of the economy (mainly financial sectors), some of which are further divided into sub-sectors. Claims have generally not been classified in sufficient detail to enable information of a "from-this-sector-to-that-sector" kind. For a detailed discussion of the statements, reference should be made to—

Staff Paper: Flow-of-Funds, Australia, 1953–54 to 1961–62 (by A. S. Holmes), (Reserve Bank of Australia, Sydney, 1965).

The first major work in the field of input-output data in Australia was done by Professor Burgess Cameron of the Australian National University. Transaction tables for 1946–47, 1953–54 and 1955–56 for an eighty, forty and twenty sector classification respectively were published in the *Economic Record* with explanation, comment and extension of the analysis. The first semi-official input-output tables for Australia were published in 1964. These tables were compiled for the year 1958–59. They were the result of a preliminary examination of the relevance of existing statistical information for the purpose of compiling a more comprehensive and accurate input-output table for a more recent year. They included transactions tables for thirty-four producing sectors. Reference should be made to—

Cameron, Burgess: "The 1946–7 Transaction Table," *Economic Record*, Vol. XXXIII, No. 66, 1957, p. 353; "New Aspects of Australia's Industrial Structure," *Economic Record*, Vol. XXXIV, No. 69, 1958, p. 362; "Inter-Sector Accounts, 1955–56," *Economic Record*, Vol. XXXVI, No. 74, 1960, p. 269.

Input-Output Tables, 1958–59 (Commonwealth Bureau of Census and Statistics, Canberra).

4. INDEX NUMBERS OF PRICES

Descriptions of the main Australian price indexes are set out below. With the exception of the Import Price Index they are all prepared by the Commonwealth Bureau of Census and Statistics.

Consumer Price Index

The Consumer Price Index was first published in 1960. It extends back to the September Quarter, 1948, and is published quarterly in special releases of the Commonwealth Bureau and in the *Monthly Review of Business Statistics*, the *Digest of Current Economic Statistics*, the *Quarterly Summary of Australian Statistics* and the *Labour Report*. It is computed for the six State capital cities separately and in combination, and for Canberra. Indexes for the major commodity groups within the complete index are also computed. The Consumer Price Index was preceded by the "C" Series Retail Price Index and the Interim Retail Price Index. The "C" Series was, until 1954, the principal Australian retail price index, being a continuous series extending back to 1914.[1] In its last years its regimen and weighting became increasingly out of date,[2] and it was supplemented in 1954 by the computation of the Interim Retail Price Index, which was extended back to 1950–51.[3]

The Consumer Price Index, like its predecessors, is designed, in the words of the Commonwealth Statistician, "to measure quarterly variations in retail prices of goods and services representing a high proportion of the expenditure of wage earner households." With the frequent and marked changes which occurred in the quality and pattern of expenditure since the end of the Second World War, the determination of a fixed regimen, which would have remained representative for any length of time, was found to be virtually impossible. Consequently the Consumer Price Index was designed as a series of short-term indexes linked together to form a continuous chain (see p. 361 above). Between 1948 and 1968 five separate indexes have been employed (1948–52, 1952–56, 1956–60, 1960–63 and 1963 onwards). Further revisions in the content of the index involving additional short-term indexes will be made as occasion demands.

Each short-term index is a fixed weight aggregative type, namely

$$P_{ot} = \frac{\Sigma p_t q}{\Sigma p_o q}$$

However, it is probably more convenient to think of this formula as a weighted arithmetic mean of price relatives

$$P_{ot} = \sum \left[\frac{p_t}{p_o} \left(\frac{p_o q}{\Sigma p_o q} \right) \right]$$

in order to examine the relative weighting of the items covered by the index.

The *regimen* of the index is divided into five major groups: food, clothing and drapery, housing, household supplies and equipment, miscellaneous. These major groups are themselves broken up into sections. Within sections there are 276 distinct items and a much greater number of grades, types, brands, etc., for which prices are obtained. The following table sets out the regimen, together with the

---

[1] See *Labour Report*, No. 40, 1951 (Commonwealth Bureau of Census and Statistics, Canberra) Appendix Section V, p. 160, and *Labour Report* No. 41, 1952 (Commonwealth Bureau of Census and Statistics, Canberra) Chapter I, p. 1 and Appendix Section V, p. 164.

[2] See Karmel, P. H.: *Applied Statistics for Economists* (Pitman, Melbourne, 1957) (1st edition), pp. 317–320 and 431–433.

[3] See *Labour Report*, No. 46, 1958 (Commonwealth Bureau of Census and Statistics, Canberra) Chapter I, p. 1.

APPENDIX B

The Consumer Price Index

Composition and Weighting Pattern as at December Quarter, 1963, for the Six State Capital Cities Combined

Group, Section, etc.	Percentage Weight	
	Section	Group
Food—		32·1
Cereal Products: Bread, flour, biscuits, rice and breakfast foods	4·0	
Dairy Produce: Milk, cheese, butter and eggs . .	7·1	
Potatoes, Onions, Preserved Fruit and Vegetables: Potatoes and onions, canned and dried fruits and canned and frozen vegetables	1·9	
Soft Drink, Ice Cream and Confectionery . .	4·0	
Other (except Meat): Sugar, jam, margarine, tea, coffee, baby foods and sundry canned and other foods	4·1	
Meat:		
Butcher's (Beef, mutton, lamb and pork) . .	9·1	
Processed (Bacon, small goods and canned meat) .	1·9	
Clothing and Drapery—		16·9
Men's Clothing	4·1	
Women's Clothing	6·5	
Boys' Clothing	0·6	
Girls' Clothing	1·0	
Piecegoods, etc.: Wool, cotton and rayon cloth, nursery squares and knitting wool . . .	1·0	
Footwear: Men's, women's and children's . .	2·7	
Household Drapery: Bedclothes, towels, tablecloths, etc.	1·0	
Housing—		12·6
Rent:		
Privately-owned houses	2·8	
Government-owned houses	0·8	
Home Ownership:		
House price	5·2	
Rates	2·6	
Repairs and maintenance	1·2	
Household Supplies and Equipment— . . .		14·5
Fuel and Light:		
Electricity.	2·4	
Gas	1·3	
Other (Firewood and kerosene)	0·9	
Household Appliances: Refrigerator, washing machine, stoves, radio set, television set, vacuum cleaner, electric iron, etc.	3·6	
Other Household Articles:		
Furniture and floor coverings	2·2	
Kitchen and other utensils, gardening and small tools	0·9	
Household sundries (Household soaps, etc.) . .	1·0	
Personal requisites (Toilet soap, cosmetics, etc.) .	1·1	
Proprietary medicines	1·0	
School requisites	0·1	

Group, Section, etc.	Percentage Weight	
	Section	Group
MISCELLANEOUS—		23·9
Transport		
Fares:		
Train	1·2	
Tram and bus	1·9	
Private motoring:		
Car purchase	3·0	
Car operation	4·4	
Tobacco and Cigarettes	3·9	
Beer	3·8	
Services:		
Hairdressing (Haircuts, wave, etc.)	0·7	
Drycleaning	0·5	
Shoe repairs	0·3	
Postal and telephone services	0·9	
Other:		
Radio and television operation	1·3	
Cinema admission	0·7	
Newspapers and weekly magazines	1·3	
	100·0	100·0

Source: Commonwealth Bureau of Census and Statistics, Australia: *Labour Report*, No. 52, 1965 and 1966, p. 8

relative weighting as at the December Quarter, 1963. This is the weighting relevant to the short-term index currently (1968) being used.

The *prices* for the groups other than housing are the averages of prices obtained from representative and reputable retailers and service establishments in each city, for each class of commodity and each service covered by the index. Prices are collected from vendors and are those actually being charged for normal cash purchases of new articles. "Bargain" or "Sale" prices of imperfect goods or discontinued lines are not used. Otherwise actual transaction prices (e.g. those reflecting grocery "specials") are used in all cases. Prices are collected for specified standards of goods and services. When qualities change, adjustments are made (by splicing, where appropriate, see pp. 362–364 above) to ensure that changes in prices are alone reflected in the index. The actual collection of prices is carried out by qualified field officers, who check the returns and make certain that the price quotations are for items of the type and standard required for the index.

The main price elements in the housing group are rents of privately-owned houses, rents of government-owned houses, house prices and rates, repairs and maintenance. The first of these are obtained from returns furnished by house agents in each city, relating to weekly rents

APPENDIX B 491

for four-, five- and six-roomed houses classified according to whether constructed of wood or brick. These returns cover an extensive sample of houses (currently numbering about 3,000) selected by field officers as being of appropriate standards. The rents of government-owned houses are obtained from housing authorities. The prices of houses are collected from private and governmental bodies engaged in constructing or financing houses for home ownership. The prices are for new houses and exclude the value of land. They comprise contract prices, sale prices and estimated building costs per hundred square feet. They are obtained for houses in selected representative categories classified by size, type of construction and material of walls. To smooth out random fluctuations in price data, four quarter moving averages are used, so that the house prices used currently in the index relate to prices over the preceding year.

As has been pointed out above, the regimen and weights of the index have been revised four times in the series extending back to 1948. The *weighting* of the index is currently (1968) based on the 1961–62 pattern of expenditure. The weights are derived from analyses of statistics of production and consumption, the Population Census of 1961, the Census of Retail Establishments of 1961–62 and the continuing Survey of Retail Establishments, and from special information obtained from trade sources and sample surveys.

Broadly speaking, the weights can be taken to reflect the relative distribution of average household expenditure between the various items in the regimen. But they only *reflect* this, and must not be taken as estimates of this relative distribution. This is so, for a number of reasons—

(i) Whereas the average household expenditure figures for most items relate to the community as a whole, the relative weighting of the Housing Group, private motoring, and several other items has been modified to reflect the pattern of wage and salary earner households.

(ii) Some items are used to represent omitted items and, for this reason, carry additional weight.

(iii) Some lines of expenditure are not fully represented (e.g. fresh fruit and vegetables, other than potatoes and onions, are not included) and others are not represented at all (e.g. medical, dental and hospital fees).

There are some differences between regimens used for the individual capital cities. The individual cities are combined to produce the Six Capital Cities index by weighting by the populations of the cities as at the 1961 Census.

The *computation* of the index follows the formula, and is relatively straightforward, except for the Housing Group. Households fall into three main groups: those renting from a private owner, those renting from a government authority and those who own or are purchasing their own houses. These three groups are represented in the index. For the first two groups the weighting is simply the estimated expenditure on rent for privately-owned and government-owned houses respectively in the base year relative to total household expenditure on items included in the index. For the third group items covering house price, municipal, water and sewerage rates, and repairs are included. The weighting for the house price is obtained by estimating a normal rate of purchase of new houses per household over the intercensal period 1947 to 1961, and multiplying this by a basic average price to obtain a basic average expenditure. This is then expressed relative to total household expenditure on items included in the index. The normal rate of purchase is estimated as what the ratio of the number of new houses acquired per annum to the stock of houses would have been if the population had been stationary. To the extent to which new houses in a stationary population are purely for replacement purposes, the ratio of new houses to the stock of houses is, in fact, the reciprocal of the average life of the houses,[1] so that this ratio is a measure of the average rate of depreciation per annum. To this extent, then, the weighting method adopted for owner-occupied houses implies that each owner "consumes" a certain fraction of a house per year, and this fraction is priced in the index at current prices. (However, no doubt some new houses in the intercensal period were second houses (e.g. holiday homes), and, thus, some allowance for these is included.) This is consistent with the treatment of other items in the regimen, which essentially relates to households' expenditure. No weight is included in the owner-occupied houses section for interest charges on moneys borrowed in connexion with house-purchase. In the same way, hire-purchase charges on motor vehicles and consumer durable goods are omitted.

The measurement of the price changes relevant to the Housing Group is somewhat complicated. A basic requirement in the compilation of any index number is that the quality of all items must be kept constant over periods for which price comparisons are made; otherwise the index will cease to measure price changes only and will include changes due to quality changes (see p. 362 above). In the case of houses, this requirement is especially difficult to achieve. Houses by their nature change in quality. Existing houses age as time passes. They are reconstructed and extended. New houses differ in style and amenities. Consequently, a comparison of the average rent or price paid for houses in two periods will inevitably contain elements of a

[1] i.e. the true death rate of the population of houses. See pp. 442–43 above.

comparison of quality as well as of price. But it is only the price element which should be included in the index.

The following are the methods used in the Consumer Price Index in an attempt to overcome this difficulty. For privately-rented houses, as has been pointed out above, the rents used are ones collected from house agents. The average rise in rent from quarter to quarter is calculated for those houses which are on the agents' books for *both* quarters, for each of the six classes of four-, five- and six-roomed, wood or brick houses. For example, suppose that between March Quarter, 1968, and June Quarter, 1968, the average rent of the sample of four-roomed brick houses (privately rented) of the appropriate quality which were on the agents' books in both quarters rose by 3 per cent; and suppose that between June Quarter, 1968, and September Quarter, 1968, the average rent of similar (but not necessarily identical) houses which were on the agents' books in both these quarters rose by 4 per cent; then we should argue that, compared with the base, March Quarter, $1968 = 100$, average rents of this type of house for the September Quarter, 1968, stood at $100 \times 1 \cdot 03 \times 1 \cdot 04 = 107 \cdot 12$. Moreover this represents a pure price change, since all comparisons have been made between identical houses.[1] The sample of houses used for these computations is carefully watched by field officers to keep its quality as constant as possible and revised from time to time.

As far as the rents charged by government authorities are concerned, the above method would hardly be suitable, since these rents are only infrequently varied after the houses are initially occupied. Under a regime of rising building costs, measurement of rent changes by this method would lead to a serious understatement of rent changes. Thus, if the rents of these houses were rigidly fixed at levels based on building costs, the rent of no single house would change, although the average level of rent actually paid would rise more and more as new houses were built at higher costs. The method used for this section of the Housing Group is to average all rents actually paid in appropriate categories and to weight these categories to provide a measure of average changes in these rents. Since the newly built houses are of much the same quality as the old ones, the quality of the houses is fairly constant. However, some quality changes are certainly taking

[1] This contention is not strictly valid. If rent tends to vary inversely with age of house, other things being equal, this method inevitably introduces a cumulative bias in rent variation due to the ageing of houses. Thus, if over a period of years the rent of no house changed except by reason of its increasing age, this method would produce an index showing a continuous decline in rents, in spite of the fact that there were no "price" variations in the rents of houses. Moreover this would be so, even if the sample of houses used were kept constant in respect of age distribution, since the comparisons are between the *same* houses in two consecutive quarters and hence some ageing is unavoidable. On the other hand, improvements are made to many houses from time to time, so that perhaps these can be assumed to offset the ageing.

place and these are inevitably included. Thus the measure of changes in the rents of government-owned houses relates to a changing stock of houses.

Reference has already been made to the collection of data for measuring changes in house prices. An attempt is made to keep quality reasonably constant by classifying houses into certain quality categories, measuring price changes within those categories, and combining these price changes in fixed proportions reflecting the relative importance of the categories. This achieves reasonable constancy of quality for short-period comparisons. For long-period comparisons it may be impossible to hold quality constant in the face of trends in house design.

COMPARISON OF WEIGHTING OF "C" SERIES RETAIL PRICE INDEX AND CONSUMER PRICE INDEX AT MARCH QUARTER, 1960

Components (b)	Percentage Contribution (a)	
	"C" Series	Consumer Price Index
Food Group		
Meat (Butcher's)	16·5 ⎫	8·8 ⎫
Potatoes and Onions	1·9 ⎬ 41·9	1·0 ⎬ 32·1
Soft Drink, Ice Cream and Confectionery	(c) ⎥	4·0 ⎥
Other Food	23·5 ⎭	18·3 ⎭
Clothing and Drapery Group	29·6	19·0
Housing Group		
Home Ownership	(c) ⎫	7·8 ⎫
Rent of Privately-owned Houses	13·3 ⎬ 13·3	2·0 ⎬ 10·7
Rent of Government-owned Houses	(c) ⎭	0·9 ⎭
Household Supplies and Equipment Group		
Fuel and Light	4·2	4·2 ⎫
Household Appliances ⎫	0·7	4·5 ⎬ 13·2
Other Household Supplies ⎭		4·5 ⎭
Miscellaneous Group		
Transport:		
Fares (Rail, tram and bus)	3·5 ⎫ 15·2	4·4 ⎫
Private Motoring	(c) ⎥	6·9 ⎥
Tobacco and Cigarettes	2·1 ⎬	3·9 ⎬ 25·0
Beer	(c) ⎥	4·1 ⎥
Services, Cinema, Radio and Television Licences and Newspapers	4·7 ⎭	5·7 ⎭
	100·0	100·0

(a) Weighted average of six capital cities.
(b) Listed under Consumer Price Index headings.
(c) Not included.

Source: Commonwealth Bureau of Census and Statistics, Australia: *Labour Report*, No. 48, 1960, p. 39

A comparison of the Consumer Price Index with the old "C" Series illustrates quite strikingly the problems of out-of-dateness in regimen and weights. The Consumer Price Index contains nearly twice as many distinct items. In particular a wide variety of consumer durables and services, which were completely unrepresented in the "C" series, has been included. The change in the relative weighting of broad groups is shown in the table above.

In the Consumer Price Index less relative weight is attached to food (particularly meat), clothing and housing expenditure, and considerably more weight to household appliances, private motoring and services (few of which were included in the old index). This reflects not only substitution against food and clothing, the prices of which have risen much more than the average since the regimen for the "C" Series was determined, but also the effect of rising consumption standards leading to a smaller proportion of expenditure being devoted to necessities.

In the table below is set out a comparison of movements in the Consumer Price Index and the "C" Series. The indexes are for the six capital cities with base 1952–53 = 100·0.

RETAIL PRICE INDEXES
AUSTRALIA, 1949–1960

June Quarter	"C" Series	Consumer Price Index
1949	62·1	62·9
1950	67·8	68·4
1951	81·1	80·7
1952	97·6	97·2
1953	101·4	101·4
1954	102·8	102·1
1955	105·0	103·6
1956	111·8	110·2
1957	113·8	113·7
1958	115·3	114·8
1959	118·4	116·8
1960	125·5	121·1

Source: Commonwealth Bureau of Census and Statistics, Australia: *Labour Report*, No. 48, 1960, p. 40

The Consumer Price Index has shown a smaller rise in prices since 1952–53 than the "C" Series. This is mainly due to the relatively lighter weight in the former index of meat and privately-rented houses, which have risen more in price over this period than the average, and to the inclusion of consumer-durables, whose prices have risen less over this period than the average.

Finally, three comments on the Consumer Price Index. First, since the index consists of a chain of relatively short links, one cannot strictly speaking describe it as measuring the changes in the cost of a given collection of goods and services. Unavoidably, precision in interpretation has had to be sacrificed for up-to-dateness in weighting (see p. 349 above).

Secondly, the regimen and weighting of the index is vastly improved compared with the "C" Series. Indeed the only important gap appears to be fresh fruit and vegetables, other than potatoes and onions. The inclusion of fresh fruit and vegetables would present difficulties, both because it is difficult to ascertain the prices of fruit and vegetables of a constant quality and because at certain times of the year some of the items are not available and consequently have no price. In addition, the prices of fruit and vegetables exhibit a great deal of seasonal variation so that the question would arise whether this variation would obscure quarterly variations in the rest of the index or whether the index should show the deseasonalized trend in price movements. These are real difficulties. They have been tackled in various ways in the countries which include some fruit and vegetable items in their retail price indexes. In Australia, the prices of fruit and vegetables vary erratically in the short period, and their movements often differ violently in degree and direction from those of other commodities.

Thirdly, while the Housing Group is much more satisfactory in concept and extent than its counterpart in the "C" Series, there are still some outstanding questions, the answers to which must depend on the purpose of the index being constructed. In particular what should be done with interest payable on houses being purchased by instalments? It would seem that if the Consumer Price Index is to be conceived in terms of a measure of the price component of households' expenditure, some account should be taken of such interest payments and possibly also of the cost of land. Their omission results in an underweighting of the Housing Group. On the other hand, variations in interest rates would have to be suitably averaged since many interest payments are not adjusted with changes in current interest rates and the measurement of variation in the price of land could present considerable difficulties.

REFERENCE

Labour Report, No. 52, 1965 and 1966, p. 8 (Commonwealth Bureau of Census and Statistics, Canberra), Chapter 1, pp. 1–37.

Wholesale Price (Basic Materials and Foodstuffs) Index

This index is calculated monthly, starting with January, 1928 (although a now obsolete index carries back to 1861), and is published in the *Labour Report*, the *Monthly Review of Business Statistics* and in other publications. Its object is to measure the price movements of basic

APPENDIX B   497

commodities entering into production for home consumption, and the commodities are priced in their primary or basic form. Its regimen contains seventy-six commodities. The regimen is divided into seven groups: metals and coal; oils, fats and waxes; textiles; chemicals; rubber and hides; building materials; foodstuffs and tobacco. Index numbers are computed for each group and for the combined seven groups. In addition, two indexes are published referring to those goods entering the regimen which are principally home-produced and those which are principally imported. The latter is not a measure of changes in import prices generally, since it includes only imported basic materials. Prices are mainly Melbourne prices and are collected from manufacturers and merchants. The weights are based on average annual consumption from 1928–29 to 1934–35 inclusive, and the formula is of the fixed-weight aggregative type.

Although the regimen contains seventy-six items, it is dominated by fourteen items. These items are iron and steel, coal, petrol, wool, timber (soft and hard), tea, tobacco, wheat, sugar, beef, mutton, butter-fat and milk, and they accounted for 80 per cent of the aggregate cost of the regimen in 1960. In particular coal and beef alone accounted for 25 per cent. The effect of such dominance is discussed on pp. 359–360 above.

The validity of the weighting and the representativeness of the index have become increasingly affected by changes in usage and industrial structure; and work is currently (1968) proceeding on the preparation of new series of wholesale price index numbers relating to materials used and articles produced by sectors of industry.

REFERENCE

Labour Report, No. 52, 1965 and 1966 (Commonwealth Bureau of Census and Statistics, Canberra), Chapter 2, pp. 38–44.

Export Price Indexes

The current export price index uses a fixed-weight aggregative formula, with reference base 1959–60. It is published in special releases and in the *Digest of Current Economic Statistics*, in the *Monthly Review of Business Statistics* and the *Quarterly Summary of Australian Statistics*.

The regimen contains twenty-nine items, constituting about 83 per cent of the value of exports. As is proper in this index, wool is dominant. In 1959–60, wool made up about 50 per cent of the aggregate cost of the regimen. The weights are based on average annual exports 1956–57 to 1960–61. The prices used relate to specified standards for each commodity and in most cases are combinations of prices for a number of representative grades, types, etc. The indexes are published for individual commodities (wool, sugar, gold) and groups of commodities (meats, dairy produce, cereals, dried and canned fruits, hides and tallow, metals and coal) as well as for the regimen as a whole.

REFERENCE

Official Year Book of the Commonwealth, No. 54, 1968 (Commonwealth Bureau of Census and Statistics, Canberra), Chapter 10, pp. 273–6.

Import Price Index

This index, which is prepared in the Research Department of the Reserve Bank of Australia and published in the *Reserve Bank of Australia Statistical Bulletin*, is a fixed-weight aggregative type. The index is available from 1928 onwards. It is at present calculated monthly but quarterly figures only are available prior to July 1955.

The weighting system has been revised periodically. The latest revision, using 1962–63 as a base, was calculated back to 1960–61 and linked with earlier data compiled using different weights. Until the latest revision, the index measured f.o.b. prices at the time of exit from the country of origin, but the index now attempts to measure prices f.o.b. at the time of entry into Australia. The effects of changes in freight and insurance charges are excluded from the index. Imports of live animals, aircraft and parts, ships, passengers' effects, military equipment and imports due for re-export are not covered.

A classification of the total index by commodity groups is published monthly. The following categories are distinguished: food, drink and tobacco; basic materials; fuels and lubricants; textiles; base metals; metal manufactures; motor vehicles; electrical machinery; other machines and machinery; and other manufactures. Details of a classification of the total index by source of origin are also published monthly. This classification distinguishes: United Kingdom; U.S.A. and Canada; European Economic Community; Japan; and Other Countries.

In principle, the index is a fixed-weight aggregative type, but the clarity of this definition is reduced by the fact that certain of the "prices" included are, in fact, index numbers themselves. Because of the difficulty in obtaining figures of actual prices paid by importers, the index is based largely on overseas price indicators which are lagged by varying amounts depending on average shipping times from the sources of origin. Thus, the price index for goods imported from the United Kingdom is based almost entirely on unit value indexes for exports, while that for imports from the U.S.A. and Canada is mainly based on wholesale price indexes.

REFERENCE

Statistical Bulletin, Economic Supplement, December 1967 (Reserve Bank of Australia) pp. 41–43.

Wage Rates Indexes

These indexes are calculated monthly and date back to 1939 for adult males and 1951 for adult females (although a now obsolete index carries back to 1891 for males and 1914 for females). They are published in the monthly bulletin *Wage Rates and Earnings*, the *Monthly Review of Business Statistics*, the *Digest of Current Economic Statistics*, the *Labour Report* and other publications.

The indexes are of the fixed-weight aggregative type and their object is to measure movements in minimum weekly and hourly wages as prescribed in the awards and determinations of industrial tribunals or as specified in formal agreements. The regimen consists of specified representative occupations in fifteen distinct industry groups, covering all industries except rural. The weights for the industries are based on the number of employees covered by awards, determinations and agreements in 1954. Within industry groups, the weights for occupations are based on the number of employees within selected awards.

The minimum wage rates used in the indexes are derived from awards, determinations and agreements in force at the end of each month. For adult males 3,415 award designations, covering 2,313 distinct award occupations are included. For females the figures are 1,100 and 515 respectively. The regimen is kept up to date by including representative occupations from awards in new industries, where these are important in particular industry groups.

The indexes are derived from the averages of awards, weighted in the way indicated above, and are published both as weighted average minimum weekly wage rates (in dollars and cents) and as index numbers with base $1954 = 100 \cdot 0$. The aggregate weighted averages and indexes (for adult males and adult females separately) are published for Australia and the six States separately. Weighted averages and indexes are also published for the fifteen industry groups separately.

The Commonwealth Statistician also publishes weighted average weekly wage rates, split into Commonwealth and State awards separately for Australia and for each State, in the monthly bulletin, *Wage Rates and Earnings*.

These indexes relate to prescribed *minimum* wage rates for a full week's work, for a given structure of occupations, and they must not be used as indicators of changes in average earnings. Movements in average earnings can diverge from movements in minimum adult wage rates by reason of changes in industrial structure, in age structure, in amount of overtime worked, in amount of bonuses or over-award payments made, and in occupational grading. A series of average weekly wage and salary earnings per employed male unit, based on pay-roll tax and other statistics, is published in the *Monthly Review of Business Statistics* and the monthly bulletin, *Wage Rates and Earnings*.

REFERENCES

Labour Report, No. 52, 1965 and 1966 (Commonwealth Bureau of Census and Statistics, Canberra), Chapter 3, pp. 58–84.

Minimum Weekly Wage Rates, 1939 *to* 1965 (Commonwealth Bureau of Census and Statistics, Canberra).

Wage Rates and Earnings, latest issue (Commonwealth Bureau of Census and Statistics, Canberra).

5. MEASURES OF QUANTUM

Until comparatively recently there were no official measures of quantum of gross national product or industrial production in Australia. However, in the early 1950s, several private research workers produced indexes of quantum of industrial production of varying degrees of quality, and in 1954 the Australia and New Zealand Bank began publishing an index of factory production.

This index was originally based on 1948–49 and carried back to that year. However, in 1958, 1960, 1964 and 1967, it was revised substantially by extending its coverage and bringing its weighting up to date. The method of construction of the index is that of indicators with fixed weights given by 1963–64 values added (see section 14.5, p. 386 above). There are 448 series included in the index and its coverage is stated to be 73 per cent of the value added in all factory industry, including fuel and power. Movements in actual production are used as indicators in the case of about 90 per cent of the items, while input indicators are used for the majority of the other items. Other indicators include value series and employment series. The index is calculated monthly and published in the *Quarterly Survey* of the Australia and New Zealand Bank and in special monthly releases. The index is corrected for variations in the length of the working month and for public holidays, and is seasonally adjusted. Broadly speaking, the index corresponds to the usual concept of an index of industrial production. It is published for total factory production, including and excluding power, and for nine major industrial groupings and for a number of sub-groupings. The major groupings are: metals, machinery and apparatus; transport equipment; building and construction materials; furniture and household goods; textiles, clothing and footwear; food, drink and tobacco; chemicals and allied industries; miscellaneous industries; fuel and power.

Since 1963, annual estimates of gross national product at constant prices have been published in *Australian National Accounts—National Income and Expenditure* and in the budget paper *National Income and Expenditure*. These are derived by revaluing the components of gross national expenditure (see section 14.2, p. 370 above). The period 1948–49 to 1967–68 has been covered by two overlapping series, one

APPENDIX B

Gross National Product at Current and Average 1959-60 Prices
($ million)

—	1963-64	1964-65	1965-66	1966-67	1967-68 (a)

At Current Prices

	1963-64	1964-65	1965-66	1966-67	1967-68 (a)
Personal consumption expenditure	11,076	11,943	12,629	13,498	14,583
Current expenditure by financial enterprises and public authorities	2,011	2,296	2,666	3,006	3,374
Gross fixed capital expenditure—					
Public	1,565	1,796	1,984	2,105	2,293
Private—Dwellings	754	890	886	959	1,066
Other	2,130	2,481	2,768	(2,724)	(2,902)
Increase in value of stocks	174	681	232	494	(311)
Statistical discrepancy	12	195	176	252	303
Total Gross National Expenditure	17,722	20,282	21,341	23,038	24,832
Gross National Expenditure after stock valuation adjustment	17,601	20,111	21,138	22,763	24,621
Plus Exports of goods and services	3,162	3,048	3,137	3,474	3,541
Less Imports of goods and services	2,864	3,478	3,627	3,702	4,159
Gross National Product after stock valuation adjustment	17,899	19,681	20,648	22,535	24,003

At Average 1959-60 Prices

	1963-64	1964-65	1965-66	1966-67	1967-68 (a)
Personal consumption expenditure	10,351	10,825	11,072	11,538	12,102
Current expenditure by financial enterprises and public authorities	1,761	1,917	2,165	2,300	2,450
Gross fixed capital expenditure—					
Public	1,387	1,542	1,653	1,686	1,784
Private—Dwellings	712	817	788	830	900
Other	2,050	2,319	2,520	(2,420)	(2,526)
Statistical discrepancy	11	175	153	212	(247)
Gross National Expenditure after stock valuation adjustment	16,319	18,066	18,371	19,186	20,085
Plus Exports of goods and services	2,930	2,928	2,967	3,337	3,549
Less Imports of goods and services	2,969	3,503	3,639	3,689	4,111
Gross National Product after stock valuation adjustment	16,280	17,491	17,699	18,834	19,523

(a) Preliminary.

Source: Commonwealth of Australia, *National Income and Expenditure 1967-68* (Government Printer, Canberra 1968), p. 16.

at average prices of 1953–54 (1948–49 to 1959–60) and the other at average prices of 1959–60 (1953–54 to 1967–68).

A number of methods are used in preparing these estimates. For some components (e.g. for a considerable part of personal consumption expenditure) the method adopted is to revalue directly quantities at base-year prices. For other components it is often not possible to express the values in successive years as the product of prices and homogeneous units of quantity. These components are revalued by means of price or volume indexes for final use goods, or where this is impracticable (e.g. for current expenditure by financial enterprises and public authorities, and for parts of gross fixed capital expenditure) values are divided by indexes of input costs (i.e. prices of labour and materials). The table above illustrates the revaluation of Australian gross national product at constant prices.

The Commonwealth Statistician also publishes Indexes of Factory Production, Indexes of Quantum of Farm Production, and Indexes of the Value of Exports and Imports of Merchandise at Constant Prices. The annual Indexes of Factory Production were first published in 1965. They are published annually with a lag of some eighteen months after the end of the year to which they relate. There are two overlapping series. The first covers the period 1949–50 to 1959–60, at constant 1955–56 prices; the second covers 1955–56 and the period commencing 1959–60 at constant 1959–60 prices. Monthly or quarterly indexes of factory production are not yet available. The annual indexes comprise indexes of output at constant prices for 64 individual sub-classes, which are combined by the use of value-added weights into indexes for thirteen classes and all classes combined (see section 14.5, pp. 386–91 above). The coverage of the index at 1959–60 prices was 72 per cent of the value of output of factory production. A detailed description of concepts and methods is contained in each issue of the annual bulletin.

Indexes of quantum of farm production are published each year in the statistical bulletin *Value of Production and Indexes of Price and Quantum of Farm Production*, and are available back to 1911–12. The indexes are of the fixed-weight aggregative type, the weights currently in use being average unit gross values for the period 1936–37 to 1938–39. Separate indexes are published for agriculture (including an index for wheat production), pastoral (including an index for wool production) and dairying, poultry and bee-farming, as well as an index for all farm production. The indexes are also published in the *Official Year Book*, together with indexes of the quantum of production, exports and consumption of farm products for food use.

Indexes of the value of exports and imports at constant prices have been published for the period from 1947–48 onwards. The period

APPENDIX B

from 1947–48 to 1966–67 is covered by two overlapping series, one at constant 1955–56 prices (for the period from 1947–48 to 1959–60) and one at constant 1959–60 prices (1959–60 to 1966–67). Annual indexes have been published for the whole period, while indexes have been published for quarters from the September quarter, 1957 onwards. These series were published in the half-yearly *Balance of Payments* bulletins and in the *Monthly Review of Business Statistics*. From 1966–67 onwards, a new series of indexes have been published at constant 1966–67 prices, and a detailed description of the series is contained in the initial bulletin *Exports and Imports of Merchandise at Constant Prices*. The indexes are now published quarterly in a bulletin of this title, as well as in the other publications referred to above. Broadly, the indexes are derived by expressing the value of exports (or imports) at the prices of the base year, either by direct revaluation of quantities at base-year prices or by division of values by price indexes.

REFERENCES

Australian National Accounts, National Income and Expenditure, latest issue (Commonwealth Bureau of Census and Statistics, Canberra).

Indexes of Factory Production, latest issue (Commonwealth Bureau of Census and Statistics, Canberra).

Exports and Imports of Merchandise at Constant Prices, latest issue (Commonwealth Bureau of Census and Statistics, Canberra).

Value of Production and Indexes of Price and Quantum of Farm Production, latest issue (Commonwealth Bureau of Census and Statistics, Canberra).

Official Year Book of the Commonwealth, No. 53, 1967 (Commonwealth Bureau of Census and Statistics, Canberra), Chapter 30, pp. 1210–12.

Quarterly Survey, Vol. IV, No. 1, October, 1954, p. 14; No. 2, January, 1955, p. 15; Vol. VII, No. 2, January, 1958, p. 12; No. 3, April, 1958, p. 18; No. 4, July, 1958, p. 18; Vol. X, No. 2, January, 1961, p. 5; Vol. XIII, No. 3, April, 1964, p. 4; Vol. XVII, No. 1, October, 1967, p. 5; and following issues (Australia and New Zealand Bank Ltd., Melbourne).

Stevens, S. P.: "An Index of Australian Manufacturing Production," *Economic Record*, Vol. 30, No. 58, 1954, p. 89.

Carver, S. R.: "Indexes of Volume and Quantum," *Economic Record*, Vol. 30, No. 59, 1954, p. 274.

Horner, F. B.: "The Meaning of Production Indexes," *Economic Record*, Vol. 37, No. 77, 1961, p. 82.

Mathews, R. L.: "The Australian Flow-of-Funds Accounts," *Economic Record*, Vol. 38, No. 81, 1962, p. 94.

Haig, B. D.: "Input–Output Relationships, 1958–59," *Economic Record*, Vol. 41, No. 93, p. 118.

Haig, B. D.: "Indexes of Australian Factory Production, 1949–50 to 1962–63," *Economic Record*, Vol. 41, No. 95, p. 451.

6. Demography

Data on population are published in detail periodically in census volumes and annually in the bulletin *Demography*. Summary information is published in the *Official Year Book of the Commonwealth*, the *Quarterly Summary of Australian Statistics*, the *Monthly Review of Business Statistics*, and in the series *Australian Demographic Review*. For textual comment on Australian demographic statistics references should be made to—

Official Year Book of the Commonwealth, No. 54, 1968 (Commonwealth Bureau of Census and Statistics, Canberra), Chapters 7 and 8, pp. 117–218.

Census of the Commonwealth of Australia, 30th *June*, 1961, *Statistician's Report* (Commonwealth Bureau of Census and Statistics, Canberra).

Australian Life Tables, 1960–62 (Commonwealth Bureau of Census and Statistics, Canberra).

APPENDIX C

A SHORT LIST OF REFERENCES

The following is a small selection from the very extensive literature which exists on statistical methods and theory and on the various specialized fields covered in the text. The books listed represent convenient avenues for further study and themselves contain reading lists.

1. Statistical Methods

Allen, R. G. D.: *Statistics for Economists* (Hutchinson's University Library, 1963)

Croxton, F. E., and Cowden, D. J.: *Practical Business Statistics* (Prentice-Hall, 1960)

Croxton, F. E., and Cowden, D. J.: *Applied General Statistics* (Prentice-Hall, 1963)

Davies, O. L. (ed.): *Statistical Methods in Research and Production* (Oliver and Boyd, 1961)

Ezekiel, M., and Fox, K. A.: *Methods of Correlation and Regression Analysis* (John Wiley and Sons, 1963)

Fisher, R. A.: *Statistical Methods for Research Workers* (Oliver and Boyd, 1958)

Hoel, P. G.: *Elementary Statistics* (John Wiley and Sons, 1966)

Neiswanger, W. A.: *Elementary Statistical Methods* (Macmillan, 1963)

Suits, D. B.: *Statistics; An Introduction to Quantitative Economic Research* (Rand McNally, 1963)

Tippett, L. H. C.: *Statistics* (Home University Library, 1956)

Williams, E. J.: *Regression Analysis* (John Wiley and Sons, 1959)

Yule, G. U., and Kendall, M. G.: *An Introduction to the Theory of Statistics* (Griffin, 1958)

2. Mathematical Statistics

Cramér, H.: *Mathematical Methods of Statistics* (Princeton University Press, 1963)

Freund, J. E.: *Mathematical Statistics* (Prentice-Hall, 1962)

Goldberg, S.: *Probability—an Introduction* (Prentice-Hall, 1960)

Hoel, P. G.: *Introduction to Mathematical Statistics* (John Wiley and Sons and Chapman and Hall, 1964)

Kendall, M. G., and Stuart, A.: *The Advanced Theory of Statistics*, Vols. I, II and III (Griffin, 1961 and 1963)

Mood, A. M. and Graybill, F. A.: *Introduction to the Theory of Statistics* (McGraw-Hill, 1963)

Schlaifer, R.: *Probability and Statistics for Business Decisions* (McGraw-Hill, 1959)

Weatherburn, C. E.: *A First Course in Mathematical Statistics* (Cambridge University Press, 1961)

Wilks, S. S.: *Mathematical Statistics* (Princeton University Press, 1947)

3. Econometrics

Christ, C. F.: *Econometric Models and Methods* (John Wiley and Sons, 1966)
Goldberger, A. S.: *Econometric Theory* (John Wiley and Sons, 1964)
Hood, W. C., and Koopmans, T. C. (ed.): *Studies in Econometric Method*, Cowles Commission Monograph No. 14 (John Wiley and Sons, 1953)
Johnston, J.: *Econometric Methods* (McGraw-Hill, 1963)
Klein, L.: *Economic Fluctuations in the United States 1921–1941*, Cowles Commission Monograph No. 11 (John Wiley and Sons, 1957)
Klein, L. R.: *An Introduction to Econometrics* (Prentice-Hall, 1962)
Koopmans, T. C. (ed.): *Statistical Inference in Dynamic Economic Models*, Cowles Commission Monograph No. 10 (John Wiley and Sons, 1950)
Stone, R.: *The Measurement of Consumers' Expenditure and Behaviour in the United Kingdom 1920–38*, Vols. I–II (Cambridge University Press, 1954–66)
Tinbergen, J.: *Econometrics* (George Allen and Unwin, 1961)
Tintner, G.: *Econometrics* (John Wiley and Sons, 1963)
Wold, H., and Juréen, L.: *Demand Analysis* (John Wiley and Sons, 1964)

4. Sample Surveys

Cochran, W. G.: *Sampling Techniques* (John Wiley and Sons, 1963)
Deming, W. E.: *Some Theory of Sampling* (John Wiley and Sons and Chapman and Hall, 1950)
Deming, W. E.: *Sampling Design in Business Research* (John Wiley and Sons, 1960)
Hansen, M. H., Hurwitz, W. N., and Madow, W. G.: *Sample Survey Methods and Theory*, Vols. I and II (John Wiley and Sons, 1966)
Jones, D. C.: *Social Surveys* (Hutchinson's University Library, 1950)
Moser, C. A.: *Survey Methods in Social Investigation* (Heinemann, 1963)
United States Dept. of Commerce: *The Current Population Survey: A Report on Methodology* (U.S. Government Printing Office, Washington, 1963), Technical Paper No. 7
Waksberg, J., and Hanson, R. H.: *Sampling Applications in Censuses of Population and Housing* (U.S. Government Printing Office, Washington, 1965), U.S. Bureau of the Census Technical Paper No. 13
Yates, F.: *Sampling Methods for Censuses and Surveys* (Hafner Pub. Co., 1963)

5. Quality Control

American Defence Emergency Standards: *Guide for Quality Control and Control Chart Method of Analysing Data* (Reproduced by courtesy of the American Standards Association), B.S. 1008: 1942 (British Standards Institution, 1942)
Davies, O. L. (ed.): *Statistical Methods in Research and Production* (Oliver and Boyd, 1961), Chapters 9 and 10
Grant, E. L.: *Statistical Quality Control* (McGraw-Hill, 1964)
Huitson, Alan, and Keen, J.: *Essentials of Quality Control* (Heinemann, 1965)
Pearson, E. S.: *The Application of Statistical Methods to Industrial Standardization and Quality Control*, B.S. 600: 1935 (British Standards Institution, 1935)

6. TIME SERIES

Croxton, F. E., and Cowden, D. J.: *Applied General Statistics* (Prentice-Hall, 1963), Chapters 11–16 and 22

Kendall, M. G., and Stuart, A.: *The Advanced Theory of Statistics*, Vol. III (Griffin, 1963), Chapters 45–47

Tinbergen, J., and Polak, J. J.: *The Dynamics of Business Cycles* (Routledge and Kegan Paul, 1950), Parts I and II

7. NATIONAL INCOME AND SOCIAL ACCOUNTS

Board of Governors of the Federal Reserve System: *Flow of Funds in the United States 1939–1953* (Washington, 1955)

Central Statistical Office: *National Income Statistics—Sources and Methods* (H.M.S.O., 1956)

Copeland, M. A.: *A Study of Moneyflows in the United States* (National Bureau of Economic Research, 1952)

Downing, R. I.: *National Income and Social Accounts* (Melbourne University Press, 1966)

Edey, H. L., and Peacock, A. T.: *National Income and Social Accounting* (Hutchinson's University Library, 1954)

Goldsmith, R. W., and Lipsey, R. E.: *Studies in the National Balance Sheet of the United States*, Vol. I (Princeton University Press, 1963)

Goldsmith, R. W., and Mendelson, M.: *Studies in the National Balance Sheet of the United States*, Vol. II (Princeton University Press, 1963)

Leontief, W. W.: *The Structure of the American Economy* (Oxford University Press, 1951)

Mathews, Russell: *Accounting for Economists* (Cheshire, 1965)

National Bureau of Economic Research: *The National Economic Accounts of the United States* (Washington, 1958)

Organization for European Economic Co-operation: *A Standardized System of National Accounts* (Paris, 1952)

Powelson, J. P.: *National Income and Flow-of-Funds Analysis* (McGraw-Hill, 1960)

Reserve Bank of Australia Staff Paper (A. S. Holmes): *Flow-of-Funds, Australia, 1953–54 to 1961–62* (Sydney, 1965)

United Nations: *A System of National Accounts and Supporting Tables* (Studies in Methods, No. 2, New York, 1953)

United Nations: *Measurement of National Income and the Construction of Social Accounts* (Studies and Reports on Statistical Methods, No. 7, Geneva, 1947)

United States Department of Commerce: *U.S. Income and Output* (U.S. Government Printing Office, Washington, 1958)

8. PRICE INDEX NUMBERS

Croxton, F. E., and Cowden, D. J.: *Applied General Statistics* (Prentice-Hall, 1963), Chapters 17 and 18

Fisher, I.: *The Making of Index Numbers* (Kelley, 1967)

Hofsten, E. v.: *Price Indexes and Quality Changes* (George Allen and Unwin, 1952)

International Labour Office: *Computation of Consumer Price Indices* (*Special Problems*) (Geneva, 1962)
Joint Economic Committee, 87th Congress, 1st Session, *Government Price Statistics* (U.S. Government Printing Office, Washington, 1961)
Ministry of Labour: *Method of Construction and Calculation of the Index of Retail Prices*, Studies in Official Statistics No. 6 (H.M.S.O., 1964)
Mitchell, W. C.: *The Making and Using of Index Numbers* (Kelley, 1965)
Mudgett, B. D.: *Index Numbers* (John Wiley and Sons and Chapman and Hall, 1951)

9. Quantum Index Numbers

Carter, C. F., Reddaway, W. B., and Stone, R.: *The Measurement of Production Movements* (Cambridge University Press, 1965)
Central Statistical Office: *The Index of Industrial Production*, Studies in Official Statistics No. 2 (H.M.S.O., 1952)
Stone, R.: *Quantity and Price Indexes in National Accounts* (O.E.E.C., Paris, 1956 and H.M.S.O., 1957)
United Nations: *Index Numbers of Industrial Production* (Studies in Methods No. 1, New York, 1961)

10. Demography

Cox, P. R.: *Demography* (Cambridge University Press, 1959)
Dublin, L. I., Lotka, A. J., and Spiegelman, M.: *Length of Life* (Ronald Press 1949)
Glass, D. V.: *Population Policies and Movements in Europe* (Cass and Co. Ltd. 1967)
Kuczynski, R. R.: *The Measurement of Population Growth* (Sidgwick and Jackson 1935)
Office of Population Research: *Population Index*, Quarterly (Princeton University)
Organization for Economic Co-operation and Development: *Demographic Trends, 1965–80, in Western Europe and North America* (Paris, 1966)
Royal Commission on Population: *Reports and Selected Papers of the Statistics Committee*, Papers, Vol. II (H.M.S.O., 1950)
United Nations: *The Determinants and Consequences of Population Trends* (New York, 1953)

INDEX

AGE distribution of population—
adjusted, 401-3
and migration, 409-10
and wars, 409-10
census results, 401, 403
factors determining, 408-11
graduated, 403
importance of, 411-13
intercensal estimates of, 404-8
misstatement of age, 401
recorded, 401
Allen, R. G. D., 369n
Alternative hypothesis, 123
Analysis of statistical data, 2, 4-6
Analysis of variance, 175n, 205n
Appropriation account, 289n
Arbitrary origin, 54, 201-2, 242
Area sampling, 183-4
Arithmetic mean—
capable of algebraic treatment, 66
defined, 53
formula for, 53, 56
interpretation of, 65
of grouped data, 55-7
of linear combination, 116-17
of population, 116
of sample means, 120
of ungrouped data, 53
properties of, 53-4, 66
relation with median and mode, 65
reliability of, 66
significance of, 124-33, 148-51
use of arbitrary origin, 54
use of class interval units, 56-7
uses of, 66-7
weighted, 58
Arithmetic scale, 32
Array, 42
Articulated accounts, 293
Attribute, 7
Australia—
abridged life table, 432
age distribution of population, 401, 403, 413
crude birth rates, 423
crude death rates, 423
crude marriage rates, 423
crude rates of natural increase of population, 423, 458
gross national product at constant prices, 372-6, 500-2
gross reproduction rates, 449
infant mortality rates, 423
life table survivors at various ages, 441

Australia (*contd.*)—
mean expectation of life at various ages, 439
national income and expenditure, 481-87
net reproduction rates, 453
number of births, 396, 397
number of confinements, 396
probabilities of dying at various ages, 440
rates of net migration, 423
rates of population growth, 423
sex distribution of population, 401-3, 411-12
size of population, 398, 400, 423
specific mortality rates, 441
true rates of natural increase, 458
Australia and New Zealand Bank, 500
Australian Business Surveys, 478
Australian "C" Series Retail Price Index—
and basic wage, 365n
and changes in the cost of living, 365
as measure of retail prices, 487-8
compared with Consumer Price Index, 494-96
weighting in, 360-1
Australian Censuses, 397-8, 401-3, 413, 477
Australian Consumer Price Index—
and deflation of gross national product, 372-3
comparison with "C" Series Retail Price Index, 494-6
computation of, 488, 490-1
description of, 487-8
housing component in, 491-4
limitations of, 495-6
object of, 488
regimen of, 488-90
revision of regimen, 362
weighting in, 491
Australian Export Price Indexes—
described, 497-8
use of Ideal formula, 358
Australian Import Price Index, 498
Australian Indexes of Quantum—
exports, 502-3
farm production, 502
imports, 502-3
industrial production, 502
Australian Indexes of Wage Rates, 499-500
Australian Interim Retail Price Index, 488

Australian national income—
  historical, 481
  method of estimation of, 482–3
  tables, 483–6
Australian statistical sources, 472–6
Australian Survey of Retail Establishments, 186, 478–81
Australian Wholesale Price Index—
  and deflation of gross national product, 373
  described, 496–7
  dominance in, 359, 497
  table, 328

BALANCE of international payments, 293
Bar charts—
  composite, 29
  described, 25
  grid in, 26
  histograms, 28
  multiple, 27
  reference to source in, 27
  rules for construction of, 26–7
  scale of, 26
  simple, 26–7
  title of, 27
Base period, 11, 327, 350
Behaviour equation, 236
Bias—
  in estimation, 158–9
  in estimation of variance, 70, 121–2
  in index number formulae, 337–8
  in moving averages, 258, 272–3
  in sampling, 158–9, 162–4
Binary comparisons, 347
Binomial coefficients, 98, 100n
Binomial distribution—
  and the normal curve, 103–4
  application to proportions, 134
  characteristics of, 95–6
  defined, 95, 98
  effects of non-replacement, 95
  formula for, 98
  interpretation of, 100, 101
  mean of, 99–100
  specification of, 100
  standard deviation of, 102
  variance of, 101
Births—
  by date of birth, 396
  by date of registration, 396
  ex-nuptial, 396
  live-, 396
  masculinity of, 411
  multiple, 396
  nuptial, 396
  registration of, 395–6
  still-, 396
Bivariate normal distribution—
  characteristics of, 217–8
  defined, 216

Bivariate normal distribution (*contd.*)—
  formula for, 216
  parameters of, 217, 223
Brown, H. P., 288n
Budget deficit, published, 291

CAMERON, Burgess, 487
Capital account—
  consolidated national, 300, 301
  defined, 284, 289
  of financial enterprises, 291, 296
  of government, 290, 295
  of persons, 292, 297
  of rest of world, 292–3, 298
  of trading enterprises, 289, 294
Capital accumulation, 283
Capital formation—
  private, 301
  public, 290, 301
Carter, C. F., 391n
Carver, S. R., 503
Census, 23, 395, 397, 403
Central value, 51
Changes in financial claims account, 319–21
Chi-square—
  defined, 139
  degrees of freedom for, 139–41, 145
  distribution of, 139
  formula for, 139
  parameter of distribution of, 139
  table of values, 140, 470–1
  testing goodness of fit, 140–3
  testing independence of classification, 144–6
Circular test, 347–8
Clark, C., 481
Classification—
  defined, 7–9
  exhaustive, 9, 18–19
  in social accounts, 279–84
  mutually exclusive, 9, 18–19
Coefficient of determination—
  defined, 207
  meaning of 207, 208
Coefficient of variation, 75–6
Cohort, 424
Collection of data, methods of, 1, 14, 156–7
Collective goods and services, 282
Commonwealth Bureau of Census and Statistics, 472
Computed and observed values, 199
Confidence limits—
  explained, 153, 160
  for a mean, 153, 166, 168, 180
  for a proportion, 154, 169, 180
  for partial regression coefficients, 245
  for population numbers, 169, 181
  for population totals, 168, 180

INDEX   511

Confidence limits (*contd.*)—
  for sample regression coefficients *a* and *b*, 228, 232
Constant sampling fraction, 173
Consumers' expenditure, 292
Contingency tables, 144
Control charts, 188–9, 191
Control limits, 188–9
Correlation—
  absence of, 196–7, 219
  and causation, 224–5, 249–50, 270–1
  and empirical economic investigations, 236–40
  and time series, 269–70
  coefficient of linear, 207, 210
  coefficient of multiple, *see* Multiple correlation coefficient
  defined, 203–7
  explained and unexplained variation, 204–5, 206
  in population, 217
  meaning, 222–4
  non-linear, 223–4
  of grouped data, 212–5
  of ungrouped data, 207–8
Correlation of prices and quantities, 334, 369
Cost of living index—
  concept defined, 351–6
  measurement of changes in, 351–7
  relation between retail price index and, 356–7, 364–6
Co-variation, 117
Cox, P. R., 444n
Crawford, J. G., 481
Crude birth rate—
  defined, 413
  determined by, 414
  for selected countries, 422
  level of, 414
  misleading as an indicator of fertility, 448–9
  relation to gross reproduction rate, 448–9
Crude death rate—
  defined, 414
  determined by, 415
  for females, 414
  for males, 414
  for persons, 414
  for selected countries, 422
  level of, 415
  misleading as an indicator of mortality, 440, 442–3
  relation to true death rate, 442–3
Crude marriage rate—
  defined, 421
  determined by, 421
Crude rate of natural increase—
  defined, 415
  determined by, 415

Crude rate of natural increase (*contd.*)—
  for selected countries, 422
  level of, 415
  relation to net reproduction rate, 454
  relation to true rate of natural increase, 458
Current Population Survey, 185
Cyclical movements, 253–6

DAVIES, O. L., 187n
Death registration, 395
Deflationary gap, 317
Deflation of values, 371, 375–6
Degrees of freedom—
  for χ^2-distribution, 139, 140, 141, 145
  for t-distribution, 129, 130, 132, 228, 229, 232, 245
Demography—
  central questions of, 449
  defined, 395
Depreciation allowances, 281, 286–7, 289–90, 301, 302–3, 304
Description of data, 2, 42
Deseasonalized series, 277–8
Dichotomous population—
  defined, 133
  distribution of mean of samples from, 134
  mean of samples from, 134
  parameter of, 133–4
  standard deviation of samples from, 134
Digital computers, electronic, 23
Dispersion—
  absolute, 51, 67
  relative, 75–6
Dispersion of price movements, 328, 335, 368, 379
Dispersion of quantity movements, 331, 335, 367, 379
Disposal account—
  consolidated national, 299, 301
  defined, 284, 289
  of financial enterprises, 291, 296
  of governments, 290, 295
  of persons, 292, 297
  of rest of world, 292, 298
  of trading enterprises, 290, 294
Disposal of income, 283
Distribution account—
  consolidated national, 299, 301
  defined, 284, 289
  of financial enterprises, 291, 296
  of governments, 290, 295
  of persons, 291, 297
  of rest of world, 292, 297
  of trading enterprises, 289, 294
Distribution of income, 283
Double-entry accounting, 280, 284, 293

ECONOMETRICS, *see* Economics, empirical aspects of

Economics, empirical aspects of, 3–4, 236–8, 251–2, 279
Economic transactions—
　classification of, 280–4
　defined, 280
Edey, H. C., 325n
Edgeworth, F. Y., 333
Enumerator, 14
Episodic movements, 253–6
Erratic movements, 253–6
Errors, Type I and Type II, 146–8
Estimation—
　bias in, 70, 121–2, 159–60
　by stratification after selection, 181–2
　explained, 115, 151, 160
　from simple random samples, 168–70
　from stratified random samples, 171–2, 181
　of population mean, 151–3, 165, 168, 180
　of population numbers, 169, 181
　of population partial regression coefficients, 245
　of population proportion, 154, 169, 180
　of population regression coefficients, 228, 232
　of population total, 168, 180
　ratio estimation, 169–70, 479–80
Events—
　defined, 81
　equiprobable, 82
　independent, 86
　joint, 81, 87
　mutually exclusive, 84–5
　probability of, 82
Exact relationship, 221
Exhaustive classification, 9, 18–19
Ex-nuptial births, 396
Expected value, 89–90
Expenditure method of estimating national income, 309–10
Exports, 292, 301
Ex post identities, 301, 302

FACTOR reversal test, 337–8
Factors of production, 281
Fecundity, 445
Fertility—
　defined, 445
　measurement of, *see* Total fertility rate; Gross reproduction rate
Field-worker, 14
Final products, 285, 302
Financial enterprises expenditure, 291, 301
Financial enterprises sector—
　capital account of, 291, 296
　defined, 282, 283
　disposal account of, 291, 296
　distribution account of, 291, 296
　gross product of, 288
　production account of, 288, 296

Fisher, I., 333n, 336n
Fisher, R. A., 161, 229, 468, 470
Flow-of-funds accounting—
　described, 318–21
　for Australia, 486–7
Frame, 161
Frequency distribution, 9, 42
　classes in, 43–8
　class interval in, 43
　class limits in, 43
　class mid-points in, 45–6, 50
　comparison of, 78–9
　constructing, 47–8
　cumulative, 48–50
　description of, 78
　number of classes in, 44–5
　regularity of, 45
　relative (or percentage) form, 48, 94
　types of, 50–51
　use of class interval units, 56
Frequency polygon, 28, 43
Full count, 157

GAP analysis, 317
Geary, R. C., 379n
Generation fertility, 464
Geometric mean, 62–3, 340
Glass, D. V., 424n
Goodness of fit, 138–144
Government deficiency, 291
Government expenditure, 290, 291, 301
Government savings, 290, 301
Government sector—
　capital account of, 290, 295
　defined, 282
　disposal account of, 290, 295
　distribution account of, 290, 295
　gross product of, 288
　production account of, 288, 295
Graphical presentation, object of, 25
Great Britain, Census of Population, 184
Gross domestic expenditure, 301
Gross domestic product, 300
Gross market supplies, 300n
Gross national expenditure, 301n
Gross national product—
　at factor cost, 303n
　defined, 300
　estimation of, 308–11, 312
　relation to other aggregates, 301, 304
Gross output, 383
Gross private capital formation, 290
Gross private investment, 290, 301
Gross product—
　of a trading enterprise, 285–6, 377, 380
　of financial enterprises sector, 288
　of government sector, 288
　of trading enterprises sector, 287
Gross reproduction rate—
　calculation of, 448
　defined, 447–8

Gross reproduction rate (*contd.*)—
  interpretation of, 448, 450
  level of, 449
  limitations of, 462–4
  relation to crude birth rate, 448
  relation to net reproduction rate, 454
  use for comparisons of fertility, 448–9
Gross resident product, 300

HAIG, B. D., 486, 503
Hajnal, J., 463n
Hansen, M. H., 185n
Hanson, R. H., 184n
Harcourt, G. C., 280n, 288n
Harmonic mean, 64, 342
Hicks, J. R., 369n
Histogram, 28, 43, 94
Holmes, A. S., 487
Home-ownership—
  treatment in price index numbers, 491–4
  treatment in social accounts, 292, 305–8
Horner, F. B., 503
Households sector, *see* Personal sector
Hurwitz, W. N., 185n
Hypotheses, 115, 122–4, 146–8

IMPORTS, 292, 301
Imputation of value—
  defined, 305
  of housewives' services, 306
  of owner-occupied dwellings, 305–8
  of produce consumed on the farm, 305–6
Incomes-received method of estimating national income, 309–10
Independence of classification, 144–6
Index numbers, 327, *see also* Price index numbers; Quantum index numbers; Index numbers of industrial production
Index numbers of industrial production—
  and changes in quality, 380
  and employment indicators, 388
  and input indicators, 388
  and output indicators, 387
  and seasonal variation, 391
  and work-in-progress, 380
  comparisons over short and long periods, 391–2
  computation by method of deflation, 381–6
  computation by method of indicators, 386–91
  coverage of, 380–1, 389–90
  defined, 377
  for small and large sectors, 379, 386
  Laspeyres's, 378–9, 391
  limitations of, 381
  Paasche's, 378–9, 391

Index numbers of industrial production (*contd.*)—
  relation between indexes of output, input and net output, 385
  relation to indexes of real gross national product, 380
Index of seasonal variation, *see* Seasonal variation
Indifference curves, 353–5
Indirect taxes, 303–4
Infant mortality rate—
  approximation to probability at birth of an infant dying, 418–9
  defined, 417
  for all births, 418
  for females, 418
  for males, 418
  for selected countries, 422
  level of, 418
  refined rate, 418–9
  use in life table, 425
Inflationary gap, 317
Input-output tables—
  described, 322–3, 324
  for Australia, 487
Inquiry—
  general purpose, 14
  special purpose, 14
Interest, 288
Inter-industry analysis, 321–6
Intermediate products, 285
Inventories, 285, 286–7

KARMEL, P. H., 280n, 288n, 457n, 463n, 488n
Kuczynski, R. R., 434n, 449n, 452n

LASPEYRES, E., 332n
Law of large numbers, 83
Least squares, 199–200, 219, 241–2, 260
Length of a generation, 455
Life table—
  and life assurance, 434
  and population projections, 443–4
  and stationary population, 437–8
  construction of abridged, 427–8, 431–2
  defined, 424
  for Australia, 477
  limitations of, 433, 444
  probabilities of dying and surviving in, 424–5, 429, 432, 433–4
  survivors in, 428
  use for comparisons, 440–1
  use for measuring population replacement, 450
Linear combination of variables—
  defined, 116
  for normally distributed variables, 118
  mean of, 116–7, 120, 127, 134, 172, 227, 231, 234, 234n, 244
  variance of, 116–8, 120, 127, 134, 173, 227, 232, 234, 244

Live-births, 396
Logarithmic cycle, 35
Logarithmic deck, 35
Logarithmic scale, 33, 34–5
Lorenz curves, 39–40
Lotka, A. J., 457n
Lottery selection, 159

Madow, W. G., 185n
Manufacturers' specifications, 193
Marriage records, 395
Masculinity—
 of births, 411
 of deaths, 411
 of population, 411
Maternal mortality rate, 421
Mathematical expectation, 89
Mathematical statistics, 2
Mathews, R. L., 319n, 503
McBurney, S. S., 486
Mean, *see* Arithmetic mean
Mean age at death of infants, 406, 419
Mean deviation—
 of grouped data, 69
 of ungrouped data, 69
Mean expectation of life—
 computation of, 431
 defined, 429–30
 for selected countries, 439
 use for comparisons of mortality, 439–40, 442
 use for measuring mortality, 430
Median—
 defined, 58–9
 formula for, 59
 interpretation of, 65
 of grouped data, 59–60
 of ungrouped data, 58
 relations with mean and mode, 65
 uses of, 66–7
Migration records, 395
Mode—
 defined, 61
 formula for, 62
 interpretation of, 65
 relations with mean and median, 65
 uses of, 67
Mortality—
 comparison of mortality conditions, 438–443
 due to specific causes, 434–6
 measurement of, 423, 430. *See also* Life table
Moser, C. A., 185n
Moving averages—
 and measurement of seasonal variation, 272–4
 bias in, 258, 272–3
 defined, 257
 effects of taking, 257–8
 limitations of, 257–8

Multiple births, 396
Multiple correlation coefficient—
 defined, 242–3
 formula for, 243
 meaning of, 243, 248
Multiple regression—
 computation of constants in, 242, 245
 defined, 240–1
 relation to simple regression, 249–50
Multiplication rule for counting, 81
Multi-stage sampling, 182–4
Multi-stratification, 182
Mutually exclusive classification, 9, 18–19, 85

National budgeting, 313–8
National income, 303
National income accounts—
 content of, 285–98
 defined, 280, 284
National turnover of goods and services, 300n
Neo-natal mortality rate—
 defined, 420
 for all births, 420
 for females, 420
 for males, 420
 level of, 421
Net increase in indebtedness of government sector, 291
Net increase in indebtedness to rest of world, 293, 301
Net national product—
 at factor cost, 303, 304
 at market prices, 301, 302, 303, 303n, 304
Net output, 385
Net private investment, 303
Net product of a trading enterprise, 286
Net reproduction rate—
 calculation of, 450–1
 defined, 450
 interpretation of, 451–2
 level of, 452–3
 limitations of, 462–4
 of selected countries, 453
 relation to crude rate of natural increase, 454
 relation to gross reproduction rate, 454
 relation to true rate of natural increase, 454–5, 457
Normal curve—
 and binomial distribution, 103
 characteristics of, 103
 of errors, 103
Normal distribution—
 and expected frequencies, 113–4, 142
 and testing normality of a frequency distribution, 141–2
 application to proportions, 135
 areas under normal curve, 106–114

Normal distribution (*contd.*)—
  bivariate, *see* Bivariate normal distribution
  characteristics of, 74–5, 105
  defined, 104
  formula for, 104–5
  in standard form, 107, 125
  of differences between two sample means, 127
  of linear combination, 118
  of sample means, 120–1
  parameters of, 105–6
  table of areas, 107–8, 466–7
Normal equations—
  for linear trend, 260
  for multiple regression, 242
  for parabolic trend, 266
  for simple linear regression, 200
Null hypothesis, 123
Nuptial births, 396

OGIVE—
  construction, 48–9
  "less than", 49
  "or more", 49
  percentage, 50
Optimal allocation, 174, 479
Outcomes—
  all possible, 80–1
  compound, 81
  elementary, 81

PAASCHE, H., 332n
Parameters—
  defined, 99–100, 158–9
  in regression, 219, 237
  of bivariate normal distribution, 217–8
  of dichotomous population, 133
  of χ^2-distribution, 139
  of normal distribution, 105–6
  of t-distribution, 129
Partial regression coefficients, 241
Peacock, A. T., 325n
Percentage defective, 193
Percentage distributions, 12
Percentages, 10
Periodic movements, 253–6
Personal income, 304
Personal sector—
  capital account of, 292, 297
  defined, 282
  disposal account of, 292, 297
  distribution account of, 291, 297
  production account of, 288
Pilot survey, 20
Polak, J. J., 256n
Population (human)—
  at a date, 397
  intercensal estimates of, 404–8
  mean, 398–401
Population projections, 443–4

Population replacement—
  defined, 449
  measurement of, *see* Net reproduction rate; True rate of natural increase
Population silhouette, 410–1
Population (statistical)—
  defined, 90, 156
  finite, 90, 116, 164–5
  infinite, 90
Prediction, 222–3, 233–4
Prediction limits—
  for a mean value, 235, 240, 246
  for a single value, 236, 240, 246
Presentation of data, 1–2
Price index numbers—
  aggregative type, 329–33
  aggregative with fixed weights, 333, 347, 349, 358
  and changes in cost of living, 351–7, 365–6
  average type, 339–342
  chained, 348–9
  changes in regimen of and quality of items in, 360–4
  changing the base of, 350
  combination of, 344–5
  comparison of different formulae for, 346–7
  comparisons between many points of time, 347–51
  construction of, 358–60
  defined, 327
  dominance in, 359–60, 497
  effect of change in price of one item on, 343–4
  Fisher's Ideal, 333, 338, 349, 358
  group, 344–5
  in practice, 364–6
  Laspeyres's, 332, 336–8, 342, 346, 348
  Marshall-Edgeworth's, 333, 338, 349
  Paasche's, 332, 336–8, 342, 346, 348, 349
  regimen of, 359
  relation between aggregative and average types, 342–3
  relation between Laspeyres's and Paasche's, 333–5
  simple aggregate of prices, 330–1
  simple arithmetic mean of price relatives, 340
  simple geometric mean of price relatives, 340–1, 347
  splicing of, 363
  tests of adequacy of, 336–9, 347–8
  weighted aggregate of prices, 331–2
  weighted arithmetic mean of price relatives, 341–2, 358
  weighting in, 331–2, 359–60
Price relatives, 339
Probabilities of dying and surviving, *see* Life table

Probability—
  addition rules of, 84, 85
  a priori, 80, 83
  as relative frequency, 94
  conditional, 87
  experimental, 83–4
  measure of, 82–3
  multiplication of, 86, 88
  nature of, 80
  rules for operation of, 84
  theory of, 80
Probability curve, 94
Probability distribution—
  and expected frequencies, 92
  cumulative, 94–5
  defined, 91
  forming, 91–4
  interpretation of, 92
  specification of, 95–9
Process average, 188
Process standard deviation, 188
Production account—
  consolidated national, 299–300, 301
  defined, 284
  of a trading enterprise, 287, 293
  of financial enterprises sector, 288, 296
  of government sector, 288, 295
  of persons sector, 288
  of rest of world sector, 288
  of trading enterprises sector, 287, 293
Production method of estimating national income, 309
Production of income, 283
Productivity—
  defined, 392
  limitations of measurement of, 392–3
  measurement of, 392–4
  overall, 392
  proper, 393
Propensity to have families, 462
Propensity to marry, 462
Proportions—
  distribution of sample proportions, 134–5
  significance of, 135–7
Punch cards, 21–2
Purposive sampling, 162

QUADRATIC mean, 64
Quality control—
  advantages of, 193–4
  and destructive tests, 193, 194
  and specifications, 193
  control charts, 188–9, 191
  control limits, 188–9
  estimation of process average and standard deviation, 188–90
  explained, 187–8
  use of percentage defective, 193
  use of sampling ranges, 192

Quantity weights, 332–3, 359–60
Quantum changes, 367
Quantum index numbers—
  Laspeyres's, 370
  of input, 385
  of net output, 385
  of output, 385
  Paasche's 370
  relation between Laspeyres's and Paasche's quantum and price indexes, 371
  use of method of deflation, 371
  See also Index numbers of industrial production
Quantum of production, 376
Quartile deviation, 68
Quartiles, 60–1
Questionnaire, 14
  example of, 16
  field of inquiry of, 20–1
  form of questions in, 18–20
  ideal data from, 15
  instructions in, 20
  leading questions in, 20
  non-refutable questions in, 18
  object of, 15
  practicability of questions in, 17
  refutable questions in, 18
  setting out of, 20
Quinquennial age groups, 403, 407 426–7, 446
Quota sampling, 164

RAISING factor, 168
Randomness of a sample, 123, 142–3
Random numbers, 161
Range, 67–8
Rate of net migration—
  defined, 415
  determined by, 416
Rate of population growth—
  average annual, 417
  compared with rate of total increase, 416–7
  defined, 416
Rate of total increase—
  compared with rate of population growth, 416–7
  defined, 416
Ratio charts, *see* Semi-logarithmic charts
Ratio estimation, 169–70, 479–80
Ratios, 10, 12
Real gross national product—
  and the terms of trade, 375–6
  computed by deflation of money gross national product, 372–6
  defined, 367–8
  limitations of concept of, 369–70
  meaning of, 369
  measurement of, 370–6
Reddaway, W. B., 391n

Regimen, 359
Regression—
 and time series, 269–71
 coefficients, 199, 209, 211
 computation of constants in simple linear regression, 200–1
 defined, 197–99
 in empirical economic investigations, 236–40
 in population, 217, 218, 220
 meaning of, 219–20
 multiple, *see* Multiple regression
 normal equations for simple regression, 200
 of grouped data, 212–5
 of ungrouped data, 202–3
 relation of multiple to simple regression, 249–50
 two lines of, 209–10, 220–1
Relationship between two variables, 196–8, 221
Relative prices, 328–9
Relatives, 11
Replacement potential of human population, 452
Respondent, 14
Rest of world sector—
 capital account of, 292–3, 298
 defined, 283
 disposal account of, 292, 298
 distribution account of, 292, 297
 production account of, 288
Root mean square deviation, 70
Rugg, H. O., 466

SAMPLE design, 170, 182–4
Samples—
 adequacy of, 158
 area, 183–4
 explained, 90, 156–7
 from a finite population, 164–5
 multi-stage, 182–4
 multi-stratified, 182
 purposive, 162
 quota, 164
 random, 91, 93
 selection of random, 160–1
 simple random, 91, 115, 159, 165–70
 size of, 166
 stratified random, 170–82
 systematic, 161, 163
Sample statistics, 91, 99–100, 158
Sample surveys—
 accuracy of, 157
 advantages of, 157–8
 bias in selection of sample, 162–4
 costs of, 158, 177
 design of, 170, 182–3
 frame of, 161
 in Australia, 478–81
 in practice, 184–6, 478–81

Sample surveys (*contd.*)—
 speed of, 157
 substitutions in, 163
 with purposive selection, 162
Sampling distribution—
 defined, 115
 of a partial regression coefficient, 244
 of a proportion, 134
 of sample regression coefficient a, 231–2
 of sample regression coefficient b, 227
 of the difference between means, 127
 of the mean, 119–21
Sampling errors—
 explained, 115, 158–60
 in simple random sampling, 165–6
 in stratified random sampling, 170–2 176–7
Sampling fraction—
 constant, 173
 defined, 165
 with optimal allocation, 174
Savings—
 defined, 289
 enterprise, 290
 financial enterprise, 301
 government, 290
 personal, 292, 301, 315
 rest of world, 292
Scatter diagram, 196, 203, 211
Seasonal variation—
 and indexes of industrial production, 391
 computation of index of, 275–7
 defined, 253, 254, 271–2
 deseasonalizing a series, 277
 index of, 273
 measurement of, 271–3
 uses of index of, 277–8
Secular trend, *see* Trend
Semi-logarithmic charts—
 description of, 32–4
 grid in, 35
 rules for construction of, 34–5
 uses of, 37–8, 400
Separation factor, 419
Sex distribution of population—
 and measurement of fertility, 462–3
 and migration, 411
 and wars, 411
 factors determining, 411
 importance of, 411–2
Sharpe, F. R., 457n
Significance, level of, 123, 146
Significance, tests—
 explained, 122–3
 of coefficient of linear correlation, 226, 229–30
 of difference between p and π, 135–6
 of difference between p_1 and p_2, 136–8
 of difference between $\bar{X}$ and μ, σ known, 124–6

Significance, tests (*contd.*)—
  of difference between $\bar{X}$ and μ, σ not known, 128–31
  of difference between $\bar{X}_1$ and $\bar{X}_2$, σ known, 127–8
  of difference between $\bar{X}_1$ and $\bar{X}_2$, σ not known, 131–33
  of goodness of fit, 140–3
  of partial regression coefficients, 244–5, 248
  of sample regression coefficient a, 231–3
  of sample regression coefficient b, 226–8, 230
  one-tailed tests, 148–51
  procedure in, 123–4
  two-tailed tests, 148–51
Skewness—
  defined, 51, 76
  effect on mean, median and mode, 65
  Pearsonian measure of, 76–7
  quartile measure of, 77
  relation between Pearsonian and quartile measures, 77, 79
Social accounts—
  balancing properties of, 311, 483
  classification in, 279–84
  consolidation of, 298–300
  defined, 279
  estimation of items in, 308–11
  interpretation of, 302–5
  uses of, 279, 312–13
  See also National income accounts
Specific fertility rates, 445–6
Specific mortality rates, 425
Splicing, 363
Stable population—
  and true rate of natural increase, 457, 459
  defined, 457
  example of, 459–62
Standard deviation—
  characteristics of, 74–5
  defined, 70
  denominator of, 70, 121–2
  formula for, 70, 71, 72, 73
  in normal distribution, 74–5
  of grouped data, 73–4
  of sample means, 120
  of ungrouped data, 71–2
  pooled estimate of, 132
  unbiased estimate of, 70, 121–2, 132
  use of arbitrary origin, 71–2
  use of class interval units, 73
Standard deviation units, 76
Standard error—
  defined, 120
  of a prediction, 233–5, 245–6
  of partial regression coefficients, 244–5
  of sample regression coefficient a, 231–2
  of sample regression coefficient b, 227–8

Standard error (*contd.*)—
  of the mean, 120
  of the mean from a finite population, 164–5
  of the mean of a stratified random sample, 172–3
  of the mean of a stratified random sample with constant sampling fraction, 173
  of the mean of a stratified random sample with optimal allocation, 174
  relation between standard errors of mean computed from simple random sample and from stratified random sample, 175–6
Standard error of estimate—
  defined, 208–9, 210, 222n
  in multiple regression, 244
Standardized value, 76
Standard measure, 76
Standard of living—
  concept of, 351–7
  measurement of, 312, 369
Stationary population, 437, 457
Statistical inference, general problem of, 115–6
Statistical methods, 2–3
Statistics—
  nature of subject of, 1
  sources of Australian, 472–8
  sources of economic and social, 23–4
  theory of, 2
  See also Sample statistics
Stevens, S. P., 503
Stigler, G. J., 353n
Still-births, 396
Stochastic relationship, 221
Stone, J. R. N., 391n
Stratification, 170–82
Stratification after selection, 181–2
Structural equation, 236
Summation sign, 51–3
Survivorship ratio, 443
Systematic sampling, 161, 163

t-DISTRIBUTION—
  defined, 129
  degrees of freedom for, 129, 130, 132, 228, 232
  parameter of, 129
  table of values, 130, 151, 468–9
  uses of, 130, 132, 153–4, 228, 229, 232, 235, 245
Tables, statistical—
  aggregative type, 9
  column and row headings, 7
  comparisons in, 10–12
  frequency type, 9
  reference to source in, 7
  rules for construction of, 7

Tables, statistical (*contd.*)—
  setting out of, 7
  units of measurement, 7
Tabulation—
  cross-, 23
  explained, 21
  mechanical, 21–3
Taxation, 289, 290
The Social Survey, 185
Time reversal test, 336–7
Time series—
  assumptions in analysis of, 253–6
  characteristic behaviour of, 252–3
  correlation of, 269–70
  defined, 10, 251
  graphing of, 31–2
  limitations of analysis of, 256, 268–9
  objectives of analysis of, 251–2
  time-scale in, 31
Tinbergen, J., 256n
Tippett, L. H. C., 270
Total fertility rate—
  calculation of, 446
  defined, 445
  interpretation of, 447
Trade cycle, 251, 253
Trading enterprises sector—
  capital account of, 289, 294
  defined, 281
  disposal account of, 290, 294
  distribution account of, 289, 294
  gross product of, 285–6
  production account of, 287, 293
Transfer payments, 289
Tree diagram, 96
Trend—
  and moving averages, 272
  defined, 253
  exponential, 263–5
  freehand drawing of, 257
  interpretation of, 268–9
  linear, computation of, 260–3
  linear, defined, 259
  logistic, 266–7
  measurement of, 257–67
  origin in, 260, 262
  parabolic, 266
  removal of, 269–71
  selection of, 266–71
Trend-free series, 269–71
Trend values, 260n, 262
True death rate—
  relation to crude death rate, 442–3
  use for comparisons of mortality, 442
True rate of natural increase—
  and stable population, 457, 459–62

True rate of natural increase (*contd.*)—
  calculation of, 455–6
  interpretation of, 457–8
  limitations of, 462–3
  relation to crude rate of natural increase, 458
  relation to net reproduction rate, 454–5, 457

UNBIASED estimate—
  of mean, 151
  of partial regression coefficients, 245
  of sample regression coefficient a, 232
  of sample regression coefficient b, 227
  of standard deviation, 70, 121–2, 132, 168n
Unbiased prediction, 235
Undistributed company profits, 290, 301
Unincorporated enterprises, 289
United Nations Statistical Office, 280n, 288n, 444n
United States Census, 184
United States Department of Commerce, 185n

VALAORAS, V. G., 418n
Value added, 285, 377–8
Value of government output, 288
Value of input, 285, 378
Value of output, 285, 378
Value of production, 285, 378–9
Value weights, 341
Variable—
  continuous, 7, 46
  defined, 7
  discrete, 7, 46
  explanatory, 236, 249
  random, 91
Variance—
  formula for, 70
  of a linear combination, 116–9
  of population, 116
Variation—
  explained, 204–5, 206
  total, 117, 204, 206
  unexplained, 204–5, 206

WAKSBERG, J., 184n
Wallace, R. H., 280n, 288n
Walsh, C. M., 332n
Whelpton, P. K., 464n
Wilson, R., 379n
Work done, 285, 377, 388

YATES, F., 161, 162n, 183n, 468, 470
Youngman, D. V., 486
Yule, G. U., 224